Introduction to Fire Protection

4th Edition

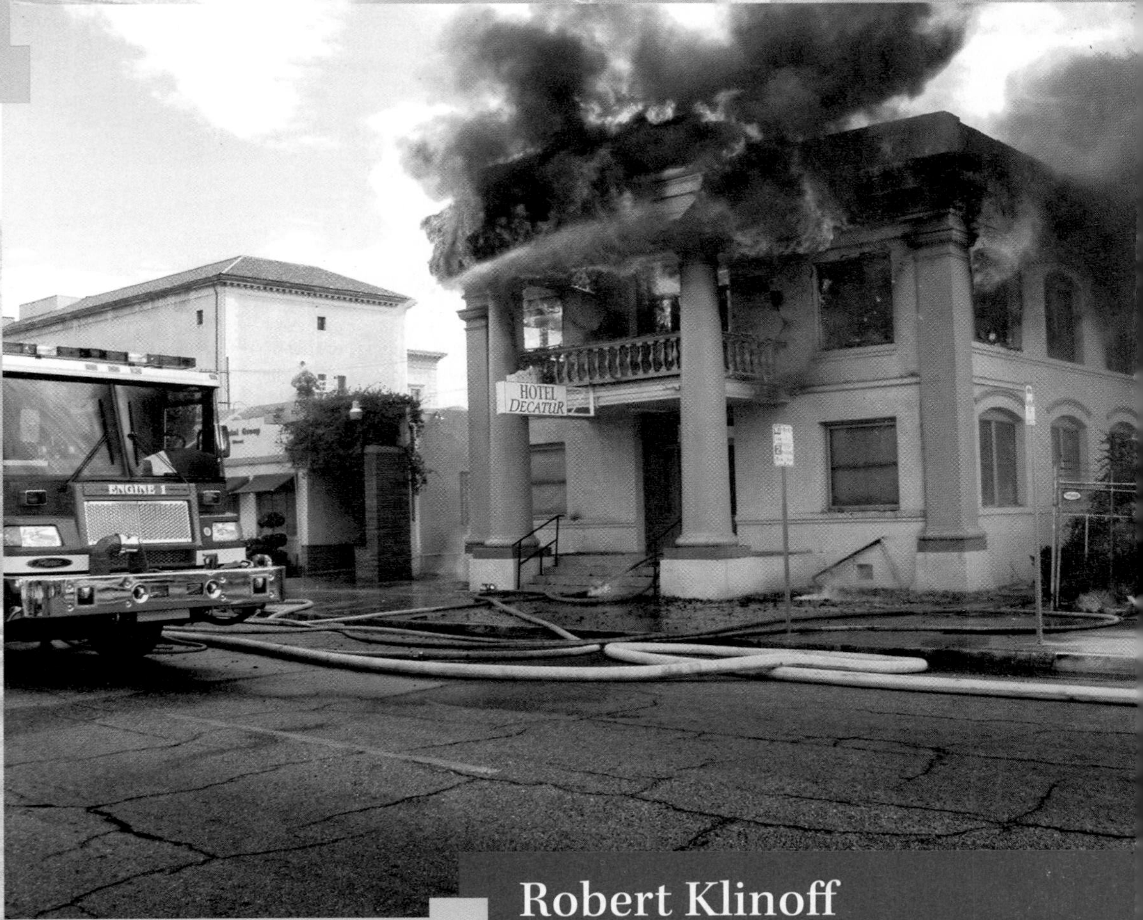

Robert Klinoff

DELMAR
CENGAGE Learning™

Australia • Brazil • Japan • Korea • Mexico • Singapore • Spain • United Kingdom • United States

Introduction to Fire Protection, 4th Edition
Robert Klinoff

Vice President, Editorial: Dave Garza

Director of Learning Solutions: Sandy Clark

Senior Acquisitions Editor: Janet Maker

Managing Editor: Larry Main

Senior Product Manager: Jennifer Starr

Editorial Assistant: Amy Wetsel

Vice President, Marketing: Jennifer Baker

Marketing Director: Deborah S. Yarnell

Senior Marketing Manager: Erin Coffin

Associate Marketing Manager: Erica Ropitzky

Production Director: Wendy Troeger

Production Manager: Mark Bernard

Senior Content Project Manager: Jennifer Hanley

Senior Art Director: Casey Kirchmayer

Library of Congress Control Number: 2010941295

ISBN-13: 978-1-4390-5842-8
ISBN-10: 1-4390-5842-3

Delmar
5 Maxwell Drive
Clifton Park, NY 12065-2919
USA

Cengage Learning is a leading provider of customized learning solutions with office locations around the globe, including Singapore, the United Kingdom, Australia, Mexico, Brazil, and Japan. Locate your local office at: **international.cengage.com/region**

Cengage Learning products are represented in Canada by Nelson Education, Ltd.

To learn more about Delmar, visit **www.cengage.com/delmar**

Purchase any of our products at your local college store or at our preferred online store **www.cengagebrain.com**

Printed in the United States of America
1 2 3 4 5 6 7 15 14 13 12 11

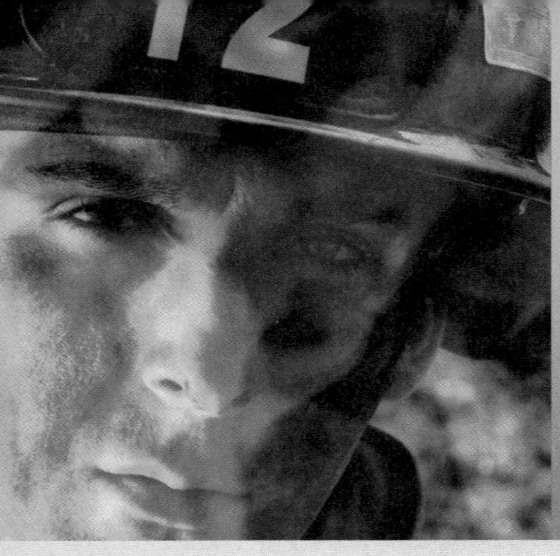

CONTENTS

Chapter 4 CHEMISTRY AND PHYSICS OF FIRE / 95

Chapter 5 PUBLIC AND PRIVATE SUPPORT ORGANIZATIONS / 120

Chapter 6 FIRE DEPARTMENT RESOURCES / 144

Chapter 7 FIRE DEPARTMENT ADMINISTRATION / 215

Chapter 8 SUPPORT FUNCTIONS / 250

Chapter 9 TRAINING / 275

Chapter 10 FIRE PREVENTION / 316

Chapter 11 CODES AND ORDINANCES / 348

Chapter 12 FIRE PROTECTION SYSTEMS AND EQUIPMENT / 377

Chapter 13 EMERGENCY INCIDENT MANAGEMENT / 424

Chapter 14 EMERGENCY OPERATIONS / 463

APPENDICES

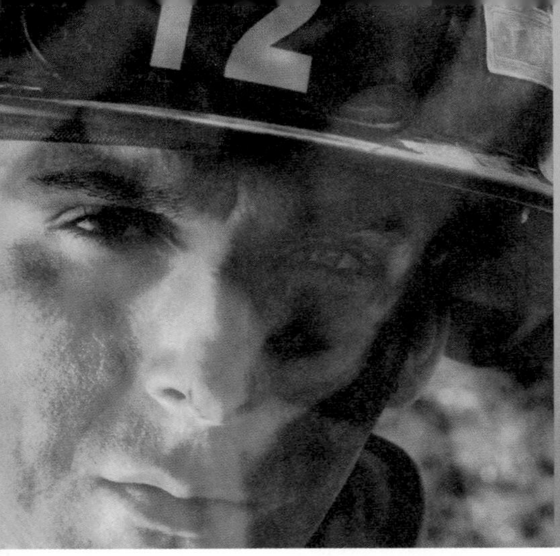

PREFACE

The study of fire science is multifaceted. It involves the study of the complete fire protection system, including fire department resources, private resources and systems, the chemistry and physics of fire, safety, fire department management, emergency management, the legal environment, and fire prevention.

ABOUT THIS BOOK

This text introduces the student to the many areas of fire protection. It is primarily intended for the person who wishes to become a firefighter. An overview of aspects of the selection process is presented to assist the student in preparing for this very competitive process. Fire protection careers other than firefighter in the public and private sector are presented as well.

This text introduces the student to the concept of the systems approach to fire protection by presenting the system components of modern fire department responsibility, including suppression, prevention, public education, emergency medical service, hazardous materials response, and urban search and rescue.

Throughout this text, safety and professionalism are stressed. Other concepts emphasized are incident effectiveness, customer service, physical fitness, training, decision making, and fire prevention.

This text meets the FESHE course outcomes for *Principles of Emergency Services* and introduces material contained in the other FESHE courses.

WHY I WROTE THIS BOOK

The first edition of *Introduction to Fire Protection* came about at the prompting of Delmar Publishers. When I began this work in 1997, I was instructing the course titled "Introduction to Fire Science" at my local community college. I received a questionnaire from

Delmar inquiring about the courses I was teaching and what books I was using. In conversation with Delmar it was determined that the texts available for this course were well out-of-date. In response to this I took on the task of writing this text. It has been a lot of work and has been both rewarding and frustrating. Now, going into the fourth edition I am pleased and proud of this accomplishment and look forward to a fifth edition in the future. Creating and revising this text has given me the opportunity to pass on some of what I have learned in my 35 years in the fire service. This is something I have always felt strongly about doing. I continue to teach Fire Science courses and strive to develop future firefighters and fire officers.

HOW TO USE THIS BOOK

First, this book presents the requirements for and advantages of a degree in fire science.

Then it describes the selection process and gives job descriptions to help the student decide if this is a career he or she is interested in and can qualify for.

The text can be divided into seven areas: fire protection as a career (Chapters 1 and 2), the history and future of the fire protection system (Chap. 3), chemistry and physics (Chap. 4), support functions (Chapters 7, 8, 9, and 10), the legal environment (Chap. 11), systems and equipment (Chap. 12), and emergency operations (Chapters 13 and 14). The areas of the text can be taught out of sequence, but each area should be kept together as a unit for better understanding. The chapters on emergency operations are presented last as they require a base of knowledge developed in the preceding chapters. The text is designed to be presented over either a quarter- or semester-length course schedule.

FEATURES OF THIS BOOK

- **Correlates to FESHE Course Outcomes for Principles of Emergency Services.** (See Correlation Grid following this Preface.)
- **Full-Color Design** features new color photos and illustrations and includes references to accompanying live-action firefighter video clips to enhance learning for the student.
- **Safety Notes** offer practical advice for keeping firefighters safe on the job.
- **Notes** highlight important information for the student and offer a quick review of critical concepts.
- **Discussion Questions Follow the Review Questions** at the end of each chapter and offer additional opportunity for students to develop their critical-thinking skills.
- **Robust Appendices** offer many helpful resources and references for the aspiring firefighter, including a report out on the U.S Fire Problem, a listing of the NFPA's National Fire Codes, common acronyms, conversion charts, fire-related websites, and more!

NEW TO THIS EDITION

In revising this book, the focus was placed on ensuring the information within the text remained current, as well as on introducing new advances or technologies in the fire services from recent years. Consequently, the fourth edition provides a comprehensive look at the world of the fire service today:

- **All-New Full Color Design** features new color photos and illustrations to enhance learning for the student.
- **Highlights Accompanying Video Clips** that offer live-action firefighter video depicting specific topics covered in the chapters—accessible FREE via www.cengagebrain.com on our *Introduction to Fire Protection* companion site. (Refer to the Note section at the end of this Preface.)
- **Information on New Positions and Organizations in the Fire Service** offer students insight on the wide range of opportunities and resources available to them.
- **A Closer Look at Safety** includes a discussion of various tactics on the fireground that contribute to or detract from firefighter safety; new sections on firefighter apparatus, retro-reflective striping, and highway scene safety; coverage of the "16 Firefighter Life Safety Initiatives"; and new safety standards and models.
- **Emphasis on Decision Making** for the fire service, with the latest thinking about how firefighters make decisions and how they should be making decisions to engage in incidents both effectively and safely.
- **Current Information on Technology**, such as geographic information systems (GIS), ensures that firefighters keep pace with the latest advances in the fire service.
- **Thoroughly-Revised Section on the National Incident Management System (NIMS)** provides an overview to this new Homeland Security Presidential Directive, which creates a nationwide approach to prevent, prepare for, respond to, and recover from domestic incidents, as well as an overview to the Incident Command System as specified in FEMA's ICS I-100a course.
- **Updated Statistics, Examples, and References** offer students a handy resource for current information in order to be better equipped to enter the fire service.

SUPPLEMENT TO THIS BOOK

To assist instructors in classroom preparation and training, Delmar Cengage Learning offers *Instructor Resources on CD-ROM* to accompany this book, which contain the following features:

- **Lesson Plans** correlate to the accompanying PowerPoint® presentations and prepare the instructor for the classroom. The Lesson Plans are also editable to meet the specific needs of the course.

- **Answers to Review and Discussion Questions** provide feedback on the questions at the end of each chapter and enable instructors to evaluate student knowledge of the content.
- **PowerPoint® Presentations** combine illustrations and photos with an outline of the important concepts in each chapter. The presentations correlate to the Lesson Plans and are editable to meet the specific needs of the course.
- **Test Banks** available in ExamView 6.0® format enable instructors to evaluate student comprehension of the concepts presented in each chapter. Question banks are editable—allowing instructors to create new tests, add or remove questions, and revise existing questions.
- **Motivating Your Students, Lesson by Lesson** provides instructors with a short introduction to the lesson that they can choose to deliver to their students at the beginning of each class. It outlines background information on the topic and explains the importance of the information contained within the chapter. These brief introductions are intended to motivate the student and engage them in the topic being presented.
- An **Image Gallery** provides photos, tables, and line art from the book. You may use this art to supplement PowerPoint® presentations or to create presentations for classroom instruction.
- A **Correlation Grid** outlines the National Fire Academy FESHE course outcomes for *Principles of Emergency Services.* The NFA is focused on providing standardized training for firefighters, and Delmar, Cengage Learning strives to provide the high-quality training materials necessary to meet that goal.

NOTE

To access the video clips referenced in this book, please visit www.cengagebrain.com. At the CengageBrain.com home page, search for this book using the search box. On the page illustrating this book, click on the "Access Now" button and this will direct you to our additional resources!

ABOUT THE AUTHOR

Robert W. Klinoff is a California Certified Chief Officer, National Fire Academy Executive Fire Officer, and retired Chief Deputy Fire Chief with the Kern County Fire Department. He is a 35-year fire service veteran. His background includes experience as a firefighter while a student at Columbia College, as an ambulance driver, as a firefighter with the U.S. Forest Service, as a firefighter with the City of San Gabriel, California, and 29 years with the Kern County Fire Department. He has served the Kern County Fire Department in Operations and as a Training Officer and Fire Marshal.

His education includes an associate of science degree in Fire Science from Columbia College and a bachelor's degree in Occupational Studies—Vocational Arts from California State University Long Beach. He is a California State Fire Service Training and Education System–certified Chief Officer, Master Instructor, Movie and Television Fire Safety Officer, and Hazardous Materials Specialist. Robert is a graduate of the Executive Fire Officer Program at the National Fire Academy.

As an instructor for the State of California, he teaches both Company Officer– and Chief Officer–level courses. He also serves as adjunct faculty for the National Fire Academy, teaching Infection Control for Emergency Response Personnel, Managing Company Tactical Operations, and Chemistry of Hazardous Materials.

Robert has further served his community as the president of the Kern SAFE Coalition, an agency dedicated to childhood injury prevention. He is a National Fire Protection

Association *Risk Watch* Champion and a National Highway Traffic Safety Administration–certified Child Passenger Safety Technician/Instructor.

Under the National Interagency Incident Qualifications System, he is certified as an Incident Commander, Safety Officer, Division Supervisor, and Strike Team/Task Force Leader. Robert served as a Safety Officer on California Interagency Management Teams. He and the team responded to the Pentagon incident on September 11, 2001 as well as numerous other major incidents throughout the country. As a result of the team's efforts at the Pentagon, they received the Group Honor Award for Excellence from the U.S. Department of Agriculture.

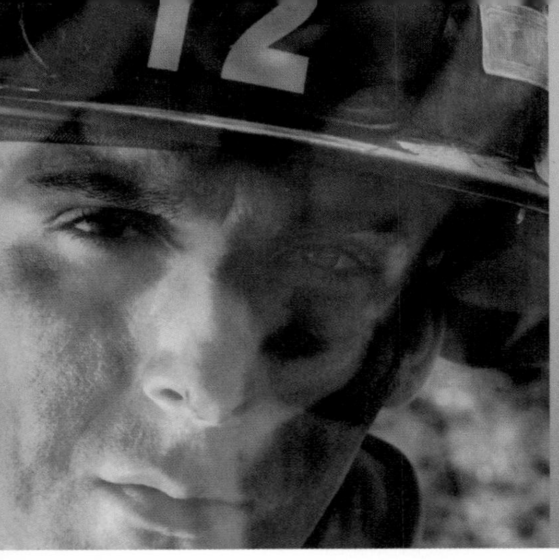

ACKNOWLEDGMENTS

I would like to personally thank Jack Amundsen, the retired chief of the Columbia College Fire Department. He is the person who first hired me as a firefighter in 1971 and showed me what a rewarding career it can be. He always encouraged me to be the best that I could be. Many other instructors and co-workers have helped me to develop myself as a professional and increased my base of knowledge. They have done much to elevate the professional standing of all firefighters in the eyes of the public.

I would also like to thank my wife, Helen, who has stood by me during my career with long periods away from home due to incidents and course work. She has always been supportive.

Many people have contributed their time and effort in assisting me with gathering the information to write this text. Their interest in developing the fire service should be recognized.

I would also like to thank the staff at Delmar, Cengage Learning for encouraging me to undertake the writing of this text and for their help in its preparation.

In addition, the author and Delmar, Cengage Learning would like to thank the reviewers who contributed to the fourth edition of this book:

Professor Gary Courtney
Lakes Region Community College
Laconia, NH

Firefighter Joe Guarnera
Massachusetts Firefighting Academy
Stowe, MA

Al Iannone
Fire Technology Coordinator
Sacramento Regional Public Safety Training Center
American River College
Sacramento, CA

Terry Koeper
Professor of Fire Science
Crafton Hills College
Yucaipa, CA

Judith Kuleta
Fire Science Program Advisor
Bellevue College
Bellevue, WA

Professor Lee Silvi
Director
Fire Science Technology and Emergency Management Programs
Lakeland Community College
Mentor, OH

Paula Simone
Fire Program Director
Central Oregon Community College
Redmond, OR

David Weirather
Fire Prevention Staff Officer
Bakersfield City Fire Department
Bakersfield, CA

Also, we would like to thank all the reviewers and contributors that participated in previous editions of the text.

FIRE AND EMERGENCY SERVICES HIGHER EDUCATION (FESHE)

In June 2001, the U.S. Fire Administration hosted the third annual Fire and Emergency Services Higher Education Conference, at the National Fire Academy campus, in Emmitsburg, Maryland. Attendees from state and local fire service training agencies, as well as colleges and universities with fire-related degree programs, attended the conference and participated in work groups. Among the significant outcomes of the working groups was the development of standard titles, outcomes, and descriptions for six core associate-level courses for the model fire science curriculum that had been developed by the group the previous year.

The six core courses are Building Construction for Fire Protection, Fire Behavior and Combustion, Fire Prevention, Fire Protection Systems, Principles of Emergency Services, and Principles of Fire and Emergency Services Safety and Survival.

FESHE CONTENT AREA COMPARISON

The following table correlates the Model Curriculum Course Outcomes for the Principles of Emergency Services requirements to this textbook's chapters.

COURSE CORRELATION GRID

Principles of Emergency Services		
Course Description:	This course provides an overview to fire protection and emergency services; career opportunities in fire protection and related fields; culture and history of emergency services; fire loss analysis; organization and function of public and private fire protection services; fire departments as part of local government; laws and regulations affecting the fire service; fire service nomenclature; specific fire protection functions; basic fire chemistry and physics; introduction to fire protection systems; introduction to fire strategy and tactics; and life safety initiatives.	***Introduction to Fire Protection 4th Edition*** Chapter Reference
Outcomes:	Illustrate and explain the history and culture of the fire service.	1
	Analyze the basic components of fire as a chemical chain reaction, the major phases of fire, and examine the main factors that influence fire spread and fire behavior.	4
	Differentiate between fire service training and education and explain the value of higher education to the professionalization of the fire service.	1, 9
	List and describe the major organizations that provide emergency response service in both the public and private sector.	5
	Identify the protection and emergency-service careers in both the public and private sector.	2
	Define the role of national, state, and local support organizations in fire and emergency services.	5
	Discuss and describe the scope, purpose, and organizational structure of fire and emergency services.	3, 7
	Describe the common types of fire and emergency service facilities, equipment, and apparatus.	6
	Compare and contrast effective management concepts for various emergency situations.	13
	Identify the primary responsibilities of fire prevention personnel including: code enforcement, public information, and public and private fire protection systems.	10
	Recognize the components of career preparation and goal setting.	1
	Describe the importance of wellness and fitness as it relates to emergency services.	14

CHAPTER 1

Fire Science Education and the Firefighter Selection Process

LEARNING OBJECTIVES

Upon completion of this chapter, you should be able to:

- Explain the differences between a community college certificate, an associate degree, and a four-year degree in fire science.
- List the advantages of obtaining a certificate or degree from a regionally accredited institution.
- Describe the availability of on-line fire science programs and training.
- Assess your career potential in the fire service.
- Give examples of work ethics.
- Explain the need for sensitivity to diversity inside and outside of the workplace.
- Describe the different levels and availability of training programs.
- Give examples of different types of personnel development programs.
- List the steps in the selection process and important aspects of each.
- List ways you can prepare for the selection process.
- Explain the purpose and importance of the probationary period.
- Identify the steps in setting SMART goals.

INTRODUCTION

The fire science **curriculum** is designed to produce a student with a comprehensive background knowledge in the **technical training** for fire suppression and prevention. **Manipulative training**, the actual hands-on firefighter training, may or may not be included as part of the curriculum, depending on the school attended.

The field of fire science–related courses includes many more people than just the firefighter. Not all people are physically or mentally qualified for the rigorous and demanding profession of firefighting. Some may choose to avoid the dangers of a strictly firefighting career. Some fire service–related jobs do not require the same level of physical ability or the ability to operate under severe stress in dangerous situations. These other jobs are closely related to, and a necessary part of, the delivery of a total fire protection system to the community.

curriculum A particular course of study.

technical training Training in the specifications and limitations of equipment or calculation of information necessary to operate the equipment.

manipulative training Training in the operation of tools and equipment.

COLLEGE FIRE SCIENCE PROGRAMS

In some colleges, fire science curricula allow the student to earn a certificate without completing all of the requirements of a degree in the program. The certificate program requires the completion of a set number of accredited core courses and additional specified courses in the area of general education. Although not a degree in and of itself, the certificate attests to the accumulation of a body of knowledge in the fire science subject area.

NOTE Research the college's catalog and meet with a college counselor to plan a course of action to achieve the desired goal.

The completion of an associate degree in fire science requires more general education units to accomplish. Some of the courses may be transferable to a four-year college; others are not.

The core courses may or may not be transferable. They may serve as prerequisites for acceptance and count as credit in an **upper division** program. Each school differs in its requirements. It definitely benefits the student to research the college's catalog and meet with a college counselor to plan a course of action to achieve the desired educational goal from a regionally accredited institution.

In 2000 a conference was held at the National Fire Academy (NFA) in Maryland. At this conference fire service leaders from throughout the country and state directors of fire service training came together to establish recommendations for fire service–related training curriculums and created the National Fire Science Curriculum Committee (now called the National Fire Science Programs Committee, or NFSPC).

The plan the NFSPC produced is called the *Fire and Emergency Services Higher Education (FESHE) Model Curriculum: Transforming to a National System*[1] The committee continues to meet on an annual basis to update and refine the model curriculum. At the 2002 conference, the 2002 FESHE IV, an experienced-based model that recommends an efficient path for fire service professional development, was produced. This model addresses how education and training should be integrated. The National Professional Development Model (**Figure 1-1**) illustrates the relationship between education and training in a professional development matrix.

upper division courses College-level courses that are applicable to a degree program for a bachelor's degree or higher. Lower division courses are those taken on the college level that are either prerequisites for higher levels of study or are used to receive an associate of arts or sciences degree.

FIGURE 1-1 National professional development model.

Courtesy of Federal Emergency Management Agency (FEMA)

The core six-course curriculum for associate degrees is as follows:

- Building Construction for Fire Protection
- Fire Behavior and Combustion
- Fire Prevention
- Fire Protection Systems
- Principles of Emergency Services
- Principles of Fire and Emergency Services Safety and Survival

In addition, curricula were developed for noncore courses that may be offered. These are:

- Introduction to Fire and Emergency Services Administration
- Fire Investigation I
- Fire Investigation II
- Fire Protection Hydraulics and Water Supply
- Hazardous Materials Chemistry
- Legal Aspects of the Emergency Services
- Occupational Health and Safety
- Strategy and Tactics

The NFA also released its 15-course upper-level Degrees at a Distance Program (DDP) curriculum to accredited baccalaureate degree programs. The DDP remains as NFA's delivery system for the courses. These courses are also accessible through programs that have signed up with the NFA. They include:

- Analytical Approaches to Public Fire Protection
- Applications of Fire Research
- Community Risk Reduction for the Fire and Emergency Services
- Disaster Planning and Control
- Fire and Emergency Services Administration
- Fire Dynamics
- Fire Investigation and Analysis
- Fire Prevention Organization and Management
- Fire Protection: Structures and Systems
- Fire-Related Human Behavior
- Managerial Issues in Hazardous Materials
- Personnel Management for the Fire and Emergency Services
- Political and Legal Foundations of Fire Protection

Two additional courses, developed by the NFSPC Bachelor's Group, are available in the model course outline format:

- Issues in Fire/EMS Management
- Advanced Principles in Fire and Emergency Services Safety and Survival

Several four-year colleges in the United States offer bachelor degree programs in fire-related fields. One of these schools is the California State University at Los Angeles,[2] offering a degree in fire protection administration. Oklahoma State University offers a degree program in fire protection and safety technology.[3] Included in this program are summer internships for students, giving them actual experience working in fire departments and industry. The University of Maryland offers a degree program in fire protection engineering.[4] The Federal Emergency Management Agency (FEMA), through the Emergency Management Institute, has a list of schools that offer higher education programs from the Associate to the Doctoral level in Emergency Management and Homeland Security.[5]

On-line Programs

A more recent development in firefighter training and education is the availability of on-line programs. These programs allow firefighters and prospective firefighters to earn degrees in fire science and related training without having to attend courses at a traditional "brick and mortar" facility. These programs are available from numerous community colleges and schools of higher learning throughout the country. Associates through Masters Degrees may be earned in these programs. Some of the advantages are that one does not need to live in close proximity to the school being attended to complete the courses. Another is that firefighters with a set duty schedule may not be off duty on the days that the course is taught in a traditional classroom setting. Previously, personnel would have to arrange duty trades or other time off to attend the courses. They can now participate in the course over the computer and complete the work as time permits within the course requirements, such as completing the work for week two during the second week of the course. Primary examples of these types of courses, from a training standpoint, are the National Incident Management System (NIMS) courses offered by the Federal Emergency Management Agency through the Emergency Management Institute (EMI) in an on-line format. These courses include:

- IS-100.a—Introduction to the Incident Command System (ICS)
- IS-200.a—ICS for Single Resources and Initial Action Incidents
- IS-700.a—National Incident Management Systems (NIMS), An Introduction

For more information on these courses visit www.training.fema.gov

Northwood University in Michigan offers a distance/online accredited bachelor's degree program in fire service management that grants credit for training certifications and life-long experience.[6] Other schools, such as the International Association of Fire Fighters Virtual Academy, offer distance learning or extended university programs. These programs work much the same way as the open learning program, allowing students to complete their course work from locations away from the college campus. These schools are just a sample of those offering these types of degree programs.

Pursuing a higher education in fire service–related courses can make you a more effective member of the fire service community. It may also help you achieve promotions after you gain employment. An education may have a direct dollar value in the workplace. Many fire departments offer, as a part of their compensation package, a pay incentive for a fire science certificate or degree or for other specified types of training certification. This usually ranges from 2½% to 10%. Calculated out over a 30-year career and carried over into retirement, this amounts to quite a bit of money.

In some departments the completion of certain courses is a condition for completion of the probationary period. The probationary period is described later in this chapter.

OTHER COLLEGE PROGRAMS

Another popular course of study for fire professionals is public administration. Most fire departments operate as public agencies governed by local or state government—this makes a public administration educational background vital to the fire executive. Another reason for the popularity of this program is the availability of a master's degree in public administration.

There are other courses of specialized study beneficial to the fire professional in the fields of emergency management, risk management, industrial hygiene, law, emergency medicine, and chemistry.

An alternate and complementary course of study to the fire science technical education is the firefighter certification. Based on National Fire Protection Association *Standard 1001, Fire Fighter Professional Qualifications*,[7] this course of study is primarily manipulative in nature with technical instruction where necessary. The course of study includes instruction in fire behavior, fire extinguishers, self-contained breathing apparatus, ropes and knots, forcible entry, rescue, water supply, fire streams, ventilation, salvage and overhaul, fire cause determination, fire suppression techniques, automatic sprinklers, and fire prevention inspection.

CAREER POTENTIAL ASSESSMENT

Becoming a firefighter requires you to be a person of the highest moral and ethical character. You represent one of the proudest professions there is. When you pin on a firefighter's badge, you represent hundreds of years of tradition of selfless service and sacrifice. You are assuming the reputation of the entire fire service. You are expected to act at great personal risk to save the lives and property of others without seeking recognition or acclaim.

When watching the news on television you may have noticed that no matter what the disaster, the fire department is usually there. Fire department personnel are usually in the background, quietly performing their jobs, whether rescuing people from flood or fire, giving medical attention to victims of crimes or accidents, delivering babies, or preventing the spread of hazardous materials. The fire department does not have numerous television action shows to advertise its abilities. People do not often think much about the fire department until they need it. The public expects a high level of professionalism and competence. Political candidates do not run on fire protection platforms, as they do with law and order. Fire departments may not get much press, but they still give the public much more than their money's worth.

Some people want to become firefighters because they see it as their chance to become a hero. Anyone who seeks this career for the singular purpose of becoming a hero is misguided. There are few opportunities to become a hero, and besides, if you do something heroic, what if nobody notices? Firefighting is not about being a hero; it is about doing the best you can to save lives and property every time you are given the opportunity.

Firefighting entails a certain amount of danger and excitement. Over 100 firefighters a year make the ultimate sacrifice and give their lives in the line of duty.[8] Statistics on firefighter deaths for the years 1977 through 2008 are contained in Appendix A. Firefighting is dangerous, but it is not a profession for those who disregard their own safety or the safety of others. With proper training and care, you should be able to make it through your career and enjoy your retirement in relatively good health.

This profession requires long hours of drill and study to master the myriad tasks you are expected to perform, often in extremely stressful situations. It has been proven that people do what they were trained to do when things go badly and danger surrounds them. You must be willing to perform the preparation before you can perform the job. Training never ends, due to the constantly changing demands of the workplace. When new chemicals are developed, new industrial processes are invented, building construction techniques change, new subdivisions of homes are built in more remote locations, or changes in the public's behavior occur, the fire department ends up being involved.

NOTE You must be willing to perform the preparation before you can perform the job.

Firefighters suffer the same ills and problems as the rest of society. The divorce rate among firefighters is high; alcoholism and drug abuse do occur. To many firefighters, their co-workers are their extended family, and what affects one affects all. Firefighters spend long periods of time together **on duty** and see each other under the worst of conditions. Responding to other people's tragedies has a way of drawing firefighters closer together and forming strong bonds among them.

Firefighters are required to show compassion and become skilled in dealing with people at the worst times in their lives—at accident scenes when loved ones have been killed or seriously injured, and at fires and other incidents of devastating loss. Firefighters must be able to deal with the injury and deaths of people of all ages, from infants to elders,

on duty The time firefighters spend performing their jobs.

under horrible conditions. As firefighters become more involved in the delivery of medical aid, they see more instances of child abuse and other tragedies.

The modern fire service has started to address the problem of stress in the workplace. Many agencies have employee assistance programs that allow firefighters to talk confidentially with a counselor when they are having stress problems at work or home. There are also provisions for alcohol and drug abuse treatment programs. No one can be a firefighter and be totally unaffected by what they see in the line of duty. **Critical incident stress debriefings** are being introduced to help firefighters cope with particularly bad incidents that they have responded to. Imagine, if you can, responding to the crash of an airliner with the expectation of rescuing people. The crash is such that all 300 persons aboard are killed. Firefighters still search the wreckage to see if there is anyone left alive, the same as they did after the bombing of the Federal Building in Oklahoma City in 1995 and the attack on the World Trade Center in New York City in 2001. Another possible scenario that firefighters face is responding to a vehicle accident where a family member, personal friend, or co-worker has been killed or seriously injured. In such situations it is not uncommon to ask yourself, "What could I have done to save them?" Another common question when a co-worker is killed or injured is, "What if it had been me?" After experiencing enough of these kinds of incidents, they start to build up inside you and manifest themselves in home or work-related problems. You must be willing to ask for help to cope when necessary.

Firefighters must be team members. They must be willing to give up personal desires to benefit the team. When the team succeeds, the *whole* team succeeds.

When the team performs extraordinarily well and praise is theirs, it should be shared equally. When a baby was saved from a burning house, one firefighter carried the baby out and got his picture in the paper—but without the help and backup of the others at the scene, the rescue would not have happened. No one can perform the job alone. Whether at emergencies or on routine work assignments, the whole team needs to pitch in and help until the work is done.

Washing the equipment and doing dishes are not glamorous or fun, but they are a part of the station routine (**Figure 1-2**) and working together as a group makes the job go by much more quickly.

A career-long commitment to physical fitness is part of being a firefighter. In 2008, 45 firefighters, or almost half of the 118 firefighters who died while on duty, fell victim to sudden cardiac death (heart attack).[9] Your fellow firefighters expect you to be able to carry your share of the load in a physically demanding profession. Letting yourself get out of shape not only endangers you, but also your co-workers and the public. They have every right to expect you to stay fit (**Figure 1-3**). If one of them goes down, it may be up to you alone to get them out.

critical incident stress debriefing A discussion in which personnel are encouraged to express their feelings after responding to and operating on particularly stressful events that result in high loss of life or other significant conditions. Conducted to help personnel better deal with their emotions.

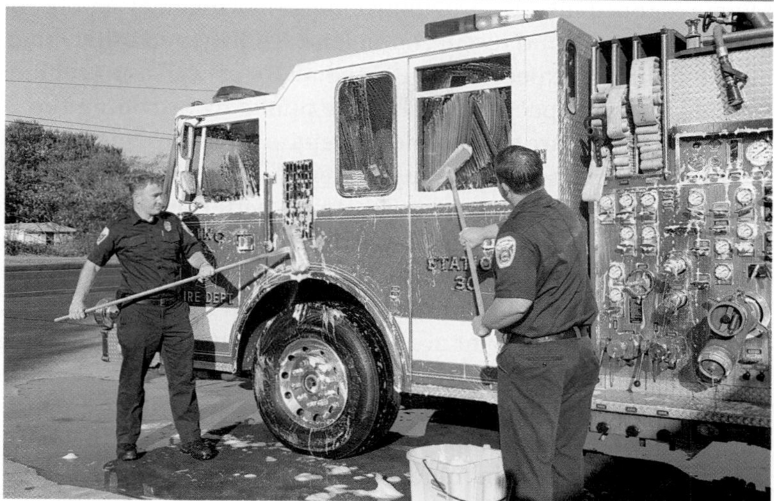
Delmar/Cengage Learning

FIGURE 1-2 Firefighters performing routine station duties.

Delmar/Cengage Learning

FIGURE 1-3 Firefighter maintaining level of physical fitness.

NOTE A career-long commitment to physical fitness is part of being a firefighter.

The role of the fire service has changed and will continue to do so. The job of firefighter has changed to meet the demands of the new fire service. It used to be that the firefighters' main job was to control and extinguish hostile fires. Nowadays the fire department has taken responsibility for providing emergency medical service, fire prevention, hazardous materials, search and rescue, homeland security, and other emergency services. The modern firefighter is expected to be an educator as well as a technician. As public employees, firefighters are expected to serve the public by providing the services it requires. If you think that all you are going to do is prepare for and extinguish fires, you are wrong. Instead of saying, "We are doing enough already," the contemporary fire chief is asking, "What more can we do and in what ways can we better serve the community?"

As a member of the fire department you may very well be asked to give of your time off to participate in community programs. These programs are beneficial because they promote the image of the fire service and firefighters as a whole. If you were to ask fire service professionals what are the most important things the fire department must have, the item at the top of the list is the support of the public. For the fire department to exist and receive the money it needs to function, the political support of the public is essential. As a firefighter, you are not solely a community employee, but a part of the community. You must always keep in mind who in fact pays your wages.

You may very well ask yourself, "Am I ready and willing to proceed into a burning building when everyone else is running out?" If you can meet all of these criteria and successfully complete the selection process, you could pursue one of the most exciting and personally satisfying careers there is.

HUMAN RELATIONS AND WORK ETHICS

When choosing a career in the fire service you not only work with other firefighters but are also often expected to live with them, sometimes 24 hours a day (with 48 hour shifts becoming more popular) in the fire station during your tour of duty. You must also perform such tasks as station and vehicle maintenance, training, meal preparation, and sleeping. An ability to get along with others is very important in the often cramped confines of the fire station. In some departments the situation is much like a military barracks.

In the area of human relations the firefighter has to be prepared to deal with diversity in the workplace. Women and minorities who were formerly excluded from service are now a large part of the force. Affirmative action programs and equal opportunity employment laws have guaranteed these groups representation in the fire department.

NOTE One of the quickest ways to lose your job as a firefighter is to become involved in harassment.

Equal Employment Opportunity law prohibits discrimination against any person in the classified service or any person seeking admission thereto because of race, national origin, sex, age, physical disability, color, medical condition, marital status, ancestry, or union activity. Discrimination on the basis of age, sex, or physical disability is prohibited except where age, sex, or physical requirements constitute a bona fide occupational requirement. Physical disability is one of the few reasons a person cannot be hired by the fire department as a firefighter. The Americans with Disabilities Employment Act specifies that there must be a clear reason why the disability excludes the person from being hired. As we will see in Chapter 2, there are jobs in the fire department other than firefighter that a person with a disability can hold.

One of the quickest ways to lose your job as a firefighter is to become involved in harassment. There is no tolerance for this type of activity in the modern fire department. Sensitivity must be shown to all groups, at the station and on the scene of an emergency. One careless remark can cost you your job and tarnish the image of all firefighters. As a firefighter, you will respond to assignments requiring you to serve people of all backgrounds. It does not matter how much money they have or how they live; all of your customers deserve to be treated with dignity and respect and provided the full benefit of your best efforts.

Many attributes make a firefighter a valued member of the fire department community. Loyalty is hard to define, but it is sticking by your fellow firefighters through good times and bad. It involves not bad-mouthing your superiors, subordinates, or co-workers. Much of the time you will not be out performing on emergencies. During this time it is very easy to fall into the habit of griping and gossiping. This only makes others think that you will talk about them when they are not around. If other firefighters do not feel that they can trust you, they will not want to work with you.

Dedication to duty is how you approach your job. As a new firefighter, you are expected to be the first one to volunteer for the dirty jobs, such as crawling under a house to retrieve a lost kitten. It does not stop there, however. Your dedication to duty should last your entire career. It is not enough to apply yourself for the first couple of years until you find yourself a comfortable station and crew that you can call home. You must constantly strive to be the best firefighter you can be.

A firefighter must have the ability to accept hardship without complaint. When the alarm sounds, the firefighter goes to work. It does not matter that you are tired from the last incident or that you haven't had a chance to eat, even though the food is on your plate. Long hours under stressful and often extremely tough conditions are included in the job.

In some departments or station assignments in larger departments, there are few calls for service and members must constantly train to maintain their skill proficiency. One of the greatest dangers to firefighters is complacency, or taking the attitude that I do not need to practice because my skills will not be necessary.

Being able to follow orders is a part of the team effort (**Figure 1-4**) required of all firefighters. On the fire ground is no time to argue with your officer or to **freelance**. Only through the total directed effort of all of the people on the scene can the problems be overcome. Around the station many orders are stated as requests: this does not mean that

freelance The act of performing operations without a coordinated effort or the knowledge of one's superior officer.

FIGURE 1-4
Firefighters attacking a structure fire.

Courtesy of Edwina Davis

they can be ignored. It is just a polite way of getting things accomplished and establishing a more relaxed, stress-free atmosphere.

You must have the ability and willingness to learn. The field of fire suppression and prevention is dynamic and ever changing. The fire service itself has changed drastically in the last few years. The fire department has an ever-expanding role in providing service to the community. Major changes have come about in the types of hazards encountered and how they are handled. The equipment available has also changed at a rapid rate. Firefighters must be ready and willing to accept and adapt to these changes or become dinosaurs—and we all know what happened to the dinosaurs.

Every firefighter must have a positive safety attitude.

You must be willing to accept personal responsibility for your actions. When given a job, it is your responsibility to complete it, not someone else's. You should not have to be closely supervised once you are trained to perform the required job. Emergency activity can and does disrupt the other work that firefighters perform on a regular basis. If this happens, it is your responsibility to return to the previously assigned work once the equipment is placed back into service after an incident. If you are unable to complete the assigned task by yourself, seek help or advise your supervisor.

Last but not least, every firefighter must have a positive safety attitude. Firefighting is an inherently dangerous profession. Numerous firefighters are disabled and others lose their lives every year in the line of duty. With a proper safety attitude you can do your best to avoid serious injury and death while still performing your job aggressively.

TRAINING PROGRAMS

Pre-service training programs in manipulative skills are available through **Explorers** (Figure 1-5), **volunteer firefighting** programs, **reserve/cadet programs**, the National Junior Firefighter Program from the National Volunteer Fire Council, colleges, and training

Explorers A program of the Boy Scouts of America for persons 15 to 21 years of age. The Explorers work in conjunction with a professional organization such as the fire or police department to learn the operation and job requirements.

volunteer firefighting Performing firefighting services without pay. In some areas a variation of this is the Paid Call Firefighter program. Under this program firefighters are paid a specified sum when they respond to incidents or attend training.

reserve/cadet programs Organized programs sponsored by paid fire departments that provide training in return for personnel volunteering their time.

FIGURE 1-5
Explorer program firefighters attending ladder training.

Delmar/Cengage Learning

associations. The goal of these programs is to teach the actual skills necessary for a firefighter to perform on the fireground. These programs start with an academy that teaches skills in handling ladders, fire extinguishers, salvage equipment, self-contained breathing apparatus (SCBA), and hose lays.

Sometimes having medical training such as emergency medical technician (EMT) or paramedic training can be of benefit when seeking employment. This varies by department and the level of medical service it provides.

Some programs are sponsored by individual fire departments, associations, or professional groups, with college credit issued. Generally you are required to attend on your own time at your expense. Often incentives are offered in the way of special treatment in the hiring process. They may even be used to establish a direct hiring pool for the department(s) involved. The completion of the fire science program on the college level along with a certificate from a certified pre-service academy can assist you in competing for the job of firefighter.

In-service training programs have been developed to train active firefighters. These programs start with the academy and move on to the station, battalion, department, area, state, and national levels. Training programs are sponsored by the departments themselves, state and local training officer's associations, the offices of State Fire Marshals, and colleges. The courses run the gamut of subject matter from hose lays to specialty courses in hazardous materials and heavy rescue, and often require department-sponsored attendance due to **worker's compensation** coverage.

The first level of training for newly hired firefighters is the academy. The new firefighter reports to the school instead of to the fire station. The purpose of the academy is to train the new firefighter in department equipment and methods, in courses that are required by law (such as emergency medical training and Hazardous Materials First Responder Operational), and to observe the new firefighter's physical and mental performance (**Figure 1-6**). During the academy the new firefighters are evaluated on their performance on written tests and during **drills**. If for some reason new firefighters do not measure up to department standards, they are dismissed. This time can be very stressful for new firefighters. They are being watched very closely for the slightest infraction. A great amount of homework and studying is required to perform well on the written exams.

NOTE If for some reason new firefighters do not measure up to department standards, they are dismissed.

Numerous schools around the country offer technical training programs. One of the best known is the National Fire Academy and the Emergency Management Institute (both

worker's compensation Money paid to persons who have been injured in the course of their employment and are unable to work either temporarily or permanently.

drill The practicing of tasks and jobs to improve performance.

FIGURE 1-6 Academy firefighters performing physical fitness training.

Delmar/Cengage Learning

located at the National Emergency Training Center), which offer courses year-round at the facility in Emmitsburg, Maryland. Instructors from all over the country are employed to present the widest viewpoint and to give the courses national appeal. The instructors are of the highest caliber and are recognized as experts in their field. There is no tuition charged to attend the National Fire Academy. For students to attend they must be sponsored by their department and be accepted after filing an application.[10] When accepted, the student's travel is reimbursed and lodging is provided. All the student pays for while attending the course is food.

A model fire training program follows the Fire Service Career Ladder as seen in **Figure 1-7.**[11] Most of the instructors are fire service professionals and relate the material to the firefighter's needs very well. Several of the courses parallel those offered by the National Fire Academy.

Throughout the United States there are colleges that sponsor fire department–related courses at their facilities. A program that covers a wide variety of fire-related subjects is offered at Texas A&M.[12] Specialized training in various types of firefighting is offered at other locations as well.

PERSONNEL DEVELOPMENT PROGRAMS

A new trend in the fire service is the personnel development program. In this program a firefighter is trained as high as two ranks above the one currently held. The purpose of this program is to develop an understanding of how the department works and to prepare

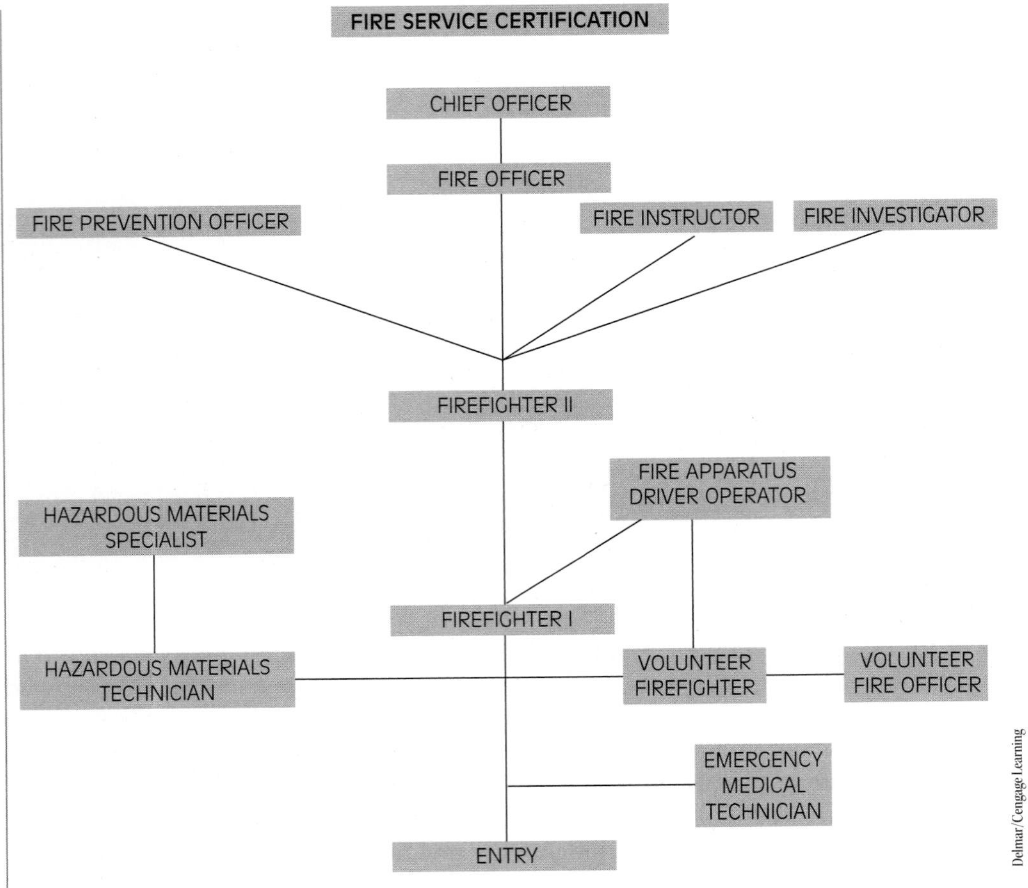

FIGURE 1-7
Example of career ladder showing training certifications.

the leaders of tomorrow. In some instances, students are assigned a **mentor** to aid in goal setting, to help monitor their progress, and to assist them as necessary.

Modern firefighters must be **generalists** and be able to perform many fire-fighting functions. They may also want to become **specialists** in one or more areas of fire

mentor A person who guides and directs toward a goal.

generalist A person with general knowledge of no great depth in many subject areas.

specialist A person with extensive training in one area of operations or information.

department operations. The modern fire service has so many responsibilities that no one can know everything or be an expert in every aspect of the job to be performed. It is the duty of every firefighter to seek training on all levels.

SELECTION PROCESS

The selection process for the fire department contains a number of steps. Different departments use various combinations of the steps presented here. It is important to research the department you are applying for to determine which of the steps are used and prepare accordingly. The steps presented here, for the purpose of illustrating the process, are: application, written examination, skills test, oral interview, physical agility/ability, medical examination, background check, final oral examination, and probationary period. A representative selection process, from start to finish, is presented here.

NOTE It is important to research the department you are applying for to determine which of the steps in the hiring process are used and prepare accordingly.

Recruitment

The selection process starts during recruitment. Fire departments are looking for the most qualified applicants they can find. When you apply for a fire department job you will be competing for a limited number of openings against others who have prepared themselves to varying degrees.

Students who are currently in or have completed fire science programs have already shown an interest in a fire department career and a willingness to invest time and money in pursuing an education in this field. They have also shown the drive and ability to study and learn in a classroom environment.

Most fire departments have prerequisites for application. The most basic prerequisites are a valid driver's license, a high school diploma or general equivalency diploma (GED), and no felony convictions. Another prerequisite often used by smaller departments that do not have the money or the staffing to place a newly hired firefighter in an academy is a Firefighter I certification.[13] Smaller departments are usually funded to have only the minimum staffing required to keep their apparatus responding; they need new personnel on the equipment and ready to go the first day they report for work. When new personnel are hired and placed in an academy, they are getting paid and the department does not receive any direct benefit from their employment. Some departments promise employment at the completion of an academy program of their choosing, but you must attend on your own time. By using instructors from their own department, it also allows them to get a good look at you before you are hired. Another common prerequisite may be **emergency medical technician** or **paramedic** certification.

emergency medical technician A specified level of medical training that usually consists of approximately 100 hours of classroom and practical training.
paramedic An advanced level of medical training. Paramedics can perform invasive procedures on the patient, such as starting intravenous lines.

Application Process

Administering examinations to large numbers of applicants is time consuming and therefore expensive for public agencies. There are several ways of limiting applications. One way is to limit the number of applications given out. It does not make much sense to give out 600 applications when the department's anticipated need is for one to ten new firefighters. Another method is to give out applications for one day only. Limited advertising is also used by publishing the job announcement only in the local newspaper. In some departments that have reserve/cadet firefighter programs, those in the program get a guaranteed application out of an already limited number.

With the rise in health care costs and as a long-term cost-saving measure, many departments have gone to a nonsmoking policy. Applicants agree to not use tobacco in any form for the term of their employment. Other departments are even more restrictive in that they require applicants to sign an affidavit at the time of applying that they have not used tobacco for a year prior to being hired. Violation of the nonsmoking agreement can result in disciplinary action and possible termination.

A limit affecting applicants is sometimes imposed as a result of court action. In some cases the fire department has been found to not represent the **demographics** of the community. In these cases the court may impose an order requiring an affirmative action program to be put into place.

Some jurisdictions also have a residency requirement. A part of the reason for these requirements is political—to hire local residents for public jobs. Another reason is to have at least some of the force available on a **call-back** basis in case of a major disaster or large fire. Some of these residency requirements require the applicant to reside in the jurisdiction prior to their application being accepted. Others require the new firefighter to reside in the jurisdiction or within a certain radius for a specified period of time after being hired. Fire departments, like any other public agency or business, are trying to attract only the best applicants and at the same time keep costs down in the hiring process.

demographics The statistical characteristics (e.g., age, race, gender, income) of the population of an area.
call-back A recall of personnel to on-duty status, usually due to an emergency situation.

The application process consists of first, finding out when the application will be available, and second, when it must be completed and returned. One way to avoid the problems with finding out about application dates is to subscribe to an application notification service. These services will, for an annual fee, send you a postcard with information on upcoming firefighter examinations and application filing periods along with prerequisites such as Firefighter I, EMT, or Paramedic. Some departments are now posting job openings on the Internet. Another way to receive notification of application filing dates is to visit the personnel department of the jurisdiction of your choice and fill out a job interest card (**Figure 1-8**). This postcard will include your name and address and is left with the personnel department. When the application filing date is announced, the department will mail the card to you. Be sure to take your own stamps for the cards. Most personnel departments will not mail out applications—they must be picked up in person.

NOTE Most personnel departments will not mail out applications—they must be picked up in person.

NOTE The job announcement is very important as it contains much of the information you need to fill out the application.

When you have picked up your application, make a copy. Use the copy for practice and return the original. The application should come with a job announcement (flyer) (**Figure 1-9**). The job announcement is very important as it contains the information you need to fill out the application. Most applications are a standard form for any job in that jurisdiction. Always ask if there are any supplemental materials with the application. There may be additional materials in the form of a pretest guide and/or study booklet. There may also be a preparation class offered to assist you in performing well on the exam.

Before leaving the personnel department and starting to fill out the application, read the job announcement carefully. Make sure you have the right job announcement for the position you are applying for. They are usually all the same color. There is nothing like driving home and then having to return because you have the job announcement for Painter. Be sure to pay particular attention to the date the application must be returned. Applications turned in late are not accepted.

A lot of information on the job announcement will assist you in filling out the application. The first item is the test number and title. This is entered on the application for personnel department sorting purposes. The job announcement also contains information about pay, working hours, and locations and also specifies the minimum requirements of the position you are applying for.

The best, and only way, a job application should be turned in is typed. The application must, at the minimum, be printed neatly in ink. This has changed somewhat as many departments now require applications be filed on-line. Your first impression on the oral panel may be your application. The members of the panel usually have your application

NOTIFICATION OF RECRUITMENT Item No. __________

(Print) Last Name First Name Middle Initial

Title of Job ______________________________

ARE YOU WILLING TO WORK SHIFTS ____Yes ___ No

DO YOU SPEAK SPANISH FLUENTLY ___Yes ___No BAKERSFIELD ONLY ☐

Telephone No. ______________________ Date ______________

The KERN COUNTY Civil Service Commission is presently recruiting for the above named job. Applications are obtainable at the office of the PERSONNEL DEPARTMENT, 1115 Truxtun Avenue, BAKERSFIELD, and must be filed on or before ______________.

If you fail to file an application for this job, it will be necessary to complete another of these forms in the event you wish to be notified in the future for this type of work.

Advise us of changes of address. Notification is a courtesy. The Personnel Dept. is not responsible for cards that aren't mailed or mailed and not delivered.

IF NO RECRUITMENT OCCURS FOR THIS JOB WITHIN TWO YEARS THIS CARD WILL BE DISCARDED.

THIS IS NOT AN APPLICATION

Personnel 580 1310 054 (Rev.7/92)

Place
Postage
Here

PRINT

NAME ______________________________

ADDRESS ______________________________

CITY and STATE ______________________________

ZIP CODE ______________________________

FIGURE 1-8
Job interest card showing front and back.

FIREFIGHTER #1900

HOW TO APPLY: Applicants interested in participating in this examination must complete an official City of Bakersfield Application For Employment (no copies). Applications must be **received and stamped** in the Human Resources Office, City Hall, 1501 Truxtun Avenue, Bakersfield, CA 93301 during the filing period listed below.

FILING PERIOD:

Tuesday, August 8, 2001	8:00am - 5:00pm
Wednesday, August 9, 2001	8:00am - 5:00pm
Thursday, August 10, 2001	8:00am - 5:00pm

The Human Resources Office will NOT accept applications prior to or after the filing period. Applications which are postmarked or FAXed will not be accepted.

NOTE: EMPLOYMENT APPLICATIONS MUST BE PROPERLY COMPLETED IN ACCORDANCE WITH INSTRUCTIONS ON FACE OF APPLICATION FORM. ALL PERTINENT INFORMATION NEEDED TO DETERMINE THAT THE APPLICANT MEETS THE MINIMUM QUALIFICATIONS MUST BE SHOWN ON THE APPLICATION; OTHERWISE THE APPLICATION WILL BE REJECTED. RESUMES WILL NOT BE ACCEPTED IN LIEU OF COMPLETED APPLICATION.

MINIMUM QUALIFICATIONS:

AGE: Must be 18 years of age at time of written exam.

EDUCATION: Must possess a high school diploma or G.E.D.

LICENSE: Valid driver's license required. Possession of a valid California driver's license at time of appointment is required.

VISION: Visual acuity in each eye not less than 20/40 vision without correction and must have normal color vision.

PHYSICAL CONDITION: Good physical condition. Weight must be in proportion to height.

SALARY: $2,902 - $3,537 per month

EXAMINATIONS: ALL APPLICANTS WILL BE NOTIFIED BY MAIL OF DATE, TIME, AND PLACE OF EXAMS.

Written Exam: (Pass/Fail) The written exam is the first phase of the examination process and may measure comprehension of oral and written material, mathematical ability, and mechanical aptitude. ***Note: Only those applicants with the top 100 written scores who achieve a minimum score of 70% will be invited to the Oral Appraisal Interview.***

Oral Appraisal Interview: (Weighted: 100%) Appraisal will be made of applicant's personal qualifications, education/training, and experience. A minimum rating of 70% is required to qualify for the eligible list. To qualify for placement on the Civil Service Eligible List, applicants must pass both the Written Exam and the Oral Appraisal Interview.

Physical Agility Exam: (Pass/Fail) The Physical Agility Exam will be administered AFTER the Eligible List has been certified. The top 50 ranking candidates on the Eligible List will be invited to participate in the physical agility exam. If additional candidates from the eligible list are needed throughout the effective period of the list, eligibles will be notified to appear for the Physical Agility Exam. Failure to pass the Agility Exam will disqualify an applicant from further consideration.

Background Investigation: (Pass/Fail) Prior to appointment, applicants must successfully complete an investigation of their personal history and background to determine suitability for the position of Firefighter with the Bakersfield Fire Department.

Nonsmoking Policy: Newly hired employees must be nonsmokers. Prospective employees will be required to sign an affidavit indicating that they have not smoked during the twelve (12) month period prior to hiring by the City. Further, they shall agree that they will not smoke, either on or off duty, during the term of their employment with the City. Violation of the nonsmoking agreement shall result in disciplinary action and possible termination of employment.

AN EQUAL OPPORTUNITY/AFFIRMATIVE ACTION EMPLOYER
WOMEN, MINORITIES, AND INDIVIDUALS WITH DISABILITIES ARE ENCOURAGED TO APPLY

The provisions on this bulletin do not constitute a contract expressed or implied and any provisions contained in this bulletin may be modified or revoked without notice.

FIGURE 1-9
Job announcement flyer for firefighter position.

before them as they conduct the interview. They may also review it after you are interviewed. Spelling and grammar are important in your answers. Be extremely careful to answer all questions required completely and honestly. Incomplete applications and those that are not signed are unacceptable.

When filling out the portion of the application that asks for job history and duties, use the job flyer as a guide. Specify how your current and past jobs meet the duties of the new job you are applying for.

NOTE The best, and only way, a job application should be turned in is typed.

If you have an expanded job history or description that will not fit on the application, attached sheets are allowed. For these to be accepted they must be turned in at the same time as the application. If you plan to bring a **resume** to your oral exam, attach a copy to your application as well. Attach a copy of your high school diploma or GED and if you have a college certificate or degree or any other special job-related training certificates, attach them as well. If the jurisdiction awards **veteran's points**, you must submit acceptable proof of discharge along with the application.

resume A listing of a person's areas of experience and education.

veteran's points Points added to a person's final score on a competitive examination process. Given to persons who have satisfactorily performed military service.

Written Examination

The written examination is designed to test the candidates' ability to learn firefighting procedures and techniques. The test evaluates **mechanical aptitude**, general intelligence, mathematical ability, behavioral reactions to given situations or events, mental alertness, adaptability to the work of firefighting, and the ability to understand orders and written material (reading comprehension). Since firefighting experience is not a prerequisite for employment, questions on the written examination are geared to the inexperienced candidate. This does not mean that there cannot be a reading comprehension question with a firefighting situation and related questions. It just means that the answer will be found in the reading material. Questions are multiple choice and computer scored. Written examinations are weighted differently by different departments. In some, the written exam will be worth 40% to 50% of your overall score. In others, the written is pass/fail.

Several resources are available in preparing for the written exam. One of these is to go to the local library and check out firefighter exam preparation texts and videos.

mechanical aptitude The ability to figure out the operation and construction of equipment from drawings.

Another source for these materials is the Firefighters Bookstore.[14] These manuals have exercises in mechanical aptitude, reading comprehension, mathematics, and so forth. If you have not done any long division or have not been taking multiple choice tests for a while, the experience you will gain by practicing with these texts is very worthwhile.

Skills Test

The skills test portion of the exam is not used by all departments. When it is, the top performers from the written exam are invited to participate. The skills test simulates real-life occurrences likely to be encountered on the job. The applicants' mental ability to deal with these situations is graded. A few examples of this sort of test are as follows:

- Applicants are given a written procedure to study for a specified period of time, then are asked questions about the material.
- Applicants are shown a video tape of operations at a fire scene and then are asked questions about the video, such as "How many firefighters were on the roof?" or "Where was the engine parked?"
- Applicants listen to an audio tape of a dispatcher giving directions and then are asked questions, such as "What street were you on after the third turn?"

Oral Examination/Interview

The next step is an interview with the **oral interview panel** (**Figure 1-10**). There is usually a representative from the personnel department in attendance and the interview is recorded on tape. The purpose of this is twofold: to ensure consistency in the way the interviews are conducted should they be challenged, and to avoid disagreement on what the answers were. This also establishes a record if you claim education or experience that is untrue. Oral panels commonly consist of three interviewers, usually of fire officer's rank. They sit on one side of a table and the applicant sits on the other.

The oral exam is designed to evaluate education and work experience and to measure the personal attributes required of firefighting personnel. It may evaluate the following characteristics: ability to act under stress, ability to accept authority, ability to get along with fellow firefighters, ability to deal with the public, and motivation to be a firefighter.

The panel of interviewers will attempt to put you at ease in discussing your qualifications as a potential firefighter. This portion of the examination is competitive and may be rated as much as 100% of your final score when a pass/fail written exam is used.

oral interview panel An interview technique in which the interviewers ask questions and evaluate the answers given by job candidates. They assign a score to the candidate's responses for ranking purposes during the selection process.

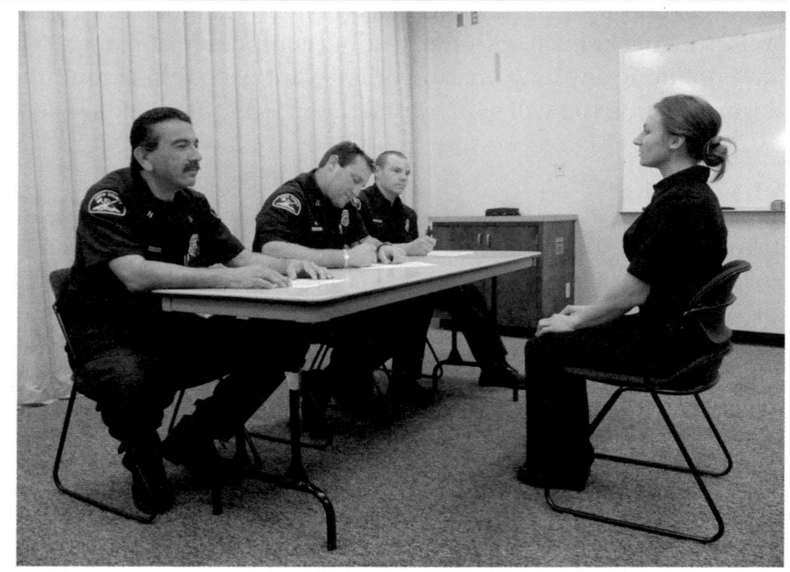

FIGURE 1-10 Candidate being interviewed by oral examination panel.

Delmar/Cengage Learning

Punctuality and appearance are important. Neat grooming and attire will give a good first impression of you to the panel. The old saying, "Look in the mirror, would you hire this person?" applies here.

Project an image of self-confidence by thinking you are the best person for the job. Be courteous to the interviewers; speak clearly and loudly enough to be heard by all members of the panel. When responding to a question, look directly at the questioner. Avoid a harsh or hasty answer. If necessary, take time to think out your answer before responding. In a positive way, stress the value of your abilities and how you can contribute to the fire service.

Avoid distracting mannerisms that might draw the interviewer's attention away from your statements. Maintain good habits, posture, and poise during the entire interview process. If you are seated in a chair that rocks, sit still. Place your hands in your lap and try to relax.

You can prepare for the oral examination by first assessing your delivery. Write down likely questions and then answer them while looking into a mirror. You may want to record your answers to see how you sound. A very common request is, "Tell us about yourself," or a close variation. This response is an excellent one to practice. Another method is to contact the local fire department and see if you can set up a "mock oral." In a mock oral, the room is set up much as it would be in a real oral and firefighters ask you questions just as an oral panel would. Sources for video tapes and written material on preparing for oral examinations can be found by looking through fire service magazines.

The oral interview exam is the final step in competing for placement on the fire department list of certified eligibles. Your score on the oral is determined by averaging the scores given to you by the individual panel members. A minimum average score of 70% is required to pass the oral exam. Applicants receiving less than 70% will be disqualified.

In situations where the written and oral examinations are both weighted, the average of your two scores will determine your ranking on the list of eligibles. In an examination process where the two scores are weighted 50/50, a written score of 80 and an oral of 90 will give you an overall score of 85. This score will be compared to all the others and you will be ranked 1, 2, 3, . . . accordingly. Being eligible for veteran's points can raise your score, so read your job announcement carefully or ask the personnel department for information on how to receive credit. Another method of ranking candidates is called *banding*. In this process, candidates with scores between specified percentages are placed in groups. When positions become available, the candidates in the first band are offered the job. If the number in the group exceeds the number of available positions, a lottery is conducted to offer a position to a certain number of candidates in the band being used.

Physical Ability/Agility

Physical ability/agility tests (**Figure 1-11**) are administered to judge the candidates' overall physical conditioning and ability to perform firefighting-related tasks. This test is expensive and time consuming to administer and usually only a select number of the top-performing candidates are asked to participate. The test consists primarily of climbing, hoisting, carrying, lifting, and dragging. The emphasis is on overall body strength and endurance. There is no set standard for this type of test, but the basic events are much the same in most tests because they are required to be fire fighting related. Appendix B describes a sample physical ability test.

As with the written and oral tests, preparation is important. The people administering the test will be your future employers and they may very well remember your performance. It is a good idea to contact someone already on the fire department and have them help you practice the basic types of activities.

NOTE As with the written and oral tests, preparation is important.

Just like any other physical endeavor, technique is involved in many of the test activities. Dragging a hose around several obstacles may appear very simple, but until you have done it several times you are not likely to be very good at it.

Many departments with ladder trucks will require you to ascend and descend a raised ladder. This event can be very frightening the first time you perform it but you can become more comfortable with practice (**Figure 1-12**). When you are taking the physical agility test and being timed is not the occasion to find out you have not prepared properly.

The physical ability/agility test may or may not be competitive and part of your final score. Most are usually pass/fail. It is good to know this and prepare accordingly.

FIGURE 1-11
Physical ability/agility testing.

Delmar/Cengage Learning

FIGURE 1-12
Candidate climbing aerial ladder.

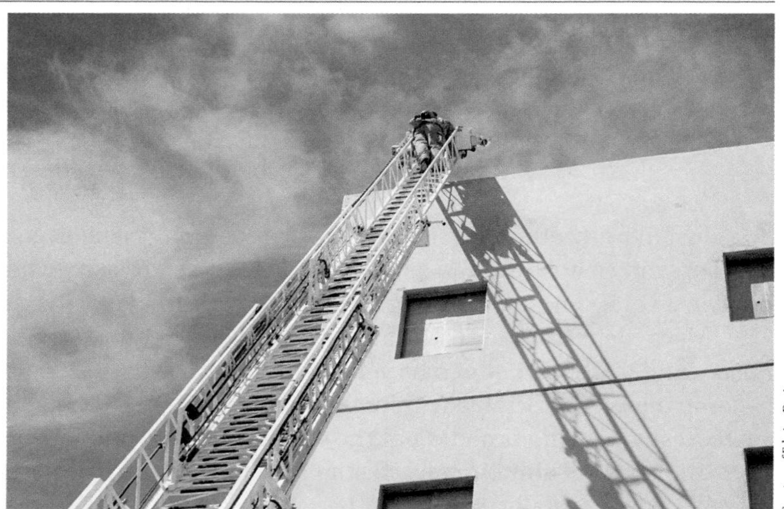

Courtesy of Edwina Davis

Firefighter Combat Challenge

Another physical agility test that is sometimes used is the Firefighter Combat Challenge developed by ARA/Human Factors, Inc. The participants are required to perform the test wearing full turnout gear while breathing from a self-contained breathing apparatus. The test starts with carrying a pack made up of two 50-foot lengths of 1½-inch hose to the fifth floor of a drill tower or other structure. Then, using the hand-over-hand method and a rope, the candidate hoists a 50-foot roll of 2½-inch hose from the ground into the fifth floor window.

The next step is to drive a 165-foot steel I beam a distance of 5 feet using a 9-pound shot-filled sledge hammer. This step is performed on a specially designed device called a "Keiser Force Machine." The fourth step is to advance a 1½-inch charged hose line a distance of 75 feet, then crack the nozzle and squirt a small amount of water. The last step is to drag a 175-pound dummy a distance of 100 feet. This is a condensed description of the Firefighter Combat Challenge. For a more detailed description contact Health Metrics at http://arahumanfactors.com.

Work Capacity Test for Wildland Firefighters

Federal wildland firefighters are required to pass the work capacity test (Pack Test) at the "arduous" fitness level.[15] Some fire departments in areas with wildland firefighting responsibilities require this test as well. The test used in the Wildland Fire Qualification Subsystem for positions requiring an arduous fitness level is as follows: the person being tested is to carry a 45-pound pack a distance of 3 miles in 45 minutes or less (equivalent to a pace of 4 miles per hour). Elevation corrections are applied for elevations over 4,000 feet.

Altitude	**Correction added to time allowed**
4,000 to 4,999 feet	30 seconds
5,000 to 5,999 feet	45 seconds
6,000 to 6,999 feet	60 seconds
7,000 to 7,999 feet	75 seconds
8,000 to 8,999 feet	90 seconds

Background Investigation

One of the components of the hiring process that is becoming more prevalent is the background investigation. Before offering candidates a job, many departments conduct a comprehensive check of the prospective employee's background.

Components of the background investigation may include a review of the application for errors and omissions, a personal information check, a fingerprint check, and a polygraph examination. Department investigators may also visit social networking sites to see if the candidate has posted any information or photos which may compromise their application for employment.

The personal information component consists of a questionnaire to be filled out by the applicant. One of these questionnaires can easily extend to 20 or 30 pages. It may include: all residences for the past 10 years; names and addresses of all relatives and several references; all educational experience, including high school; any experience or employment, including voluntary and temporary, for the past 10 years; military information, including all locations and supervisors; legal information regarding perjury, convictions for felonies or misdemeanors, lawsuits, and traffic violations; ownership, carriage, and use of firearms, including concealed weapons permit; prejudice against any groups and any discriminatory actions; drug use or any involvement with drugs directly or indirectly; motor vehicle operation, including driver's license and insurance information; and financial information, including accounts, mortgages, credit cards and loans, late payments, judgments, child support, and bankruptcy. There may also be questions regarding any physical altercations or domestic abuse incidents that you may have been involved in. The final question may be worded as: "Did you in any way cheat, lie, or commit fraud during the application or evaluation process or during any of the background processes?"

All relatives, employers, roommates, current and former spouses, and references may be sent a personal confidential inquiry about you. Typical questions are:

- Does the applicant have good communications skills?
- Do you feel the applicant can make logical decisions and use common sense?
- Do you feel the applicant retains information?
- Do you feel the applicant has the willingness to confront problems?
- Is the applicant dependable and well motivated?
- Is the applicant generous and willing to help others even at his/her inconvenience?
- Do you think the applicant has the desire for self-improvement?
- Please describe the applicant's demeanor, grooming, and personal care.
- Do you feel the applicant is of strong moral character and is honest?
- Do you feel the applicant is physically able to perform the duties of a firefighter?
- How do you feel about the applicant working with the public?

These are subjective questions and not all responses are going to be weighted the same. You may be asked to explain why someone said a particular thing about you in his or her response.

Once the packet is reviewed by the employer and all of the references that reply have been reviewed, a polygraph examination may be administered. Its purpose is to verify that all of the information presented on the personal information questionnaire is correct. It is considered the best practice to not misrepresent any of the information. Every question allows explanation and one of the worst things that can be done is to leave something out or lie about it. Doing so is grounds for failure of the background portion of the examination. In addition, you should keep records of the information that the employer is likely to ask for so you can truthfully and completely fill out the background investigation packet.

Final Interview

The last interview is with the fire chief or designated representative(s). Conducted in a manner similar to the oral examination interview, the primary purpose is to discuss various phases of the fire service career. This interview takes place when your name appears within the top positions on the **eligible list**.

The interviewer(s) on this panel can change your ranking among the people they are interviewing. More than one person may be interviewed for a vacancy.

Stay alert and pay attention. This interview may make the difference in whether you get hired or not. You could walk into the interview as the top-rated candidate on the list and walk out as the bottom-rated person of the group interviewed.

eligible list A certified list of persons who have successfully completed the testing process.

Medical Examination

A step in your evaluation process prior to appointment as a **probationary firefighter** is a complete medical examination, which may include drug screening.[16] It is given by a physician appointed by the department at department expense. Each applicant must be in good general physical condition, free from disease or defects that would interfere with the satisfactory performance of the duties of the position. The results of the pre-employment medical examination are used to determine whether applicants possess the prescribed standards of physical health and physique required for the position of firefighter. Applicants who fail to pass this examination have their names removed from the eligible list.

probationary firefighter A person hired by the fire department who has not been granted permanent status.

Probationary Period

The probationary period is the last step in the selection process. The academy program can be considered a part of the probationary period. At the end of the academy program there may be a final examination that covers all of the skills and technical material presented. The firefighter then goes to an assignment at a fire station. Probationary firefighters are expected to perform independent study to learn the required departmental policies and procedures. The probationary period may last up to 12 months because a firefighter working a 24-hour shift schedule works only an average of 10 days a month. Twelve months is often considered to be enough time to observe how the new firefighter adjusts to fire department life and performs at emergencies and regular work assignments.

During the probationary period, new firefighters may not enjoy full civil service protection of their employment. This allows the department to remove firefighters who cannot adjust or are found to be unable to perform their duties. Removals from service are not done without cause. By this time the department has invested a lot of money in the selection process and training.

Many departments have a comprehensive combination written/manipulative test at the end of the probationary period. The purpose of this test is to assess the knowledge and skill of the firefighter. The candidate is expected to correctly perform the skills taught in the academy as well as those learned during the probationary period. The written test consists of knowledge of policies and procedures and technical information about the department and equipment. If at this point the firefighter cannot pass these tests, dismissal may occur.

Goal Setting

To achieve your goal of becoming a skilled professional firefighter, you must first clearly define what your goals are. You must create a road map to success with a well-defined destination. If you wander aimlessly, no matter how fast you go you will not arrive at the desired destination. Planning and working hard are required to achieve your goal. Without both you will not succeed.

The time to start preparing and planning is now. No one is going to wait for you. The job you desire is going to come open, and unless you are truly prepared you are not going to be ready to get it. Firefighter entrance examinations are competitive. Not all who apply are going to be accepted. With few exceptions those who are the best prepared are going to perform the best. In civil service hiring, lists are established and those at the top of the list get first consideration for vacancies.

Goal Setting Process

When you develop your goals, keep the acronym SMART in mind. Goals should be Simple, Measurable, Accountable, Realistic, and Timely.

The first step is to visualize what your goals are and write them down. Keeping a list in your head is ineffective. The interference from everyday events will soon have you losing sight of your goals unless they are clearly written and easily referred to on a regular basis. Writing your goals down assists you in determining whether they are truly realistic.

By writing your goals down you are making a contract with yourself. This personalizes the goals to you and no one else. Goal accomplishment requires commitment. Establishing a contract with yourself helps to create that commitment.

Long-term planning is based on current decisions. What you do now will have consequences—positive or negative—later. The background investigation requires you to stay out of trouble and not do certain things. If you choose to start or continue doing things that will preclude you from becoming a firefighter in the present, they are going to cost you in the future. Look at the news and count the number of people who derail themselves through current actions. They were on the road to success and some decision they made caused them to lose the ability to accomplish their career goals. It is the same with physical fitness. You will be required to be in good physical condition to pass the physical ability test. You cannot get in shape overnight. It takes months to prepare for a physical ability test such as the Candidate Physical Ability Test (CPAT). The CPAT involves both physical ability and skills at performing operations. Both of these can and should be practiced ahead of time.

Preparation is important, but so is taking action to achieve your goals. Preparation paralysis can set in and you may never really take the action required to achieve the goal. Setting a timeline on certain goals may assist you in doing this. Projecting goals into the future with no completion date lets you take forever to achieve them.

Look at the job requirements and once you meet the minimum, fill out the application and take the entrance exam. You can continue to prepare as you continue to take entrance exams. Do not wait until some unspecified time in the future to start taking entrance exams. The more often you take them the better you can prepare to take them in the future. Testing can be practiced just like other skills. You may not do well at first because you are not used to the testing process. Do not be discouraged or give up. You were not able to run when you first learned to walk. The intimidation factor of that first oral interview panel can be overcome with practice. If at first you lack poise, work on it.

You need not do this all by yourself. Do not be afraid to ask for help. Many of the instructors in fire science programs are working or retired firefighters. Ask for assistance. For the most part they are involved in this type of instruction because they wish to "give something back" to the service they love. They can do this through you. If they are unable to assist you, they probably know someone who can. Do not be hesitant about asking. Firefighters tend to be "can do" people and they respect this in others.

Your goals have been established and now something has happened that causes you to be unable to achieve one or more of them. That does not mean that they are forever unattainable. You just have to make a course correction and move on. If you did not do well on the exam in your hometown, go out there and take other exams. Focus on the positive. Everyone has a bad day now and then. Life has a way of taking strange twists and turns that adaptable people use to their advantage. To adapt, overcome, and continue on is the way successful people react to setbacks and obstacles.

Celebrate success. Take some time to reflect when you have succeeded in achieving a personal goal—the completion of your degree, Firefighter I certification or EMT, for example. A feeling of success on a regular basis will help energize you to accomplish the goals that are left.[17]

Goals are accomplished in three steps. The first is to visualize the goals. Second, clearly define the goals and write them down. Finally, take the actions required to achieve them. The absence of one of these three steps does not guarantee failure, but it will hinder accomplishment.

SUMMARY

The process of becoming a firefighter starts, for many people, by seeking an education in the fire science field. The level of education sought is mostly up to the person involved. The pursuit of an education is only one step in the process.

The prospective firefighter must prepare for all the areas of the selection process. Even the highest written and oral examination scores will not get you hired if you fail the physical agility or medical examinations. Before seeking a career as a firefighter, be sure that you meet the minimum medical requirements. The rigorous medical examination required for a position on a municipal fire department is not required of those seeking employment with the U.S. Department of Agriculture Forest Service or the Department of the Interior Bureau of Land Management. These positions are discussed in Chapter 2.

Always be sure that whenever you attempt the selection process you are properly prepared for all phases, and remember: if you do not think you are the best person for the job, neither will anyone else.

REVIEW QUESTIONS

1. The fire science curriculum is aimed at providing the student with what types of skills?
2. Before enrolling in classes the student should meet with which college official?
3. What is the basic college degree in fire science?
4. List several advantages of attending on-line courses over traditional course settings.
5. List two ways to attend fire academies.
6. What manipulative certification should you have before applying for a position with the fire department?
7. Training programs are offered at various levels, such as state and local. List two others.
8. List two pre-service opportunities for gaining firefighting experience.

9. What is the minimum age requirement for firefighters in your state?
10. List two basic prerequisites in applying for the firefighter exam.
11. What is the first step in the selection process?
12. List two ways to find out when application periods are open for firefighter examinations.
13. List a source of material used to prepare for the written examination.
14. List two ways to prepare for the physical ability/agility test.
15. List two ways to prepare for the oral examination.
16. What is the purpose of the probationary period?
17. What are the steps in preparing SMART goals for yourself?

DISCUSSION QUESTIONS

1. What are some ways you can prepare yourself to perform well in the firefighter selection process?
2. Why is hiring for diversity (e.g., women and minorities) an important issue in public agencies such as the fire service?
3. Prepare answers to the following questions you may be required to answer in an oral interview: Why are you the best person for the job? What have you done to prepare yourself to be a firefighter? What is your goal in the fire service?
4. Why is the National Professional Development Model based on education and training?

NOTES

1. Fire and Emergency Services Higher Education (FESHE) Model Curriculum, www.usfa.dhs.gov/nfa/higher_ed/feshe/feshe_model.shtm
2. California State University, Los Angeles, Fire Protection Administration and Technology, 5151 State University Drive, Los Angeles, CA 90032.
3. Oklahoma State University, Fire Service Training, Stillwater, OK 74078-0114.
4. University of Maryland, Department of Fire Protection Engineering, College Park, MD 20742.
5. Federal Emergency Management Agency, Emergency Management Institute, 16825 S. Seton Ave., Emmitsburg, MD 21727. http://www.training.fema.gov/EMIWeb/edu/collegelist/
6. Northwood University, Midland Program Manager, 400 Whiting Drive, Midland, MI 48640. Email firesci@delta.edu.
7. National Fire Protection Association, *Standard 1001, Fire Fighter Professional Qualifications* (Quincy, MA: National Fire Protection Association, 2007).
8. National Fire Protection Association, *The U.S. Fire Problem* (Quincy, MA: National Fire Protection Association, 2003).
9. Unites States Fire Administration, *Annual Report on Firefighter Fatalities in the United States*, retrieved from www.usfa.dhs/gov/media/press/2009releases/092409.shtm
10. National Fire Academy, 16825 S. Seton Ave., Emmitsburg, MD 21727.
11. California Fire Service Training and Education System, 7171 Bowling Drive, Suite 600, Sacramento, CA 95823-2034.

12. Fire Protection Training Division, Texas Engineering Extension Service, Texas A&M University System, College Station, TX 77843-8000.
13. National Fire Protection Association, *Standard 1001, Fire Fighter Professional Qualifications* (Quincy, MA: National Fire Protection Association, 2007).
14. Firefighter's Bookstore, 18281 Gothard #5, Huntington Beach, CA 92648-2719. Phone 800-727-3327. www.firebooks.com
15. U.S. Department of Agriculture Forest Service, Washington, DC 20250.
16. National Fire Protection Association, *Standard 1582, Medical Programs for Fire Departments* (Quincy, MA: National Fire Protection Association, 2007).
17. Successories Library, *Goals Guidelines for Designing an Extraordinary Life* (Successories Inc., retrieved from www.successories.com, 1999).

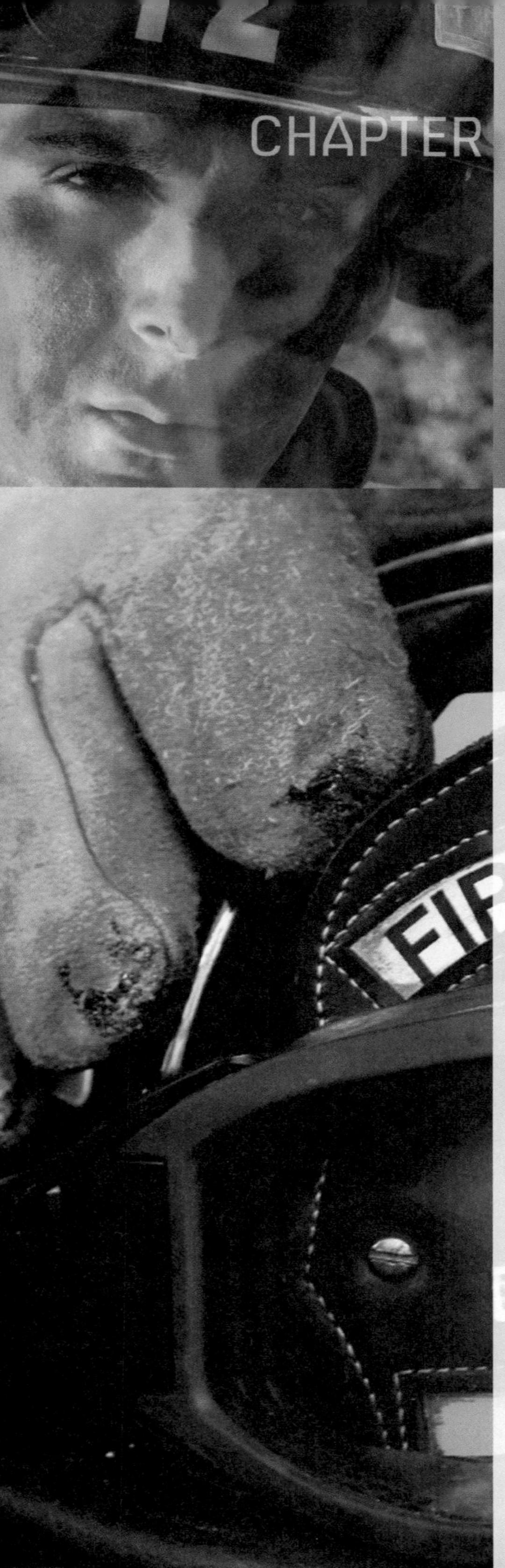

CHAPTER 2

Fire Protection Career Opportunities

LEARNING OBJECTIVES

Upon completion of this chapter, you should be able to:

- Identify fire protection jobs in the public and private fire service.
- List duties and requirements of the position of firefighter trainee and firefighter.
- List duties and requirements of the position of firefighter/paramedic.
- List duties and requirements of fire heavy equipment operator.
- List duties and requirements of firefighter forestry aid.
- Give examples of fire service jobs other than firefighter.

INTRODUCTION

The study of fire science opens up a new world of opportunity for the student. There are many different job descriptions in the fire protection field with jobs available in both the public and private sectors. There are avenues of promotion within the fire department that require advanced training in the fire science field. The process and requirements for getting hired in the average municipal fire department were presented in Chapter 1. This chapter presents the jobs themselves, their duties, and other requirements.

PUBLIC FIRE PROTECTION CAREERS

The first selection of jobs we will look at directly involve firefighting. These job descriptions are representative of those in municipal and rural fire departments.

There are several fire career–related job posting websites listed in Appendix H.

Firefighter Trainee–Fire Department

In cases in which the fire department employs new firefighters under specific trainee programs, this job title would be representative. A person working in this capacity typically receives less pay and fewer benefits than a regular firefighter.

After a specified period of time and satisfactory completion of the program, the person is promoted to Firefighter–Probationary. This program would not have a prerequisite of Firefighter I certification but would instead lead to this certification.

Definition

Under close supervision and in a learning capacity, to assist in the various phases of fire suppression and prevention; to learn the functions carried on by the department; and to do related work as required.

Typical Tasks

Performs a variety of work assignments, including responding to fire alarms and other emergency incidents to protect life and property; assists fire personnel by performing selected duties of significant learning value; participates in continuing training and instruction program by individual study of technical material and attendance at scheduled drills and classes; and may drive and operate fire engines and similar equipment.

Employment Standards

Knowledge of modern fire prevention and suppression methods or ability to acquire such knowledge; ability to learn technical firefighting techniques and principles of **hydraulics** applied to fire suppression; ability to understand and follow oral directions; ability to establish and maintain cooperative relationships with fellow employees and the public; ability to keep simple records and prepare reports; mechanical aptitude; physical endurance and agility.[1]

As you can see, the emphasis is on close supervision in a learning capacity. These positions are often used to target specific groups for entry into the fire department due to affirmative action requirements.

hydraulics The computation of the required pressure to be applied to water to overcome the effects of pressure loss due to friction in piping and fire hose.

Firefighter–Fire Department

This position is the standard entry level position for the fire department and may or may not require the completion of a Firefighter I academy or certification to apply.

Definition

Under supervision, to respond to fire alarms and other emergency incidents to protect life and property; and to do related work as required.

Typical Tasks

Responds to alarms and assists in suppression of fires; cleans up and performs **salvage** operations after fires; assists in maintaining and caring for fire apparatus, equipment, fire station, and grounds; responds to emergency incidents (**Figure 2-1**); operates **resuscitator** and automatic external defibrillator, and administers first aid; makes residential and business inspections to discover and eliminate potential fire hazards and to educate the public in fire prevention; participates in continuing training and instruction program by individual study of technical material and attendance at scheduled drills and classes; may drive and operate fire engines and similar equipment; may train or assist in training auxiliary firefighters; and may act as relief for a **driver/operator** or **company officer**.[2]

salvage Operations performed to prevent or reduce fire, smoke, or water damage to items of value.

resuscitator A mechanical device that can perform forced ventilation or provide oxygen to a victim or patient.

driver/operator The primary responsibility of this position is to operate the pumping or aerial apparatus assigned to the fire department. Depending on the jurisdiction involved, this position may be identified by various titles, such as engineer, chauffeur, or truck operator.

company officer The first line supervisor in the fire department. Depending on the jurisdiction involved, this position may be identified by various titles, such as captain, lieutenant, sergeant, station manager, module leader, or unit manager.

FIGURE 2-1 Firefighters performing extrication of trapped victim.

Courtesy of Edwina Davis

Employment Standards

The same as for firefighter trainee.

Promotional Opportunity

In departments with several ranks, the firefighter may become eligible to test for the position of driver/operator or company officer after a specified period of time in rank, and completion of any department-specified prerequisites.

The time in rank and prerequisites vary depending on the department.

In departments where the driver/operator position is of a higher rank than firefighter, the position usually has more responsibility to ensure that company functions, such as program work, are completed. When it is a promotional position there is an increase in pay.

Firefighter–Fire Department Federal

The federal government has numerous positions for firefighters at federal installations.

These installations are mostly located at military bases. The job descriptions, requirements, and promotional opportunities are much the same as for municipal firefighters.

Firefighter Paramedic

As the traditional role of the fire department has changed, as we will see in Chapter 3, the position of firefighter paramedic has become much more prevalent. The position of firefighter paramedic has increased responsibility over that of firefighter and requires the completion of advanced medical training.

Definition

The same as firefighter with additional responsibility of the delivery of advanced life support.

Typical Tasks

Depending on the department, the firefighter paramedic may respond as part of an engine crew and provide medical aid as needed as an adjunct to fire fighting duties. In other departments, the firefighter paramedic responds in either a special squad vehicle or ambulance. If responding in a squad vehicle, patient transportation is handled by others. If responding in an ambulance, the firefighter paramedic treats victims at the scene and transports them to the hospital (**Figure 2-2**).

Employment Standards

In addition to the standards of firefighter, the firefighter paramedic must possess or be capable of acquiring a paramedic certification valid for the jurisdiction served.

Promotional Opportunity

The same as for firefighter. As with any other preparation for promotion, a willingness to accept additional responsibility is considered an asset.

FIGURE 2-2 Firefighter and paramedic checking victim for injuries.

An advantage to becoming a firefighter paramedic is that persons with these qualifications are in great demand and **lateral transfer** to another department may be available. Another advantage is that along with the increased responsibility and ability to assist the public in medical emergencies, there is usually a pay incentive.

lateral transfer The act of changing jobs from one fire department to another without coming on at the bottom of the rank structure.

Fire Heavy Equipment Operator

The fire heavy equipment operator position is mostly limited to departments that provide fire protection for wildland areas (**Figure 2-3**).

Definition

Under direction, to operate heavy motorized equipment in fire control work and to construct and maintain fire breaks and roads, and to do related work as required.

Typical Tasks

Operates a bulldozer in forest fire areas over steep, rough terrain in establishing fire **control lines** or in fire hazard reduction and conservation work; operates motor graders, heavy duty transports, fire trucks, and other types of equipment used in fire suppression

control lines Removing fuel, applying water, or using natural barriers to stop a wildland fire from spreading.

FIGURE 2-3 Fire heavy equipment operator constructing fire line.

Courtesy of Edwina Davis

and maintenance work; services and assists in making mechanical repairs to equipment, including welding and limited body repair; as assigned, works with or supervises others on fire line, road construction, conservation, and in-camp work projects or fire incidents. Maintains records of work accomplished and equipment service and makes reports; as assigned, responds to alarms and assists in suppression of fires; cleans up and performs salvage operations after fires; assists in maintaining and caring for fire apparatus, equipment, fire station, and grounds; responds to emergency incidents, operates resuscitator, and administers first aid; participates in continuing training and instruction program by individual study of technical material and attendance at scheduled drills and classes; and assists in roadside burning and in building or clearing fire breaks or fire roads.

Employment Standards

Graduation from high school or equivalency and experience in operating heavy motorized equipment, including bulldozers and heavy transport trucks, some of which has been in rugged terrain.[3]

Promotional Opportunity

This may be a one-class position. To promote would require transferring to the firefighter or driver/operator rank.

Benefits

The preceding positions, with the exception of firefighter trainee, are all part of the regular fire department structure and receive the standard benefits, as far as pay, sick leave, vacation, and medical coverage, as provided to other positions within the public agency.

Safety Section Retirement

In some states, firefighting positions are classified as **safety section**, along with law enforcement, and receive a higher level of retirement compensation than those employees classified as general (nonsafety) members. For example, Employee A operates heavy equipment for the roads and bridges department for the jurisdiction. Upon retirement, after 30 years of service, at 55 years of age, his percentage of final compensation is 44%. Employee B operates heavy equipment as a fire heavy equipment operator for the same jurisdiction. Upon retirement, after 30 years, at 55 years of age, his percentage of final compensation is 78%. This works out to Employee B's pension being 1¾ times as much as Employee A's. This increased pension amount is based on the fact that fire fighting is an extremely dangerous and stressful occupation.

safety section A body of law that sets the retirement benefit rate for certain professions, primarily those that are high hazard and deal with public safety, namely, fire and police.

Another difference under this law is that safety section members are eligible for service retirement after 20 years, regardless of age, versus 30 years, regardless of age, for general members.[4]

The next group of positions we review are those of the federal wildland fire agencies, including the U.S. Forest Service, Bureau of Indian Affairs, National Park Service, Bureau of Land Management, and state Departments of Forestry, which tend to be seasonal employment for entry-level personnel.

It may take several summers as a seasonal employee to secure a permanent position.

Firefighter (Forestry Aid) Wildland GS 3

Definition

Position is that of an advanced trainee on a **wildland** fire suppression engine or hand crew and/or a fuels management crew (**Figure 2-4**). The work is performed in a forest environment in steep terrain, where surfaces may be extremely uneven, rocky, covered

wildland Open land in its natural state.

Courtesy of Edwina Davis

FIGURE 2-4
Forestry aids constructing fire line with assistance of water-dropping helicopter.

with thick, tangled vegetation, and so forth. Temperatures are frequently extreme, both from the weather or from the fire. Smoke and dust conditions are frequently severe. This position is usually considered to be entry level and the training required may be provided on the job or from a local community college or high school program prior to being allowed to engage in firefighting operations.

Typical Tasks

Performs assignments to develop knowledge of **fuels management** and fire suppression techniques and practices, such as fire line construction, use of pumps and engines, **hose lays**, **foam** and **retardant**, working around aircraft, safety rules, and fire and fuels terminology.

Searches out and extinguishes burning materials by moving dirt, applying water by hose or backpack pump, and other methods. Chops brush, or fells small trees to build fire line using various tools such as axe, shovel, **Pulaski**, **McLeod**, and power saws to control spreading wildland fire and to prepare lines prior to **prescribed burning**. Chops, carries, and piles **logging slash**. Patrols the fire line to locate and extinguish sparks, flare ups, and hot spot fires that may threaten developed fire lines. Cleans, reconditions, and stores fire tools and equipment.

May be assigned to other resource management activities such as recreation, timber, or **reforestation** when not performing fire suppression or fuels management duties.

The purpose of the work is to carry out assigned tasks in the reduction of ground fuels and the suppression of wildland fires. The scope varies from that of a small prescribed fire that may be managed by the crew, to that of a large fire involving several thousand people. The effect of the work performed is to manage fire under controlled conditions to accomplish work or to minimize resource loss under wildfire conditions.

fuels management A program where naturally growing fuels, such as brush, are reduced to lessen fire intensity or to open up areas for wildlife and cattle.

hose lay The method of laying out hose at a fire scene.

foam The finished product of water combined with certain agents that aid in the water's ability to extinguish fires. These agents are further discussed in Chapter 12.

retardant A material spread on fuels that inhibits their burning. This agent is further discussed in Chapter 12.

Pulaski A tool for fighting wildland fires with an axe on one side of the head and a grub hoe on the other.

McLeod A tool for fighting wildland fires with a scraping blade on one side of the head and a rake on the other.

prescribed burning Planned application of fire under specified conditions in a predetermined area to achieve management objectives, which include: removal or modification of fuels; clearing paths through brush; and killing unwanted plant growth.

logging slash The remnants of logging operations, including limbs trimmed from downed trees and broken tree trunks.

reforestation Planting seedling trees in areas destroyed by fire or logging.

Employment Standards

Knowledge of forestry practices and techniques, including accepted fire suppression and prescribed burning methods to be used in various types of fuels and under a variety of weather and terrain. Working knowledge of fire behavior and fire control techniques to carry out assigned fuels and wildland fire tasks. Skills in the use of hand tools such as axe, shovel, Pulaski, McLeod, portable pumps, and chain saws to build fire lines and extinguish burning materials. Knowledge of safety practices to prevent injury and loss of life.

The work requires strenuous physical exertion for extended periods including walking, climbing, shoveling, chopping, throwing, lifting, and frequently carrying objects weighing 50 pounds or more. The duties of the position require that the candidate meet prescribed physical requirements as measured by the Work Capacity Test for Wildland Firefighters (Chapter 1).[5]

Promotional Opportunity

With satisfactory performance and available positions, the firefighter may be promoted to a higher pay position, Firefighter, or transfer into a position requiring greater knowledge and proven ability such as **helitack** crew person, **smoke jumper**, fire prevention technician, or fire engine operator trainee.

Many college students fill these positions due to their seasonal nature. It gives them the ability to make money during the summer for educational and living expenses while gaining experience in fire suppression. A person trained in wildland fire suppression is a great asset to fire departments that have wildland responsibilities. This experience can be a valuable addition to your resume when trying to secure a firefighting job. If you recall the work ethics and human relations portions of Chapter 1, it is obvious that experience of this type is directly applicable.

There are numerous civilian (nonsafety) positions in the public fire service as well. These jobs fall into several areas and are an integral part of the delivery of a fire protection system. The jobs we now consider do not have the rigorous physical requirements of firefighter, although many are stressful and mentally challenging in their own right.

helitack Personnel whose primary means of transportation to fires is by helicopter. They also assist in helicopter operations when the helicopter is being used for water drops or for crew and equipment transportation.

smoke jumper Highly trained personnel who parachute in to suppress fires in remote areas.

Fire Prevention Specialist

Definition

Under direct supervision, to perform a variety of routine fire prevention, training, and hazardous materials disclosure work.

Typical Tasks

Performs routine field checks of fire prevention systems including, but not limited to, hydrant flow test, fire sprinkler system check, and access requirements (**Figure 2-5**). Assists in hazard reduction/weed abatement inspections to ensure compliance with federal, state, and local standards. Participates in fire safety, emergency medical services, and hazardous materials disclosure, training, and education programs. Researches federal, state, and local policies, procedures, codes, and ordinances. Gathers and correlates hydrant information such as location, size, type, water pressure and flow rates; and updates water maps as directed. Provides fire prevention and hazardous materials disclosure information to the public, department personnel, and other related agencies. Assists in developing procedures for, and coordination of, hazardous materials disclosure program. Evaluates information provided on disclosure forms and assigns fees. Assists in developing, producing, coordinating, and maintaining hazardous material data on facilities in jurisdiction area. Assists in gathering and correlating statistical information. Writes reports based on field notes. Assists in file and record keeping. Assists fire prevention officers and inspectors with office coverage.[6]

Employment Standards

Determined by the employing jurisdiction.

Promotional Opportunity

Upon obtaining experience and proficiency the person may be promoted to Fire Prevention Specialist II. People have also used this position to gain a foothold in the fire department and have used the work experience, background, and knowledge gained to perform well on the firefighter exam.

FIGURE 2-5 Fire prevention specialist inspecting fire sprinkler system.

Delmar/Cengage Learning

Fire Hazardous Materials Program Specialist

Definition

Under direction, to evaluate potential hazards of unusual chemicals and materials and determine which chemicals are subject to the hazardous materials disclosure sections of the health and safety code; to serve as the technical advisor to the hazardous materials control unit of the fire department; and to do related work as required (**Figure 2-6**).

Typical Tasks

Evaluates potential hazards of unusual chemicals and materials; determines which chemicals are subject to the hazardous materials disclosure sections of the health and safety code; develops, maintains, and updates list of hazardous materials; analyzes industrial manufacturing processes to identify those potential chemical usages requiring regulations; provides technical support to fire department staff regarding regulations; reviews and analyzes requirements for chemical storage facilities, hazardous material inventory statements, and business emergency plans; assists in the development, implementation, and review of the area plan; reviews completed disclosure forms and emergency plans involving the most complex substances; interprets sections of the laws pertaining to hazardous materials and provides information to emergency personnel, health officials, elected officials, businesses, and the general public; coordinates hazardous materials disclosure to the fire code; reviews and comments on hazardous material preparedness

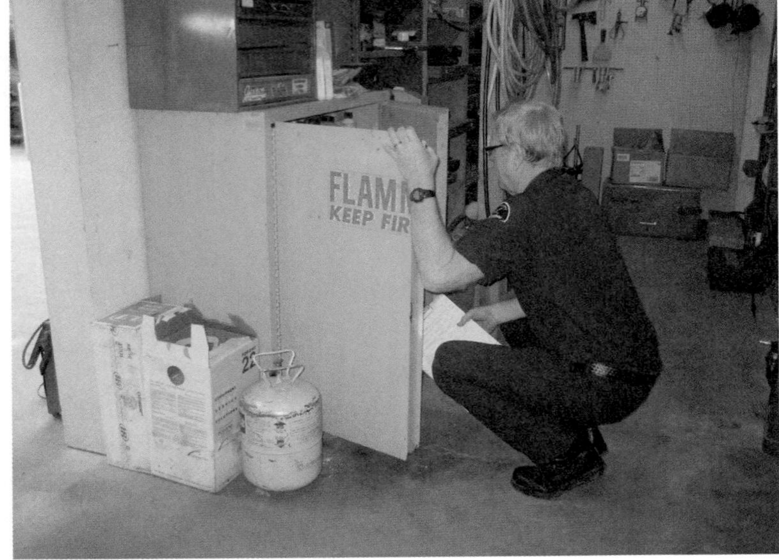

FIGURE 2-6 Hazardous materials specialist performing business inspection.

Delmar/Cengage Learning

programs; assists in technical evaluation of materials at emergency scenes as required; assists other agencies as required regarding the effects of hazardous materials and their effects on the environment; provides information and assistance to industries, other agencies, and the public on compliance with the hazardous materials disclosure regulations; works with other local, state, and federal agencies involved in determining the effects of hazardous materials on the environment; provides technical support to the fire department personnel conducting on-site inspections and participates in inspections of facilities involving complex and volatile chemicals; writes reports based on field notes; and prepares and reviews technical documents, reports, correspondence, documentation, and evidentiary material.

Employment Standards

These are determined by the jurisdiction offering the position.

The following standards are offered as an example: Bachelor's degree in a physical or biological science, environmental engineering, industrial hygiene, or related field (fire protection technology, fire protection engineering); and the requisite years of increasingly responsible experience in hazardous materials control, industrial hygiene, industrial processes, or related areas. Or a master's degree in industrial hygiene, chemistry, environmental engineering, industrial engineering or related field and two years' experience in hazardous materials control, industrial hygiene, industrial processes, or related areas.

Knowledge of the principles of physical, organic, and inorganic chemistry, including qualitative and quantitative analysis; knowledge of chemistry and laboratory techniques, including **toxicology**, hazardous materials identification, and biology; knowledge of effects of hazardous materials and their interaction on the environment; knowledge of local, state, and federal regulations and laws relating to hazardous materials; ability to perform research work on technical problems; ability to analyze manufacturing processes; ability to analyze and interpret data; ability to critically review chemical reports and documents; ability to interpret local, state, and federal laws and regulations relating to hazardous materials; ability to prepare clear, accurate, and concise written technical reports; ability to present evidence in court.[7]

toxicology The science of materials that are poisonous to living things, especially humans and animals.

Fire Department Training Specialist

Definition

Under supervision, to plan, develop, and produce training, informational, and educational materials (**Figure 2-7**); as assigned, to supervise or participate in the preparation of multimedia instructional and informational materials.

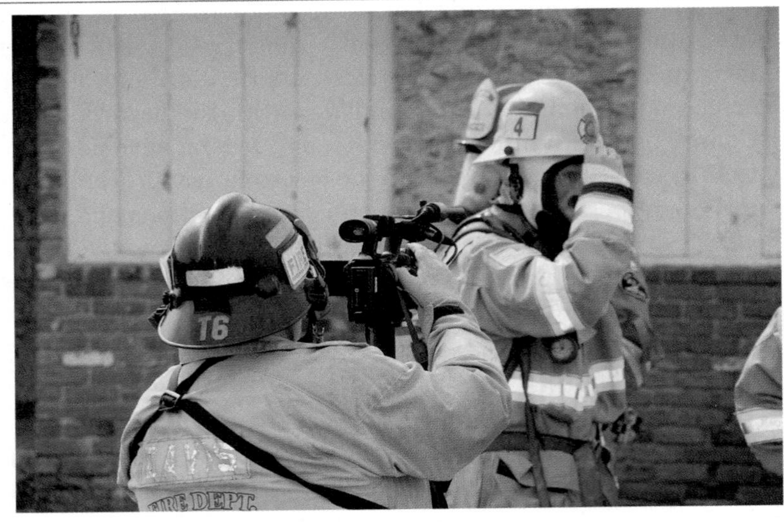

FIGURE 2-7
Training specialist shooting video for training.

Courtesy of Edwina Davis

Typical Tasks

Plan and develop materials for use in training programs. Develop training materials for fire department and public. Present training programs to fire department personnel and public. Assess training needs through testing and evaluation of personnel. Evaluate effectiveness of current and future training programs. Research training needs in regard to department needs and state and federal laws. Implement training programs. Develop and maintain record-keeping system to ensure and verify compliance with legal requirements. Schedule in-service training programs to maintain company efficiency. Review standard operating procedures for accuracy and consistency with fire department operations.

Ensure department member ability to utilize new equipment. Evaluate and recommend use of outside-developed training programs and materials. Prepare training budget. Research and recommend new procedures.

Employment Standards

College-level study in instruction and evaluation. Ability to communicate effectively both orally and in writing. Ability to organize and schedule required training.[8]

Public Fire Safety/Education Specialist

Definition

Provides public education within the community in all aspects of life/fire safety, including providing instruction in fire prevention, fire escape planning, fire extinguishment methods, and other related aspects. Coordinates department-sponsored community awareness, community relations, and community improvement programs, services, and activities related to public fire and burn safety education. Assists public educators with fire safety education (**Figure 2-8**).

Typical Tasks

Plans, conducts, and develops safety, fire, and burn prevention programs; investigates and evaluates various community needs, and implements programs to address these needs; presents programs to homeowners' groups, public and private schools, large groups, civic organizations, industry, businesses, and other segments of the community; provides instruction in identifying and correcting potential fire and burn hazards in the home; reviews data to determine areas of the community requiring an emphasis in fire prevention instruction; coordinates engine company activities in implementing fire safety programs; provides instruction to citizens regarding escape planning in the event of fire and other disasters; describes function and benefits of various types of smoke and fire detectors; instructs citizens regarding extinguishing minor fires pending the arrival of the fire department; coordinates Juvenile Fire Setters program; prepares press releases and public service announcements in coordination with community programs; performs a variety of administrative and research assignments in response to requests from management staff; conducts special studies of organizational policies, procedures, and practices relating to departmental and state-mandated policies; compiles and prepares oral and written reports; compiles data and prepares grant proposals; organizes fund-raising activities; assists in budget preparation; may be required to attend continuing education classes; and performs other duties as required.[9]

FIGURE 2-8
Public safety/education specialist presenting fire extinguisher training.

Courtesy of Marsha Giesler

Employment Standards

College graduation with studies emphasizing fire science, public relations, communications, or education; and work experience, or an equivalent combination of training and experience that provides the capabilities to perform the desired duties. Ability to communicate effectively both orally and in writing; establish and maintain effective working relationship with staff, public, and private representatives; compile, analyze, and summarize statistical and technical data; prepare and present reports and programs related to public fire and burn safety education; create brochures and informational packets; and organize, coordinate, and schedule work projects efficiently.

Dispatcher/Telecommunicator

In some departments the position of dispatcher is held by firefighters.

Definition

Under direction of the shift supervisor, to receive emergency and nonemergency telephone and radio calls and dispatch equipment; to operate a computer-aided dispatch system; and to do related work as required (**Figure 2-9**).

Typical Tasks

In a learning capacity and on an assigned shift, receives and acts upon emergency and nonemergency incidents in accordance with established policies and procedures; operates communications equipment; instructs the public on proper medical techniques under

FIGURE 2-9 Fire department telecommunicator (dispatcher) at console.

emergency situations; secures and records incident information; refers inquiries to appropriate public and private agencies; keeps fire control officers advised of situations and dispatches additional personnel or equipment when so advised by the incident commander; continually monitors the status of fire units; logs all departmental emergency activity; operates other emergency service radios; operates computer systems, including video display terminals; operates telephone equipment; sends and receives telephone messages; prepares and types reports; keeps necessary records; and performs related work as required.

Employment Standards

Graduation from high school or successful completion of GED examination. Minimum typing proficiency of required net words per minute. Knowledge of basic principles and techniques of communications equipment and computer terminal operations; knowledge of local geography, communities, and location of streets and highways; ability to learn the operation of emergency communication equipment and computer-aided dispatch systems; ability to remain calm, act quickly, and exercise good judgment in emergency situations; ability to speak clearly and concisely in a well-modulated voice, perform basic mathematics, follow oral and written instructions, develop reports, and keep records.[10]

Emergency Services Planner/Manager

A field that is gaining in usage across the country is Emergency Management. With the increase in Homeland Security funding and increased focus on disaster management, more and more personnel are being recruited and seeking careers in this field. These are primarily civilian (non-safety section) positions.

Definition

This position is responsible for the development of plans for meeting emergencies, leading the development of operational concepts and orders, and directing coordination with heads of departments, agencies, and cities. This position requires a high degree of initiative, resourcefulness, diplomacy, and organizational ability.

Typical Tasks

- Formulates plans for the organization and implementation of a jurisdiction's emergency services program.
- Plans for mobilization of personnel and resources.
- Acts as an assistant to the Director of Emergency Services during times of declared emergencies.
- Integrates and coordinates activities and programs with emergency officials and local and area emergency offices.
- Completes studies and inventories of resources in the jurisdiction relating to emergency services.

- Develops and prepares public service announcements, press releases, and other media-oriented reports.
- Supervises public informational broadcasts during emergencies.
- Conducts tests to determine adequacy of preparation.
- Other Tasks
    - Prepares correspondence and comprehensive reports.
    - Maintains necessary records.
    - Supervises the coordination of the county-wide shelter program.
    - Supervises training programs to acquaint workers with their emergency roles.
    - Performs other tasks as required for the management of the jurisdiction's emergency services.

Employment Standards

Graduation from an accredited four (4) year college with a degree in Business Administration, Public Administration, or closely related field and two (2) years of increasingly responsible administrative experience in emergency service or other civil defense activities. Two (2) years of directly related experience may be substituted for education requirement on a year-for-year basis up to a maximum of two (2) years. Applicants must possess the physical capacity to perform all essential tasks. Possession of a valid Class C Driver's License prior to the time of appointment is required.

Another requirement is knowledge of the Federal Disaster Relief and Emergency Act and the applicable state Emergency Services Act; knowledge of fiscal disaster assistance programs of local, state and federal governments; knowledge of policies related to emergency action.

Also required are an ability to analyze emergency situations accurately and adopt effective courses of action; ability to develop and manage interagency emergency service programs; ability to speak effectively before public gatherings; ability to work with others and gain their respect and cooperation; ability to organize and coordinate training programs; ability to prepare informational and directive materials and formulate correspondence; and ability to supervise others engaged in emergency services activities.[11]

PRIVATE FIRE PROTECTION CAREERS

Firefighter

There are opportunities in the private sector as well for the position of firefighter. An example of a large corporation that hires its own firefighters for plant protection is the private fire protection provided at the Northrop Grumman B-2 Division Fire Department. This fire department is wholly subsidized by the Northrop Grumman Corporation and provides plant protection for their aircraft production facilities in Palmdale and Pico Rivera, California.

Two of the more well-known private firefighting companies are Red Adair Wild Well Control and Fire Fighting and Boots and Coots Fire Fighting, both from Texas. The Boots and Coots Company was very much involved in firefighting in Kuwait after the Gulf War.

Some companies have their employees form fire brigades for plant protection. The personnel are not employed primarily as firefighters. They receive training in SCBA and basic firefighting methods and perform firefighting duties when required. There are also private wildland firefighting companies that contract out to public agencies on large wildland fires, several of which operate in the Northwest. Contracting with the federal government through the National Interagency Fire Center in Boise, Idaho, these contractors provide engines and other wildland firefighting apparatus along with the crews to operate them. Some of these companies are branching out into structure protection from wildland fires, disaster response, and hazardous materials response as well to give them a year-round operation.

Insurance Companies

Insurance companies require people with fire science backgrounds in the area of loss prevention, as inspectors, emergency plan developers, claims adjusters, and investigators, especially in arson-related cases.

Industry

As well as employing firefighters, industry utilizes fire science graduates as loss prevention specialists and safety consultants. These positions inspect properties for fire and other hazards and develop and present employee training programs.

Fire Protection Systems Engineer

In any building or structure with an installed fire protection system, the system had to be engineered. These engineers design the specifications and plan the installation of systems (**Figure 2-10**). As manufacturing processes become more exotic and higher in dollar value, special systems need to be developed to prevent damage from system operation as well as fire. These jobs require advanced degrees, usually a bachelor's or master's degree.

Fire Protection System Maintenance Specialist

Numerous contractors across the country sell, install, and maintain fire protection systems as their primary business. Where required by law, they service fire extinguishers annually. They also inspect and maintain fixed fire protection systems and equipment, including fire sprinklers and special systems to protect computer rooms and other special applications (**Figure 2-11**).

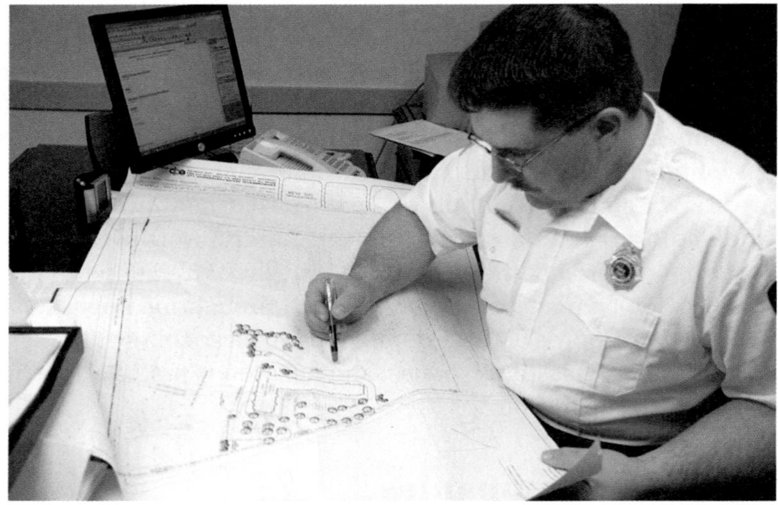

Delmar/Cengage Learning

FIGURE 2-10 Fire protection systems engineer plans check of water supply system.

Delmar/Cengage Learning

FIGURE 2-11 Fire protection equipment company that employs fire extinguisher and fire protection system maintenance specialists.

Invention and Innovation

Over the years many firefighters have contributed their inventions to the fire service.

As they were actually doing the job they saw the need for tools and techniques to make operations more efficient. The following are several examples of the ingenuity displayed by firefighters to improve their ability to do their jobs.

Around 1800, George Smith, a New York firefighter, invented the fire hydrant. In 1913, Edward Pulaski of the U.S. Forest Service invented the grub hoe and axe combination tool that carries his name and is still used by wildland firefighters.[12] The FIRESCOPE incident command system, the basis of the National Incident Command System (NIMS–see Chap. 13) used by firefighting agencies across the country was developed by firefighters. Many other tools, techniques, and devices have been developed by firefighters and will continue be developed by firefighters.

As the business of fire fighting becomes more complex, there is always a need for new and innovative equipment and techniques. A background in fire science can aid a person in assessing these needs and assisting in the development of new firefighting equipment and methods.

The following story is an example of a firefighter—Captain Scott Park of the Kern County Fire Department—recognizing a need and using his background and experience to invent a device to fill the need.

On October 17, 1989, at 5:03 P.M., a magnitude 6.8 earthquake hit the San Francisco Bay Area. Later named the Loma Prieta earthquake for the location of its epicenter, this event is well remembered by many people because they saw it live on television. The earthquake hit just as game three of the 1989 World Series between the San Francisco Giants and the Oakland Athletics was about to begin. The game was being played at Candlestick Park in San Francisco. People all over the world had tuned in to the pregame show when disaster struck.

The extent of the disaster was not immediately known. The Goodyear Blimp was overhead for the game and was broadcasting live shots of the numerous large fires in the area as the sun was setting. At first the focus was on the fires because they were the most evident and communications lines were disrupted. As reports finally got through, it became evident that there was widespread structural collapse throughout the area (**Figure 2-12**). The collapses were not only homes and other businesses, but also a section of the Bay Bridge and Interstate 880, the Cypress Freeway.

As the hours went by it became evident that the worst rescue problems involved the freeway collapse. The freeway was built as a double deck with post and lintel construction, which left a void space between the underside of the upper deck and the lower roadbed. An undetermined number of cars and their occupants were trapped in this void space.

A massive search effort was immediately launched to try to rescue these victims. The problem was to locate those trapped and assess their condition. The operation was severely impeded by the falling darkness and sheer mass of the freeway sections.

FIGURE 2-12
Structure collapse due to earthquake.

Courtesy of Edwina Davis

Where there was a sufficient opening, brave rescuers crawled into the void space to search for trapped victims. In other areas there was not enough space for a person to squeeze between the lower part of the upper deck and the wall on the lower deck. To compound the problem, rescuers were faced with numerous aftershocks and the very real possibility of further collapse of the weakened structure.

A way needed to be found, and quickly, to locate the trapped vehicles and their occupants. To remove the top layer of freeway from the whole collapsed structure would take too long. Metal detectors would not work to locate the cars because of the thickness of the concrete and the presence of steel reinforcing rod.

The decision was made to dig through from the top, which entailed cutting concrete and reinforcing rod and removing material as they went. There was no way to know if the hole was anywhere near a trapped vehicle. The danger of further structural collapse had to be evaluated before heavy equipment could be brought in on the top deck to dig through. These holes were needed to check out areas that were not accessible in any other way. In the end, the last person to be found alive was not extricated for four days.

Captain Park watched what was going on and realized the need for a faster way to check out void spaces, which occur in almost any structural collapse. To make a man-sized hole in reinforced concrete six inches thick takes up to two hours. It often requires the use of heavy equipment that may not be available or feasible to use.

To save lives in any structural collapse situation, speed is essential. The first thing that must be done is to locate the victims. Assessing where to dig and the mechanism of entrapment is necessary to make a decision as to the best way to proceed. The use of heavy equipment is often not possible due to the problem of further endangering the victims.

Captain Park was already an active member of the Kern County Fire Department's Urban Search and Rescue Team and had a long-time interest in this sort of work and its related problems. Captain Park's intention was to invent a device that could gain audio/visual access to a void space in a minimum amount of time. After much thought he came up with the idea of the Searchcam® (**Figure 2-13**).

The Searchcam is a compact unit with headphones and a video screen worn at waist level. It has a hand-held probe with a miniature video camera, lights, and microphone in the end. The probe is inserted into a hole drilled through the concrete or other surface to check out the area on the other side.

Drilling the hole and inserting the probe are much faster and easier than trying to cut a man-sized hole. It also requires no special or heavy equipment. A hole big enough to insert the probe can be drilled through six-inch concrete in two to three minutes with a hand-held drill versus the already stated two hours to cut a man-sized hole with heavy equipment. By properly determining where the victims are and what is holding them, precision digging can take place, speeding up the extrication process tremendously.

FIGURE 2-13
Searchcam® in use at a structure collapse.

Courtesy of Edwina Davis

By using the video/lighting capability, the presence or absence of trapped victims can rapidly be determined. The audio capability helps to assess if the person is still alive. Overall the device aids in the live rescue of trapped victims.

These kinds of inventions do not come into production easily. It took two and a half years from the concept to the first prototype. It took another year from prototype to production. Captain Park used his education, experience, and contacts as part of the Urban Search and Rescue Team to aid in the development of the Searchcam. This device has been very successful, with each of the active Federal Emergency Management Agency Urban Search Task Forces having at least one. There have been sales to other countries as well and the Searchcam is now being used worldwide.[13]

In the early morning hours of January 17, 1993, a magnitude 6.6 earthquake hit the Northridge area of the San Fernando Valley in Los Angeles, California. Considered the most expensive disaster in the history of the United States, the earthquake caused an estimated $20 billion in damage. This quake, much like the one in San Francisco, toppled freeway overpasses, caused widespread structural collapse, and ignited numerous fires. Miraculously there were only 33 deaths attributed to the quake.

A year later, almost to the day, a devastating magnitude 7.2 quake struck the city of Kobe, Japan. This quake occurred in a densely populated area, causing almost 5,000 deaths, and it virtually destroyed the city.

These disasters occurred in two of the most technologically advanced and best prepared countries in the world. As the population density of the major cities continues to increase, disasters of this level are bound to happen, illustrating the need for continued efforts to design new and better equipment to deal with emergency situations.

SUMMARY

There are opportunities for motivated and qualified individuals with fire science degrees in the public and private workplace. Many of these provide good pay and benefits as well as providing a satisfying career. By pursuing a fire science degree, you are showing your prospective employer the ability to learn and complete a course of study.

It is extremely difficult to become a paid firefighter because of the stiff competition. Those you will be competing against are preparing themselves by seeking an education and related work experience. By properly preparing yourself, you can compete on the level required. Everything you do to prepare yourself to pursue a job in the field of fire science can pay off when applying for a job as firefighter. Modern fire departments require personnel with a wide range of fire protection specialties to perform their duties. A good education and experience in a related field are viewed as assets and should be brought out on the application and during the interview. If you do not meet the medical requirements for firefighter, you can still have a job in a closely related field.

REVIEW QUESTIONS

1. What is the difference between the firefighter trainee and firefighter positions?
2. What is the meaning of "under supervision" as it relates to being a firefighter?
3. What additional requirements are placed on a firefighter paramedic over those of a regular firefighter?
4. What particular skills are required of a fire heavy equipment operator?
5. What are the differences between a safety section and a general retirement?
6. As a Firefighter (Forestry Aid) GS 3 can you expect to have a full-time job?
7. What are the benefits of having a summer job as a firefighter (forestry aid)?
8. What jobs are available in the fire service that do not require actual firefighting to be performed?
9. What are some of the private sector jobs you could have that would help you prepare for the position of firefighter–fire department?

DISCUSSION QUESTIONS

1. How would you go about identifying the need for a new device or procedure for the fire service?
2. What is the job description for firefighter in the area in which you reside?
3. How do you fit the requirements for the position you are seeking in the fire service?

NOTES

1. Kern County Personnel Department, *Firefighter Trainee* (Bakersfield, CA: Kern County Personnel Department).
2. National Fire Protection Association, *Standard 1001, Fire Fighter Professional Qualifications* (Quincy, MA: National Fire Protection Association, 2008).
3. Kern County Personnel Department, *Fire Heavy Equipment Operator* (Bakersfield, CA: Kern County Personnel Department).
4. State of California, *Retirement Act of 1937* (Sacramento, CA: State of California).
5. US Forest Service, *Forestry Aid GS-0462-03, Standard Job No. N2011* (Washington, DC: US Department of Agriculture Forest Service). Retrieved from www.usajobs.gov
6. National Fire Protection Association, *Standard 1031, Professional Qualifications For Fire Inspector and Plan Examiner* (Quincy, MA: National Fire Protection Association, 2009).
7. Kern County Personnel Department, *Fire Hazardous Materials Program Specialist* (Bakersfield, CA: Kern County Personnel Department).
8. National Fire Protection Association, *Standard 1041, Fire Service Instructor Professional Qualifications* (Quincy, MA: National Fire Protection Association, 2007).
9. National Fire Protection Association, *Standard 1035, Public Fire and Life Safety Educator Professional Qualifications* (Quincy, MA: National Fire Protection Association, 2005).
10. City of Bakersfield Personnel Department, *Fire Dispatcher* (Bakersfield, CA: City of Bakersfield Personnel Department).
11. Kern County Personnel Department, *Emergency Services Manager* (Bakersfield, CA: Kern County Personnel Department)
12. DK Publishing, *Firefighting* (New York, NY 10014).
13. Park, Scott, Captain, c/o Kern County Fire Department, 5642 Victor St., Bakersfield, CA 93308.

CHAPTER 3

Public Fire Protection

LEARNING OBJECTIVES

Upon completion of this chapter, you should be able to:

- Identify the origins of modern fire protection.
- Describe the evolution of fire protection.
- List the causes of the demise of the volunteer fire companies in the major cities.
- Identify the U.S. fire problem.
- List the general responsibilities of the modern fire service.
- Describe the evolution of modern firefighting equipment.
- Describe the evolution of protective clothing and equipment.
- Describe how major fire losses have affected the modern fire service.
- List the reasons for fire defense planning.
- Define risk and risk management.
- Illustrate the fire department's role in community risk reduction.

INTRODUCTION

The only creature in the world that has learned how to initiate and utilize fire is man. Also, humans seem to be unique in that they do not immediately flee from flame. This is apparent when visiting a campground and campers are sitting around the campfire as well as in the fact that people have fireplaces in their homes. However, the very earliest humans apparently feared fire. It could destroy their homes and food sources. If they were caught in the open in a fire on a grassy plain or in a forest fire, it could overtake and kill them. It is known that fire existed for thousands of years before people learned to use it. It is apparent fires were initiated by natural phenomena: Lightning struck trees and started fires; flaming meteors fell to earth; and volcanoes spewed fire. These fires probably frightened people because they did not understand what fire was or how it could either harm or help them. People of the time thought that fire was a product of the gods.

We do not know exactly when people learned to use fire, but there is reported evidence that it was between 200,000 and 400,000 years ago. In caves near Beijing, China, the remains of a hearth and charred animal bones were discovered along with the bones of a humanlike creature known as Peking Man. It is not certain if this individual had learned to make fire or just learned to carry it to the cave, because the charred animal bones only suggest the fire was used for cooking.

Once people learned how to create and control fire it lost much of its mystique. When people first learned to use fire, their culture and society changed dramatically and fire became humanity's constant companion. Previously unexplored, colder areas of the world could be inhabited, and people gained more control over their environment. As the world population expanded, new land could be explored in the constant hunt for food.

Later, people used fire to make tools, implements, and pottery, which could be used to store foodstuffs. This important discovery of the use of fire to fashion implements is the very basis of our civilization. Before people learned to farm and store grains, they had to spend most of their time hunting for fresh game and fresh wild food. When they learned how to grow and store food, it meant they could develop permanent settlements. Populations no longer had to be nomadic.

The Industrial Revolution of the eighteenth century was initiated with the development of steam power—steam generated by heat from controlled fire. Steam was used for transportation, manufacturing, and the generation of electricity (which it still is). Fire also became a weapon of war. Explosives and boiling oil both involve the use of fire.[1]

We have become a nation that relies heavily on energy, and much of that energy is a by-product of fire. We still use fire for the same processes that early humans did. We use heat generated by fire to keep us warm and cook our food, to make glass and shape metal, and to propel our means of transportation. Most homes have at least two essential fires—the stove for cooking and the furnace for heating. Some houses also have fireplaces and/or wood stoves for warmth and comfort. As with any tool, fire can be used correctly or incorrectly. As we have seen over the centuries, fire used incorrectly can have tragic and disastrous results.

The three main causes of hostile fires are considered to be men, women, and children. When people become too comfortable with fire and lose their respect for it, problems arise. In the United States, people who are adversely affected by fires are considered to be victims, even if the fire is the result of their own negligence. It has not always been this way. Over the centuries, as man has become more of a city dweller, fire has become more of a destructive force. The fire service, as we will see in this chapter, has evolved to deal with the problems this presents.

EVOLUTION OF FIRE PROTECTION

The first recognized firefighting force was organized in Rome by the Emperor Augustus in A.D. 6. After numerous disastrous fires, the Corps of Vigiles was introduced. The Vigiles were divided into battalions of 1,000 men each. Each battalion was commanded by a man responsible to the Emperor himself. The Vigiles, equipped with buckets and axes, patrolled the streets and fought fires. They also performed fire prevention duties by warning the citizens to be careful with their fires and to keep water in their homes for fire extinguishment. The cost of maintaining the Vigiles was paid out of public funds. Fires were investigated and those found responsible were subject to corporal punishment.[2]

NOTE The first recognized firefighting force was organized in Rome by the Emperor Augustus in A.D. 6.

The first settlement in America, Jamestown, was almost a failure due to a fire. Upon landing in the New World, settlers were forced to hastily construct shelters of readily available materials. These materials were brush, sticks, clay, and grasses used as roof thatching. These homes were equipped with large central fireplaces for warmth and cooking. The chimneys were constructed of brush and sticks or wooden planks with a thick coating of mud or clay. After a while, the mud or clay would come loose and the chimneys, well dried from numerous fires, would burst into flame. Combined with the thatched roofs and the closeness of the structures to each other, the colony was a design for disaster (**Figure 3-1**).

The first permanent colony was founded in 1607. The first conflagration followed in the winter of 1608. This conflagration destroyed almost all of the colonists' homes and provisions. Because of the severity of the winters in Virginia, many of the colonists died from exposure or starvation.[3] To this day, the combination of structures closely spaced, either to each other or to combustible vegetation, still leads to conflagration-scale fires.

Fire protection at the time of the first settlers mostly consisted of creating fire breaks by pulling down structures. The bucket brigade also came into use. The townspeople would gather and form two lines between the water source and the fire (**Figure 3-2**). The men would pass the full buckets toward the fire and the women and children would return the empty buckets to be refilled.[4]

In 1647 the governor of New Amsterdam (later renamed New York), Peter Stuyvesant, took the first steps to save the new city from fire. He drew up a building code prohibiting

FIGURE 3-1
Thatch roof house with wooden chimney.

FIGURE 3-2
Bucket brigade.

Delmar/Cengage Learning

wood or plaster chimneys. He appointed four volunteer fire wardens, whose primary function was fire prevention by enforcing the law and making sure that chimneys were properly maintained. The wardens were allowed to levy fines when violations were found. The money accrued from the fines was used to purchase fire equipment.

The greatest fire threat was that fires would get a good head start while people slept. New Amsterdam, Boston, and other towns adopted a curfew. At 9:00 P.M. a bell was rung and all fires were to be extinguished or covered until 4:30 A.M. Stuyvesant went a step further and appointed a group of young men to patrol the streets at night carrying wooden noisemakers that were twirled to sound an alarm if a fire was discovered. They became known as the "Rattle Watch."[5]

In 1666, the Great Fire of London destroyed almost two-thirds of the city. The fire went on for five straight days, destroying the city's primarily wood frame structures. It was fought with the implements available at the time—buckets, hooks, and primitive fire engines.[6]

As a result of this tremendous fire, the city passed a code of building regulations. Many new firefighting methods and architectural improvements were proposed but were not enacted. These included wider streets, green spaces, building setbacks, and the use of noncombustible materials.[7] One major effect of this fire was the creation of fire insurance companies. This idea, like so many others, soon spread to the New World.

At this time, the cities had no well-organized firefighting forces of their own. The owners of the fire insurance companies saw the need for, and benefit of, an organized firefighting force. They were soon advertising the availability of their own firefighting forces to protect the properties they insured. The insurance company would place a plaque, called a fire mark, on the buildings that they protected (**Figure 3-3**). Because this was a business and not a public service, the responding fire brigades would fight fires only in the businesses they insured. If they arrived at the scene and it was a business that was uninsured or with another provider, they let it burn.[8]

Courtesy of National Museum of American History, Kenneth E. Behring Center, Smithsonian Institution

FIGURE 3-3
Fire marks used to identify which insurance company protected a home or building.

In 1679, Boston established the first publicly funded, paid fire department in America. Volunteer organizations called mutual fire societies were organized to assist the fire department. Ben Franklin observed the operation of the mutual fire societies and in 1736 organized the Union Volunteer Fire Company in Philadelphia.[9] He is considered to have conceived the idea of the volunteer fire companies that continues today.

NOTE In 1679, Boston established the first publicly funded, paid fire department in America.

In England, the fire insurance companies had their own firefighting forces. In America, the insurance companies performed salvage work, but the firefighting was left up to the volunteers. The insurance companies would pay the first company to arrive to put water on the fire.[10] This, along with the great pride the firefighters had in their company and its abilities, led to fierce and often violent competition between the companies of volunteers.

Membership in a volunteer fire company was a great source of pride for many Americans. There were long waiting lists of adventurous young men who wanted to join. Many famous Americans served on volunteer fire companies, including George Washington, John Hancock, Alexander Hamilton, Samuel Adams, and Paul Revere.[11] After the Revolutionary War, the concept of volunteer fire companies spread across the nation. Men joined the local volunteer company for social reasons as well as to serve the public and to protect their towns.

City water systems consisted of wooden logs hollowed out to carry water (**Figure 3-4**). Every half block or so there was a wooden plug placed into the log. This plug was removed to access the water for firefighting, thus giving us the term "fire plug." There was great competition to see which company could get "first water" on the fire. Some companies would pay boys to race ahead to the scene of the fire and hide the plug from other companies. These boys were called "plug uglies." Another source of water was cisterns dug under the streets whose lids were removed to allow access during fires (**Figure 3-5**).[12]

The volunteer fire companies continued to develop and grow in strength. Pride in the companies grew to the point that the equipment became very ornate, with the advent of uniforms, fancy painting on the engines, bells, and a constant desire to have the latest and best equipment.

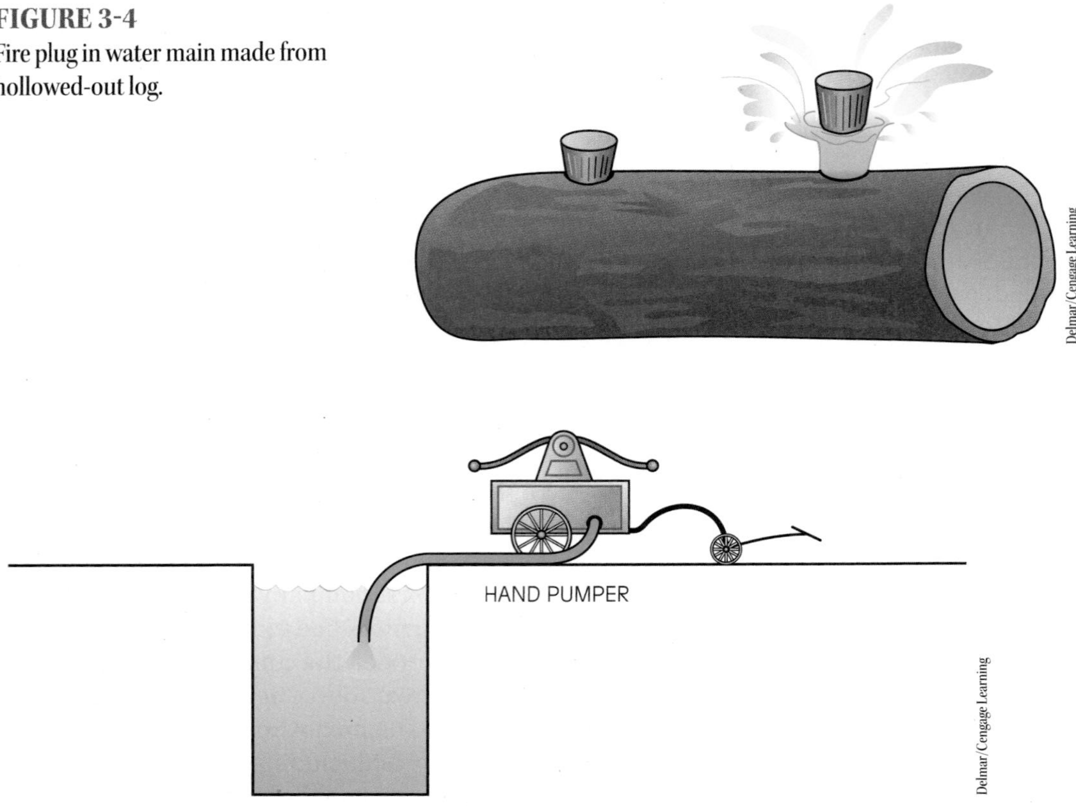

FIGURE 3-4
Fire plug in water main made from hollowed-out log.

FIGURE 3-5
Cistern for firefighting water supply. This method of fire water storage still applies as modern pumpers are capable of drafting water from a static source.

These were rough men and rough times. The competition got to the point that companies arriving on the scene at the same time would fist-fight to see who could claim "first water." Some companies would send a man with a bucket of water to the scene as fast as possible so he could throw it on the fire and claim first water for his company.[13]

These early firefighters reveled in the excitement and danger of attacking fires. They were readily prepared to be injured in the line of duty. They often took bad falls, were run over by the equipment when they pulled it to the fire, or were singed and blistered by the heat. They wore their scars and bandages proudly, their injuries proclaiming to all that they were real men. It was common to go out in the dead of winter and have their wet clothes freeze to their bodies as they brought a blaze under control. Perhaps a warehouse was storing black powder or chemicals or it presented the hazards of falling walls and roofs.[14] None of the hazards daunted the brave firefighters in the pursuit of their task. The fire was their enemy and they would not surrender.

The competition among the companies hastened their demise in the large cities. The fighting and rowdyism got to the point that the city leaders decided to organize paid fire departments. In 1853, Cincinnati became the first city with a fully paid fire department.[15]

NOTE In 1853, Cincinnati became the first city with a fully paid fire department.

Since these early beginnings, the fire department has continued to evolve and expand its role in the United States. The time of waiting around the firehouse for the alarm to go off is gone. The modern fire department is proactive in the community in trying to minimize loss from fire and other disasters. Fire departments now routinely provide such diverse services as fire prevention, public safety education, fire suppression, medical aid, rescue services, and hazardous materials response. The age of the firefighter being perceived as a professional is here.

The fire service is still full of traditions. New firefighters are still required to prove themselves on the fireground. Most fire engines are still painted red, even though other colors are more visible. People are still required to come in at the bottom of the organization and work their way up through the ranks. This is not all bad. Tradition and a sense of history help to instill pride in the organization and the people who work there. Many of the traditions, like that of service to the public, are what make the fire department one of the most respected public agencies. After especially large and destructive fires, the public has been known to come out in force and show its appreciation for the job done by the firefighters. It is a great feeling to watch the evening news and see signs hung out showing the public's appreciation for a tough battle fought and won (**Figure 3-6**).

A negative statement sometimes made in reference to the fire service is "one hundred years of tradition unimpeded by progress." Fire departments and firefighters must be able to change with the times. The fire service should not lose touch with its long and honorable history, but it must be able to see where change is necessary and adapt for the future.

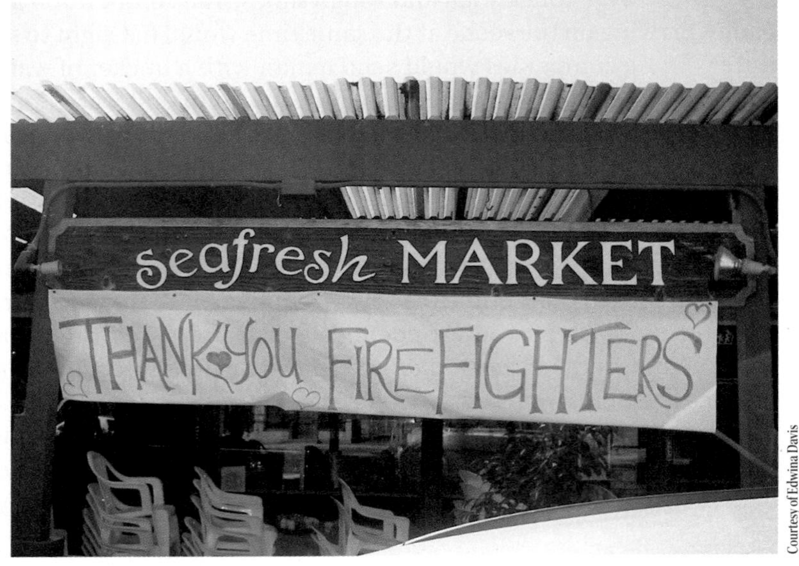

Courtesy of Edwina Davis

FIGURE 3-6
Sign erected by citizens to thank firefighters for saving their homes and businesses.

Actual fire suppression activities occupy very little of the firefighters' time in most departments. The fire service has become more involved in medical aid. Some departments are reporting requests for medical aid to be 70% of their incident volume. Others provide ambulances and paramedic programs to their communities. Television shows such as *Emergency*, which aired from 1972 to 1977 about the Los Angeles County Fire Department's paramedic program, have raised the public's expectations of the fire service. The public is no longer content to pay for firefighters to sit in the station—it wants service for its tax dollar.

This demand for increased service at least partly fueled the evolution of the fire service education system. As illustrated in Chapter 1, the educational system has progressed to the point that many firefighters are no longer content with time on the job as their only educational opportunity. As in many other professions, national and state standards of qualification have been adopted. Certifications and degrees are available in a wide variety of fire service specialties. Possibly in the future the name of the position will change from firefighter to public safety specialist or emergency services technician.

EQUIPMENT

Firefighting equipment evolved because of the need for more firefighting capability. As cities grew and the fire load increased, new ways needed to be found to apply water where and when it was needed. The first known fire pump, called a *siphona*, was invented in the fourth century B.C. (**Figure 3-7**).[16] It was comprised of two cylinders with pistons that

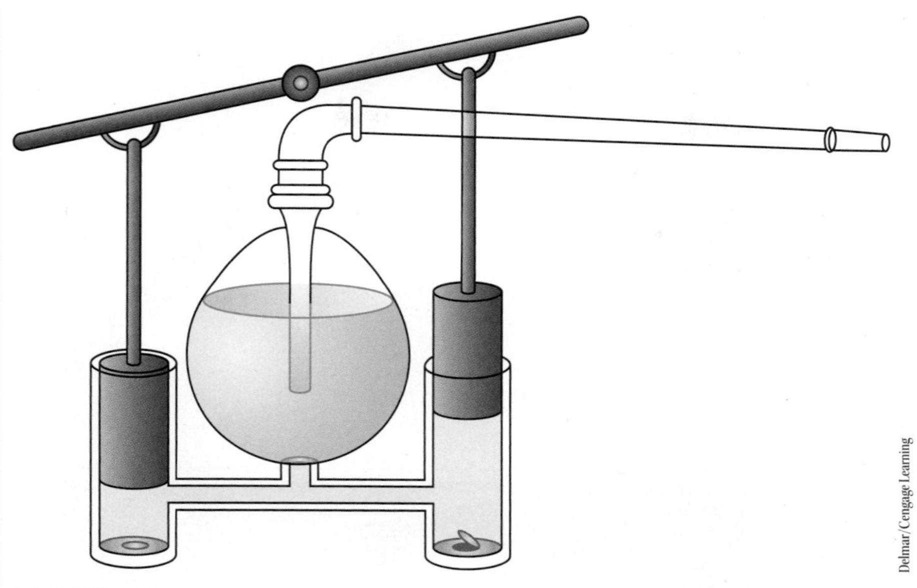

FIGURE 3-7
Siphona using double action piston pump. This principle was used to create hand-operated fire pumps (hand pumpers).

sucked water in at the base as each piston rose and expelled it as the piston went down. The water was fed into an air-filled chamber, compressing the air in the chamber and forcing the water out of a nozzle on the top. The purpose of the air space in the chamber was to give a steady stream. For centuries this information was lost. When rediscovered, it was used to design the engine commonly referred to as the *hand pumper*. This concept carried on into the hand-pumped engines that followed and continued to be used into the twentieth century. Fire hooks were designed to pull burning thatch from roofs and to pull down houses to create fire breaks (**Figure 3-8**). In the sixteenth century, large syringes were developed to squirt water onto the fire. As you may well imagine, they lacked effectiveness.

NOTE Firefighting equipment evolved because of the need for more firefighting capability.

Many different designs of hand-pumped engines were tried and many were discarded due to inefficiency or poor operation. The most effective and longest lasting design used the principle of the siphona, with long handles that could be swung out from the sides and latched into place. The handles were hinged to allow the pumper to pass through narrow streets and alleys. These handles were called brakes. Up to 15 feet long, the brakes were manned by volunteers and the engine was pumped between 60 and 170 strokes a minute

FIGURE 3-8
Fire hook used for tearing down buildings to create a fire break.

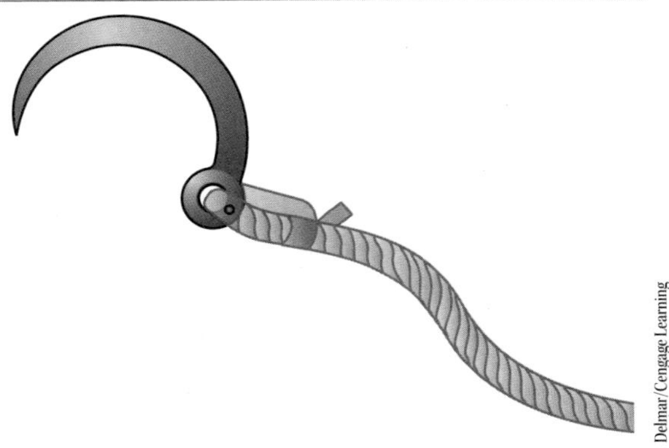

Delmar/Cengage Learning

FIGURE 3-9
Hand pumper fire engine.

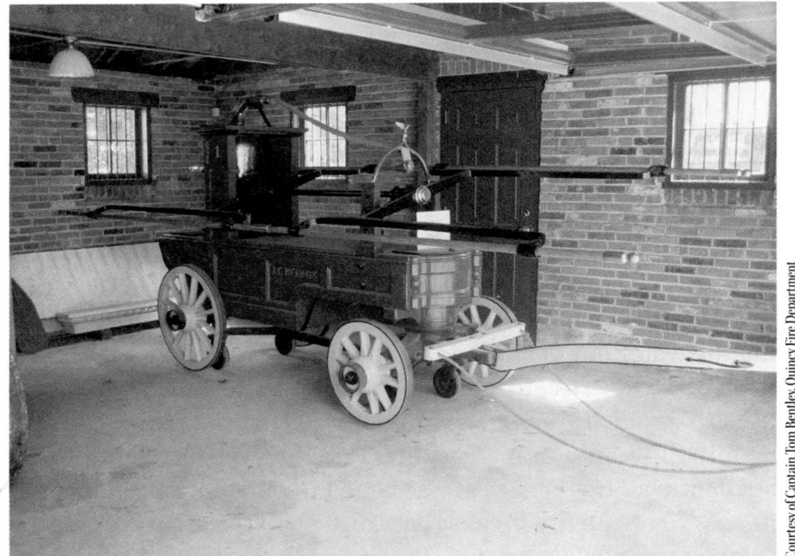

Courtesy of Captain Tom Bentley, Quincy Fire Department

(**Figure 3-9**). Pumping the engines was hard work, and at high rates of speed the persons pumping had to be relieved at regular intervals.

The basic design and principle of the engines were the same, but the execution varied greatly. Most were of the two-cylinder design, but at least one four-cylinder engine exists. Some were of the endstroke design with the brakes perpendicular to the tongue; others were of the sidestroke design. To gain pumping efficiency in a short length, even double-deck pumpers were made that used men both on the ground and up on a platform to operate the brakes.

The first hand pumpers discharged water through a large nozzle mounted on top called a gooseneck. The limitation of this design was that the engine had to be positioned close to the fire and this limited the ability to aim the discharged water. Firefighters, wanting to be able to better position their water streams, experimented with extending a hose from the gooseneck. This gave them the capability of attacking the fire from different angles without having to move the engine. It also gave them the capability of having the engine closer to the water source and further from the heat, smoke, and embers from the fire. From this improvement, the concept of the interior attack was born. Firefighters could now apply the water directly on the seat of the fire, not just through a window or doorway.

Another limitation of the first hand pumpers was that they had to have the tub around the intake of the pump filled by a bucket brigade. When the design was changed so the engines could take suction (draft) from cisterns under the streets or be filled from hydrants, the efficiency increased greatly. All hands were needed to man the brakes, whether the problem was a long hose lay to the fire, a need for a hose stream to be aimed into an upper floor window, or a large fire that required long hours of pumping. When this suction hose was permanently attached to the body of the pumper it was called a squirrel tail.

The first hoses were made of sewn leather, which had a tendency to badly leak, reducing the hose stream at the nozzle. The next improvement was riveted leather hose, which was much tougher and leaked considerably less.[17] These improvements continued on through linen hose, rubber-lined cotton hose, and today's synthetic hose.

As water systems improved and hose became more popular, hose companies were formed. The hose was carried on a wheeled carriage called a hose cart (**Figure 3-10**). If

FIGURE 3-10
Hose cart used for transporting hose to the fire scene. These are still used in some settings.

the water system was good enough, the hose companies could use the pressure from the water mains to place nozzle streams used to attack the fire into service. If the water system was insufficient, the hose companies would respond with, and were often affiliated with, the hand pumper companies.

Until this time, almost every piece of wheeled firefighting equipment was pulled by hand—which could be very dangerous, especially in hilly areas. A fully rigged hand pumper is very heavy and the common practice was to run with it, to the fire, as fast as possible. When a pumper went out of control due to hills or sharp turns, the only thing to do was to get out of the way. This was very hard on the equipment and the firefighters. Many a volunteer was run over or pinned to a building by the very apparatus he was pulling to the fire.[18]

The next major improvement in fire fighting apparatus was the steamer (**Figure 3-11**). The first steamers were pulled by hand but were soon discovered to be too heavy so horses were used to pull them. The steamer took only a three-person crew to get it to the fire and place it into operation. The steamer could pump for as long as there was coal to keep it running, without having large numbers of volunteers to man the brakes. At first the volunteers saw the steamers as a threat to their existence and fought hard against their adoption. The steamer gave the cities the opportunity to have a regular paid fire department at low cost due to the small number of men required to operate it. This, along with the problems of fighting among the volunteer companies at fires, led to the extinction of the volunteer fire departments in many of the larger cities.[19]

As buildings grew in height, a way had to be found to effect rescues from windows when the access through the inside of the building was untenable. This led to the introduction of the ladder company. The responsibility of the ladder company was to bring

FIGURE 3-11 Steam-powered fire pumper.

Courtesy of National Museum of American History, Kenneth E. Behring Center, Smithsonian Institution

the ladder wagon to the fire. The ladder wagon was also equipped with hooks, ropes, and other equipment, giving us the term still in use "hook and ladder." The pike pole, looking much like a lancer's pike, was developed to pull down ceilings to get at concealed spaces. The pike axe, a single blade axe with a point on the back side of the head, was designed for forcible entry (**Figure 3-12**). With these tools carried on the ladder wagons, the job of the ladder company evolved into one of forcible entry, rescue, and ventilation, which it still is today.[20]

Extension ladders up to 75 feet in length were brought to fires. These allowed access as high as the fifth floor but were extremely heavy and cumbersome. By 1870, the aerial ladder apparatus was developed. Its ladder was extended by hand cranks. The next improvements were spring-assisted raising, compressed air, and finally, hydraulic pressure.[21]

The chemical wagon, invented in the late 1800s, carried two tanks—one of soda and water, and the other of acid—which, when mixed, created carbon dioxide gas. The gas created pressure in the vessel, forcing the water out through the hoses. Their effectiveness was limited to the amount of gas created to expel the water. These engines were smaller than steamers, weighed much less, and were easier to place into operation, but their effectiveness was limited to the water they carried and the amount of gas created to expel the water.

Because of its speed and ease of operation, the chemical engine extinguished many fires in their first stages. However, a severe problem developed if the hose became kinked or plugged, because an explosion could occur.

FIGURE 3-12
Pike pole and pike axe firefighting tools.

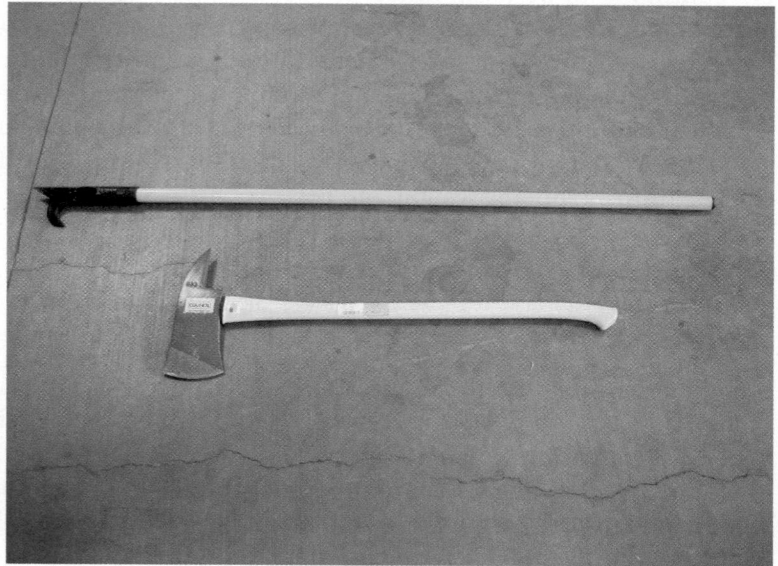

Delmar/Cengage Learning

FIGURE 3-13 Restored 1928 Moreland gasoline-powered fire engine.

Courtesy of Edwina Davis

The next major improvement in fire apparatus was the invention and use of the internal combustion engine (**Figure 3-13**). Unlike horses, the gasoline engines did not get tired on long runs nor were they subject to disease or being frightened by dogs or people. The first gasoline-powered engines were not very reliable and steam pumpers continued to be used. Since the departments already had steam engines in use, gasoline-powered tractors were built to pull the steamers to the fire, replacing the horses. As improvements were made in design and reliability, the motorized apparatus finally dominated the scene and other types of pumping apparatus and steam and chemical engines were eventually phased out. In the engine design in use today, the same motor that propels the apparatus drives the pump. This design further simplified the apparatus, reducing weight and cost.

FIRE SERVICE SYMBOLS

Maltese Cross

A common symbol in the fire service, the Maltese cross, is considered a symbol of protection and a badge of honor. The Maltese cross dates back to the days of the Crusades. When the Christian knights approached the cities held by the Saracens, they were assaulted by glass bombs containing naphtha, a flammable liquid. Once the naphtha had penetrated the knights' armor, the Saracens would throw flaming torches at them. Hundreds of the knights were burned alive, and many others perished or were injured coming to their aid. To recognize them for their bravery, the survivors were awarded a special cross as a badge of honor. Many of the knights were members of the Knights of St. John, which was headquartered on Malta. The badge became known as the Maltese cross. The symbolism continues in that a firefighter will lay down his life for others. The Maltese cross remains the firefighter's badge of honor and in some departments the badges are this shape.

Dalmatians

When horses were introduced into firehouses, Dalmatian dogs were also introduced as fire service mascots (**Figure 3-14**). For centuries they were used as followers and guardians of horse-drawn vehicles. They were kept with the horses to keep them calm. When responding to fires, the dog ran ahead, behind, or under the fire apparatus and chased off other dogs and animals that would bother the horses. The Dalmatian is still the only recognized carriage dog in the world. The Dalmatian is also known for its friendliness and ability to work.

FIGURE 3-14
Dalmatian dog fire service mascot.

Courtesy of Mark/Raycroft/Minden Pictures/Getty Images

FIRE STATIONS

As paid firefighters came more into use, the fire station needed to have sleeping quarters, because the firefighters were there 24 hours a day. Previously, a shed for the apparatus was all that was needed. A stable for the horses was added as horse-drawn steamers were brought into service. Many old fire stations still have a hole in the ceiling that went up to the hay loft. In multistory fire stations, the sliding pole was introduced to give quick access to the apparatus floor when the sleeping quarters were upstairs. The first poles were wood; then brass was used. In some stations, due to injuries from sliding down the poles, slides were introduced. The design of the modern fire station is presented in Chapter 6.

PERSONAL PROTECTIVE EQUIPMENT

In the early days of the volunteers, the firefighters fought fire in whatever they happened to be wearing at the time. As pride in the companies grew, uniforms were used as a means of identifying the companies' members (**Figure 3-15**). The uniforms were mostly worn

FIGURE 3-15
Old-style uniform with firefighter holding speaking trumpet.

Courtesy of National Museum of American History, Kenneth E. Behring Center, Smithsonian Institution

at parades and gatherings. They were trimmed in white or blue and had large letters or numbers on the front of the shirt. A thick leather belt with the company number on the buckle was also worn. The belt came in handy as a place to hang tools.[22] Standard uniforms were adopted by all of the larger paid departments during the early 1800s. The badge identifying the member's rank and department took the place of the large numbers and letters.

The fire helmet was a product of necessity (**Figure 3-16**). Over the years, many different designs, from cloth and felt hats to leather, metal, fiberglass, and plastic, were tried. The traditional shape was created with a short brim on the front and a long brim on the back to keep embers and hot water from going down the back of the firefighter's neck. The shield was mounted on the front with the company number and the firefighter's or department name. The ribbed shape was added to provide rigidity. This design improvement became especially important as interior attack became more prevalent.

Bunker gear, or turnout clothing, went through the same development process as the rest of the firefighting equipment. As the methods of attack became more aggressive, the firefighters' personal protective equipment had to keep up. Long canvas coats were worn with boots that came up to the thighs. These protected the firefighter from falling embers. Water barriers were added to the coat in an attempt to keep the firefighters dry; insulation was added to protect them from heat. The long boots were eventually replaced with knee-length boots and insulated pants. These better protected the firefighter from fire underneath them (refer to Figure 6-58).

FIGURE 3-16 Assortment of fire helmets, including leather, metal, fiberglass and plastic.

Delmar/Cengage Learning

SAFETY Today, self-contained breathing apparatus is used on almost all fires.

Today, self-contained breathing apparatus is used on almost all fires. The firefighters of old wrapped wet rags around their faces to protect themselves from smoke and ash. The next improvement was the canister (gas mask), which protected the firefighters from smoke particles but did nothing to protect them from toxic fire gases and low oxygen concentrations. When the self-contained breathing apparatus was invented and adapted to firefighting use, the ability to perform interior attack was greatly enhanced (**Figure 3-17**). The firefighters no longer had to back off from the fire and "take a blow" to fill their lungs with fresh air. One method used for getting fresh air in a smoky building was to place the face down by the nozzle because the nozzle draws air alongside the fire stream. The use of breathing apparatus should also reduce the number of firefighters suffering cancer from fire-related exposures to toxic gases and soot.

FIRE LOSSES

Throughout U.S. history, fires have resulted in tremendous loss of life as well as property. We first look at fires that resulted in property loss.

NOTE Throughout U.S. history, fires have resulted in tremendous loss of life as well as property.

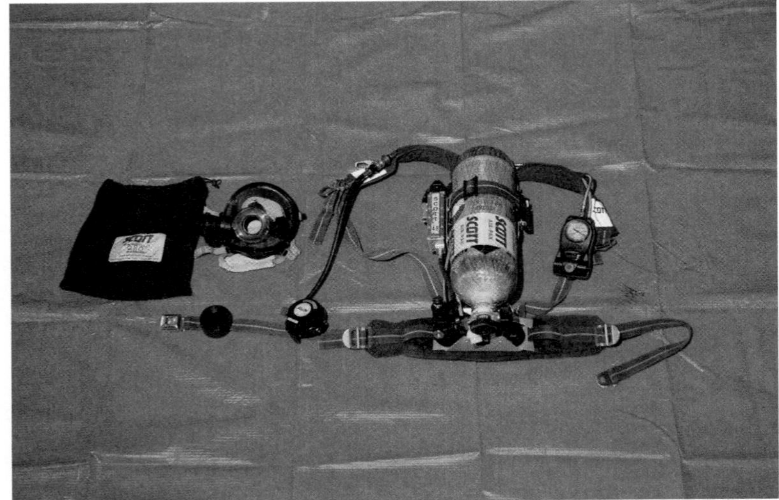

FIGURE 3-17 Modern self-contained breathing apparatus (SCBA).

Delmar/Cengage Learning

One of the most disastrous fires ever to occur in the United States was the so-called Great Chicago Fire, which occurred on October 8–10, 1871. This fire burned for three days and destroyed 3⅓ square miles of the city of Chicago, Illinois. The fire left 100,000 people homeless and killed approximately 300. As a result of this and other fires, President Calvin Coolidge proclaimed the first National Fire Prevention Week to be October 4–10, 1925. This tradition continues today with Fire Prevention Week occurring each October.

At the same time as the Chicago fire, October 8, 1871, another disastrous but lesser known fire started in the vicinity of Peshtigo, Wisconsin. The fire began as a forest fire and soon overran the town of Peshtigo. In this fire 2,400 square miles were burned and approximately 1,000 people were killed.

The city of Baltimore, Maryland, was stricken by a **conflagration** on February 7, 1904. A fire started in a storage area in the basement of the six-story Hurst Building. Soon the fire leapt up an unenclosed shaft and spread to the top floor. Nearby buildings with unprotected openings were soon afire. As the conflagration spread, the mayor called for help from other major cities in the area. Fire equipment and firefighters were shipped in by train. Upon their arrival, they found that the threads on their hoses would not fit those on the Baltimore hydrants. The out-of-town firefighters were forced to rip up the streets and create ponds of water around the hydrants to supply their engines. This problem eventually led to the development of standardized hose threads for firefighting equipment. By the time the fire was stopped, 155 acres of mercantile property valued at $50 million was destroyed and 50,000 people lost their jobs.[23]

On April 18, 1906, an earthquake rocked the city of San Francisco. The earthquake lasted only 1½ minutes, but the resulting fires lasted for two days. The fires destroyed 4.7 square miles of the city and burned over 25,000 buildings. Four hundred fifty people were killed in the disaster. The estimated loss was from $350 million to $1 billion.[24]

On April 12, 1908, Chelsea, Massachusetts, was devastated by a fire started in rags. High winds whipped the flames out of control and destroyed approximately one-half of the improved area of the city. When the fire was extinguished, 3,500 buildings over an area of almost a half mile were destroyed. In October 1973, 65 years later, the same thing happened: Chelsea had another conflagration with the same problems of fire spread and control.[25]

In the early part of the twentieth century, when combustible construction was prevalent, conflagrations occurred at an alarming rate. In 1913, a major fire hit Hot Springs, Arkansas, destroying over 500 buildings. Salem, Massachusetts, was ravaged by fire on June 25, 1914; 1,600 buildings were destroyed.[26]

The list goes on and on. The common denominator in these conflagrations is **combustible construction**. The streets were narrow and there were ineffective building

conflagration Very large fires that defy control efforts and cause extensive damage over a large area.

combustible construction The use of unprotected wood and wood byproducts in building construction.

codes in place regarding **unprotected vertical shafts**, which led to rapid fire spread throughout the structures.

More recently a primary cause of large fires resulting in great property loss and numerous deaths is terrorism. On one day, September 11, 2001, the threat of large-scale terrorist attacks in the United States became reality. Two hijacked airliners were purposefully flown into the twin towers of the World Trade Center in New York City and a third airliner hit the Pentagon in Arlington County, Virginia. As a result of these incidents, the life toll was over 3,000 persons, including 343 firefighters. The property loss was estimated to be in the tens of billions of dollars. There is a constant threat that these types of incidents will continue to occur.

Part of the reason southern California widland fires reach conflagration proportions is the extreme weather experienced in the fall. The weather pattern builds up high pressure over the Great Basin, the area of eastern Washington, Idaho, and Utah, coupled with a low pressure off the coast of Southern California. This difference in pressure causes hot, dry Santa Ana winds to blow from the Great Basin across Southern California to the Pacific Ocean.[27] Powered by these winds and fueled by dry brush, wildfires grow quickly to conflagration size. With combustible exterior construction in the form of wooden decks, wood shake roofs, and wood exteriors, the fire soon rages out of control. In the great fire siege of October 26 through November 7, 1993, 1,152 structures and more than 200,000 acres were consumed by fire. Many of these structures were homes. Two of the fires alone were responsible for massive destruction: The Laguna fire burned 366 structures and the Malibu fire burned 323.[28] Two factors that finally let the firefighting forces get the upper hand were the winds dying down and the fire burning to the Pacific Ocean.

NOTE Urban interface and intermix fires will continue to grow in frequency and devastation.

These types of fires are not limited to California. Similar fire experiences are occurring across the nation. In all, at least 39 of the 50 states have received federal disaster declarations due to wildland fires.[29] This problem is not limited to the U.S.—Australia, Italy, Greece and Israel are experiencing these problems as well. As more and more people leave the cities and build their homes in the hills and mountains, these **urban interface** and **urban intermix** fires will continue to grow in frequency and devastation. People are building homes at the heads of canyons and on ridge tops, offering homeowners a

unprotected vertical shafts Laundry chutes, elevator shafts. When not enclosed in fire-resistive construction, these openings act as chimneys, allowing rapid fire spread in multilevel buildings.

urban interface The area where built-up areas of homes and businesses have little separation from the natural-growing wildland area.

urban intermix Buildings interspersed in wildland areas. Homes built in the woods.

beautiful view but creating a firefighting nightmare. Often built at the end of narrow streets and cul-de-sacs, these home sites are death traps for firefighting crews if the fire gains speed as it races uphill. As firefighting is done further down the hill and people water their roofs to protect their homes, the water supply at the higher elevations becomes depleted. Soon there is not enough pressure for engines to operate from the hydrant system, if there is one. This leaves the firefighters with only the water in the tank on the apparatus to deal with an onrushing wall of 50-foot or higher flames.

Combustible roofs are a cause of conflagration fires. In the western United States, particularly in California, conflagration fires have started out as brush fires spreading to structures. This happened in the Berkeley hills in 1923 and again in 1991. On October 20, 1991, the East Bay Hills fire took the lives of 23 civilians, one firefighter, and one policeman. This fire destroyed an average of one home every 13 seconds for the first nine hours of its existence, for a total of 2,103 structures. Similarly, in 1990 the Paint fire burned 545 structures in the Santa Barbara area.

Arson is another cause of high-loss fires. Many people have been killed as the result of arsonists setting fires to occupied buildings. Usually motivated by spite or revenge, the arsonist has no feeling for the consequence of his actions. Arson is especially common during riots. In periods of civil unrest, it is not uncommon for firefighters and their equipment to be attacked. A Los Angeles city firefighter, driving a fire engine down the street, was shot in the neck by a passing motorist during the 1992 riots. The 862 structure fires set during the Los Angeles riots caused three deaths, and damage was estimated at $1 billion.[30]

NOTE The United States has one of the highest fire death rates per capita in the industrialized world.

THE U.S. FIRE PROBLEM

Annually the National Fire Protection Association collects data from the National Fire Incident Reporting System (NFIRS) and develops a snapshot view of the national fire problem in the United States.

- The United States has one of the highest fire death rates per capita in the industrialized world. The fire death rate is 13.1 per million of population.
- Approximately 4,000 people die every year in the United States as the result of fires and another 22,000 are injured.
- About 100 U.S. firefighters are killed annually in duty-related incidents and another 87,000 are injured.
- Fire is the third-leading cause of accidental death in the home.
- An average of 1.9 million fires are reported each year. Keep in mind that not all fires are reported.
- Direct property loss due to fires is estimated at more than $10 billion each year.

- The South has the highest fire incident and death rates in the nation. This statistic is related to income level, with poverty-stricken areas having the highest fire and death rates.
- Eighty-two percent of all civilian fire fatalities occur in the home.
- Cooking is the leading cause of home fires and injuries in the United States.
- Careless smoking is the leading cause of fire deaths.
- Careless smoking is the leading cause of residential fire deaths.
- Heating equipment was the second leading cause of home fires and home fire deaths.
- In commercial properties, arson is the leading cause of deaths, injuries, and dollar loss.
- The fire death risk among senior citizens is more than double the average population.
- The fire death risk for children under age five is nearly double that of the average population.
- Children playing with fire start more than 50% of the fires that kill young children.
- Men die or are injured in fires twice as often as women.
- Approximately 90% of U.S. homes have at least one smoke alarm.
- More than 40% of residential fires and 60% of residential fatalities occur in homes with no smoke alarm.
- The U.S. fire service responds to a fire every 17 seconds.
- Twice as many fires occur in the kitchen than in any other area of the home.
- Nationwide, there is a fire death every 147 minutes and a fire injury every 24 minutes.
- The top five areas of fire origin in residences are: kitchen, 26%; bedroom, 13%; living room/den, 8%; chimney, 8%; and laundry area, 5%.
- Residential sprinklers decrease the home fire death rate per 100 fires by 74%[31]

NOTE Senior citizens are at the highest risk of being killed in a fire.

NOTE The southeastern United States has the highest fire death rate per capita.

NOTE Careless smoking is the leading cause of residential fire deaths.

NOTE A working smoke detector doubles a person's chance of surviving a fire.

To put all this in perspective, if something is stolen in a crime it is lost, not destroyed. Usually only items considered to be of resale value are taken. If a house burns down, many of the family's belongings are destroyed and unrecoverable. Anyone who has ever had a home burn can tell you it is not the loss of the DVD player or the television that really matters, as these can be replace with money provided by insurance, but the loss of the irreplaceable photos and family heirlooms. When a business burns or is severely damaged by fire, jobs and tax revenue are lost. Other businesses take over the burned-out business's customers.

With the cost of rebuilding and regaining its previous customers, the burned-out business cannot compete and only a few reopen, which has a serious negative impact on the community as a whole.

In the national forests and other wildlands, lightning is the cause of many large-acreage fires each year. One of the reasons lightning accounts for so much of the acreage is that firefighting resources are soon overwhelmed when a large storm comes through and ignites numerous fires. In the summer, dry lightning storms occur and **ground strikes** start fires.

In the year 2008 there were 78,979 wildland fires (the average number of fires for the years 1992–2001 was 103,112). There were 5,292,468 acres burned and suppression costs were over $1.3 billion. The number of people involved in fighting the fires, including firefighters and support personnel, was more than 30,000.[32] When fire seasons such as 1990, 1994, 1997, 2000, 2003, and 2007 occur, resources from across the country, Canada, and Mexico have been mobilized.

In the fall of 2003 Southern California was stricken by a fire siege unequaled in previous experience. It lasted for ten days, from October 21–30, and consisted of 14 large fires that occurred in six counties: San Diego, Riverside, Los Angeles, San Bernardino, Ventura, and Santa Barbara. By the time the fires were controlled, they had burned 745,933 acres (1,165 square miles). Included in the losses were 3,646 residences, 36 commercial structures, and 1,169 outbuildings. The human toll included 22 fatalities: one firefighter and 21 civilians. There were an additional 239 reportable injuries. The fires were fought by 15,631 personnel, 1,898 engines, 296 hand crews, 181 bulldozers, 105 helicopters, and 43 air tankers.[33] A similar situation impacted Southern California in 2007, resulting in the deaths of five U.S. Forest Service firefighters performing structure protection on the Esperanza Fire. Nationally, there were 85,705 wildland fires in 2007 with 9,328,045 acres lost to fire.[34] Despite the availability of resources and improvements in firefighting equipment and technique, Los Angeles County experienced the largest fire in its recorded history (160,5557 acres) in 2009, which also cost the lives of two firefighters.[35]

Another common cause of wildfires is humans, sometimes referred to as "two-legged lightning." Careless campers who leave campfires unattended, burn trash carelessly, smoke in high–fire danger areas, operate faulty equipment, or drive in dry grass with catalytic convertors on their vehicles are the source of fire starts. Human causes, other than arson, accounted for 60% of the fire starts and 22% of the acres burned. Arson accounted for 26% of the fire starts and 10% of the acres burned.[36]

ground strikes When lightning bolts reach objects on the ground, often starting fires in trees when there is no rain received with the storm.

Once these fires gain headway, vast areas of natural resources are lost, animals are killed, and valuable **watersheds** are destroyed. In time these areas will return to their previous condition, but this may take well over a hundred years.

watersheds Complex geographic, geologic, and vegetative components that control runoff of rain water and support varied ecosystems.

PURPOSE AND SCOPE OF FIRE AGENCIES

The main purpose of the fire service is to manage community risk. By having trained personnel and specialized equipment on hand, emergency responders exemplify a community's response to risk. Most departments now focus on the control of fires in individual properties and the rescue of endangered occupants. Departments exist to limit the probable loss to the community when a fire occurs. Imagine the consequences if there were no fire departments.

In a department with a mission statement such as "protecting life and property through effective public education, fire prevention, and emergency services," we are talking about risk reduction. Managing risk is done through reducing risk, the weighing of cost versus benefit. It would be impossible and too expensive to totally eliminate risk.

A community expresses its assessment (perceived threat) of its overall fire risk through the resources it is willing to commit to its fire department. If the fire department is unable to perform its fire control mission, the community's fire risk balance could be compromised.

The modern fire service manages more community risk than just that of fires. The fire service now provides services in the areas of emergency medical services (EMS) to reduce the consequences of medical emergencies. It also provides emergency management, providing services to reduce the effect and consequences of disasters such as earthquakes, floods, tornadoes, and ice storms. Rescue teams safely remove citizens from situations such as structural collapse, swift water, ice, trench rescue, and vehicle crashes. Hazardous materials ("hazmat") response teams protect the population and the environment from the effects of uncontrolled releases of hazardous materials.

As mentioned in the mission statement, emergency response is only one factor in the reduction of risk. Placing a fire station on every corner would still not prevent incidents from occurring. Often, after the incident has occurred is too late to save lives; therefore departments engage in public education, prevention, and code enforcement (discussed in Chapter 10).

With this tremendous opportunity to be seen as heroes comes responsibility as well. Emergency response organizations are viewed as essential public services.

As such, members need to ensure that they are always ready to respond and carry out their sometimes life-saving functions. When departments cannot respond, public outcry can be intense. Managers must always be aware of possible disruptions to service (equipment failure, unavailable personnel, and other reasons). It may not be your fault, but no one wants to hear excuses.

Another part of this responsibility is that safety organizations are also recipients of public funds. Resources must be managed carefully in a cost-effective manner to allow the utilization of funds where they are most needed. If firefighters are careless and lose or destroy equipment, money must be diverted from other areas to replace it. If personnel are not safety conscious, injuries rise and costs for worker's compensation, medical treatment, and replacement personnel rise as well.

Fire departments must manage financial, liability, and safety risks within three major categories:

1. Risk to the community—community risk (the NFA offers a course in community risk reduction)
2. Risk to the fire department organization—organizational risk
3. Risk during emergency operations—operational risk[37]

The original purpose of the firefighters was to "save lives and property from fire." Generally that meant pulling people from burning buildings and spraying water until the fire was out. In doing so, much property was lost from water damage and the demolition of buildings to stop the spread of fire through the creation of fire breaks. As time went on and the fire service became more sophisticated, salvage work became a fire department responsibility. It was no longer acceptable to extinguish fires by "washing it off the lot." The term "saving property from fire" took on a new meaning.

NOTE The original purpose of the firefighters was to "save lives and property from fire."

Firefighters have always saved lives at fires when they could. As equipment has improved, this capability has been greatly enhanced. This mission of saving lives has expanded to the point that the modern fire department is responsible for almost all rescue activities, including dive teams, urban search and rescue, technical rescue, swift water rescue, and cliff rescue, as well as auto accident extrication capabilities along with advanced life support capability (**Figure 3-18**). The fire department is a readily available, trained professional force, making them the logical choice to assume increased rescue responsibility. As the population increases and people become bolder in their searches for excitement, the importance of the rescue capability of the fire service is sure to grow.

In the area of fire prevention, the original firefighters were limited to inspecting chimneys and making sure fires were extinguished at curfew. Today fire prevention activities have increased to the point in which the fire prevention bureau is involved in the design process of new buildings through codes, water flow requirements, and inspection. This process does not stop when the building is finished and occupied by the tenant. Periodic inspections are made to ensure compliance with codes and ordinances.

FIGURE 3-18
Firefighters performing technical rescue training.

Courtesy of Edwina Davis

NOTE Public education for fire safety and prevention has become a fire department responsibility in many jurisdictions.

Public education for fire safety and prevention has become a fire department responsibility in many jurisdictions. Demonstrations are put on at shopping malls, service clubs, businesses, schools, and fairs. This education is not limited to fire safety; problems particular to an area are addressed as well. In the western United States, earthquake safety lectures and demonstrations are common. Phoenix, Arizona, has a fire department-sponsored home swimming pool safety program to prevent children from drowning. Fire departments in the Midwest present information on tornado safety, and on the East Coast programs are presented on hurricane safety. Fire departments are now being requested to become involved with the proper installation of child passenger-safety seats.

As hazardous materials have become more of a concern, fire departments have taken over responsibility for hazardous materials control functions including inspecting businesses and keeping inventories. In some areas, when old underground fuel tanks are removed, the fire department issues the required permits and inspects the operation.

Arson investigation has become an important function of the fire department.

In conjunction with law enforcement personnel, fire investigators search for the cause of fires and gather evidence to prosecute criminals. As budgets have tightened, cost recovery has become more popular as a means of retrieving money spent to combat fires caused by negligence or illegal activities. Some departments extend this concept to pursuing cost recovery on traffic incidents involving drunk drivers. Cost recovery requires the fire department, law enforcement, and prosecution services to work together in a close relationship.

NOTE To provide an effective, integrated fire defense system at the least cost, planning must take place.

FIRE DEFENSE PLANNING

To provide an effective, integrated fire defense system at the least cost, planning must take place. The fire chief must take into account all public and private resources at his disposal to achieve the desired goals. One of the first questions that must be asked is, "What is the acceptable level of loss due to fire?" Losses due to fire can be divided into three general categories: deaths, injuries, and property.[38] The fire chief and the fire department staff decide what they envision the community's fire defenses should be. In many jurisdictions, civilian advisory boards are included in this process.

NOTE The goals of fire protection are to prevent fire from starting, to prevent loss of life in case fire does start, to confine fire to its place of origin, and to extinguish fire once it does start.

The goals of fire protection are to prevent fire from starting, prevent loss of life in case fire does start, confine fire to its place of origin, and extinguish fire once it does start. The goals state what level of fire protection is desired. Once the goals are stated, the objectives can be set. Objectives are steps to be taken to achieve the goals. The objectives must be clearly stated and measurable. We must determine if what we are doing is accomplishing the objectives.

In order to formulate any plan, we must first know where we are. Statistics are kept and analyzed to determine what is presently occurring. Reports are generated specifying type of loss, type of occupancy, time of day, ignition source, item first ignited, and direct cause of loss.[39] Using these facts, the fire department can look for trends and major factors. Once these factors are identified, the department determines which areas to target, in the form of objectives in the overall plan.

Goals must be matched with jurisdiction and department philosophy. It must be determined if the goals are economically and politically attainable. There are measures that, if taken, would not be acceptable to the public as a whole. An example of this is home fire prevention inspections. Although it seems like a good idea, people have the Fourth Amendment right, under the U.S. Constitution, not to have their homes invaded without reasonable cause. This program can only be conducted with the homeowner's approval. Another factor is the cost: How many firefighters would it take to inspect a city of 40,000 homes on an annual basis?

The next step is to establish policies. Policies are general guidelines of how things will be done. An example of a policy would be that all firefighters would wear breathing apparatus during **overhaul** operations, or until declared unnecessary by the officer in charge. The policy clearly states the department's position on wearing breathing apparatus, while still leaving room for decision making on a case-by-case basis. The objective is to reduce firefighter injuries; the policy provides a guideline to achieve the objective.

Personnel must be provided to attain the objectives. If reduced property loss is the objective, enough people need to be assigned to the fire in a timely fashion to control, extinguish, and perform salvage work. Personnel are the single greatest cost in a fully paid fire department. Often up to 90% of the fire department budget is taken up in personnel costs.[40] The decision on equipment staffing is one of the most often–contested in the nation. Some groups feel that the minimum effective engine company staffing is four people; others think three are sufficient. Costs must be balanced against benefits and available resources to come to a decision.

The National Fire Protection Association has developed two standards to address the issue of adequate staffing: *Standard 1710, Career Fire Departments, Organization and Deployment* and *Standard 1720, Volunteer Fire Departments, Organization and Deployment.* These may be used by fire departments to justify staffing patterns and deployment of resources.

Procedures are established to describe how employees will perform recurrent and important functions. Procedures differ from policies in their specific nature. A typical procedure is for firefighters to **inventory** all of the equipment on their apparatus on a daily basis.

Facilities for training and housing employees must be provided. Facilities also include apparatus and equipment. Without tools, firefighters cannot perform their jobs. Budgeting must take into account that tools become worn out, lost, and broken and must be replaced. A fire station may last 100 years or more, an engine for 20, turnouts for three to five years, and flashlight batteries, a matter of months.

Providing funding for the personnel, facilities, and equipment requires an excursion into the political arena. The chief and his staff must determine how much the

overhaul The operation performed to ensure that all embers are extinguished after a fire is controlled. The final extinguishment actions.

inventory The act of accounting for all of the equipment and tools assigned to a piece of apparatus.

system will cost. Cost analysis is performed from within and from outside the department. The department's business manager determines cost factors and these are reviewed by the jurisdiction's budget analyst. After the cost analysis is completed, the city council or other elected body is lobbied for the budget to be approved. This is done mostly behind the scenes. When the chief appears before the governing body to seek approval of the budget, it is mostly a formality. The appearance before the governing body gives the public a chance to review the requested budget and make comment. One must always keep in mind that there never seems to be enough money to satisfy the needs of all of the groups competing for funds. As a firefighter, it may be obvious to you that the department needs new equipment and more personnel, but this is not so obvious to the people who want longer library hours. In tight fiscal times, everyone has to make sacrifices.

Once the budget is approved, the chief is charged with applying the available resources to achieve the desired results. At every step of the way, results must be evaluated to see if the goals are being achieved. The process does not stop here. Planning is a never-ending cycle of setting goals, determining objectives, and evaluating results. The fire department exists in a dynamic political and social climate with ever-changing demands and expectations.

ALL-HAZARD PLANNING

With the modern fire department responding to many incident types, it is important to look at all of them when planning. The fire chief and his staff must prepare the department to respond to incidents such as fires, floods, earthquakes, terrorism, medical incidents, swift water, and any other incident they may be charged with. Resources must be obtained, trained, and staged appropriately. Considerations must be made to comply with legal requirements for certain incident types. These are discussed in Chapter 11.

Much like fire defense planning, risk assessments and financial decisions must be made as to the likelihood and consequences of such incidents. It is difficult to maintain public focus on supporting the costs of all of these specialty resources when an incident of this type may not have occurred for a while. It is also difficult to maintain readiness on the part of the firefighters when they may believe there is little likelihood of an occurrence.

Agreements must be reached as to which will be the lead among the agencies responding before one of these incident types occurs. Training on interagency cooperation and establishing communications **interoperability** must be completed beforehand.

interoperability The ability of different departments to communicate on common radio frequencies at incidents is called interoperability. This includes assisting fire departments and other departments, such as public works and law enforcement.

RISK MANAGEMENT

Risk is both a noun and a verb. Expressed as a noun, risk is defined as "exposure to harm or loss"; as a verb, risk is defined as "to expose to the chance of injury or loss." The determining factors in both of these definitions are the probability that an undesired event might occur, with a harmful or undesirable consequence, and the severity of the harm that might result.

In describing risk, the probability of an occurrence can be in subjective terms, such as rare or high, or in numerical terms, such as 20% or one in three. Harmful consequences are often expressed in descriptive terms such as death, injury, and disaster. Probability and consequences can be combined and expressed mathematically as the product of loss and probability. An insurance company, for example, might describe a facility as a $10 million risk that has a 2% probability of loss. The probability of risk is composed of two factors. The chance that something undesirable might happen and also the probable outcome as rated on a scale of negative consequences.

Consider how the two factors of probability and consequence aid in planning. Some risks are low probability and high consequence (e.g., a helicopter crash); others are high probability and low consequence (e.g., burning your fingers changing the oil in your car). What we have to really watch for are the situations that are high in both categories. One of these is firefighting, as evidenced by firefighter deaths and injuries provided in Appendix A.

Risk Management

The term "risk management" refers to any activity that involves the evaluation or comparison of risks and the development of approaches that change the probability or the consequences of a harmful action. Risk management comprises the entire process of identification and evaluation of risks as well as the identification, selection, and implementation of control measures that might alter risk.[41]

The three control measures can be categorized as: administrative, engineering, and personnel protection. Administrative controls consist of guidelines, policies, and procedures established to limit losses. Their intention is to make the task safer for the worker. These include standard operating procedures (SOPs), training requirements, safe work practices, and regulations and standards.

Engineering measures are systems engineered to remove or limit hazards. Engineering measures include equipment design, traffic engineering (traffic signal preemptive devices), and mechanical ventilation. These measures are also designed to make tasks safer for the worker.

Personal protection consists of equipment, clothing, and devices designed to protect the worker. Examples of personal protective equipment (PPE) include helmets, gloves, turnouts, hoods, SCBA, and tools. The purpose of these items is to protect the worker from the hazard(s).

Risk Management Plan

Revisions in NFPA *Standard 1500, Fire Department Occupations Safety and Health Program* require that a written risk management plan be prepared by fire departments and be made part of their official policies and procedures. A risk management plan establishes policy. The plan serves as documentation that risks have been identified and evaluated and that a reasonable control plan has been implemented and followed. In the federal government, this process is referred to as a Job Hazard Analysis. The components of a risk management plan required by NFPA 1500 are:

- Risk identification
- Risk evaluation
- Risk control techniques
- Program evaluation and review

The elements of the plan are intended to apply to all aspects of a fire department's operations and activities, not just emergency activities.

Community Risk Reduction Planning

When planning to protect the community from risk, the process can be divided into four steps. Step one is preparation. Preparation includes assessing what types of risks are to be faced (e.g., fires, aircraft crashes, tornadoes, flooding, earthquakes, terrorism, etc.). Step two is mitigation. Once the threats have been determined, mitigation steps are taken to reduce the threat. Mitigation may include training hospital and city personnel in emergency management concepts, applying engineering solutions to prevent minor disasters from becoming major disasters, implementing changes in building and fire codes, and locating and equipping firefighters to deal with threats to the community. The third step is response. This is when the fire department and other agencies apply their personnel, training, and equipment to lessen the damage from the incident. In the recovery phase, things are returned to normal as quickly and as much as possible.[42]

THE FUTURE OF FIRE PROTECTION

The future holds many changes for those in the firefighting profession. The term "fire department" has become outmoded. More properly, the term would be "emergency services agency." With the sole exception of police functions, the fire department is responsible for most public safety emergency functions and many that are nonemergency. The organization, its responsibilities, and equipment are much different than they were in the past. The only thing that has not changed is the public's need for emergency services and the firefighter's willingness to do what it takes to serve and protect. The fire department of tomorrow will be vastly different than the one of today.

NOTE Overall the trend is expected to be toward fire prevention and away from manual fire suppression.

Recent focus has been on protecting workers from job hazards. Vast bodies of law and regulation cover all aspects of the work environment. Many of these regulations are contained in Code of Federal Regulations 29, Part 1910, the body of law that establishes the Federal Occupational Safety and Health Administration (OSHA). As these regulations are more closely applied to the fire department, the risks taken by firefighters in their jobs will be scrutinized for safety and necessity. This is just another area that modern firefighters are going to have to adapt to.

Overall the trend is expected to be toward fire prevention and away from manual fire suppression. Some people feel that the fire department of the future will be more of a regulatory agency for built-in protection systems. The case can be made for the fact that fires have become more difficult to extinguish because of building design changes and construction components. It is becoming harder and harder to apply water to a fire in a **high-rise building**. In a political atmosphere of user fees versus general taxation, it stands to reason that built-in suppression systems will take the place of a large firefighting force. This more directly places the cost of protection on the owner.

high-rise building Multistory buildings in which the upper floors are beyond the reach of aerial equipment. Fires must be attacked from the inside due to the height of the building.

SUMMARY

The fire service and firefighters have a long history rich in tradition and honorable sacrifice. No less is expected of the firefighters of today. It is important to review what has happened in the past so mistakes can be avoided in the future. As the fire service progresses, there will be new challenges to be met. The best firefighters will not only accept the challenges but also embrace them.

The acceptable level of fire protection must be determined by the officials of every jurisdiction.

Their decision is based on political and economic factors. It usually takes a major loss to focus the public's attention on the fire suppression needs of the community. It is the firefighter's job to keep the public, despite its apathy, as safe as possible. There is truth in the saying that "the longer it has been since your last major fire, the closer you are to the next one."

The modern fire service is a delivery system that contains many parts integrated into a whole. Only by understanding the parts and their relationship to one another can perspective be gained on the fire protection picture. Without an understanding of the big picture, it is impossible to envision the future of fire protection and to plan accordingly.

REVIEW QUESTIONS

1. What was the first organized force of firefighters called?
2. Why was organized firefighting developed?
3. List several of the measures developed to prevent fires in the early United States.
4. List reasons for joining a volunteer fire company.
5. What factors led to the demise of the volunteer fire companies in the large cities?
6. Trace the development of fire department pumping apparatus.
7. Which groups are at the highest risk of being killed in fires?
8. Where do most fatal fires occur?
9. What is the leading cause of residential fires?
10. What factors have caused the public to expect more services from the fire department?
11. What services are provided by the fire department in your area?
12. What developments have taken place in the area of personal protective equipment for firefighters?
13. What are the common denominators in most fires of conflagration proportion?
14. What are the four goals of fire protection?
15. What is the difference between a policy and a procedure?
16. What are the two components of risk?
17. What are the three control measures in risk management?

DISCUSSION QUESTIONS

1. What is the fire department's role in community risk reduction?
2. What is the one most effective thing that could be done to improve fire protection in your area?
3. The level of fire protection provided to a municipality is a political decision. What can be done to improve the political climate to increase the fire protection level?
4. Why is pre-incident planning important in the fire service?

NOTES

1. National Fire Protection Association, *Learn Not to Burn* (Quincy, MA: National Fire Protection Association, 2009).
2. Institute for Training in Municipal Administration, *Managing Fire and Rescue Services* (International City Management Association, 2002).
3. Ibid.
4. International Fire Service Training Association, *Fire Service Orientation and Terminology* (Stillwater, OK: Fire Protection Publications, University of Oklahoma, 2004).
5. Ibid.
6. Institute for Training in Municipal Administration, *Managing Fire and Rescue Services* (International City Management Association, 2002).
7. Ibid.

8. International Fire Service Training Association, *Fire Service Orientation and Terminology* (Stillwater, OK: Fire Protection Publications, University of Oklahoma, 2004).
9. Ibid.
10. Ibid.
11. Ibid.
12. Ibid.
13. Steven R. Frady, *Red Shirts and Leather Helmets: Volunteer Fire Fighting on the Comstock Lode* (Reno, NV: University of Nevada Press, 1991).
14. Ibid.
15. International Fire Service Training Association, *Fire Service Orientation and Terminology* (Stillwater, OK: Fire Protection Publications, University of Oklahoma, 2004).
16. Ibid.
17. Ibid.
18. Steven R. Frady, *Red Shirts and Leather Helmets: Volunteer Fire Fighting on the Comstock Lode* (Reno, NV: University of Nevada Press, 1991).
19. International Fire Service Training Association, *Fire Service Orientation and Terminology* (Stillwater, OK: Fire Protection Publications, University of Oklahoma, 2004).
20. Ibid.
21. Ibid.
22. Institute for Training in Municipal Administration, *Managing Fire and Rescue Services* (International City Management Association, 2002).
23. Ibid.
24. Ibid.
25. Ibid.
26. Ibid.
27. United States Department of Agriculture Forest Service, *Fire Weather* (Washington, DC: U.S. Government Printing Office, 1977).
28. "The Southland Fires, A Special Report," *Los Angeles Times*, 7 November 1993.
29. Federal Emergency Management Agency, *Are You Ready?* (Jessup, MD, Federal Emergency Management Agency).
30. *Fire House Magazine* (New York, NY: Firehouse Communications, Inc., 1994).
31. National Fire Protection Association, *NFPA Journal* (Quincy, MA: National Fire Protection Association, 1993).
32. National Interagency Fire Center, *National Fire News* (National Interagency Fire Center, 2009, http://www.nifc.gov/fire_info/nfn.htm).
33. *California Fire Siege 2003: The Story* (Washington. DC: U.S. Government Printing Office, 2003).
34. National Interagency Fire Center, *National Fire News* (National Interagency Fire Center, 2009, http://www.nifc.gov/fire_info/nfn.htm).
35. InciWeb Incident Information System, Station Fire, retrieved from http://inciweb.org/incident/1856/
36. National Interagency Fire Center, *National Fire News* (National Interagency Fire Center, 2009, www.NIFC.gov/fire).
37. United States Fire Administration, *Risk Management Practices in the Fire Service* (Washington, DC: Federal Emergency Management Agency, 1995).
38. Institute for Training in Municipal Administration, *Managing Fire and Rescue Services* (International City Management Association, 2002).
39. Ibid.
40. Ibid.
41. United States Fire Administration, *Risk Management Practices in the Fire and Rescue Service* (Washington, DC: Federal Emergency Management Agency, 2002).
42. Ibid.

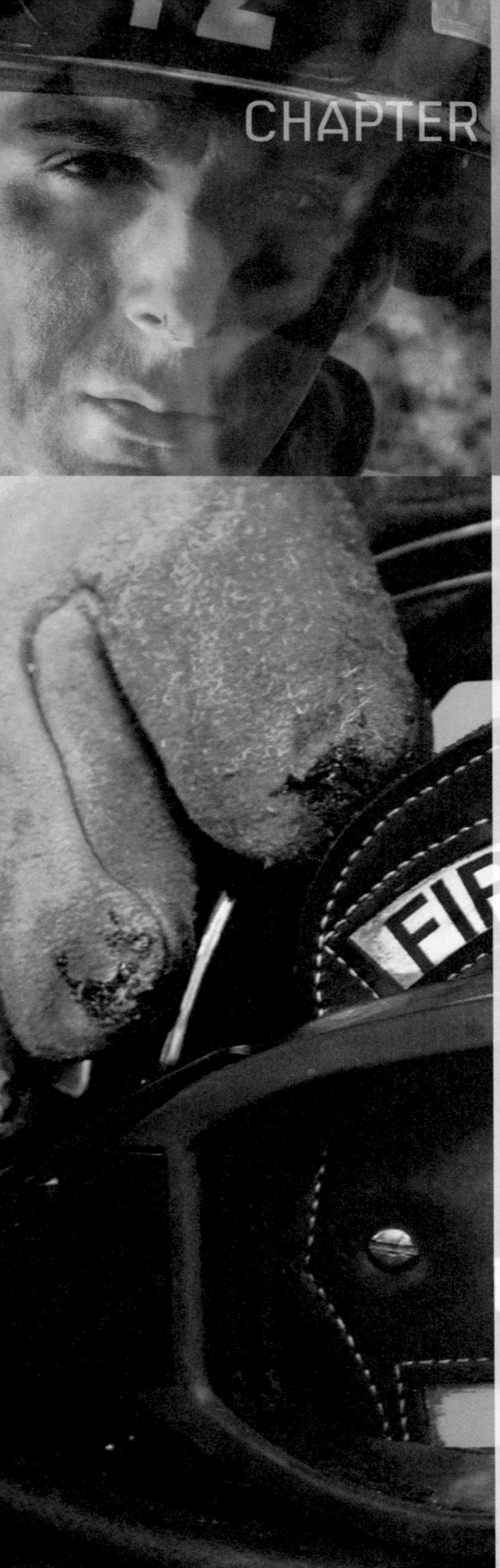

CHAPTER 4

Chemistry and Physics of Fire

LEARNING OBJECTIVES

Upon completion of this chapter, you should be able to:

- Define the difference between the fire triangle and the fire tetrahedron.
- Describe what constitutes an oxidizer.
- Describe what constitutes a fuel.
- Illustrate the states of matter.
- Explain the process of pyrolysis.
- Describe the properties affecting solid fuels.
- Describe the properties affecting liquid fuels.
- Describe the properties affecting gas fuels.
- Differentiate heat and temperature.
- Illustrate the four methods of heat transfer.
- Illustrate the five classifications of fire.
- Describe the four stages of fire.

INTRODUCTION

To appreciate how fire is controlled, we must first understand the chemical and physical properties of fire itself. This information can be used to predict what the fire will do with the available fuel and where it is headed. With this knowledge, we are able to choose the proper extinguishing agent and its method of application.

FIRE DEFINED

Fire is a rapid, self-sustaining **oxidation** process accompanied by the evolution of heat and light in varying intensities.[1] Combustion is described as a chemical reaction that releases energy as heat and usually as light.[2] When a substance is undergoing combustion we usually refer to it as being "on fire." Another very common oxidation reaction is the rusting of iron. This is not considered to be combustion because it evolves insignificant amounts of heat, without light, and proceeds at a slow rate.

oxidation The chemical combination of any substance with an oxidizer.

FIRE TRIANGLE

Fire was originally believed to be based on three elements being present: fuel, air, and heat (**Figure 4-1**). Fires could be controlled or prevented by removing one of these three elements. To be more accurate, in the discussion of fires and combustion from a chemical and physical standpoint, the air leg of the triangle will be replaced with **oxidizer** and the heat leg will be replaced with **energy** (**Figure 4-2**).

oxidizer A substance that gains electrons in a chemical reaction.

energy The capacity for doing work.

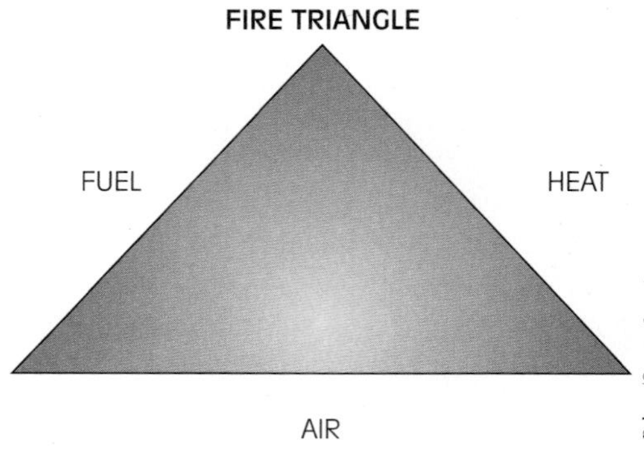

FIGURE 4-1
Original fire triangle.

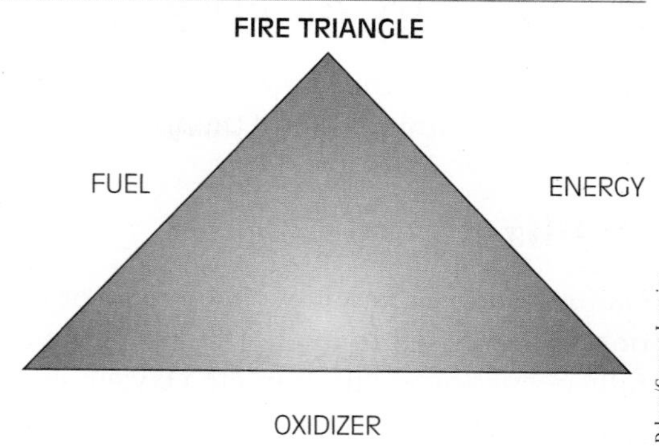

FIGURE 4-2
New fire triangle.

FIRE TETRAHEDRON

Scientists have determined that a fourth component, called the chemical chain reaction, is present in the combustion process (**Figure 4-3**).[3] The chain reaction occurs when the fuel is broken down by heat. When paper burns, the **molecules** in the paper (fuel) are broken down by heat, producing chemically reactive species called **free radicals**, which then combine with the oxidizer. This recombination process releases more heat and chemical reactants causing the fuel to break down further and the process continues.

As long as there are fuel, oxidizer, and energy, in the appropriate amounts, and the chemical chain reaction is not interrupted, the combustion process will continue.

molecule Combined groups of atoms. Molecules composed of two or more different kinds of atoms are called compounds.

free radicals An atom or group of atoms that is unstable and must combine with other atoms to achieve stability.

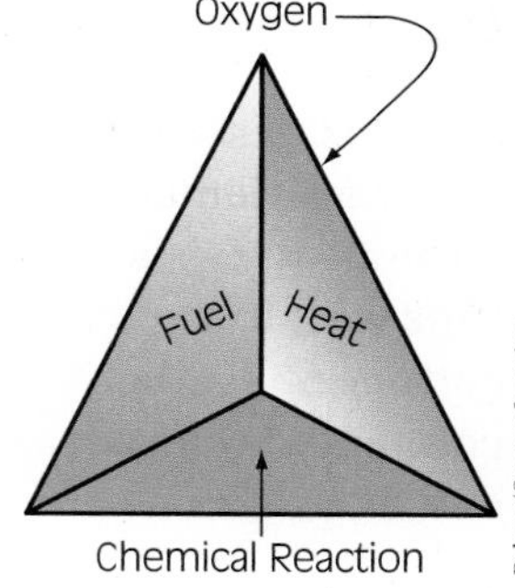

FIGURE 4-3
Fire tetrahedron.

CHEMISTRY OF FIRE

When looking at the chemistry of fire there are two basic necessary components, oxidizer and fuel. Without either of them the fire cannot occur.

Oxidizer

Oxidizers are substances that evolve or generate oxygen, either at ambient temperatures or when exposed to heat.[4] Oxygen is the most commonly occurring oxidizer. It is a naturally occurring element in air. Oxygen in its natural state is a gas. Air is a **mixture** of approximately 21% oxygen and 78% nitrogen with the remaining 1% other elements. Air is a mixture, not a compound, which leaves the oxygen readily available. Most fires we come in contact with are burning in air at 21% oxygen. In an oxygen-enriched atmosphere (over 21%), the fire will intensify. These atmospheres can occur when certain chemical compounds are mixed with, or in close proximity, to the fuel. An example of a substance that can release oxygen when heated is a common swimming pool chlorinating compound, calcium hypochlorite. When energy is added to this compound, usually from the heat of a fire, the oxygen it contains is released and can enrich the atmosphere. Another oxidizer is ammonium nitrate, which is used in fertilizers and as an oxidizer in blasting agents. Conversely, in an oxygen-deficient atmosphere, the rate of combustion will decrease and the fire will possibly go out on its own.

NOTE In an oxygen-enriched atmosphere (over 21%), the fire will intensify.

Other elements can take the place of oxygen in the combustion reaction. Two of these, fluorine and chlorine, are found on the periodic table of the elements listed under the halogen family. Both naturally occur as gases, as does oxygen. Fluorine is a much stronger oxidizer than oxygen, and a fire in a fluorine atmosphere would burn more rapidly than one occurring in air.

mixture A substance made up of two or more substances physically mixed together.

Fuel

A fuel is described as anything that will burn. Carbon and hydrogen are the two most common elements in fuels. These fuels are referred to as hydrocarbons. Carbon in its natural state is a solid. It is the main element in organic fuels, that is, fuels that at one time were living things, such as petroleum products (gasoline, natural gas), wood, paper, cotton, and other natural fibers. Hydrogen is a flammable gas. Other elements and compounds can act as fuels, including metals that will burn, such as sodium, aluminum, and magnesium.

NOTE Carbon and hydrogen are the two most common elements in fuels.

The most common fire involves a fuel composed of mostly carbon and hydrogen in an atmosphere with oxygen present. The oxygen combines with hydrogen to produce water vapor and with the carbon to form carbon dioxide. These are the two by-products of complete combustion. Complete combustion rarely occurs due to outside factors. These factors may include, but are not limited to, fuel size, arrangement, contaminants, and the availability of the oxidizer. This gives us the by-products of incomplete combustion such as smoke, carbon monoxide (CO), carbon dioxide (CO_2) and other fire gases.

PHYSICS OF FIRE

Fuel

Fuel may occur in one of the three states of matter: solid, liquid, and gas (**Figure 4-4**).

All molecules vibrate at normal temperatures. The molecules in liquids vibrate faster than those in solids, and gas molecules vibrate the fastest of the three. The state of the fuel is temperature dependent. As we raise the amount of energy in the molecule by applying heat, and raise the temperature, the molecules vibrate faster and faster. As this vibration increases, solids become liquids and liquids become gases.

Combustion usually occurs when the fuel has been converted to the vapor or gaseous state because the oxidizer occurs as a gas, and it takes both oxidizer and fuel in the gaseous state for the recombination to occur. Solid and liquid fuels are converted to the gaseous state by the application of energy. This can be observed by carefully watching a candle burn. The flame seems to float a small distance from the wick. This process is called **pyrolysis** (**Figure 4-5**). Pyrolysis is defined as the chemical decomposition

pyrolysis The chemical decomposition of matter through the action of heat.

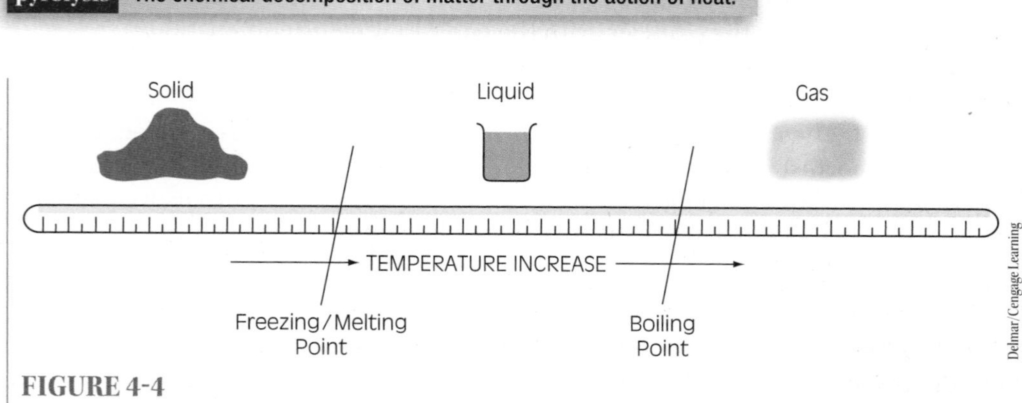

FIGURE 4-4
States of matter, in many cases, are temperature dependent.

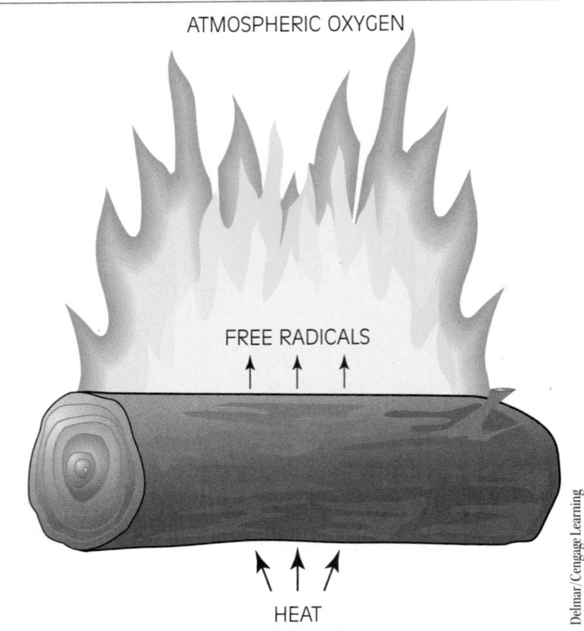

FIGURE 4-5
Pyrolysis, the release of free radicals due to input heat.

of matter through the action of heat. Some fuels, such as pure carbon or the combustible metals, do not have to convert to the vapor state to burn. They do not have to decompose or vaporize; the union with oxygen takes place directly.[5] When the fuel is brought to the temperature at which it continues combustion without any external input of heat and becomes self-sustaining, it is called the **ignition temperature**.[6]

As heat is added to a substance, the molecules are broken down into shorter and shorter chains. The longer molecules break into shorter molecules and the shorter ones break into what are called free radicals. The free radicals are the byproducts of the fuel that combine directly with the oxidizer, with the end result being combustion.[7]

Let us look at the process in a piece of wood, as an example. Wood is made up of long chains of molecules that contain hydrogen and carbon. As heat is applied, the molecules receiving the heat start to break down into shorter and shorter molecules. The end result is the shortest hydrocarbon molecule, methane, a flammable gas. Several other short chain hydrocarbon molecules resulting from this process will burn as well, including ethane and propane. When these molecules are further broken down into free radicals by heat, they combine with the oxidizer and combustion takes place. Sufficient amounts

ignition temperature The minimum temperature to which a substance must be raised before it will ignite. The piloted ignition temperature is usually much lower than the autoignition temperature. Piloted ignition may be provided by a spark or flame or by raising the general temperature.

of oxidizer, usually oxygen, and an ignition source of sufficient temperature must be present. In the absence of either of these, pyrolysis can produce flammable gases that are only waiting for the missing component, sufficient oxygen or other oxidizer. When the oxidizer is added, combustion begins.

Solid Fuels

Several factors affect the rate at which solid fuels are pyrolized: mass, arrangement, continuity, and moisture content.

Mass affects the fuel in that the smaller or more finely divided the fuel, such as dust or chips, the less heat required to get it to pyrolize because the fuel has a greater surface area in relation to its mass (**Figure 4-6**). Ignition capability is based on the mass and surface area being ignited. A wooden 2 × 4 has greater mass vs. surface area than sawdust, which has less mass vs. surface area per particle. When lighting a campfire, paper and small twigs are easier to ignite than larger logs.

Arrangement is based on the spacing of the fuel particles, regardless of their size. A house in the framing stage of construction (**Figure 4-7**) will ignite more easily and burn faster than a stack of lumber (**Figure 4-8**), even though both are basically the same fuel. When gasoline is ignited in a container it burns rapidly on the surface; when atomized into a fine spray it burns explosively.

Continuity is the grouping of fuel over a prescribed area. This may be vertical, horizontal, or both. An example of vertical continuity is dry grass spreading fire upward into

FIGURE 4-6
Relative fuel sizes, showing grass, brush, and trees.

Courtesy of Edwina Davis

FIGURE 4-7 Structure in framing stage of construction, illustrating loose fuel arrangement. Easy to ignite and burns rapidly.

Delmar/Cengage Learning

FIGURE 4-8 Stacked lumber, illustrating tight fuel arrangement. It is difficult to ignite.

Delmar/Cengage Learning

brush and then into trees. An example of horizontal continuity is the same fire's spread being stopped by a road or fire break (**Figure 4-9**).

Moisture content is the amount of moisture contained in a fuel. The moisture content of a fuel affects its ease of ignition. On a humid or rainy day, forest fires are not much

FIGURE 4-9 Firefighters constructing fire break to break fuel continuity.

Courtesy of Edwina Davis

of a problem. The fuels involved can absorb moisture directly from the air. The fuel has too high a moisture content to burn readily. For pyrolysis of the fuel to take place and burning to begin, the moisture must be turned to vapor and removed. This process absorbs large amounts of energy. If a large enough fire with sufficient energy is started, it can overcome the high moisture content of the fuel and burn rapidly.

Flame Spread

The Steiner Tunnel (ASTM E-84) is frequently referenced as a method to assess flame spread and smoke density and is a mandated test for many commercial building materials. The three attributes measured are flame spread, temperature, and smoke density. The test consists of a 25' vented tunnel, lined with firebrick, with the test material mounted to the top of the chamber. At one end of the chamber, the sample is subjected to a high-energy flame for ten minutes. A fan draws the flame across the surface of the material being tested. Flame spread is determined visually through windows built into the tunnel. An optical cell mounted at the tunnel exhaust measures smoke density. Provision may also be made for the measurement and analysis of combustion gases. The standards used for comparison of flame spread characteristics are asbestos-cement board, rated at 0, and red oak flooring, rated at 100.[8]

NOTE Most flammable liquids, such as gasoline, have a specific gravity less than 1.0 and will float on water.

Many fires have spread very rapidly across interior finishes leading to loss of life and property. Although not the only factor, flame spread was a contributing factor in these fires. Smoke is a factor in life safety due to its toxic characteristics and its ability to obscure vision.[9]

Liquid Fuels

Molecules of liquids move freely and flow like water but do not readily separate. Liquids will assume the shape of their container. Liquid fuels have physical properties that make them difficult to extinguish and they increase the hazards to persons and property.

When a liquid fuel is spilled, it flows across the ground, increasing the size of the spill. It also flows downhill, pooling in low-lying areas. Several other properties of liquids are important to firefighters:

- *Specific gravity (liquid).* The weight of a liquid as compared to the weight of an equal volume of water. Water is assigned a value of 1.0. **Nonmiscible** liquids with a specific gravity less than 1.0 will float on water. Nonmiscible liquids with a specific gravity higher than 1.0 will sink in water (**Figure 4-10**). Most flammable liquids, such as gasoline, have a specific gravity less than 1.0 and will float, and burn, on water.
- *Volatility.* Volatility is the ease with which a fuel gives off vapors at **ambient temperature**. This property has a definite effect on the flammability, ease of extinguishment, and fire prevention considerations of the fuel.
- *Vapor pressure.* All liquids have a vapor pressure, which is the pressure exerted by vapor molecules on the sides of a container, at equilibrium. Vapor pressure is

nonmiscible Not capable of mixing, will separate.

ambient temperature The temperature surrounding an object. Air temperature.

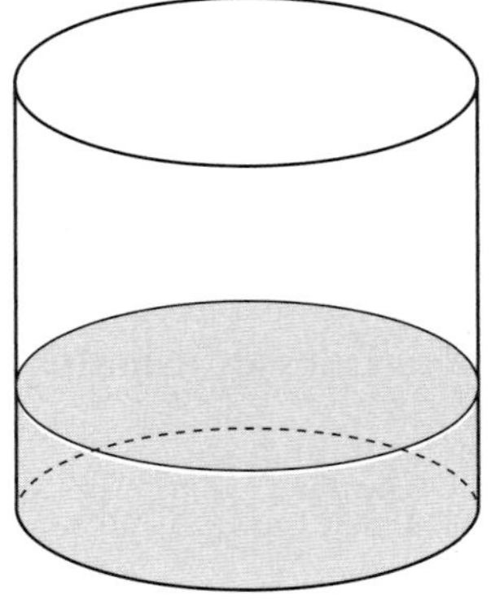

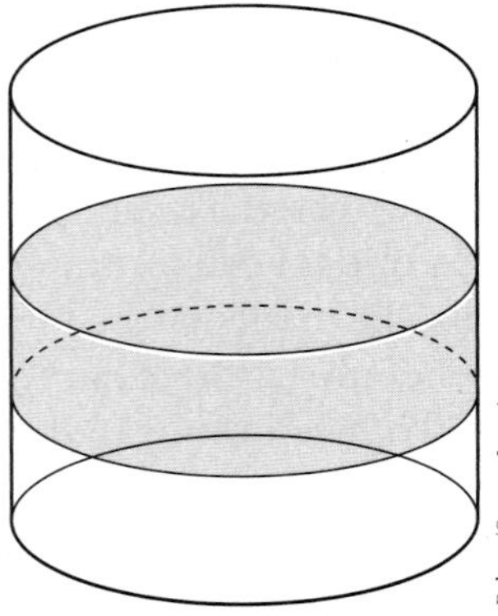

Delmar/Cengage Learning

FIGURE 4-10 Specific gravity. Liquid in container on left is heavier than water and will sink in water. Liquid on right is lighter than water and will float on water.

FIGURE 4-11
Vapor pressure. Container on left has less vapor pressure than container on right.

caused by the molecules of the liquid escaping the surface through **evaporation**. When the number of molecules leaving the surface of the liquid is the same as the number returning, the liquid is at equilibrium. Vapor pressure is temperature (energy) dependent. As the temperature of the liquid rises, the molecules have more energy to overcome atmospheric pressure and the vapor pressure of the liquid increases (**Figure 4-11**).

- *Boiling point.* When the vapor pressure equals **atmospheric pressure** at the surface of the liquid, the boiling point is reached. At this point, the molecules of the liquid have enough energy to actively leave the surface of the liquid. Liquids at their boiling point evolve vapors at a much more rapid pace than through evaporation. This can easily be observed with a pan of water on the stove. As energy is added in the form of heat, the water begins to boil and leaves the pan as vapor (steam). If the pan has a lid, the steam will soon develop enough vapor pressure to lift the lid off the pan. If the pan of water were left unheated sitting on the counter, without a lid, the water would eventually evaporate, but at a much slower rate.
- *Vapor density.* The relative density of a vapor or gas as compared to air is the vapor density. Air is assigned a value of 1.0. This is important in that a vapor with a higher density than air will sink and lie along the ground, seeking low spots in which to pool. A vapor with a lower density than air will tend to rise and dissipate (**Figure 4-12**). Gasoline vapors, which are very heavy, will flow downhill and pool, like a liquid. Liquefied petroleum gas vapors act in much the same way. Natural gas vapors will rise and dissipate in the slightest breeze. The exceptions to this occur when the ambient

evaporation The changing of liquid to a vapor.

atmospheric pressure The pressure of the atmosphere exerted on any point. 14.7 psi at sea level.

FIGURE 4-12
Vapor density. Vapors heavier than air will sink and those lighter than air will rise.

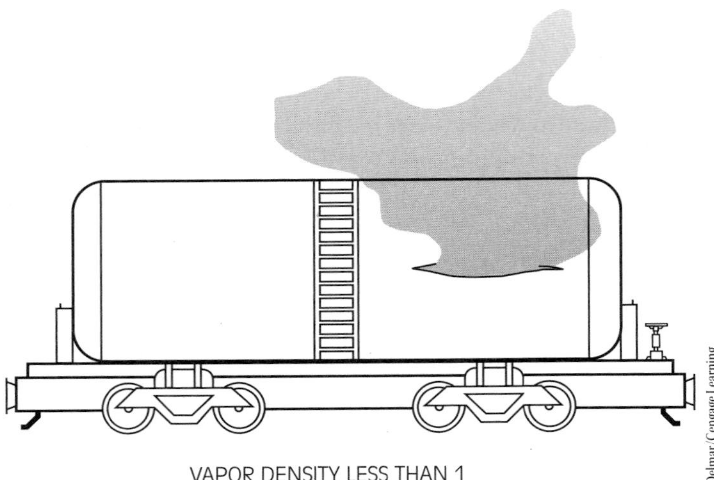

Delmar/Cengage Learning

temperature is very high or very low. If it is cold enough, a vapor that is lighter than air at normal temperatures may act as if it is heavier than air. On a hot day, especially when spilled on hot asphalt, heavy vapors can be heated to the point that they will rise. Relative humidity can also have an effect on how a vapor reacts. This effect occurs because humid air contains more moisture than dry air and is therefore heavier.

NOTE A vapor with a higher density than air will sink and lie along the ground, seeking low spots in which to pool.

- *Flash point.* The flash point is the minimum temperature of a liquid at which it gives off vapors sufficient to form an ignitable mixture with air, but ignition is not sustained (in other words it will flash and then quit burning). Above the surface of a liquid, due to evaporation or boiling, there is a vapor space. If an ignition source is introduced into this vapor space and ignition occurs, the liquid is above its flash point. When liquids have a flash point below 100°F, they are called *flammable* liquids (see **Table 4-1**). When the flash point is at or above 100°F, they are called *combustible* liquids.[10]

TABLE 4-1 Physical Properties of Some Flammable and Combustible Substances

Name	Boiling Point (F°)	Flash Point (F°)	Flamm. Range % in Air	Specific Gravity	Vapor Density
Acetone	133	0	2.5–12.8	0.79	
Acetylene[a]			2.5–100		0.91
Ammonia	−28		15–28		0.60
Benzene	176	12	1.2–7.8	0.88	
Butane	31		1.6–8.4	0.6 (31°)	2.11
Carbon monoxide	−313		12.5–74		0.97
Ethyl alcohol	173	55	3.3–19	0.79	
Ethyl ether	94	−49	1.9–36	0.71	
Ethylene oxide[a]	51	−20 (51°)	3.0–100	0.82	1.49
Formaldehyde	−6		7.0–73		1.04
Gasoline (average)	102	−45	1.4–7.6	0.72–0.76 (60°)	
Hydrogen sulfide	−77		4.0–44		1.19
Isopropyl alcohol	181	53	2.0–12.7	0.79	
Isopropyl ether	154	−18	1.4–7.9	0.73	
Kerosene	347–617	100–162	0.7–5.0	0.81	
Methyl alcohol	147	52	6.0–36	0.79	
Methyl bromide	38		10–16	1.73 (32°)	3.36
Toluene	232	40	1.1–7.1	0.87	
Turpentine	309–338	95	0.8–?	0.86	

Source: The NIOSH Pocket Guide to Chemical Hazards, 2005.

[a]Referring to the table notice that some of the substances have flammable ranges that go as high as a 100% concentration. This makes them extremely dangerous, even when in their containers. All of them can be classified as fuels.

- *Miscibility.* Miscibility is the ability of a substance to mix with water. Most flammable liquids do not mix with water, making them harder to extinguish. Because they cannot be diluted, they float on the surface of the water and continue burning. Fuels that mix with water can be diluted to the point that they will no longer burn. An example of a **miscible** fuel is alcohol.

miscible Capable of mixing without separation.

Gas/Vapor Fuels

Gas is the state of matter defined as a fluid that has neither independent shape nor volume but tends to expand indefinitely. Gases always fill the container in which they are stored. Materials that are normally gases, when contained as liquids, are above their boiling points at ambient temperature and convert to gas upon their release. For example, when butane and propane are released from their containers, they revert to the gaseous state.

- *Upper flammable limit.* The maximum concentration of gas or vapor in air above which it is not possible to ignite the vapors is its upper flammable limit. A vapor that is above its upper flammable limit is said to be "too rich" to burn.
- *Lower flammable limit.* The minimum concentration of gas or vapor in air below which it is not possible to ignite the vapors is its lower flammable limit. A vapor that is below its lower flammable limit is said to be "too lean" to burn.
- *Flammable range.* The proportion of gas or vapor in air between the upper and lower flammable limits (**Figure 4-13**). This proportion of vapor in air is measured in

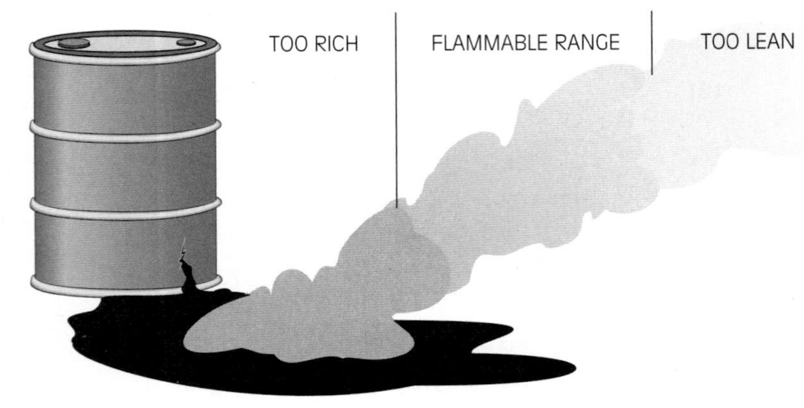

Delmar/Cengage Learning

FIGURE 4-13
Flammable range. Too rich to burn at liquid's surface, within flammable range, and too lean to burn at increased distance from liquid's surface

percent. At the liquid surface, the vapor will be at 100% concentration. As we move further from the surface, the concentration will lessen until it reaches 0%.

Classification of Gases

Gases can be categorized as flammable and nonflammable. There are gases that are not flammable that support combustion, the best example of which is oxygen. Oxygen itself does not burn, but if the oxygen concentration is increased, most fires will burn more rapidly.

Once the vapor warms up, it disappears from view but may still be lingering in the area.

A word of caution about gaseous fuels: When a compressed gas, like butane, is released, the vapor cloud you see indicates that the gas is colder than the air temperature, condensing the moisture in the air. It appears much like fog. This is not the extent of the vapor cloud. Once the vapor warms up, it disappears from view but may still be lingering in the area. It is possible to stand in a vapor cloud with a concentration that is within its flammable range. If the cloud were to ignite, the person would be burned severely, if not killed. If a vehicle with a diesel engine is driven through a vapor cloud, it can act just like stepping on the accelerator. Letting off the throttle would have no effect. The governor would not limit engine speed, because it limits fuel flow, not air flow. The engine would rev to the point that it would probably come apart. There would more than likely be ignition of the vapor cloud from the heat of the turbocharger or pieces of hot carbon being blown out of the exhaust.

HEAT AND TEMPERATURE

Heat is defined as a form of energy. There are several possible sources of heat: chemical (the breaking down and recombination of molecules), mechanical (friction, friction sparks, compression), electrical (arc or spark, static electricity, and lightning), and nuclear (fission and fusion). When heat energy is absorbed by a substance, its temperature rises.

In the fire service, heat energy's ability to change the temperature of a substance is expressed in several ways. The English system uses two units of measurement. One is the British thermal unit (BTU). This is defined as the amount of energy required to raise the temperature of one pound of water one degree Fahrenheit. The other expression is in metric terms and is the calorie. A calorie is the amount of heat required to raise one gram of water one degree Celsius. The SI system uses the joule. A calorie equals 4.187 joules. There are 1,054 joules in one BTU.[11]

Temperature is the measure of the hot or coldness of an object. A more scientific description is the measurement of the average speed of vibration of the molecules in a substance. Temperature is measured on four scales: Fahrenheit, Celsius, Kelvin, and Rankine (**Figure 4-14**). On the Fahrenheit scale, water freezes at 32°F and boils at 212°F. On the Celsius scale, water freezes at 0°C and boils at 100°C. On the Kelvin scale, water freezes at 273K and boils at 373K. (Note: The Kelvin scale does not use the symbol ° to represent degrees.) On the Rankine scale, water freezes at 492°R and boils at 672°R. On the Kelvin and Rankine scales, there are no negative numbers since 0K or 0°R represent the lowest attainable temperature. Scientific theory and experiment show that there is a limit below

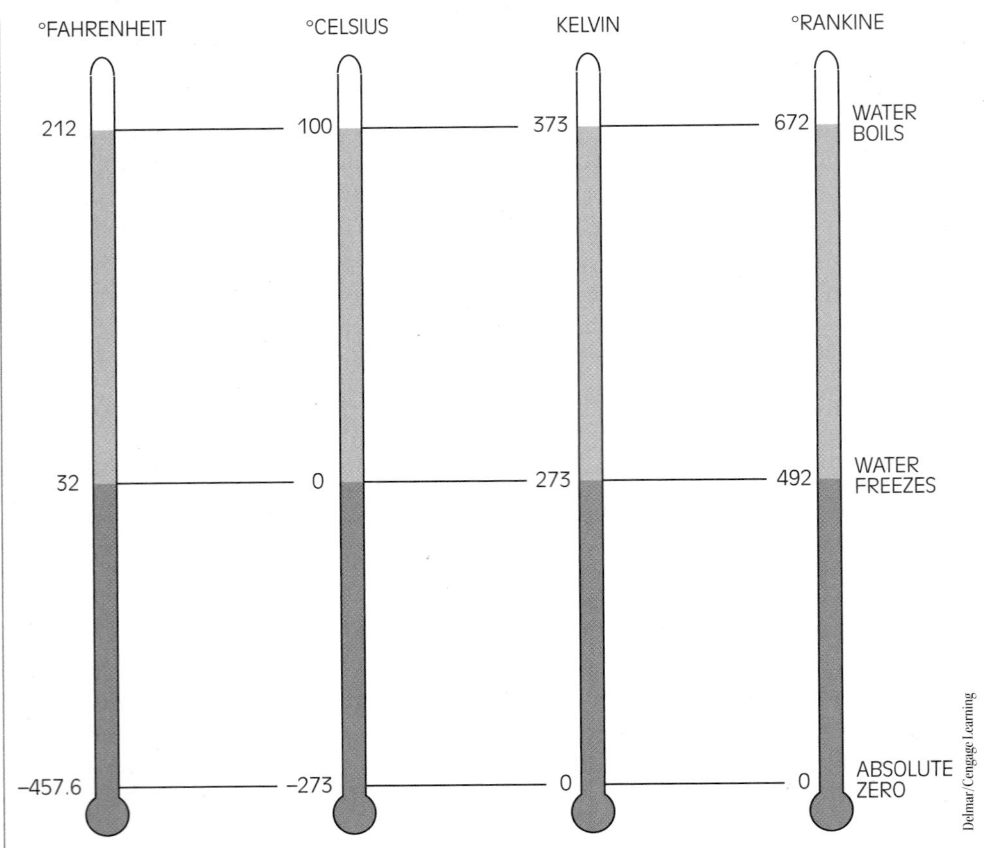

FIGURE 4-14
Illustration of relationship of temperature measurement scales.

which it is impossible to cool matter.[12] For the purpose of this book, temperatures will be given in degrees Fahrenheit (°F) and degrees Celsius (°C).

Heat and temperature are not to be confused. Heat is a measurement of energy and temperature is a measurement of how much energy a material retains. To illustrate this point, go outside on a sunny day. The roof of a dark-colored car will be much warmer than the roof of a light-colored car. Both are subjected to the same amount of heat energy radiated by the sun. The dark roof absorbs and retains more of that energy and feels warmer and, consequently, higher in temperature.

HEAT TRANSFER

For pyrolysis to take place there must be a way for the heat (energy) to be transferred from the heat source to the fuel. There are three methods of heat transfer:

1. *Conduction.* The transfer of heat through a medium without visible motion (**Figure 4-15**). This can be observed by placing a pan on a hot stove. The pan soon becomes heated (a rise in temperature), but there is no visible movement of the pan. The more dense a material is, the better conductor it makes. Metals are good conductors of heat or electrical energy. The more dense the metal, the better conductor it is. This is why gold is used for electrical contacts in computers and space craft. Wood is a poor conductor of energy. Air is a very poor conductor of energy.
2. *Convection.* The transfer of heat through a circulating medium, such as liquids and gases. When a fire occurs, the smoke and heated gases rise because of convection (**Figure 4-16**). As a gas is heated, it expands and becomes less dense, making it

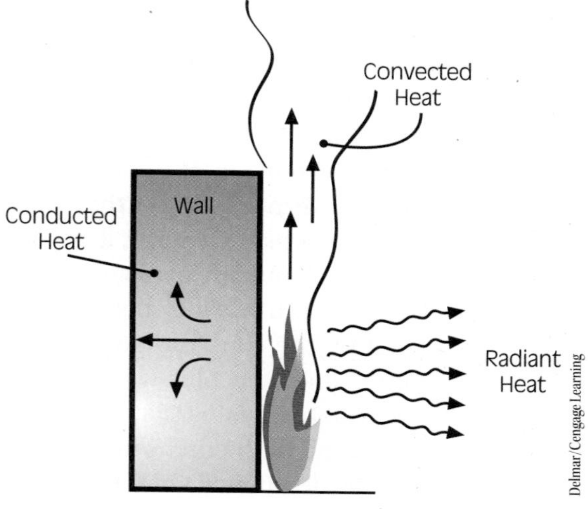

FIGURE 4-15
Transfer of heat through conduction, convection, and radiation.

FIGURE 4-16
Convection column developing over wildland fire due to rising heated air produced by the fire.

Delmar/Cengage Learning

lighter than the surrounding gases, and it rises. The hotter the gases, the faster they rise. This is evident in structure fires. As they burn with more intensity, the smoke rises at a more rapid rate. As the gas cools, it returns to its former elevation, completing the circulation.

3. *Radiation.* The transfer of heat by wavelengths of energy. Heat in the form of rays is the infrared portion of the electromagnetic spectrum.[13] As these wavelengths of energy travel through space, they eventually come into contact with an object. As the object absorbs the energy, it becomes heated, resulting in a rise in temperature. When you walk outside on a sunny day, the air temperature is warm, but objects in the sun are much warmer because they have absorbed the wavelengths of energy generated by the sun. The darker the object, the more of the wavelengths it can absorb, and it will become warmer. One thing that must be kept in mind when dealing with radiated heat is that it is not affected by wind. Radiated heat can set fire to objects upwind of the fire, whereas convection-caused fires are much more likely to occur on the downwind side (**Figure 4-17**).

FIGURE 4-17
Transfer of heat through radiation, causing fire to spread.

Courtesy of Edwina Davis

Some fire service professionals describe a fourth method of heat transfer, known as direct flame impingement or autoexposure, which is the transfer of heat through direct flame contact. This method of heat transfer can be observed in high-rise building fires. As the windows on the lower floors are broken out, the flames lap onto the windows of the upper floors, breaking them also. The flames then come into direct contact with materials on the upper floor, spreading the fire.

This fourth method of heat transfer combines elements of convection and radiation.

The convection current makes the flames rise, and when materials are bathed in flame, heat is radiated onto them. This method of heat transfer is not accepted by everyone in the scientific community but illustrates fire spread in many instances (**Figure 4-18**).

The example of a structure fire can illustrate the four methods of heat transfer. A fire starts in a stack of boxes in a warehouse, and the heat and smoke rise due to convection. When enough convected heat builds up, the underside of the roof or ceiling catches fire. The boxes at the bottom of the stack were the first to ignite and the fire spreads upward through the stack due to direct flame impingement. After a while, radiated heat spreads the fire to piled stock several feet away. As the fire gains headway, a pipe running through the wall is heated to the point that stock piled around it on the other side of the wall ignites. Rarely does a fire spread due to only one method of heat transfer; usually several methods are involved.

FIGURE 4-18 Transfer of heat through direct flame impingement. Fire lapping onto porch structure from openings.

Courtesy of Edwina Davis

CLASSIFICATION OF FIRES

Fires have been divided into five basic classifications, based on the type of fuel involved (**Figure 4-19**).[14] In some instances more than one type of fuel may be involved in a single incident.

Class A. Ordinary combustibles. This class includes materials composed of primarily natural fibers, such as wood, paper, and cotton.

Class B. Flammable liquids and combustible liquids. These include materials that can flow while burning, such as gasoline, kerosene, alcohol, and cooking oil. This

Ordinary Combustibles

Flammable Liquids

Electrical Equipment

Combustible Metals

Combustible Cooking

Delmar/Cengage Learning

FIGURE 4-19 Classification of fire symbols.

definition is often shortened to flammable liquids and should not be confused with the rating of flammable versus combustible, based on flash point.

Class C. Energized electrical. The key word here is *energized*. If the power is turned off, the fire is reclassified into what other materials are burning. One must be cautious as some appliances, such as older television sets and other electrical equipment, may retain stored energy, even years after their last use.

Class D. Combustible metals. These include magnesium, titanium, zirconium, sodium, and potassium. Most metals burn with a brilliant white flame that can damage the eyes.

Class K. Cooking materials (cooking oils/fats/grease). This is a newer classification of fire. Class K extinguishers are specifically designed to supplement fire suppression systems in commercial kitchens. These extinguishers are for cooking oil, fat, and grease fires.

It is not uncommon for more than one class of fire to be involved in an incident. In a typical automobile fire, there are Class A fuels (upholstery) and Class B fuels (gasoline), and some autos have Class D components (usually magnesium and/or aluminum). If the fire has resulted from the auto crashing into a structure or power pole, there may be Class C components also involved in the fire.

STAGES OF FIRE

Previously, fire development was divided into three phases: incipient, free burning, and smoldering. This has evolved into the concept of four stages: ignition, growth, fully developed, and decay.

In the ignition stage the fire is ignited and reaches a point where it no longer needs input heat from outside sources to continue burning (**Figure 4-20**). In this stage oxygen in the surrounding air is approximately 21%, heat is starting to be produced, and the combustion reaction begins to accelerate.

In the growth stage the fire releases heat, bringing more fuel to its ignition temperature. As the additional fuels start to burn, the fire spreads and gains in intensity. Heat is transferred onto nearby surfaces though radiation. Convected heat rises and preheats fuels above the fire, bringing them to their ignition temperature. In this manner the fire begins to spread more and more rapidly.

If a fire is burning in a confined area, the ceiling temperature increases rapidly as heated air rises. Ceiling temperatures can easily reach 1,000°F.[15] As this process continues, insufficient amounts of oxygen are available, causing incomplete combustion. Residual fire gases start to accumulate at the ceiling level. This accumulation of fire gases from time to time will ignite if sufficient oxygen is present, causing a flameover or rollover situation to occur. As soon as these gases are consumed, the flame is no longer seen and the process will repeat itself as temperatures in the enclosed areas continue to increase.

FIGURE 4-20 Fire in fully developed stage.

Courtesy of Edwina Davis

The presence of this condition lets firefighters know that they are approaching the seat of the fire and need to start applying water at the ceiling level to reduce temperature and prevent a flashover. If they fail to cool the environment, conduction, radiation, and direct flame contact will cause other combustibles in the room to pyrolize.

SAFETY Firefighters cannot survive in a room with a flashover.

When the contents in the room (this may include PPE) are brought to their ignition temperature and sufficient oxygen is present, flashover can occur. If a flashover were to take place, temperatures, even at floor level, would rise dramatically. Firefighters cannot survive in a room with a flashover, even wearing full PPE and SCBA. The only chance they have for survival is to be in very close proximity to an exit that they can access quickly. Escape up and over a windowsill is unlikely, as this will place firefighters into an even higher temperature zone. Should they escape, their SCBA will most likely have failed and they will sustain serious burn injuries. Without breathing apparatus, firefighters can survive in a room with temperatures of 300°F of dry heat for only a short period of time; without breathing apparatus they are fully exposed to fire gases, smoke particles, elevated temperatures, and low oxygen concentrations.[16]

When oxygen is introduced to the room, the fire gases present and all of the fuels above their ignition temperature can burn with explosive force.

In the fully developed stage all available fuels in the fire's perimeter are burning. In a structure fire, this may be a room and contents or the whole structure. In a wildland fire situation the fire is spreading across the countryside.

The decay stage occurs when the fire has run out of available fuel or oxygen. In an automobile or structure fire, for example, the fuel can be consumed to the point where the fire runs out of things to burn (fuel), causing it to go out on its own. The fire can also go into the decay stage when suppression action has reduced the fire to smoldering embers (**Figure 4-21**).

In a sealed environment the fire may run out of sufficient oxygen to sustain combustion. If the room involved is well sealed from outside air, the oxygen in the room is consumed. As the oxygen content drops below 15%, combustion is slowed and the fire enters the fourth phase. In this example of the fourth phase, flame may die out and glowing combustion takes place. Pyrolysis continues to occur with amounts of combustible gases produced. The room is now superheated and charged with smoke and combustible fire gases. The gases and room contents are above their ignition temperatures and the only item lacking is sufficient oxygen. The fire gases and smoke are alternately forced out and sucked back into the structure. The windows become blackened by smoke stain baked on

FIGURE 4-21
Fire in decay stage.

Courtesy of Edwina Davis

by the heat of the fire. When oxygen is introduced to the room, the fire gases present and all of the fuels above their ignition temperature can burn with explosive force. This condition is referred to as a **back draft** or smoke explosion.

back draft A type of explosion caused by the sudden influx of air into a mixture of gases, which have been heated above the ignition temperature of at least one of them.

SUMMARY

This chapter introduces the chemical and physical properties of the elements found in fires and what actually occurs during the combustion reaction. A solid understanding of the combustion process and the stages of fire must be in place before we discuss the selection, application, and tactics of extinguishing agents.

REVIEW QUESTIONS

1. List the three legs of the old and new fire triangles.
2. Making the triangle a tetrahedron, what is the fourth side?
3. What is the most commonly occurring oxidizer?
4. What are the two most common elements in fuel?
5. Fuel may occur in any of the three states of matter. What are these states?
6. Define pyrolysis.
7. Define auto- and pilot ignition temperatures.
8. List the four factors affecting the burning rate of solid fuels.
9. Would a nonsoluble liquid fuel with a specific gravity of 0.8 sink or float when water is added?
10. What happens when a liquid's vapor pressure has reached atmospheric pressure?
11. Will a gas with a vapor density of 1.2 float or sink in air?
12. What effect does the ambient temperature have on the ignitability of a liquid?
13. Which properties make flammable gases such as acetylene so dangerous in fire situations?
14. List the four sources of heat energy.
15. Water freezes at what temperature Celsius? What temperature Fahrenheit?
16. List the four methods of heat transfer. How do they affect firefighting operations?
17. When a building is burning and it sets the one across the street on fire, the fire is spread by: (two possibilities.)
18. A fire in a large trash container could contain which classifications of combustibles?

19. A forest fire, out of control, is burning in which phase of fire?
20. When a fire is burning quietly because it has consumed most of the available oxygen, it is in which phase?
21. Taking into account the chemistry and physics of fire, what can be done to reduce the risk of fire spread in a building?
22. Why is knowledge of what is currently burning and what will soon be burning important to the firefighter arriving at scene?
23. Looking at Table 4-1, which gas is the most dangerous under normal conditions? Why?

DISCUSSION QUESTIONS

1. Is direct flame impingement truly a fourth method of heat transfer?
2. Why is the burning rate of fuel dependent on its physical state?
3. Which method of heat transfer is the most important in a structure fire? Justify your answer.
4. Which of the four phases of fire poses the greatest threat to a firefighter's safety?

NOTES

1. Richard L. Tuve, *Principles of Fire Protection Chemistry* (Quincy, MA: National Fire Protection Association, 1976).
2. Eugene Meyer, *Chemistry of Hazardous Materials* (Upper Saddle River, NJ: Brady, 2005).
3. A. B. Guise, The Chemical Aspects of Fire Extinguishment, *NFPA Journal* (Quincy, MA: National Fire Protection Association, 1960).
4. Eugene Meyer, *Chemistry of Hazardous Materials* (Upper Saddle River, NJ: Prentice Hall Inc., 2005).
5. Richard L. Tuve, *Principles of Fire Protection Chemistry* (Quincy, MA: National Fire Protection Association, 1976).
6. Ibid.
7. Frank L. Fire, *The Common Sense Approach to Hazardous Materials* (New York, NY: Fire Engineering, 2004).
8. National Fire Protection Association, *Standard 255, Test of Surface Burning Characteristics of Building Materials* (Quincy, MA: National Fire Protection Association, 2006).
9. National Fire Protection Association, *Fire Protection Handbook*, Twentieth Edition (Quincy, MA: National Fire Protection Association, 2008).
10. Richard L. Tuve, *Principles of Fire Protection Chemistry* (Quincy, MA: National Fire Protection Association, 1976).
11. Ibid.
12. Eugene Meyer, *Chemistry of Hazardous Materials* (Upper Saddle River, NJ: Prentice Hall Inc., 2005).
13. Richard L. Tuve, *Principles of Fire Protection Chemistry* (Quincy, MA: National Fire Protection Association, 1976).
14. National Fire Protection Association, *Standard 10, Portable Fire Extinguishers* (Quincy, MA: National Fire Protection Association, 2007).
15. Delmar, Cengage Learning, *Firefighter's Handbook*, Third Edition (Clifton park, NY, 2008).
16. National Fire Protection Association, *Fire Protection Handbook*, Twentieth Edition (Quincy, MA: National Fire Protection Association, 2008).

CHAPTER 5

Public and Private Support Organizations

LEARNING OBJECTIVES

Upon completion of this chapter, you should be able to:

- Identify types of support organizations.
- Identify the purpose of specific organizations.
- List how these organizations assist the fire service.
- Identify the organization to contact when information regarding a specific subject is required.

INTRODUCTION

Numerous and varied organizations on national, state, and local levels are related to the fire service. Their purposes vary widely as do their memberships. In this chapter we look at these organizations and their reasons for existence as well as their effect on the fire service. Not all of the associations and agencies that affect the fire service and the delivery of fire protection are mentioned here as they are too numerous to list.

NATIONAL AND INTERNATIONAL ORGANIZATIONS

- *Association of American Railroads (AAR).* An association of the railroad companies in the United States, this organization provides training and reference materials for the fire service. Publishers of the *Pocket Guide to Tank Cars*, a handy guide for training and reference on tank car types and their fittings. Address: Association of American Railroads, 50 F Street NW, Washington, DC 20001.
- *American Fire Sprinkler Association (AFSA).* Membership consists of contractors, manufacturers, dealers, and distributors of automatic fire sprinklers. Publishers of *Sprinkler Age*. Address: 12959 Jupiter Road, Ste. 142, Dallas, TX 75238.
- *American National Standards Institute (ANSI).* Identifies public requirements for national safety, engineering, and industrial standards and coordinates voluntary standardization activities of concerned organizations. Address: American National Standards Institute, 1819 L Street NW, Washington, DC 20036.
- *American Petroleum Institute (API).* Provides wildlife clean up after an oil-related emergency. Address: 1220 L Street NW, Washington, DC 20005. Phone: 202-682-8000.
- *American Red Cross.* Provides assistance to the victims of disasters on a large scale, setting up shelters for evacuated persons, or on a small scale, assisting one family that has lost its home due to fire or other disaster (**Figure 5-1**). The Red Cross is also involved in providing training and certification to emergency responders in the area of infectious disease control through its course "AIDS Education for Emergency Workers." Address: 18th and E Streets NW, Washington, DC 20006.
- *American Rescue Dog Association (ARDA).* Provides volunteer search and rescue dogs and handlers free of charge to requesting public agencies to search for lost or missing persons. Address: PO Box 151, Chester, NY 10918.
- *Automatic Fire Alarm Association (AFAA).* Members include manufacturers, distributors, state/regional associations, engineers, users, and fire and building officials. Improves life safety through use of properly designed, installed, and maintained

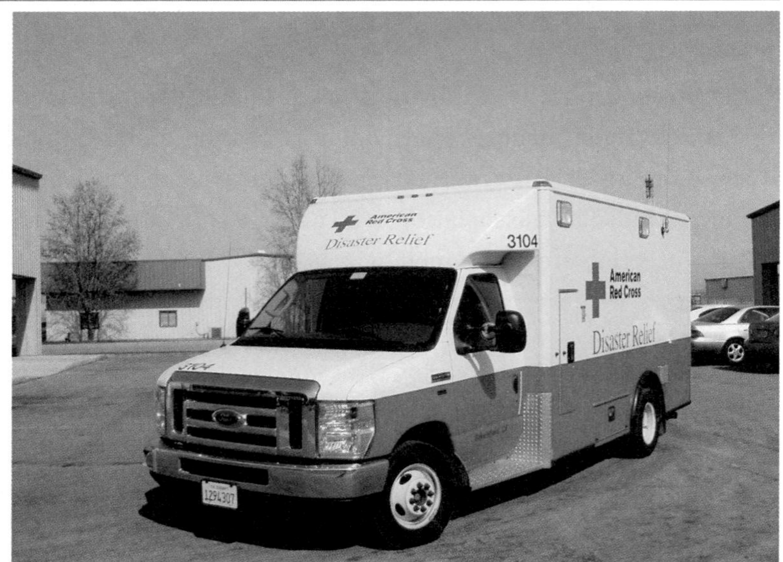

FIGURE 5-1
Red Cross emergency response vehicle (ERV).

automatic fire detectors and early warning systems. Provides training programs, awards, and seminars. Address: PO Box 951807, Lake Mary, FL 32795.

- *Board of Certified Safety Professionals (BCSP).* An association for safety engineers, industrial hygienists, safety managers, and fire protection engineers. Address: 208 Burwash Ave., Savoy, IL 61874.
- *Building Officials and Code Administrators (BOCA).* An organization that produces model codes for adoption by jurisdictions throughout the United States. The organization also sponsors training, testing, and certification for building officials and code administrators. Address: 4051 W. Flossmoor Rd., Country Club Hills, IL 60477.
- *Building and Fire Research Laboratory (BFRL).* Provides the scientific and technical basis for reducing fire losses and the costs of fire protection. Address: National Institute of Standards and Technology, 100 Bureau Drive, Stop 8600, Gaithersburg, MD 20899.
- *Chemical Transportation Emergency Center (CHEMTREC).* A service of the chemical industry providing a 24-hour emergency number to call in case of chemical emergency. Upon receipt of the initial call, with the name of the product, CHEMTREC provides immediate advice on the nature of the product and the steps to be taken in handling the early stages of a problem. CHEMTREC then promptly contacts the shipper of the material involved for more detailed information and on-scene assistance when available. Calls to CHEMTREC are to be limited to emergencies only. Emergency phone: 800-424-9300.

- *Chlorine Emergency Plan (CHLOREP).* Accessed through CHEMTREC. When called, the nearest manufacturer is notified and a representative makes contact with the agency in charge of the emergency. Emergency phone: 800-424-9300.
- *FM Global.* Resources include consulting services, property inspection, water supply and sprinkler system evaluation, safe operation of industrial processes, research, and numerous other property-related factors relating to insurance. The FM Global Laboratories were developed for evaluating fire protection devices and equipment, primarily in industrial fire protection. The laboratories are involved with the evaluation of fire protection equipment used by fire service personnel, including the systems installed in buildings, such as sprinkler systems, detection and alarm systems, and portable equipment, including extinguishers. Publishers of *A Pocket Guide to Automatic Sprinklers*, a reference source for firefighting personnel for training, inspections, and prefire planning. Address: 1151 Boston Providence Turnpike, Norwood, MA 02062.
- *Fire Apparatus Manufacturer's Association (FAMA).* Composed of manufacturers of fire apparatus and equipment. Works for the betterment of the fire service industry through the manufacture of safe and efficient fire apparatus and equipment. Address: PO Box 397, Lynnfield, MA 01940.
- *National Association of State Fire Marshals (NASFM).* Comprised of fire marshals, heads of fire prevention bureaus, fire investigators, and their staffs. Promotes control of arson and prevention of fire, promotes exchange of information between members, and assists in conduct of professional duties. Address: 1319 F Street NW, Suite 301, Washington, DC 20004.
- *FIRESCOPE.* Organization formed because of the need for a standardized system of fire management on large, multiagency incidents. As a result of its efforts, the Incident Command System was developed. Members include firefighters and fire marshals from a wide range of agencies and jurisdictions. Part of their task is to produce standard job descriptions for positions and command structure at an incident. The incident types addressed so far are wildland, hazardous materials, multicasualty, and urban search and rescue. For more information contact your state fire marshal's office.
- *Fire Suppression Systems Association (FSSA).* Suppliers, manufacturers, and installers of automatic fire systems. Promotes the use of fire suppression systems in the overall fire protection industry. Address: 5024-R Campbell Blvd., Baltimore, MD 21236.
- *Association of American Railroads/Bureau of Explosives.* Publishers of the *Emergency Handling of Hazardous Materials in Surface Transportation.* The Bureau of Explosives book is provided to aid emergency responders in handling hazardous materials incidents. Address: BOE Publications, PO Box 1020, Sewickley, PA 15143–1020.

- *Industrial Risk Insurers (IRI).* Established in 1890 in response to the need for an underwriting and loss prevention organization capable of insuring large **highly protected risk** (HPR) properties.* Today IRI provides property insurance and related coverage for about 40% of the Fortune 1000 corporations, covering some 20,000 properties in more than 70 countries. Major classes of HPR business include aircraft manufacturing, automotive manufacturing, basic steel, electrical manufacturing and assembly, food, glass, heavy/light metal, pulp and paper, textiles, and utilities. IRI is not an individual company but is an association backed by the financial resources of more than 20 of the world's leading insurance companies. IRI's headquarters are in Hartford, Connecticut, with offices located in 10 other major cities in the United States. Other offices are in Australia, Canada, England, Germany, Japan, and Puerto Rico. Address: 85 Woodland Street, PO Box 5010, Hartford, CT 06102-5010.
- *Insurance Committee for Arson Control (ICAC).* Property/casualty insurance companies and their trade associations. Serves as the focus group for the insurance industry's overall antiarson efforts and a liaison with other groups devoted to arson control. Provides public education and conferences and compiles statistics. Address: 110 William St., New York, NY 10038.
- *Insurance Services Office (ISO).* A voluntary, nonprofit, unincorporated association of insurers formed to gather information to assist in setting fire insurance rates. The ISO rates the fire protection capabilities of the jurisdictions its insurance company members insure. Utilizing the *Grading Schedule for Municipal Fire Protection*, the ISO provides the insurance industry with a means of identifying, for insurance purposes only, a relative analysis of what may be expected from public fire services. The classification spread of 1 to 10 is a scale of relative values whereby a municipal fire protection system may be compared with others. It is also indicative of a system's ability to defend against the major fires that may be expected in any given community. Where Class 10 is assigned, there is usually no protection. However,

highly protected risk A designation used by the insurance industry to describe business properties that meet certain special requirements as described in the footnote below.

*Most large corporations automatically qualify for the highly protected risk insurance designation, which is shaped by six guiding principles. These are (1) a concerned management interested in implementing an aggressive program of loss prevention and control; (2) acceptable construction in good repair with adequate exposure protection; (3) interior protection with appropriate automatic extinguishing or suppression systems installed wherever there is combustible occupancy; (4) special hazard protection based on the identification and evaluation of all hazards peculiar to the class of occupancy and with the provision of proper supplementary protection; (5) exterior protection with the provision for an adequate water supply for automatic extinguishing systems and ancillary equipment and the availability of a private emergency organization; (6) adequate surveillance, interior and exterior, using guard patrols, surveillance systems, continuous occupancy, or a combination of these.

very minimal protection in a community of extensive development may grade Class 10. Protection Class 1 represents a fire protection system of extreme capability. The rating process involves looking at the whole fire protection system of a given area, including the fire department, water supply, fire service communications, fire safety control, and system of guidance. Guidance can be obtained from nationally recognized standards, predominately those of the National Fire Protection Association and the American Water Works Association. Address: 545 Washington Boulevard, Jersey City, NJ 07310.

- *International Association of Arson Investigators (IAAI).* Formed to combat arson by bringing together arson investigators and related personnel. Provides worldwide training and an information center for arson investigators. Address: PO Box 91119, Louisville, KY 40291.
- *International Association of Fire Chiefs (IAFC).* Represents the interests of officers of the chief rank from throughout the United States and other countries. The purpose of this organization is to further the professional advancement of the fire service. One of the founding members of the International Fire Code Institute, created in 1991. Address: 1329 18th St. NW, Washington, DC 20036.
- *International Association of Fire Chiefs/Metropolitan Section.* Membership in the Metro/IAFC is open to fire chiefs who represent cities of over 200,000 population or 400 paid firefighters. Address: see International Association of Fire Chiefs.
- *International Association of Fire Chiefs/Western Fire Chiefs Association.* Publishers of the *Uniform Fire Code* with the International Conference of Building Officials. Address: PO Box 938, Fremont, CA 94537.
- *International Association of Firefighters (IAFF).* The largest union organization in the Northern Hemisphere representing firefighters and their interests, with approximately 200,000 members from the United States and Canada. Organized with local and state offices as well as a national organization. The IAFF is politically active on all levels of government. Concerned with fire prevention, protection, and the safety of all firefighters through research and legislation. The IAFF is also a major contributor to the Muscular Dystrophy Association and has its own Burn Foundation to promote burn care research. Membership is limited to active and retired paid firefighters. The IAFF is affiliated with the AFL-CIO/CLC. Address: 1750 New York Ave. NW, Washington, DC 20006.
- *International Association for Fire Safety Science (IAFSC).* Membership includes fire research scientists. Promotes research into the science of preventing and mitigating the adverse effects of fires and disseminates results of research. Address: Center for Fire Research, Gaithersburg, MD 20899.
- *International City/County Management Association (ICMA).* An association of local-government management professionals whose purposes are to strengthen urban government through quality professional management and to develop and

disseminate new approaches through training programs, information services, and publications. Publishers of *Managing Fire and Rescue Services*. Address: 1120 G St. NW, #300, Washington, DC 20005.

- *International Conference of Building Officials (ICBO).* An organization that develops model building codes for adoption. Creators of the Uniform Building Code (UBC). Member of the International Fire Code Institute. Address: 17926 S. Halstead St., Homewood, IL 60430.
- *The International Fire Code Institute (IFCI).* Consists of fire department personnel, fire service professionals, building officials, design engineers, architects, product manufacturers, and others who are interested in increasing fire and life safety to establish safer communities. Founded in 1991, IFCI became the first organization in the United States to focus on the development and publication of a model fire code, namely the Uniform Fire Code. IFCI now publishes the *Urban–Wildland Interface Code* and is a participant in the International Fire Code process. The organization provides training seminars and certification targeted toward fire department and building department personnel, architects, and engineers involved in building design, plan review, and fire/life safety inspections. Address: 5360 Workman Mill Rd., Whittier, CA 90601.
- *International Fire Service Accreditation Congress (IFSAC).* IFSAC is a peer-driven, self-governing body that accredits both public fire service certification programs and higher education fire-related degree programs. The mission of the IFSAC is to increase the level of professionalism of the fire service through the accreditation of certifying entities and degree-granting institutions. Some departments, agencies, states, and countries mandate that certification of individual firefighters come from an IFSAC-accredited entity. The Congress is comprised of the Degree Assembly, the Certificate Assembly, the Degree Assembly Board of Governors, the Certificate Assembly Board of Governors, the Council of Governors (which includes members of both Assemblies), and an Administrative Office that conducts the daily business of the organization. The IFSAC Administrative Office is located on the campus of Oklahoma State University. More information is available at http://www.IFSAC.org.
- *International Municipal Signal Association (IMSA).* Organized to assist its members with technical knowledge and information regarding fire and police alarms and traffic control signals. Address: 1115 North Main St., Newark, NY 14513.
- *International Rescue and Emergency Care Association (IRECA).* Membership consists of prehospital care providers including fire, ambulance, and rescue squad members. Address: 8107 Ensign Curve, Bloomington, MN 55438.
- *International Society of Fire Service Instructors (ISFSI).* Organized to further the exchange of educational material and techniques. Sponsor of the Fire Department

Instructors Conference. Provides training, programs, seminars, and conferences to further fire service education. Address: PO Box 2320, Stafford, VA 22555.

- *Mutual Aid Box Alarm System (MABAS).* MABAS is a mutual aid system, which has been in existence since the late 1960s. Pre–September 11, 2001, MABAS was heavily rooted throughout northern Illinois. Since 9/11, MABAS has rapidly grown throughout the states of Illinois and Wisconsin as well as parts of Indiana, Iowa and Missouri. Day-to-day MABAS extra alarms are systematically designed to provide speed of response of emergency resources to the stricken community during an ongoing emergency. Declarations of Disaster are not required for routine MABAS system activations. Today MABAS includes approximately 1,000 of the states' 1,200 fire departments and is organized within 67 divisions. MABAS divisions geographically span an area from Lake Michigan to Iowa's border and south almost into Kentucky. Twelve Wisconsin divisions also share MABAS with their Illinois counterparts. The cities of Chicago, St. Louis, and Milwaukee are also MABAS member agencies. MABAS has also expanded into Indiana and Michigan. See http://www.MABAS.org
- *National Association of Emergency Medical Technicians (NAEMT).* Promotes professionalism among EMTs and paramedics. Provides training programs. Address: 9140 Ward Parkway, Kansas City, MO 64114.
- *National Fire Protection Association (NFPA).* Organized in 1896, the NFPA has members from the fire service and private organizations. The NFPA is recognized for its efforts in developing standards on firefighter safety, equipment, and professional standards. Standards that directly affect firefighters from the first day on the job are:

  1001: Fire Fighter Professional Qualifications

  1404: Fire Fire Service Respiratory Protection Training

  1500: Fire Department Occupational Safety and Health Program

  1581: Fire Department Infection Control Program

  1901: Automotive Fire Apparatus

  1961: Fire Hose

  1971: Protective Ensembles for Structural Fire Fighting and Proximity Fire Fighting

  1975: Station/Work Uniforms for Emergency Services

  As you can see from the preceding list, the NFPA has developed standards on all manner of firefighting equipment and programs. These standards are available through the National Fire Codes Subscription Service. The NFPA is also involved in research, technical advisory services, education, and other fire-related services. The NFPA has published books on a wide range of fire-related subjects, including the *Fire Protection Handbook*, considered by many to be the ultimate source of fire protection information. Address: One Batterymarch Park, PO Box 9101, Quincy, MA 02269-9101.

- *National Fallen Firefighters Foundation (NFFF).* The United States Congress created the National Fallen Firefighters Foundation to lead a nationwide effort to remember America's fallen firefighters. Since 1992, this tax-exempt, nonprofit foundation has developed and expanded programs to honor our fallen fire heroes and assist their families and coworkers. The National Fallen Firefighters memorial is located on the grounds of the National Fire Academy in Emmitsburg, Maryland. The NFFF is also the administrator of the "16 Life Safety Initiatives," the "Courage to be Safe®," and "Everyone Goes Home.com®" programs. Address: National Fallen Firefighters Foundation, PO Drawer 498, Emmitsburg, MD 21727.
- *National Fire Sprinkler Association (NFSA).* Members include fire sprinkler contractors, manufacturers and distributors of fire sprinkler equipment, fire and building officials, and insurance authorities. Creates a market for the widespread acceptance of fire sprinkler systems in both new and existing construction. Address: Rt. 22, PO Box 1000, Patterson, NY 12563.
- *National Interagency Incident Management System (NIIMS).* Develops the Incident Command System and related components. Address: National Wildfire Coordinating Group, National Interagency Fire Center, 3905 Vista Ave., Boise, ID 83705.
- *National Response Center (NRC).* Provides a one-call notification service for hazardous materials emergencies. Notified if the property loss is over $50,000, if serious injury or death has occurred, and if there is a continuing danger to the public. The National Response Center notifies the DOT, EPA, and U.S. Coast Guard. Emergency-only phone: 800-424-8802.
- *National Safety Council (NSC).* Members include industry, labor unions, associations, hospitals, community service organizations, traffic safety associations, fire service, local safety councils, and commissions. The mission is to educate and influence society to adopt safety and health policies, practices, and procedures that mitigate human and economic losses arising from accidental causes and adverse occupational and environmental health exposures. Provides training, awards, seminars, speakers, public education, and conferences and compiles statistics. Address: 1025 Connecticut Ave. NW, Suite 1200, Washington, DC 20036.
- *National Volunteer Fire Council (NVFC).* An organization of various state firefighter associations that represent and pursue the interests of volunteer firefighters and volunteer fire departments. Formulates and promulgates programs useful to the fire service, represents the interests of its members to the U.S. Congress and with various federal agencies involved in the preservation of life and property. Provides training programs, public education, and conferences and compiles statistics. Address: 1050 17th Street NW, Suite 490, Washington, DC 20036.

- *National Wildfire Coordinating Group (NWCG).* Develops training and reference materials for wildland fire fighting. One of its publications is the *Fireline Handbook*, a reference source that contains material on safety, resource use and capabilities, organization, command, and fire behavior. Available from the National Interagency Fire Center, publication NFES 0065. Address: National Interagency Fire Center, 3905 Vista Ave., Boise, ID 83705.
- *Salvation Army.* The Salvation Army provides canteen services on an as-needed basis in some communities. This is usually limited to coffee and donuts or sandwiches on an extended incident. Address: Check local listing.
- *Society of Fire Protection Engineers (SFPE).* Professional society for fire protection engineers. Advances science of fire protection engineering and its allied fields; promotes education; and provides awards, training programs, seminars, and conferences. Address: 60 Batterymarch St., Boston, MA 02110.
- *Southern Building Code Congress International (SBCCI).* Creators of the Southern Building Code. Also sponsors training, testing, and certification for building officials. Address: 900 Montclair Rd., Birmingham, AL 35213.
- *Underwriters Laboratories Inc. (UL).* The goal of the UL is to promote public safety through scientific investigation of various materials in regard to the hazard they present by their use. After testing, the organization lists and marks the material as having passed its rigorous tests. The UL reserves the right to test a representative sample of a manufacturer's listed product at any time to assure compliance. Organized as a nonprofit corporation. Address: 333 Pfingsten Rd., Northbrook, IL 60062.

FEDERAL ORGANIZATIONS

There are numerous publicly funded agencies of the federal government concerned with protecting the public from fire. These organizations are involved in all aspects of firefighter training, research, and development. Many of the organizations have resources that can assist in emergency situations.

- *Department of Defense.* Assists fire department with antiterrorism training and response. The training is titled "Domestic Preparedness" and deals with weapons of mass destruction (WMD). It covers nuclear, biological, and chemical threats (NBC). Many military bases make resources available to local departments on a call-as-needed basis. They may also have training props that can be utilized.

  It is becoming more common to see large numbers of armed forces personnel who are trained to fight fires activated on large wildland fire incidents. Personnel from the army and reserves are now even being given this training before they are needed. With their numerous personnel, command structure, and self-contained logistical support, they are a tremendous asset in times of need. Military bases usually have their own fire departments that can respond off-base when requested by the local fire department.

They may even have specialized apparatus, such as water tenders, heavy equipment, foam units, and aircraft rescue firefighting apparatus. Helicopters from military bases may also be available to perform emergency transport of victims in rescue situations.

- *Department of Labor (DOL).* Administers and enforces the Occupational Safety and Health Act to provide safety in the workplace. The DOL compiles the National Occupational Injury and Illness data for the Bureau of Labor statistics. Address: 200 Constitution Ave. NW, Washington, DC 20001.
- *Department of Transportation (DOT).* Regulates shipping of hazardous materials in trucking and railroads and on aircraft and waterways in the United States.

  The DOT specifies placarding requirements for materials being transported. This federal agency publishes the *Emergency Response Guidebook*, which is carried on fire apparatus for reference in case of shipping accidents. Using placards and shipping papers, the *Guidebook* provides information on initial actions to be taken in case of a spill or suspected spill of hazardous materials in a transportation accident. Available from: U.S. Government Printing Office, Superintendent of Documents, Mail Stop: SSOP, Washington, DC 20402-9328.
- *Emergency Management Institute (EMI).* Authorized under the Civil Defense Act of 1950 to provide training to public sector managers to prepare for, mitigate, respond to, and recover from all types of emergencies. Provides programs in emergency management, technical development, and professional development. The parent organization for EMI is the Federal Emergency Management Agency. Address: 16825 S. Seton Ave., Emmitsburg, MD 21727.
- *National Emergency Training Center (NETC)*, sponsored by the Emergency Management Institute. At the NETC, representatives of the fire service, law enforcement, emergency medical services, civil defense, public works, state and local government, and public interest groups can meet and seek training to perform coordinated emergency response. Address: 16825 S. Seton Ave., Emmitsburg, MD 21727.
- *National Institute for Occupational Safety and Health (NIOSH).* Part of the Centers for Disease Control and Prevention (CDC) in the Department of Health and Human Services, NIOSH conducts research and provides educational functions to support OSHA. It also recommends occupational safety and health standards. Publisher of the *Pocket Guide to Hazardous Materials.* Contact NIOSH at http://www.cdc.gov/niosh/contact or 1-800-CDC-INFO.

  In fiscal year 1998, President Clinton and Congress recognized the need to address the continuing national problem of firefighter fatalities and funded NIOSH to undertake the National Fire Fighter Fatality Investigation and Prevention Program. The overall goal of this program is to better define the magnitude and characteristics of work-related deaths and severe injuries among firefighters, to develop recommendations for the prevention of these injuries and deaths, and to implement and disseminate prevention efforts. The plan consists of five parts and is centered on the field investigation of firefighter fatalities. The parts are firefighter fatality investigations; cardiovascular disease fatality investigations; the firefighter fatality database project; the intervention research project; and the information dissemination project.

- *National Incident Management System (NIMS)* The National Incident Management System (NIMS) provides a systematic, proactive approach to guide departments and agencies at all levels of government, nongovernmental organizations, and the private sector to work seamlessly to prevent, protect against, respond to, recover from, and mitigate the effects of incidents, regardless of cause, size, location, or complexity, in order to reduce the loss of life and property and harm to the environment.

  NIMS works hand in hand with the *National Response Framework (NRF).* NIMS provides the template for the management of incidents, while the NRF provides the structure and mechanisms for national-level policy for incident management.
- *National Integration Center (NIC).* The Secretary of Homeland Security, through the National Integration Center, publishes the standards, guidelines, and compliance protocols for determining whether a federal, state, tribal, or local government has implemented NIMS. Additionally, the Secretary, through the NIC, manages publication and in collaboration with other departments and agencies develops standards, guidelines, compliance procedures, and protocols for all aspects of NIMS. More information about NIMS and the NIC is available at http://www.fema.gov/emergency/nims/NIMSTrainingCourses.shtm
- *National Institute of Standards and Technology (NIST).* Founded in 1901, NIST is a nonregulatory agency of the Commerce Department that promotes U.S. innovation and industrial competitiveness by advancing measurement science, standards, and technology in ways that enhance economic security and improve our quality of life. In April of 2010, NIST's Building and Fire Research Laboratory released the *Report on Residential Fireground Field Experiments* (NIST TN – 1661), a landmark fire study on how crew sizes and arrival times influence the saving of lives and property. Information on the NIST is available at http://www.nist.gov.
- *U.S. Department of Agriculture Forest Service (USFS)* (**Figure 5-2**). A nationwide organization charged with the management of the national forests, this agency provides fire protection on national forest lands and assistance and resources to agencies protecting areas bordering on national forests. In the summer, this is one of the largest firefighting organizations in the country. Federal surplus equipment, from helicopters and road graders to hand tools, is available to local firefighting agencies through the USFS excess property program. The items are checked out to the local agency on a loan basis and must be returned when no longer needed or accounted for when no longer useable. The only cost of these items to the local agency is their transportation from where procured and any necessary changes to get them into shape for use. The availability of these items, at little or no cost, is a great asset to local agencies. Contact local National Forest Supervisor's Office.

FIGURE 5-2 USFS fire engine and firefighters at wildland fire.

Delmar/Cengage Learning

- *U.S. Department of Homeland Security (DHS).* Established to integrate resources from federal, state, and local governments to establish an agency focused on protecting the American people and their homeland. The national strategy seeks to develop a complementary system connecting all levels of government without duplicating effort.

  The DHS has five major divisions or "directorates." The directorates are: Border and Transportation Security; Emergency Preparedness and Response; Science and Technology; Information Analysis and Infrastructure Protection; and Management.

  The directorate that interacts the most with fire departments is Emergency Preparedness and Response. They assist departments by providing grant funding for equipment and training.
- *U.S. Department of the Interior Bureau of Land Management (BLM)* (**Figure 5-3**). Much like the USFS, BLM provides fire suppression services on lands under the control of the Department of the Interior, outside of the national parks, and adjoining private lands. Mostly operating in the western part of the United States. Contact local Bureau of Land Management Office.
- *U.S. Department of the Interior Park Service (USPS).* Under the Department of the Interior, the USPS, like the BLM, provides fire protection in the national parks and bordering lands where a fire may be a threat to the park. Contact local National Park Service office.
- *National Firefighting Equipment System (NFES).* Publishes reference and training manuals on all aspects of wildland firefighting. Also develops the courses used in

FIGURE 5-3
BLM engine at wildland fire.

Delmar/Cengage Learning

the qualification rating system for wildland firefighters. All of the publications available through this system are referenced by NFES number. An example of this is the book *Basic Aviation Safety*, NFES #2097. These publications are available through the National Interagency Fire Center.

- *National Interagency Fire Center (NIFC).* A cooperative effort of the Department of the Interior and the Department of Agriculture, NIFC is the central supply point for resources required on large wildland fire fighting incidents. By contacting NIFC, incident commanders can order everything from shower units to fire hose to hand crews and overhead personnel. Previously named the Boise Interagency Fire Center (BIFC). Address: National Interagency Fire Center, 3905 Vista Avenue, Boise, ID 83705.
- *Federal Emergency Management Agency (FEMA).* Created in 1978 by then-President Jimmy Carter, with the intent of placing federal disaster response coordination and services under one agency, FEMA was heavily focused on civil defense planning in the 1980s. After the Hurricane Andrew incident, FEMA was reorganized and focused more on natural disasters. FEMA's performance was highly regarded during the Loma Prieta Earthquake in the San Francisco Bay area in 1989. FEMA maintains several Type 1 Incident Management Teams across the United States. These teams are set up with personnel from various agencies and are called into action when the need arises. They are used on fires, floods, hurricanes, earthquakes, and other disasters. FEMA also maintains a list of urban search and rescue teams (USAR) activated for disasters with major structural collapse such as earthquakes and explosions.

These teams must meet specified minimum requirements for personnel and materials. When the Type 1 or USAR teams are on the top of the rotation list, the personnel assigned to the particular team must be able to respond in two hours. Address: 16825 S. Seton Ave., Emmitsburg, MD 21727.

- *United States Fire Administration (USFA).* Established by the Federal Fire Prevention and Control Act of 1974 (Public Law 93-498), its purposes are to reduce the nation's losses from fire through better fire prevention and control; to supplement existing programs of research, training, and education; and to encourage new, improved programs and activities by state and local governments. Address: 16825 S. Seton Ave., Emmitsburg, MD 21727.
- *National Fire Academy (NFA).* The NFA was also established by the Federal Fire Prevention and Control Act of 1974. The NFA is an organization dedicated to the professional development of firefighters and related professionals across the United States. The NFA has a campus in Emmitsburg, Maryland. Courses offered are in the areas of executive development, fire prevention, leadership, incident management, public education, fire service education, arson, infection control, and hazardous materials. These courses are designed to be presented on or off campus as the need arises. The NFA also maintains a large library called the Learning Resources Center (LRC). Address: 16825 S. Seton Ave., Emmitsburg, MD 21727.
- *Fire and Emergency Services Higher Education (FESHE) Conference.* The FESHE conference is a gathering of college fire service program directors, state training directors, and representatives from organizations that have an interest in fire service higher education. Out of this conference came the FESHE guidelines. These guidelines include a professional development model as well as a model curriculum for fire science degree programs at the associate's and bachelor's degree levels (see Chapter 1). More information is available at http://www.usfa.dhs.gov/nfa/higher_ed/feshe/feshe_conf.shtm.
- *Department of the Treasury Bureau of Alcohol, Tobacco, Firearms and Explosives (ATFE).* Assists in the investigation of arson and bomb incidents by gathering and processing evidence. Address: 111 Constitution Ave., Washington, DC 20224. Phone: 202-622-2000.
- *National Transportation Safety Board (NTSB).* Investigates and maintains statistics on vehicle accidents, including fire apparatus. The NTSB is involved in the process of vehicle recalls when severe problems are found. Address: 490 L'Enfant Plaza SW, Washington, DC 20594. Phone: 202-382-6800.
- *Occupational Safety and Health Administration (OSHA).* Established under federal law to ensure safe working conditions, OSHA is a part of the Department of Labor. Some states have their own state-level OSHA that enforces federal regulations. Address: Department of Labor Building, 14th St. and Constitution Ave. NW, Washington, DC 20210.

- *Environmental Protection Agency (EPA).* Responds to and acts as the coordinating agency on large hazardous materials incidents. The person responding is called the "on-scene coordinator" and has the authority to ensure the spill is contained and cleaned up properly. The EPA has money available under the Superfund to provide cleanup services at hazardous materials sites that are deemed a threat to the public. Address: 401 M St. NW, Washington, DC 20460.
- *Federal Aviation Administration (FAA).* Regulates fire protection at general aviation airports where commercial carriers operate and also aboard aircraft. It also controls the movement of hazardous materials by air. When an aviation accident occurs, the FAA investigates. Address: 800 Independence Ave. SW, Washington, DC 20591.
- *Nuclear Regulatory Commission (NRC).* Regulates all aspects of the nuclear power generating facilities in the nation, including fire protection. Address: 1717 H St. NW, Washington, DC 20555.
- *U.S. Coast Guard (USCG).* Provides on-scene coordination on hazardous materials incidents that involve or threaten navigable or coastal waterways. Publishers of the *CHRIS Manual* and other hazardous materials information sources. Address: 2100 Second St. SW, Washington, DC 20593.

NOTE If you wish to contact a federal agency, it is listed in the phone book under "United States Government Offices."

STATE ORGANIZATIONS

- *Office of State Fire Marshal (OSFM or SFM).* The OSFM receives its authority to enforce the state's fire laws from the legislature. In most states the state fire marshal is appointed by the governor. The state fire marshal's responsibilities vary by state. The usual responsibilities include the review and approval of construction plans for fire safety, the investigation and determination of fire cause, the investigation of fire deaths, the aggressive attack on arson through investigation and prosecution, regulation and control of the storage and use of explosives and other hazardous materials, preparation and promotion of fire-related legislation, and promotion and enforcement of the state's fire codes. Contact the state fire marshal's office in your state.
- *State Fire Training.* Many states have a state-level training organization. The purpose of this organization is to develop, coordinate, and deliver training throughout the state by full-time management staff and full- and part-time instructors who are state certified to present the curriculum. Often the training programs are presented under the sponsorship of a local college, which provides a way to present certifications and college credits and provides a means to pay the instructors. Contact the state fire marshal's office in your state.

- *State Emergency Management Agency.* Provides training and assistance to local governments in the preparation, response, recovery, and mitigation of disasters. May also coordinate the state Master Mutual Aid agreement. Look under "State Government Offices" in the phone book.
- *Fire Commissions.* In some states fire commissions are set up to act as advisory panels to the state fire training organization. They oversee the expenditure of public funds for training purposes. They also adopt the standards used in certification.
- *State Fire Chiefs Associations.* Fire chiefs organize on the state level to gather and find solutions to common problems. They also promote professional development and legislation.
- *State Firefighter Associations.* Members from various firefighting organizations in the state organize to lobby for legislation to further their interests. Some legislation that has been passed in certain states concerns cancer presumption and heart presumption. These items of legislation guarantee that if a firefighter has cancer or heart disease, it is presumed to be job connected, which allows for a job-connected disability pension. These organizations also provide training to their members in the areas of contract negotiation and wage and hour matters.
- *State Police.* The police and fire department join forces in their efforts to investigate arson and apprehend and convict arsonists. The police have the systems in place to check license plates on burned vehicles, have access to crime labs, and are experts on gathering and preserving evidence. If the state has a state-level highway patrol, the state and federal highway system is under its jurisdiction.
- *State Environmental Protection Agency.* Agencies become involved when hazardous material incidents threaten the environment. This responsibility is placed under the wildlife management agency in some states. When there is a state-level Super Fund, money may be available to clean up spills in some jurisdictions.
- *State Occupational Safety and Health Administration.* In the states that have them, state OSHA enforces state and federal regulations pertaining to worker safety in the workplace.
- *State Health Department.* After fires or incidents that may affect the public health, the health department may become involved in determining what must be done to correct the problem. It may condemn a load of fruit spilled on the road in a vehicle accident or condemn a freezer of food at a restaurant after a fire. Health departments also become involved in determining whether a hazardous materials spill has been cleaned up sufficiently.
- *State Forestry Department.* Lands in the state that are not incorporated into cities or under federal jurisdiction, although private property, are often the responsibility of the state for fire protection. By using state department of forestry resources and contracting with local fire protection agencies, fires are suppressed.

- *Special Task Forces.* When there are specific fire-related problems that need to be addressed, it is common for the governor to set up a special task force to deal with them. A good example of this would be an arson task force composed of firefighters, law enforcement, insurance companies, and state agency representatives to address the issue. The task force would perform a study and make recommendations based on their findings.
- *Office of Emergency Management (OEM).* This agency coordinates with fire agencies during disasters. The OEM can provide money, personnel, vehicles, and equipment to assist local agencies in combating the problems due to the emergency.
- *National Guard.* In some areas National Guard units are trained in wildland firefighting and activated when local resources are overwhelmed. In times of major wildland fires National Guard planes may be set up as air tankers, called Mobile Aerial Fire Fighting System (MAFFS) Units, to combat fires. The National Guard also has all sorts of heavy equipment and trucks that can be used in combating fires.
- *Other State Agencies.* Much like the federal government, each state has numerous agencies and departments that interact with the fire service either directly or indirectly.

LOCAL ORGANIZATIONS

- *Burn Foundations.* Burn foundations are set up to raise and disseminate funds to the victims of burn injuries and their families. The money is also used for burn injury research and to purchase burn treatment equipment. Many fire department personnel are involved, on their off-duty time, in these activities. Burn foundations are not only good public relations for firefighters but may also make the difference in having the technology available for treating you if you are injured.
- *Local Government.* Most fire departments are entities of some level of local government. The locally elected officials establish the fire department and delegate the authority to fight fires and enforce fire protection laws through ordinances. The local government raises the money necessary to purchase equipment and pay the firefighters. This may be done through general taxation, special districts, or fire taxes.
- *Law Enforcement.* Fire and law enforcement departments find themselves working together on many emergencies. When evacuation is necessary due to hazardous materials incidents, police usually evacuate people and deny entry to the affected area. In some states the County Sheriff serves as the officer in charge of wildland fire incidents.

  Police provide traffic control at emergency scenes and may be trained in emergency first aid. The goal of the police and fire departments are much the same—to save lives and protect property. A positive working relationship with local law enforcement is a definite plus in accomplishing the fire department's mission.

- *Building Department.* The fire department and the building department work together to ensure fire safety in buildings. The building department is more involved when the building is being constructed and when major remodeling that requires a building permit takes place. The building department should also notify the fire department when new buildings, other than homes, are certified ready to occupy. By working closely with the building department, the fire department can ensure that fire-related codes are enforced during the construction phase.
- *Water Department.* The water department controls the jurisdiction's water system to a great degree. If there is to be adequate water available for firefighting purposes, there must be a working relationship with the water department. If damage or repairs have affected part of the water system, it is to the fire department's advantage to be notified of the location, duration, and extent of the area affected. In times of high water demand, due to required flow to combat fires, a representative of the water company may have to be contacted to make adjustments to the water system.
- *Zoning/Planning Commission.* The local planning commission decides what types of occupancies will be built where. It decides where the residential areas will be and what the density will be. These areas can be divided into single- and multiple-family structures. It decides where commercial and industrial centers will be located. In the planning of future fire station sites this information becomes extremely important. A fire station may be in one place for 50 years or more. The surrounding area and its fire protection needs will dictate the size of the station in terms of room needed for a large or small crew and the type of apparatus it must house. The fire department should be included in the planning process in the interest of providing optimum fire protection.
- *Street Department.* This department may also be called Highways and Bridges. This agency designs and maintains streets and bridges in the jurisdiction, which concerns the fire department in terms of being able to get the apparatus into all areas and across all of the bridges (**Figure 5-4**). If cul-de-sacs are too tight or bridges cannot handle the weight of apparatus, it severely limits response routes and times as well as being dangerous to firefighters. In the case of a large brush fire, a negotiable escape route for fire apparatus must be available. Many fire departments have had their apparatus and the traffic lights equipped with a system that allows them to change the lights to green in their direction of travel, called traffic preemption systems. Another reason for maintaining a good relationship with the road department is that it has heavy equipment and operators that may be required at an emergency scene. A skip loader and a dump truck are very handy to have when cleaning up after a semitrailer wreck or when sand needs to be laid to contain a hazardous materials

FIGURE 5-4 Public works/roads department water tenders.

Delmar/Cengage Learning

spill. The road department should also notify the fire department if roads or bridges are closed due to maintenance or repair.

- *Local Judicial System.* In the U.S. legal system people are innocent until proven guilty and entitled to due process. A working relationship must be maintained with the judicial system, both the courts and the prosecutor's office, to ensure that these rights are protected. If they are not, every citation for violation of the fire code or arson case will be thrown out of court.

  It is not uncommon for firefighters to be subpoenaed into court to testify in lawsuits and criminal prosecutions arising from emergencies that they responded to. They may just be witnesses to what the scene looked like when they got there or may be defendants due to the actions they took. A positive working relationship with the local prosecutor's office needs to be maintained to assist in preparing firefighters when they are called on to present testimony.
- *Local Office of Emergency Management.* The local civil defense organization can come to the aid of the fire department in times of major emergency by providing resources in the form of personnel and equipment.
- *Local Emergency Medical Service Agency.* This agency ensures that ambulance and fire department emergency medical technicians and paramedics are provided training and meet licensing requirements. It also regulates private ambulance companies in its jurisdiction. In the case of a mass casualty incident or other specific circumstances, it may be responsible to poll the local hospitals to determine the availability of beds in their emergency rooms for victims.

- *Emergency Operations Center (EOC).* Many jurisdictions, on local and state levels, have an EOC to which different functional representatives respond during an emergency activation. Once activated, the center is the focal point of decision making on a regional basis. Resource needs are addressed and prioritized in cases of multiple incidents. Cooperators in an EOC may be government and private agencies including law enforcement, fire, human services, waste management, mental health, Red Cross, and others as needed. On the departmental level the center is called a Department Operations Center (DOC) and it performs such functions as staffing fire stations vacant due to commitment to an incident, accessing resources from departments outside the jurisdiction, and the like.
- *Firefighter Union Locals.* Firefighters that have joined state or national union organizations often have a local chapter (**Figure 5-5**). This allows them to address grievances, negotiate contracts, and deal with other matters on a local level. These organizations are the focal point for political activity in local elections. Many firefighter union locals are heavily involved in fund-raising activities for burn foundations and other charities.
- *Community Service Organizations.* Local service organizations (**Figure 5-6**) are sources of revenue and equipment for fire organizations. They also provide opportunities for firefighters to attend and make fire prevention presentations.

  These groups, in some instances, have bought smoke detectors to be given out to the public. By working together the service groups and the fire department can improve public relations as well as provide valuable public services.

FIGURE 5-5 Union local headquarters.

Delmar/Cengage Learning

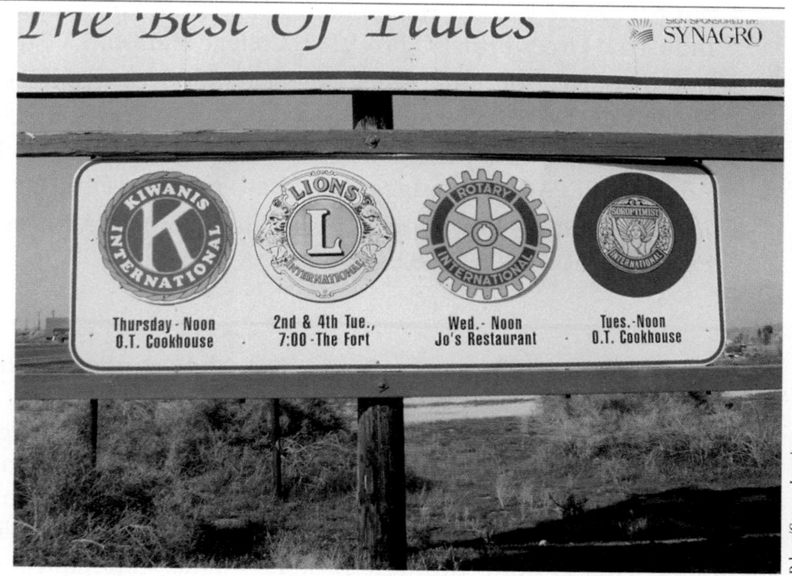

Delmar/Cengage Learning

FIGURE 5-6 Community service organizations' logos posted on city sign.

PERIODICAL PUBLICATIONS

There are numerous sources of information for those working in fire protection and emergency services. Numerous fire-related organizations are now utilizing the Internet to provide information on the World Wide Web. There is a listing of a number of these organizations in the Appendices at the end of this text.

Another source of current information is emergency service–related magazines. They aid firefighters in staying up-to-date on the latest developments in fire and emergency medical service equipment, methods, and activities.

The following is a listing of some of the leading fire service–related magazines:

- *Firehouse.* Address: Firehouse, 82 Firehouse Lane, Box 52824, Boulder, CO 80321-2824.
- *Fire Chief,* Administration/Training/Operations. Address: Communication Channels Inc., Fire Chief, PO Box 5111, Pittsfield, MA 01203-9830.
- *Fire Engineering,* "Training the Fire Service for 117 Years." Address: Fire Engineering, PO Box 1289, Tulsa, OK 74101-9853.
- *Firefighter's News,* "Information and Skills for Fire Service Decision Makers." Address: Firefighter's News, c/o Lifesaving Communications, Inc., Circulation Dept., PO Box 100, Nassau, DE 19969-9900.
- *FireRescue,* "Read It Today, Use It Tomorrow." Address: JEMS Communications, 525 B St., Ste. 1900, San Diego, CA 92101-4495.

- *Industrial Fire World*, "The Foremost Industrial Fire Protection Authority." Address: Industrial Fire World, PO Box 9161, College Station, TX 77842-9984.
- *JEMS*, Journal of Emergency Medical Services. Address: JEMS, PO Box 3730, Escondido, CA 92033-9984.
- *NFPA Journal*, "The Official Magazine of the National Fire Protection Association." Address: NFPA Journal Magazine, 1 Batterymarch Park, Quincy, MA 02669-9101.
- *Rescue*, "Uniting Rescue and Basic Life Support." Address: Rescue, PO Box 3730, Escondido, CA 92033-9982.

SUMMARY

Numerous resources are available to help the fire service achieve its goals. Many of the organizations have the goal of raising the professional standards of their members. This is done through gatherings of the members to create standards and provide training information. The professional organizations also have the goal of promoting legislation that makes providing fire protection safer and more effective. Some of the private organizations, such as CHEMTREC, are only to be used for emergencies. They provide much needed information to assist in at-scene decision making. In some instances the organizations are equipped with personnel and equipment that can respond to the scene and assist the fire department in hazard mitigation and incident control.

There are organizations that operate on a scale that varies from local to international; by doing so they can better address the needs of their members. There are federal, state, and local agencies. Their purpose for being may not be directed toward the fire department, but through the use of their resources the fire department can better achieve its purposes. They can also provide valuable resources in both emergency and nonemergency situations.

Through knowledge of these organizations and the help they can provide, information and training can be accessed on a wide variety of fire-related subjects. It is not as important to know everything there is to know as it is to know who to contact to gather information.

REVIEW QUESTIONS

1. What organization would be a source of information about fire equipment?
2. What organization would be a source of information about fire sprinklers?
3. What organization would be a source of information about labor relations?
4. What organization would be a source of information about model codes and their adoption?

5. When you arrive at a hazardous materials incident, you see a placard on a truck involved and need information as a guide as to how to proceed. What is your information source?
6. You are at the scene of a hazardous materials spill and need information on the product. Whom do you contact?
7. At the same incident the spill is now about to enter a navigable waterway. Whom do you contact to alert the proper federal agencies?
8. The responsible federal agencies have been alerted. Which ones are likely to respond?
9. Your department needs training manuals on a variety of firefighting subjects and you are asked to find a source for them. Whom would you contact?
10. There has been a large-scale natural disaster, other than a fire, in your area. Which federal agency takes responsibility for providing assistance?
11. Which state level agency is activated to assist?
12. There is a wildland fire of extreme proportions in the National Forest in the area. Which other federal agencies can be called on to assist?
13. On the preceding fire, state resources are needed; which agencies are called upon to assist?

DISCUSSION QUESTIONS

1. You are trying to raise money to purchase a rescue tool for your department. What are some of the local service clubs you can contact?
2. Arson has become a growing problem in your jurisdiction. What public and private groups can be activated to form an Arson Task Force?
3. There are on average 100 firefighter line-of-duty deaths a year. Which of the listed organizations are heavily involved in addressing this issue?
4. Utilizing the websites listed in the Appendices and the organizations listed in this chapter, list sources to be used in preparing a plan of action to search for job opportunities in the fire service.

CHAPTER 6

Fire Department Resources

LEARNING OBJECTIVES

Upon completion of this chapter, you should be able to:

- List fire department facilities.
- List advantages of a department having its own facilities.
- Describe the purpose of each of the fire department facilities.
- Describe the types of fire apparatus and their functions.
- List the types of tools and equipment carried on fire apparatus.
- Describe the uses of the various tools and equipment carried on fire apparatus.
- Describe the different types of personal protective equipment used by firefighters.
- Describe the types and uses of aircraft in firefighting.

INTRODUCTION

The modern fire department relies on many types of resources in the form of facilities, apparatus, and equipment to do its job. The facilities described in this chapter are not available at every fire department due to need and budget constraints; they represent a sample of the facilities at fire departments across the country.

The apparatus and equipment described in this chapter have evolved over a period of many years to fulfill specific functions, much of it in a particular firefighting situation or method. Not all of the apparatus and equipment is operated or carried by every fire department, as situations and types of incidents vary. It is important to be aware of the differing types of apparatus and equipment when operating in conjunction with other departments and agencies on large or complex incidents.

Not all of the apparatus and equipment available to the firefighter is listed here, as that would take up numerous volumes. As you study this chapter, try to think of some apparatus and equipment available in your area that could be adapted to firefighting use. As you look at the chapter, you should realize that this is exactly what has been done with many of the tools in use today.

FIRE DEPARTMENT FACILITIES

The modern fire department requires numerous types of facilities for response, support, and administrative functions. They are illustrated here. Not every department will need all of them due to size of response area and the size of the department.

Headquarters

The fire department headquarters is where the managerial staff of the fire department is located. The fire chief, administrative officer, and their staffs work out of the headquarters. The heads of the various bureaus—fire prevention, training, investigation, and others—have their offices at this facility. By having all of the top staff in one location, it is much easier to perform unified planning. The whole staff, or selected personnel, can be gathered on short notice to confer on items requiring immediate attention.

The headquarters may be located at the main fire station or at a separate location. There are advantages and disadvantages to either location. Having the headquarters located at the main fire station (**Figure 6-1**) helps the staff keep a finger on the pulse of the organization. It brings them closer to the troops in the field. It also provides the staff with personnel who can be used to perform tasks, such as running errands, when necessary. The disadvantages are that the firefighters at the main fire station will be assigned many of the minor jobs that the staff needs done, which is often disruptive to the routine work that the company officer has planned. Firefighters tend to be quite social and a certain amount of productive time will be lost with the station crew talking to the staff personnel and firefighters from other stations as they come and go during the day. Having apparatus responding from the headquarters is disruptive to the headquarters staff. When apparatus leaves, there is usually a certain amount of noise in the revving of engines, sirens, and air horns. The apparatus pulling out leaves behind a cloud of diesel smoke that can find its way into the office spaces unless the station is equipped with exhaust recovery systems.

There are advantages to working at the main fire station. The firefighters working at the main fire station are usually the best informed as to what is going on in the department, which can be very beneficial at promotional testing time. The other side of this is that they are the most visible to the top staff and are usually held to a higher standard just because of that visibility. In a large department with widespread stations, the old adage "out of sight, out of mind" may well apply. The headquarters firefighters may also get choice assignments. Even if this is not true, it is often the perception of the firefighters at the other stations. Headquarters firefighters are right there when things

FIGURE 6-1
Fire headquarters co-located with fire station.

become available and they can be the first to get their names on the list for classes and other events.

Having the headquarters remotely located also has advantages and disadvantages (**Figure 6-2**). A site can be chosen that allows for future expansion as the department increases in size. The role of the fire department has grown immensely during the last few years and the amount of personnel needed to administrate and perform the new functions has grown along with it. Having the headquarters located by itself reduces some of the problems noted in the previous paragraphs. The staff personnel are left alone to perform their duties without the interference and disruption of station activities. The remote location also cuts down on the production of rumors from eavesdropping and from personnel accidentally, or on purpose, seeing memos and other confidential communications or suggestions. When someone reports to headquarters for disciplinary reasons, he or she does not need a whole station crew watching and then spreading the news. Having the office away from the unofficial communication system allows the staff the luxury of brainstorming and other creative thinking without the fear that anything placed on a dry erase board or paper will get to the field as a "done deal," not just an option that may be accepted or rejected.

Some of the disadvantages are that someone has to be found to carry out errands, like heavy lifting or moving office equipment, that office personnel cannot perform themselves. This may even require having to detail a crew from a station over to headquarters to move things on an occasional basis. When new equipment—such as nozzles or turnouts—is to be tested, it needs to be taken to a fire station, not just sent downstairs for evaluation. This tends to formalize the contact of the staff officers with the field.

FIGURE 6-2
Fire headquarters building separate from fire station.

Automotive Repair Facility

Mechanics are needed to maintain a fleet of apparatus and all of the other motorized equipment used by the fire department. They are hired for their expertise in working with the types of apparatus operated by the fire department. A complete facility has hoists that can handle the weight of a fire engine for service from underneath (**Figure 6-3**). The facility should be heated and cooled for the comfort of the mechanics in winter and summer. Each mechanic requires a complete set of hand tools. The shop should be equipped with a set of specialty tools for work on certain parts of the apparatus. Heavy tools, such as lathes and presses, are needed to perform certain jobs and to fabricate parts when vehicles have a new motor installed, or when some part that can no longer be purchased needs to be replaced. A lube and oil change bay should be included as well as tire-servicing equipment. A separate area of the shop may be set up for welding and fabrication. Many manufacturers of fire apparatus have not remained in business for the service life of the apparatus, making parts, such as compartment doors, unavailable.

Training Center

One of the most important facilities a fire department can have is a training facility.[1] It need not be overly fancy or expensive. Many training props can be created from donated items.

FIGURE 6-3 Fire department vehicle repair facility.

Delmar/Cengage Learning

A drill tower (**Figure 6-4**) is effective for training personnel in the use of ground ladders and aerial apparatus. It can also be used for training in rappelling and **high-angle rescue**. If the training tower is equipped with interior stairwells and a **standpipe system** it can be used for training in high-rise firefighting. Most drill towers are constructed of concrete, brick, or steel. Training towers have been made from wood, with the uprights made out of telephone poles. Burning is not usually done in drill towers because of the damaging effects of heat and smoke.

An effective way to fill the tower with smoke is with a smoke machine. These machines leave no harmful or unsightly residue.

A burn building (**Figure 6-5**) or prop is effective for training firefighters under hot and/or smoky conditions. Demonstrations showing the first two phases of a fire and development of smoke and heat in an interior fire environment can be safely performed. These types of buildings are especially good for training firefighters in interior attack as a back draft or flashover is not likely to occur. The building should be constructed of non-combustible materials, allowing it to last through many training fires without damage. If the fires are kept to a few palettes or small amounts of ordinary combustibles, the effect of

high-angle rescue Rescue utilizing ropes and other equipment. Examples are removing persons from smokestacks, wind turbines or water towers.

standpipe system Plumbing system installed in multistory buildings for fire department use with outlets on each floor for attaching fire hose.

FIGURE 6-4
Drill tower being used for aerial ladder training.

Delmar/Cengage Learning

FIGURE 6-5
Burn building used for live fire training.

Delmar/Cengage Learning

the heat can be obtained without damage to the building. Several handfuls of damp straw with a road flare stuck in the center will make all of the smoke required. The fires should be kept as small as possible and flammable liquids, tires, or other highly flammable substances should not be used. Replaceable ceiling panels with sheetrock nailed to a wooden frame and roof panels using plywood are effective for ventilation training.

Whenever live fire is part of the drill, a safety officer, full turnouts, and self-contained breathing apparatus are requirements. The best course of action is to strictly adhere to NFPA *Standard 1403, Live Fire Training Evolutions.*[2]

Burn buildings are also useful for demonstrating the dangers of various household materials in a fire situation. A very effective demonstration is to place a dried-out Christmas tree in the living room of the burn building. The area surrounding the tree is set up with furniture and wrapped boxes. The tree is then set on fire, demonstrating the dangers of allowing the tree to dry out. A motivational drill to build speed in firefighting salvage operations in a burn building with a sprinkler system is to have one crew hooking up to the sprinkler system while another is inside spreading salvage covers. If the inside crew is fast enough, it will complete its work and get out before getting wet.

The burn building can also be used for hazardous materials training by setting up a simulated clandestine drug lab and having the team make entry. The room may contain examples of the common booby traps to promote awareness of the dangers present. Law enforcement may be interested in using these facilities to practice hostage rescue and other skills.

No training center would be complete without classrooms. These can be plain or fancy. If the money is available they can include VCRs, DVD players, televisions, satellite reception, and all of the other audiovisual training aids. Chalkboards or dry erase boards are required for drawing out hose lays or doing computations. The advantage of designated classrooms is they are designed to be used in that way. An apparatus room at a station tends to smell like diesel smoke and is not very well-heated or cooled. Tables and chairs are required, and students need to be able to take notes and sit through some classes that may last all day. Adequate lighting is necessary to lessen eye strain. If the local department cannot afford to build its own classrooms, it may be able to borrow space from recreation centers, veterans' halls, or schools on weekends. When fire department classrooms are available, other agencies are amenable to giving fire department personnel free tuition to classes in return for use of the facilities. These kinds of arrangements are often made with such agencies as the U.S. Forest Service (USFS) and the Bureau of Land Management (BLM).

The training facility requires a storage area or engine house to store apparatus and equipment used for training. The training center needs to be furnished with a variety of equipment, such as ladders, hose, and other items. To borrow this equipment from front-line engines every time a drill or academy is held is inconvenient, disruptive, and hard to manage. The engine house is also a good place for training with salvage covers and other pieces of equipment that take up large areas. If firefighters are wet from drilling, the engine house is a good place for them to gather during or after drills to get out of the weather. If they were to gather in the classroom, they would get dirty water and mud on the floors.

As an integral part of the training center, there need to be several hydrants and a **drafting pit**. The fire hydrants can be used for operator training (**Figure 6-6**). Having the hydrants within the training center allows the persons being trained to practice and be tested on **hydrant hookups** without worrying about traffic or adversely affecting an area's water supply. The hydrants can be used in performing drills at the drill tower and the burn building. When training with heavy stream appliances, there needs to be somewhere to discharge upward of a thousand gallons a minute without causing a traffic accident or other damage.

The drafting pit allows for operator training in drafting operations (**Figure 6-7**). The pit is designed so that the water discharged from the engine is directed back into the pit. This feature allows for long periods of pumping without wasting water or creating runoff problems. This pit is used for testing fire engines at draft both annually and after pump repairs.[3]

If the training center is on enough acreage, it can contain a driver training/testing course, often referred to as an Emergency Vehicle Operations Course (EVOC). It is always

drafting pit An open topped underground tank that is used for drafting operations and pump testing.

hydrant hookups Attaching the suction hose from the pumper to the hydrant.

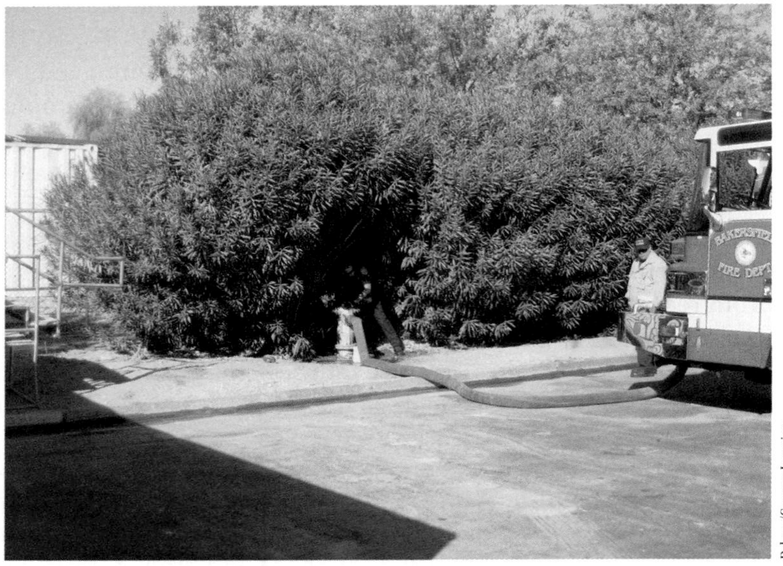
Delmar/Cengage Learning

FIGURE 6-6
Hydrant hookup practice.

Delmar/Cengage Learning

FIGURE 6-7
Drafting pit being used for operator testing.

better to train people on tasks such as emergency stopping and high-speed lane changes somewhere away from other traffic (**Figure 6-8**).

As a function of the extra space, an area can be set aside for drills in trench rescue and structural collapse props. Having these props inside a fenced and locked facility

FIGURE 6-8
Driver training on obstacle course, alley dock exercise.

Delmar/Cengage Learning

allows them to be left set up without fear of some children getting injured playing around them.

The fire department can usually come up with enough donated material to set up props for hazardous materials training (**Figure 6-9**). This would include plumbing props, railroad tank cars, and large tanks. The large tanks can be used for **confined space** rescue training as an added benefit. Many businesses will donate materials in return for access to the props to perform their own training.

In departments spread over large geographic areas, training is often accomplished by sending out a monthly video tape (now commonly placed on CD or DVD). Many fine training programs are commercially available in a video format as well. When the training center is equipped with a studio (**Figure 6-10**) and an audiovisual specialist, the department can make up its own videos. This allows department-specific training programs to be created and duplicated for distribution. Some departments even have their own closed-circuit television channels for presenting training and other information.

The training facility may have offices specifically for the training staff. These offices require copying and word-processing equipment for developing and disseminating training programs and information. Having the staff present at the facility also gives them the ability to provide instructor support. Some training facilities have complete firefighting-related libraries, which allow firefighters, instructors, and students to come to one central location to check out books, videos, and other materials.

confined space A space that is not designed to be occupied on a regular basis that is lacking in natural ventilation.

FIGURE 6-9
Hazardous materials operations training prop.

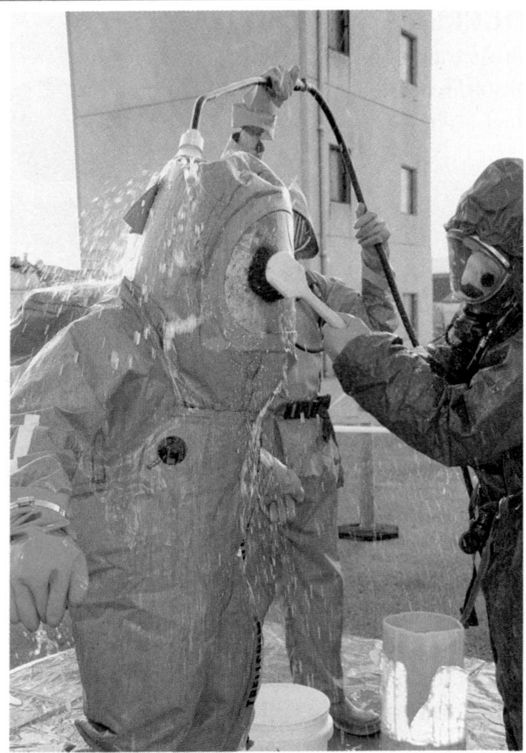

Courtesy of Edwina Davis

Warehouse/Central Stores

The fire department warehouse/central stores center is designed to stock most of the day-to-day needs of the fire department administration, fire stations, and firefighters. All of the materials required, from toilet paper to turnouts, are stored while waiting to be issued to the personnel. By having materials at hand, the department's supply orders can be filled in a timely fashion. The warehouse is also a place to store extra turnouts and other items not currently in use. A stock of these items must be maintained as the time required to obtain them from the manufacturer may be several months.

The central warehouse facility is also a good place to locate the repair facility for self-contained breathing apparatus (SCBA). In larger departments with many SCBAs, a technician is employed with this as his or her main function. This can reduce the cost of going outside for service and provide for shorter downtime of the SCBA because the fire department will be the number one priority. The SCBA for other governmental agencies that use them, such as the corrections department and environmental health department, can also be serviced at this facility. The repair facility may also contain a specialized breathing air compressor for the on-site filling of SCBA bottles

FIGURE 6-10
Electronic media studio for production of training materials.

Delmar/Cengage Learning

(**Figure 6-11**). A regular air compressor of the home or industrial type is not acceptable as it uses oil to lubricate the compressor pistons and makes the compressed air unfit to be used as a source of breathing air. The bottles for the SCUBA of the Search and Rescue Dive Team can be filled here as well.

Communications Center

The fire department receives calls for emergency assistance at the communications center (**Figure 6-12**).[4] Most of the United States now has a 9-1-1 system in place. In a typical situation, the 9-1-1 calls are received by either the local law enforcement agency or the fire department. The dispatcher asks the type of emergency. If the incident requires fire department response, it is routed to the fire department dispatcher. The dispatcher, through the use of a keyboard at the console, enters the incident's location and type into a computer-aided dispatch (CAD) computer system. The CAD system then places the information necessary to properly dispatch the required units on the dispatcher's console screen. Depending on the nature of the incident, it may require one or more

FIGURE 6-11 Specialized air compressor used for filling SCBA bottles.

Delmar/Cengage Learning

FIGURE 6-12 Emergency services dispatch center.

Delmar/Cengage Learning

pieces of apparatus. A vehicle accident may require an engine and an ambulance. An accident with pinned-in victims may be dispatched as the nearest engine, aerial ladder truck, or specialty vehicle with rescue equipment, and an ambulance. If an air ambulance (medical helicopter) or other apparatus is available, it can be dispatched as well.

An enhanced 9-1-1 system is a great improvement over the older systems in several ways. It used to be that the dispatcher received the call and looked up the location on the map. In large or complicated jurisdictions, such as large cities or departments with vast geographic areas, dispatchers were required to have an extensive knowledge of the jurisdiction. Often children and people in distress are not sure where they are and either provide the wrong address or none at all. An enhanced 9-1-1 system is programmed with the address of the land-line phone being used to make the call. This system does not work as well when a cellular phone is used to make the call. The dispatch center usually has the capability of tracking the call to the nearest cell phone transmission tower to the phone being used to make the call, which can provide them with at least a close approximation of the location of the origin of the call.

The older system also required the dispatcher to determine which apparatus/ station(s) to dispatch from "run cards." These cards had to be looked through to find the proper one for the location and type of incident. If the responsible station was out on another assignment, the dispatcher had to determine who was second in. This old system could be very time consuming, especially in a department with a large geographical area, high incident volume, and numerous stations. With the development of modern CAD systems, dispatchers already have the location and the computer tells them whose station area the incident is in and, depending on the type of incident, what apparatus to dispatch. Some systems are also equipped with GPS locators on all of the apparatus so the closest apparatus to the incident can be dispatched, regardless of station location. These GPS systems are called AVL and are described later in this chapter.

Fire Stations

All of the facilities described so far are for the support of the firefighters in the fire station. Fire stations started out as nothing more than a shed to house the fire fighting apparatus and equipment. They then evolved into a place to house the apparatus and equipment and a social hall for the volunteers to gather. With the advent of the paid fire department and permanently assigned personnel, the stations were equipped with living quarters for the firefighters. Today's fire station (**Figure 6-13**) serves many functions: There is the apparatus room for the apparatus and equipment, a kitchen for cooking meals, sleeping quarters for the crew, an office for paperwork and maintaining files, a workroom for maintaining equipment, an area with physical fitness equipment, and rest rooms and showers.

NOTE The public expects firefighters to be professionals and a run down–looking fire station does nothing to enhance our professional image.

Many of the changes being incorporated into the design of the modern fire station are due to the inclusion of women in the fire service and laws relating to handicap access. A newly constructed fire station will probably include separate bedrooms for the crew members instead of the old barracks-style format. At least one of the rest rooms will be

FIGURE 6-13
Modern fire station.

Delmar/Cengage Learning

equipped for handicapped access. The station may be located in an industrial area or in a residential neighborhood, but it is often designed to fit in as well as possible with the surrounding structures. A modern professional-looking office area is included to make the public feel more welcome when it visits the station for permits or other information. One of the main requirements is that the station be well-maintained and clean. The public expects firefighters to be professionals and a run down–looking fire station does nothing to enhance a professional image.

A well-designed fire station is situated on a large lot with enough room for maneuvering fire apparatus and performing training evolutions. The lot may have to be secured to keep people from entering when the fire crews are absent. The apparatus room should be equipped with automatic doors that can be closed by a remote control in the apparatus when the firefighters leave. The apparatus room may also have electric reels for the **motor block heaters** and air hose reels for inflating tires. Some departments are installing exhaust smoke removal systems in apparatus rooms. Ventilated storage cabinets need to be provided for the storage of turnouts of the off-duty personnel. There is also a hose rack for storing extra hose to replace that on the apparatus. There may also be a special air compressor for filling SCBA bottles. Out back is a hose tower for drying hose before storing or reloading on the apparatus.

motor block heater An electrical device that keeps oil in the motor warm and makes for easier starting and helps prevent damage when the motor is started in cold weather.

FIRE APPARATUS

The modern fire service requires many types of apparatus to perform its duties in protecting the community. These types of apparatus vary widely in their design and application. In many instances, apparatus has been modified or specially designed to better perform the required work. There are several basic designs for the specialized apparatus used today.

Cab and Chassis

Fire apparatus manufacturers start out with a cab and chassis (**Figure 6-14**). Depending on the needs and specifications of the buyer, these can vary greatly. The cab and chassis can be either two- or four-wheel drive. Fire apparatus are designed to meet NFPA specifications.[5] To meet these specifications apparatus must provide inside seating for all personnel. Wearing seat belts is law in most states and department policy as well as good safety practice. Today's new engines are likely to be of the four-door cab variety. This development in safety has mostly done away with the practice of firefighters riding on the tail step or running boards of the apparatus. Even in pumpers of the semi-closed cab type, firefighters should never release their seat belts and stand up until the apparatus is stopped and the brake is set. Whenever possible, the best practice is for the firefighters to remain seated and belted in until told to leave the apparatus by their officer. It is very easy to fall from a moving apparatus if you are standing and the driver hits a bump or swerves to miss an obstacle.

For more information on response on a motorized fire apparatus, please refer to DVD clip *Response on a Motorized Fire Apparatus*.

FIGURE 6-14
Fire engine chassis prior to buildup.

Delmar/Cengage Learning

NOTE Fire apparatus are designed to meet NFPA specifications for various apparatus types, such as pumper or aerial apparatus.

Let us now take a tour of the cab area of a standard new piece of fire apparatus. The vehicle is equipped with large mirrors to aid in safe operation. There should also be a fish-eye mirror for pulling up close alongside objects like curbs. Inside the cab is the driver's seat with a large steering wheel (**Figure 6-15**). Mounted on the dash, in front of the driver, are gauges for fuel, temperature, and air pressure in the air brake system. There is a speedometer and tachometer as well. Also on the dash is a push/pull switch that operates the air brakes and a headlight switch.

In pumpers of the semi-closed cab type, firefighters should never release their seat belts and stand up until the apparatus is stopped and the brake is set.

The vehicle will be equipped with either a manual or automatic transmission. When equipped with a manual transmission, there is a switch or lever that disengages the drive train from the rear wheels and transfers the power output of the motor to the pump. The power is redirected through a pump transfer transmission. After the power is transferred to the pump gearing, the road transmission is returned to top gear. When equipped with an automatic transmission, there is a pump switch that engages the pump transfer transmission. Once

FIGURE 6-15
Interior of fire engine cab.

engaged, the road transmission is returned to top gear. With either type of transmission, if, either through oversight or mechanical failure of the switch, the power is not redirected through the pump transmission, the vehicle may lurch forward when the clutch is engaged or the throttle is opened.

Vehicles with a manual transmission have clutch, brake, and throttle pedals. An automatic transmission vehicle has a brake and throttle pedals. If the power plant of the pumper is a diesel, it may also have an engine brake of the "**Jake brake**" type. The use of these devices greatly reduces brake fade and extends the life of brake components on a 25,000- to 40,000-pound vehicle, the weight of a typical pumper. Ladder trucks are even heavier. Fire engines and other fire vehicles lead a tough life accelerating and stopping repeatedly on the way to emergencies in metropolitan areas and when operated in hilly terrain.

In the center of the dash are the switches that control the lights, electronic siren, and/or electric siren. There are switches for the **light bar** on the roof as well as the warning lights on the rear and sides. In most pumpers there is a so-called master switch that permits all of the warning lights to be controlled by one switch, allowing the individual switches to be left in the "on" position. If there are **alley lights** mounted in the light bar, there should be two switches, one for right and one for left. Another light bar mounted device is for preemption of traffic signals. This allows the changing of traffic lights to green in the direction of travel of the apparatus. There is another switch for the lights on the rear, called "pickup" or "hose lights," which are used at night to illuminate the area around the rear of the vehicle for reloading hose or backing up. The electronic siren is equipped with an on/off switch and settings for public address, radio outside speaker, manual, yelp, high/low, and wail. In the manual position, pushing the horn button in the middle of the steering wheel activates the siren. This manual use is handy when making long rural responses in which the siren is needed only intermittently. Some sirens have an electronic air horn capability, which is usually specified on vehicles that do not have an onboard air compressor and therefore cannot support real air horns. The air horns are mounted in the front bumper, to increase their effectiveness, by placing them close to the height of most automobile windows (**Figure 6-16**).

The radio system for the apparatus to maintain contact with the dispatch center and the other apparatus is located in the cab. Modern radios have multichannel capability. The use of multiple channels allows the fire department to operate on several incidents at one time without developing overcrowding on one frequency. There may also be an intercom system, with headsets and microphones for the crew. These allow the members

Jake brake Common name for the Jacobs Engine Brake and other brakes of that type. Used on diesel motors.

light bar Roof-mounted unit containing emergency warning lights.

alley lights Lights mounted in a light bar that shine to the side of the vehicle, commonly used for spotting addresses on structures at night.

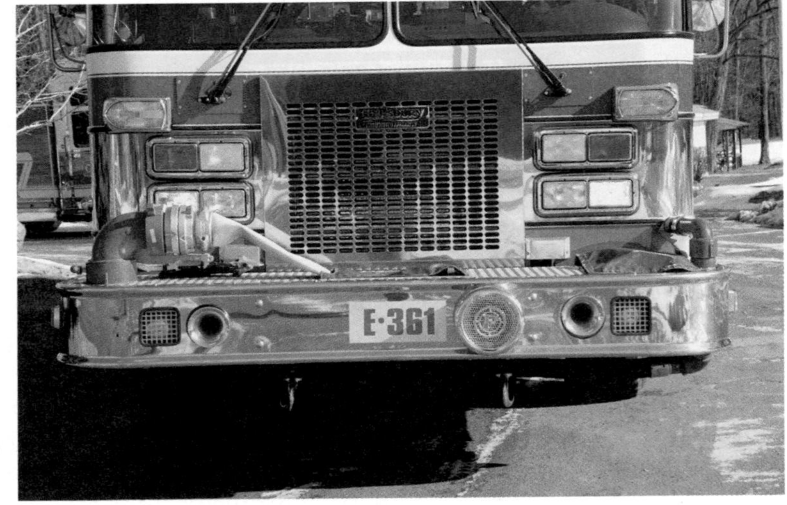

FIGURE 6-16
Front view of fire vehicle showing location of airhorns in front bumper.

Delmar/Cengage Learning

FIGURE 6-17
Operating pumper while maintaining communications through use of headphones.

Delmar/Cengage Learning

of the crew to talk to each other easily over the sound of the motor, siren, and air horn when responding. As an added benefit, they protect the firefighters' hearing. On a piece of apparatus with the firefighters separated from the officer by the back wall of the cab, the headsets allow the officer to give instructions to the firefighters and allow the firefighters to hear the radio traffic and the at-scene description given by the officer. There may be a connection for a headset on the pump panel so the pump operator can hear the radio over the roar of the motor when operating at scene (**Figure 6-17**).

Many fire apparatus are designed with a breathing apparatus mounted between the driver and the officer. This allows the officer to quickly don the breathing apparatus before leaving the cab.

In some models of enclosed cab pumpers, the firefighters ride facing rearward; in others they face forward. The seats in this area can be designed with the back cut out, allowing a breathing apparatus to be mounted where it can be quickly donned when needed.

In areas where summer heat is a factor, the apparatus may be equipped with a built-in air-conditioning system, which allows the firefighters in full turnout gear to stay cool when making long responses. It also makes riding in the apparatus much more comfortable on routine assignments.

The cab portions vary widely depending on the specifications and financial resources of the purchaser. Some fire departments have custom pumpers with an **ambulance gurney** mounted crosswise in the cab, giving the engine patient transport capability. Other vehicles have a cab with a walk-through design and enough room to contain a mobile command post.

Many fire department vehicle cabs are now equipped with **Mobile Data Computers (MDC) (Figure 6-18)**. These computers allow personnel to connect to the Computer-Aided Dispatch (CAD) system. The units provide a connection to gather and update incident information. They can provide information about addresses, such as known

ambulance gurney The wheeled cot that patients are placed on prior to transport in an ambulance.

Mobile Data Computer A computer mounted in the apparatus and connected to an antenna to provide and receive CAD information.

FIGURE 6-18 Mobile data computer (MDC) terminal for retrieving information while en route and at scene.

Delmar/Cengage Learning

hazards, owner contact information, and so forth. An on-board computer can contain a basic street map with overlays of hydrant locations, pre-incident plans, sewers and storm drains, and other information. To make the information more user friendly, the layers can be turned on and off as needed. The units are usually equipped with touch screens so the operators can provide information at scene as to availability, without using the radio—thereby minimizing voice traffic on sometimes crowded channels and reducing the possibility of a message being missed or having to be repeated.

In addition, automatic vehicle location systems (AVL) are often used in conjunction with MDCs. The unit in the apparatus receives GPS signals, and once their location is computed it is provided as a graphic display to the dispatch center CAD and in the apparatus. This allows the closest resource response to be generated. They also come with routing capability that displays the route to the scene. The route information can include and be based on one-way streets, school zones, construction zones, highway divider walls, speed limits, and other features that may restrict response to a location. When the incident is entered into the CAD, the computer utilizes the AVL and routing information to determine the closest appropriate resource.

Both MDC and AVL assist units in getting to the scene more quickly. Fires grow very rapidly and in the case of medical emergencies and rescues, time can be the difference between life and death.

Cellular phones and fax machines are just a logical forward step. With the high number of accidents that happen when backing up, video cameras aimed over the rear, like those mounted on motor homes, are coming into use. At the fire scene, one of the greatest resources the officer has is pertinent and up-to-date information. As more sophisticated devices become available for the storage and retrieval of information and communications, they will find their way into the cabs of fire apparatus.

The standard for fire engines includes two individual battery sets of the truck type, allowing for additional starting amperage as well as a backup in case one set goes dead. It also allows for increased storage capacity for the tremendous draw placed on the electrical system due to the warning lights. A switch in the cab allows the batteries to be turned off and one or both sets of batteries to be used at a time.

Motor

Fire apparatus today are mostly powered by diesel motors, noted for their long life and durability under tough conditions (**Figure 6-19**). Diesel motors are selected for their abundance of torque. It takes a lot of power to operate a large-gallon-per-minute pump and supply effective hose streams. The motor must also be able to propel a heavy vehicle to operating speed in a short period of time. When used in hilly terrain, a diesel motor with both a turbocharger and a supercharger are not uncommon.

The motor is often equipped with an oversized alternator to supply power for all of the extra lights used as warning devices. If the motor is left at idle with all of the emergency

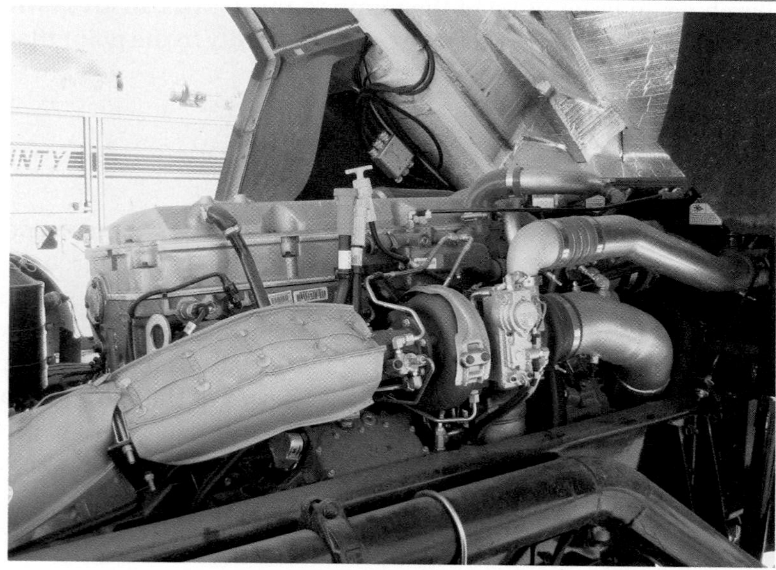

FIGURE 6-19
Diesel motor installed in fire engine. With turbocharger to boost horsepower output.

Delmar/Cengage Learning

lights operating for any length of time, it can drain the batteries. The oversize alternator must be turning at around 1,000 rpm to develop enough amperage to operate all of the additional lights and other electrical equipment on the vehicle. Apparatus are equipped with a high-idle switch that automatically increases the idle speed when engaged. Most apparatus are also supplied with an **inverter** that allows 110-volt lighting to be used without starting the onboard generator.

inverter An electrical device that converts 12-volt current to 110 volt. Used to operate lights and tools from vehicle's charging system.

Modular Apparatus

Some departments use modular apparatus. By having different modules that are mountable on the chassis, the department gains flexibility. An example would be a cab and chassis with a dismountable body stocked with hazardous-materials or heavy-rescue equipment, giving the department the capability of having less of the expensive parts of a truck—the cab and chassis—and several choices as to which bodies to mount as the need arises. The U.S. Forest Service and Bureau of Land Management have employed this concept for many years by utilizing flatbed trucks with self-contained pumper bodies mounted on them. This concept does not work out with a regular pumper because of the plumbing and mounting of the pump to the chassis. In areas where a large water tank is carried, it also tends to raise the center of gravity to the point that the apparatus is limited to on-road use only.

Pumper/Engine

The basic piece of motorized apparatus in the fire service is the pumper. (For the purposes of this text the terms "pumper" and "engine" are used interchangeably.) These apparatus are designed to meet NFPA 1901, *Pumper Fire Apparatus* specifications. Most pumpers in service are of the triple combination type. A triple-combination pumper is so named because it carries hose, a pump, and has a water tank. Other apparatus is carried as the situation dictates. All of these attributes as well as the cab and chassis can be combined in various forms, according to need. Some jurisdictions have purchased specially designed pumpers with an extendable mounted ladder or boom that gives them the capability of applying elevated streams.

NOTE Most pumpers in service are of the triple combination type.

Water Tank

Water tanks on fire pumpers vary in size. On small apparatus and metropolitan engines, where water is readily available from the hydrant system, tanks tend to be around 200 to 500 gallons. In rural areas tanks range from 750 to 1,500 gallons (**Figure 6-20**). Any more than this and the apparatus is considered to be a water tender/tanker. Once the tank exceeds 1,000 gallons, the vehicle usually rides on three axles. The tanks are equipped with **baffles** to prevent the water from shifting around and causing the vehicle to become unstable.

As water tanks increase in size and vehicle length and width stay the same, the only way to go is up. Standard-size engines with large tanks tend to have hose beds high in the air, making them harder to access (**Figure 6-21**). A roll of wet 2½-inch hose weighs approximately 60 pounds and is hard to lift into the hose bed to carry it back to the station. The higher the hose bed, the harder it is to load the wet hose. The large tank adds to the overall weight of the apparatus and going up in height raises the center of gravity, making the engines more top heavy and likely to tip over when operating in sidehill situations or when performing evasive maneuvers.

baffles Partitions placed in tanks that prevent the water from sloshing and making the vehicle unstable when turning corners.

FIGURE 6-20
Three-axle engine with 1,500 gallon water tank.

Delmar/Cengage Learning

FIGURE 6-21
Contrast of hose bed heights due to tank size.

Delmar/Cengage Learning

The large tank adds to the overall weight of the apparatus and going up in height raises the center of gravity, making the engines more top heavy and likely to tip over when operating in sidehill situations or when performing evasive maneuvers.

Plastic is a very popular material for water tank construction. Metal tanks, which have traditionally been used, tend to corrode over time and develop leaks. Plastic tanks are corrosion resistant and stand up better to wetting agents and foaming agents added to water to improve fire fighting characteristics.

Foam Systems

More and more apparatus are being equipped with built-in foam systems. These can be either class A, class B, or both. A built-in class A system provides superior knockdown on ordinary combustible materials. Class B systems are for use on hydrocarbon fuels (e.g., gasoline, diesel fuel, and jet fuel). The systems consist of a built-in foam concentrate tank, injector, and a means of adjusting the concentrate level delivered into the water stream. Many also come with a flowmeter to keep track of how much water has flowed and how much foam concentrate has been used.

Another variation of the class A foam system is the compressed air foam system (CAFS). This system utilizes an air compressor to inject air into the hose stream as it leaves the pump. Combined with class A foam this provides a light and airy foam that can stick to vertical surfaces. It is often used to pretreat structures and trees in the path of oncoming wildland fires, giving personnel the opportunity to protect structures without actually remaining at the site, directly in the path of an advancing fire. A word of caution here is that the nozzle reaction from a hose connected to a CAFS pumper is much more than that of a regular pumper at the same pump pressure due to the release of air pressure in the hoseline when the nozzle is opened.

Pumps

The main purpose of any pump is to lift water or to add pressure to the water so it can flow through hose and nozzles and be applied to the fire away from the pump. To deliver the contents of a reservoir or water tank to the third floor of a building requires some device to force the water through the hose and appliances.

NOTE The main purpose of any pump is to lift water.

Centrifugal Pump

The most commonly used main pump on fire apparatus is the centrifugal pump. It has many desirable features from the firefighting standpoint. The centrifugal pump consists of one or more vaned wheels, called "impellers," mounted on a shaft (**Figure 6-22**). Power is supplied to the pump from the motor of the pumper through a pump transmission. The transmission can consist of a transfer case or a power takeoff unit. If equipped with a transfer case, the power is redirected from the rear wheels to the pump and the pumper

FIGURE 6-22
Fire pump cutaway mounted for display.

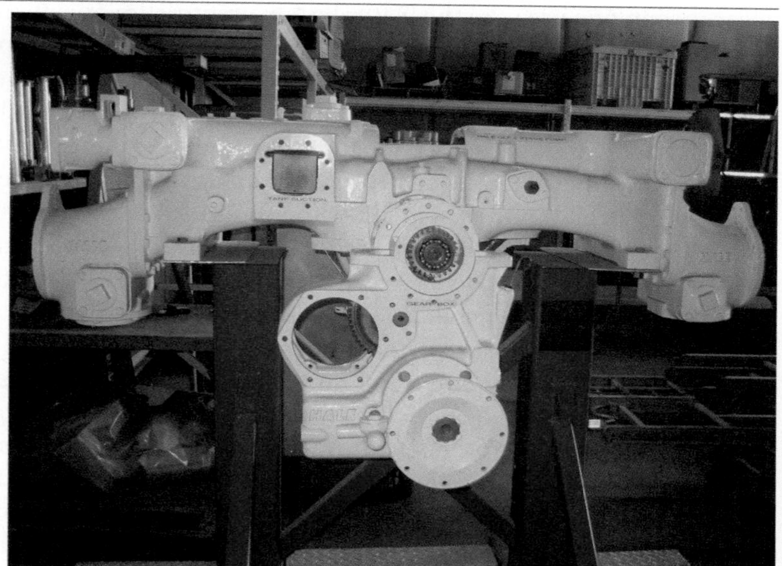

Delmar/Cengage Learning

stays stationary when the pump is operated. In the case of a power takeoff (PTO) unit, the pumper can pump and roll at the same time. This type of construction is often found on wildland fire pumpers so mobile attacks can be made. The limitation with this set up is that the pump operates at engine speed so driving at low speeds reduces pump pressure as RPM is kept low. A third option often found on engines used for wildland firefighting is a separate motor carried for operating the pump. This option also allows for the use of the pump while moving. The pump speed is determined by a separate throttle allowing for high pressure at low vehicle speed.

NOTE The most commonly used main pump on fire apparatus is the centrifugal pump.

The pump casing has one or more suction inlets where water can enter the pump. The water is then directed into the center of the impeller, called the "eye." As the pump impeller spins on its axle, it directs the water to the outside of the casing, imparting centrifugal energy to the water, hence the term "centrifugal pump." The principle is the same as a merry-go-round. As you stand in the middle of the disk and it spins faster and faster, you are pushed toward the outside. The centrifugal pump does the same thing. The impeller consists of a two-sided disk with vanes between the disks. This design better aids the impeller in transferring the centrifugal energy to the water. When the water leaves the impeller, it collides with the pump casing, resulting in increased pressure in the pump. The casing, called the "volute," is designed as a modified circle with a wider clearance in one area. As the water builds up pressure, it is forced out of the volute and into the discharge plumbing. The outflow of the water creates a partial vacuum at the eye of the impeller, drawing more water into the pump through the suction plumbing. The discharge plumbing extends out through the pump panel on the side of the apparatus and has valves and threaded ends for the connection of fire hoses (**Figure 6-23**).

The advantages of the centrifugal pump are numerous. The clearances between the impeller and the pump casing allow the pump to do several things. It can spin at high rates of speed and build up large amounts of pressure without discharging any water. This situation occurs when discharge valves and nozzles are turned off and on at the fire scene. There is not always someone available to stand by the pumper and operate the throttle as the volume and pressure demand increase and decrease. One must be careful because the water can become quite hot if no water is circulated through the pump for a while, depending on the speed, possibly damaging the pump or causing a scald burn to personnel when a nozzle is first opened. Another feature is that a centrifugal pump can take advantage of any pressure coming in on the suction side, either from another pumper or a hydrant, effectively allowing the motor driving the pump to work less hard. Centrifugal pumps can also tolerate the pumping of trash and dirty water to a certain extent. They are equipped with suction screens to keep out debris and rocks that can damage the pump (**Figure 6-24**).

One must be careful because the water can become quite hot if no water is circulated through the pump for a while, depending on the speed, possibly damaging the pump or causing a scald burn to personnel when a nozzle is first opened.

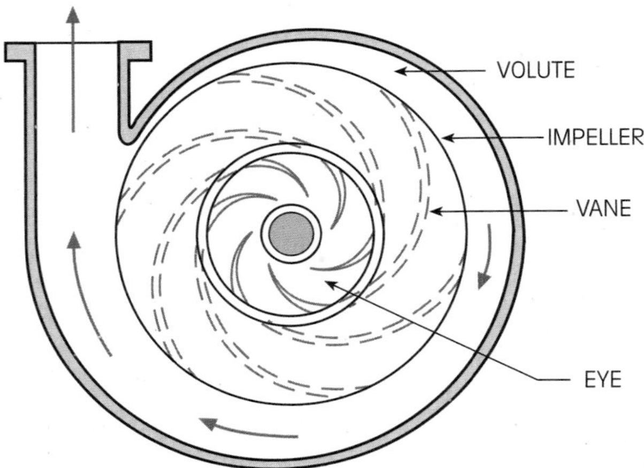

FIGURE 6-23
Interior design of centrifugal pump.

Delmar/Cengage Learning

FIGURE 6-24 Pump suction inlet screen to prevent ingestion of rocks and other debris into pump.

Attached to the pump is a device provided for the firefighter's safety known as the *relief valve* or *pressure governor*. The pressure governor reads the pressure provided to the hose lines. The maximum pressure is preset at the pump panel. If a nozzle is shut down or flow is otherwise restricted, the pressure governor adjusts the throttle setting to the pump motor to keep the remaining lines from exceeding the preset pressure.

NOTE A centrifugal pump can take advantage of any pressure coming in on the suction side.

The relief valve is another type of semiautomatic pressure-regulating device. This spring-operated valve is set at the desired pressure by a handwheel on the pump operator's panel. When the handwheel is turned to the right, the spring is compressed. The more it is compressed, the higher the actuating pressure is raised. When actuated, some of the water is redirected from the discharge side of the pump back to the suction side. When the pump is operated with two or more lines coming from it, if one of the lines were to be shut down, all of the water would try to exit through the line remaining open, which would cause a sudden pressure surge in the open line. A pressure surge could cause a firefighter to lose footing or fall from a ladder. The relief valve effectively reduces this pressure surge.

The relief valve is provided for the firefighter's safety.

The disadvantage of the centrifugal pump is that it can only act on the water that enters it. It cannot draw water into itself from a **static water source**. The water must be introduced under slight pressure, from a hydrant or other pumper, or another type of pump must be used to create a vacuum in the pump casing, causing the water to enter. Once the water does enter, the discharge of water from the pump can keep the vacuum going and the suction is self-sustaining. If the pump is driven too hard and the suction capability is exceeded, the pump will start to cavitate (to form small vapor bubbles in the interior), causing damage to the impeller.

Main fire pumps of the centrifugal type come in sizes ranging from 250 to 2,500 gallons per minute, in increments of 250 gallons per minute. The minimum recognized fire pump is 500 gallons per minute.

NOTE The centrifugal pump can only act on the water that enters it.

static water source Pond, lake, or tank used to supply fire engines.

Positive Displacement Pumps

The other type of pump mounted on fire apparatus is the positive displacement pump (**Figures 6-25 A and B**). This type of pump can come in several forms: gear pumps, piston pumps (like the hand pumpers), and diaphragm pumps. The principle is that every time the pump cycles, a specified amount of fluid is taken in and discharged. If one gallon of water enters on the suction side at the start of a cycle, one gallon will be discharged from the pressure side at the end of the cycle. As the rate of speed increases, the volume will increase in direct proportion.

NOTE The other type of pump mounted on fire apparatus is the positive displacement pump.

These pumps have the advantage of being able to pump air, thus making them self-priming. When piggybacked onto a centrifugal pump, the positive displacement pump can evacuate the air in the centrifugal pump. When used for this purpose, it is called a priming pump because it "primes" the centrifugal pump.

This action creates a reduction of pressure in the centrifugal pump, which allows water to be drawn up the suction hose and into the pump. Once the water enters the pump, pressure is added by the pump and a fire stream can be developed (**Figure 6-26**). This occurs because the weight of the atmosphere over the earth exerts approximately 14.7 pounds of pressure per square inch on the surface at sea level. When the atmospheric pressure inside the pump is reduced below 14.7 pounds per square inch, the water is forced up the suction hose and into the pump.

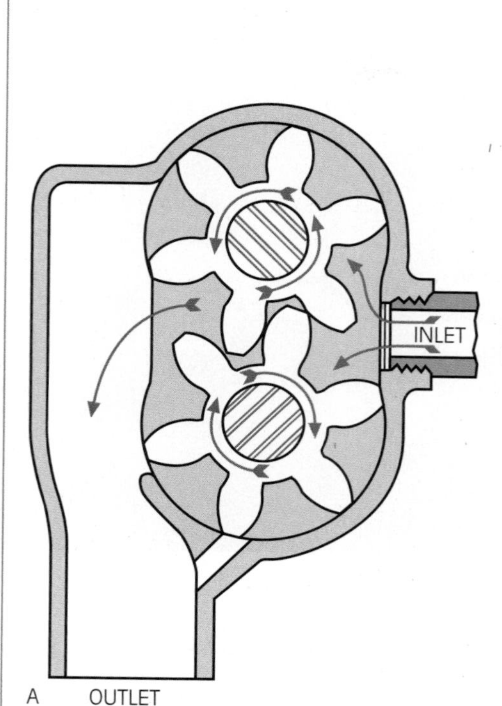

FIGURE 6-25
Gear-type positive displacement pump: (A) interior components, (B) exterior view.

NOTE These pumps have the advantage of being able to pump air, thus making them self-priming.

Because the centrifugal pump cannot pump air due to its loose tolerances, the positive displacement pump is needed to create the vacuum. The positive displacement pump is very small in relation to the centrifugal pump and is typically driven by an electric motor the size of an automobile starter motor. On some wildland firefighting engines, the priming pump is hand operated. Another advantage of positive displacement pumps is that they can create tremendous pressure when pumping fluids. For applications where high pressure is necessary, such as pressure washers, they are the pump of choice.

The disadvantages of positive displacement pumps are enough that they are not used as main fire pumps. If a positive displacement pump was being used and the nozzle were turned off, the pressure in the hose would increase until something blew out. The tolerances are so close that very small debris can jam the pump. They also do not gain any benefit from water forced into them. Because of its design, a positive displacement pump of the same gallons per minute would be much heavier and more expensive than a centrifugal pump.

FIGURE 6-26
Pumper operating at draft, taking suction from static water source.

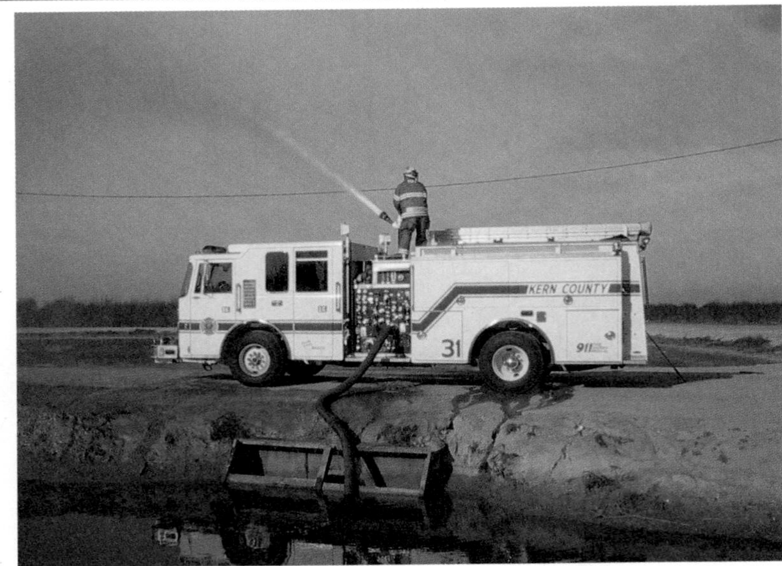

Delmar/Cengage Learning

Aerial Ladder and Elevating Platform Apparatus

Aerial ladder apparatus comes in two configurations.[6] There is the tractor-trailer type with tiller steering, with the tiller operator sitting in a small cab at the rear of the apparatus and able to steer the rear wheels (**Figure 6-4**). The advantages of the tractor trailer and tiller types are that they can maneuver in tight spaces and make sharp corners. The disadvantages are that if the driver goes too fast, the tiller operator can lose control and collide with parked cars or, even worse, people standing on the sidewalk. The other configuration is the straight chassis (**Figure 6-27**).

The aerial portion of the apparatus can come in various configurations as well. There are the extendable ladder types, in which the ladder raises from the bed and a set of fly sections are extended. These ladders are advantageous in that personnel can ascend and descend the ladder when it is raised. On this type of apparatus, the ladder placement is controlled from the operator's platform at the base of the ladder. Some of these apparatus are equipped with an enclosed platform at the ladder tip from which personnel operate. This type is known as an aerial ladder platform apparatus. The **articulated boom** type is another design, in which the boom is raised hydraulically and extended through adjusting the angle of a knuckle joint (**Figure 6-28**). This type of aerial apparatus, while very strong and easy to place from elevated platform controls, is limited in that personnel have to move the basket to the ground to enter or exit.

articulated boom Elevating device consisting of a boom that is hinged in the middle.

FIGURE 6-27
Aerial ladder truck with ladder extended.

Delmar/Cengage Learning

FIGURE 6-28
Articulating boom aircraft firefighting vehicle.

Courtesy of Jeff Riechmann

Elevated platforms are good for using heavy tools from the basket, such as a modified jackhammer for breaching walls. In addition to the aerial ladder carried on the truck, carrying 108 feet of ground ladders is required for the apparatus to be classified as a ladder truck.

If the driver goes too fast, the tiller operator can lose control and collide with parked cars or, even worse, people standing on the sidewalk.

The ladder truck is often equipped with an intercom system connecting the person at the tip or in the basket with the operator. Some apparatus come with breathing air cylinders mounted to the basket so the personnel can use them as a supply instead of SCBA bottles, which gives them a much longer supply of air. Some aerials have plumbing supplied so water can be pumped to the tip and applied through elevated streams. This design is called a *water tower.*

Aerials may be equipped with a fire pump on the apparatus and on others an engine is used to pump the fire stream.

Quint

An apparatus equipped with pump, water tank, ground ladders, hose bed, and an aerial device is called a *quint*. Such apparatus are used as a pumper and ladder truck combination.

Squads

Squad vehicles are the specialty vehicles of the fire service. Just about any time some special configuration is needed for a specific purpose, the vehicle is called a squad. Squads are usually strategically located and respond upon request. In departments that provide advanced life support medical functions without the transportation capability of an ambulance, the vehicle may very well be called a medical squad.

NOTE Squad vehicles are the specialty vehicles of the fire service.

Another form of squad is the special lighting vehicle equipped with a high-wattage generator and numerous removable lights and extension cords. Hazardous materials vans and vehicles, designed for a specific purpose and outfitted with equipment for a specific function, fit the description of squads (**Figure 6-29**).

Special air units are also squads. They are equipped with extra SCBA bottles and equipment. Some are equipped with the special compressors for breathing air or **cascade systems**. A squad with a compressor or cascade system allows the filling of SCBA bottles at the scene. These vehicles are now often combined with a large wattage electrical generator and the combination is often called an "air and light" squad or unit.

cascade system A system of large compressed gas cylinders connected to a manifold.

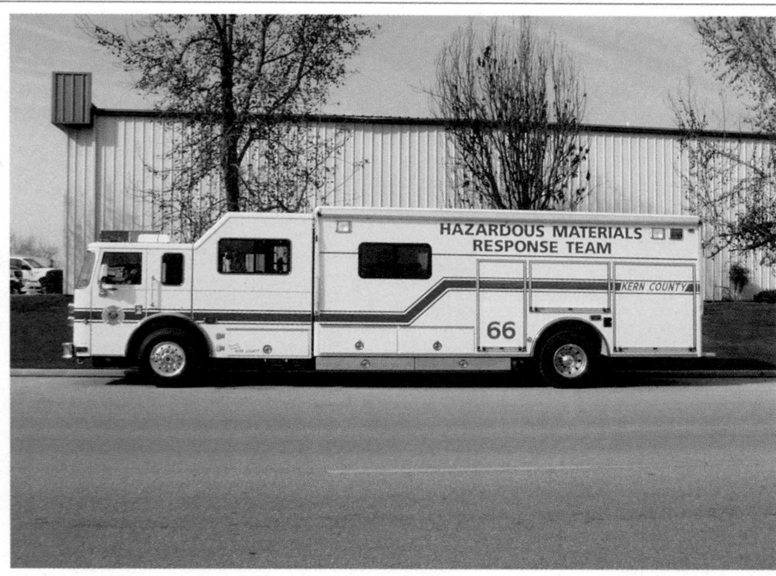

FIGURE 6-29 Hazardous materials response team vehicle.

Courtesy of Edwina Davis

A mobile command post, activated on large assignments, can be called a squad. **Tactical support** and **rehab** vehicles fit the same criteria (**Figure 6-30**).

A time in which firefighters rest, cool off, and drink liquids to replenish their body fluids

A squad type vehicle used in wildland fire fighting is a terra torch (**Figure 6-31**). This unit has a tank full of jellied gasoline, similar to napalm, that is squirted from a special nozzle equipped with an igniter. The terra torch is used to light **backfires** and perform **burning out** operations.

tactical support A vehicle equipped to provide the needs of firefighters at the emergency scene. See rehab.

rehab Short for rehabilitation.

backfires A fire lit in front of an advancing fire to remove fuel and widen control lines.

burning out Lighting a fire to remove fuel along the flanks of a fire. Also used to remove unburned islands that remain as the fire advances.

Aircraft Rescue Fire Fighting Apparatus (ARFF)

A type of apparatus specially designed for airport firefighting is the ARFF unit, also called crash fire rescue (CFR) apparatus (**Figure 6-32**).[7] These apparatus are designed with large water tanks and built-in foam tanks and systems, and are all-wheel drive for going off

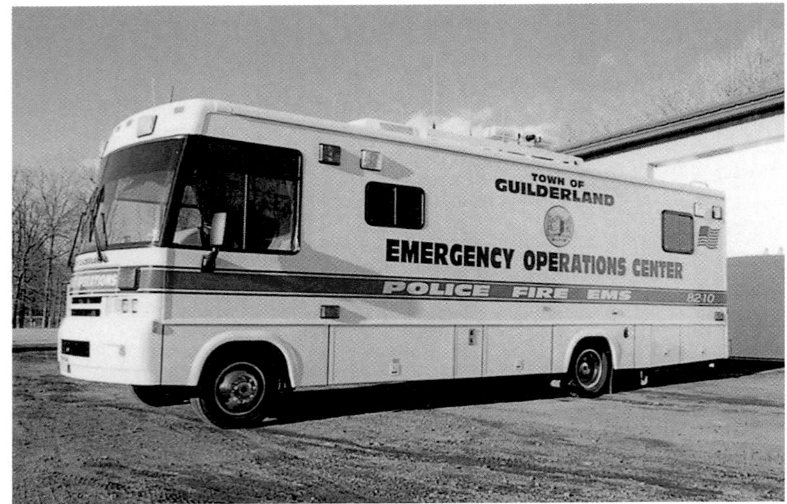

FIGURE 6-30
Tactical support vehicle.

Delmar/Cengage Learning

FIGURE 6-31
Terra torch igniting ground fuels at wildland fire.

Courtesy of Edwina Davis

runways. They are equipped with **turret nozzles** on the roof, forward-facing nozzles on the front bumper, and **ground sweep nozzles** to keep fire from beneath them when driving through burning fuel. All of these nozzles can be remotely controlled from inside the cab, making them a very effective firefighting combination with only one person on board. They

turret nozzle Roof or bumper mounted nozzle remotely controlled from inside the cab.

ground sweep nozzles Nozzles mounted underneath apparatus to sweep fire from under the vehicle.

FIGURE 6-32
Aircraft rescue firefighting specialty vehicle (ARFF).

Courtesy of Jeff Riechmann

are also equipped with attack lines that can be pulled for firefighting away from the vehicle, such as interior attack in a large aircraft. The apparatus has the capability of pumping and rolling at the same time. There are suction inlets on the side of the vehicle that allow them to connect to fire hydrants or fire engines as the need arises. They are equipped with a minimum of ladders that are used to gain access to aircraft wing surfaces and interiors.

Another type of aircraft firefighting apparatus is mounted on a standard truck chassis and is equipped with a "twinned system" that allows foam and/or dry chemical extinguishing agent to be applied at the same time. The system has large tanks of dry chemical agent and expellant gas mounted on the apparatus. There is a water tank, foam concentrate tank, and pump for the foam system. These two extinguishing agents are discharged through hose mounted on a reel with two nozzles connected.

FIRE TOOLS AND APPLIANCES

The fire service utilizes many types of tools and appliances to combat fires and perform rescues and other tasks. Those illustrated here are the more common items, but not a complete listing.

Hose

Fire hose is used for getting the required water from the source of supply to where it is needed to control the fire. Fire hose is constructed in several different ways.[8] The older type is rubber lined with one or more cotton jackets. The rubber liner prevents leaks while the cotton jacket(s) give the hose resistance to rupture under high pressure, provide abrasion resistance, and allow it to maintain flexibility. The problem is that hose of this type is heavy and requires thorough drying inside and out after use (**Figure 6-33**). When used, the hose must be rolled up, loaded on the engine and returned to the station to be dried to prevent acids forming on the inside and mildew on the outside. This condition makes the hose labor intensive and requires a complete hose change for the engine to be held in reserve to replace the hose that is drying.

For more information on fire hose, please refer to DVD clips *Fire Hose & Appliances* and *Service Testing Fire Hose.*

Now in common use is synthetic hose that has been developed that has no rubber liner and will not mildew. This hose is lighter in weight, more flexible, and can be reloaded on the engine at the fire scene. Using this type of hose reduces hose inventories required,

FIGURE 6-33
Hose drying tower.

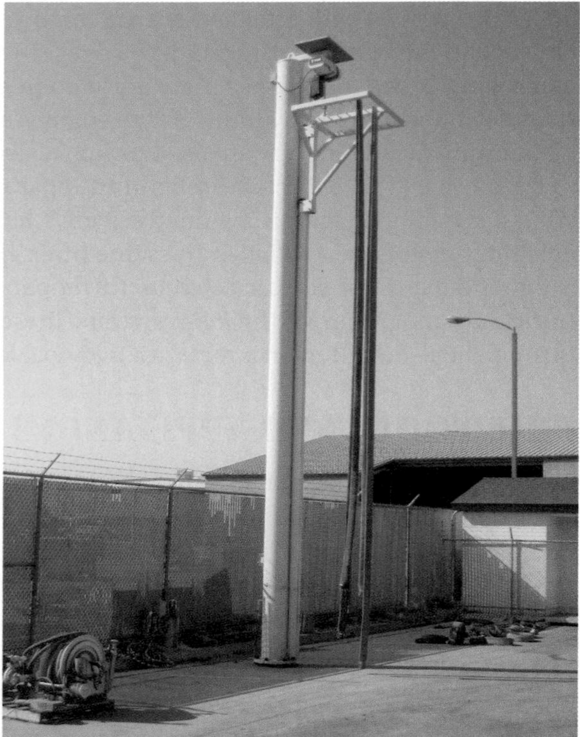

Delmar/Cengage Learning

weighs less, takes up less space in the hose bed of the engine (allowing more to be carried), and saves time. When returning from a fire with cotton hose, it is in rolls in the hose bed; synthetic hose is reloaded and ready for the next assignment. Many times an engine has been diverted to another incident before it was able to return to the station for a fresh hose load.

To connect the hose together each end has a coupling.[9] One end has exposed threads and is called the male end. The other has a swivel with interior threads and is called the female end. Couplings have traditionally been made of brass but are being replaced by pyrolite. Pyrolite is a lighter-weight metal than brass and more resistant to bending. Couplings are most commonly of either the threaded type with national standard thread or of the quick connect (Storz) type (**Figure 6-34**). The exception to this is the one-inch hose used in forestry firefighting: It has either national standard or, most commonly, iron pipe thread. Hose comes in either 50- or 100-foot lengths.

NOTE Not all departments use the national standard as their thread type. Some use thread types unique to their jurisdiction. This can lead to difficulty connecting pumpers and hose from other jurisdictions to their apparatus.

The hose load carried on a pumper is determined by the type and size of fires expected to be encountered. The hose is carried in the area of the pumper known as the hose bed. It is designed for easy access and laid out according to purchaser specification. The hose bed is often equipped with a cover to keep water, burning embers, and other debris off the hose.

FIGURE 6-34
Storz quick connect fire hose coupling on large diameter hose (LDH).

NOTE The hose load carried on a pumper is determined by the type and size of fires expected to be encountered.

Attack lines must be able to supply sufficient amounts of water (gallons per minute), while not being so heavy or rigid that they cannot be maneuvered (**Figure 6-35**). Standard attack lines for structural firefighting are 1½- or 1¾-inch with 1½-inch couplings. Hose of this size offers good flow, in the range of 100 to 200 gallons per minute, while retaining ease of mobility. Attack lines are laid on the pumper so that they can be pulled by one person, advanced, and placed into operation as rapidly as possible. Supply hose is laid in the hose bed so that it can be fed onto the ground as the engine drives forward (**Figure 6-36**). This makes it easier to establish a hose lay in a short period of time.

Supply line hose is designed to be laid out and not moved around very much, especially when full of water. The past standard was the 2½-inch hose. Its attributes were that it could flow respectable amounts of water as supply line as well as having the capability of being used as an attack line when necessary. A jurisdiction with mostly rural areas would more likely carry 2½-inch supply lines. This allows for more linear feet of hose to be carried to facilitate the longer hose lays needed in an area with long

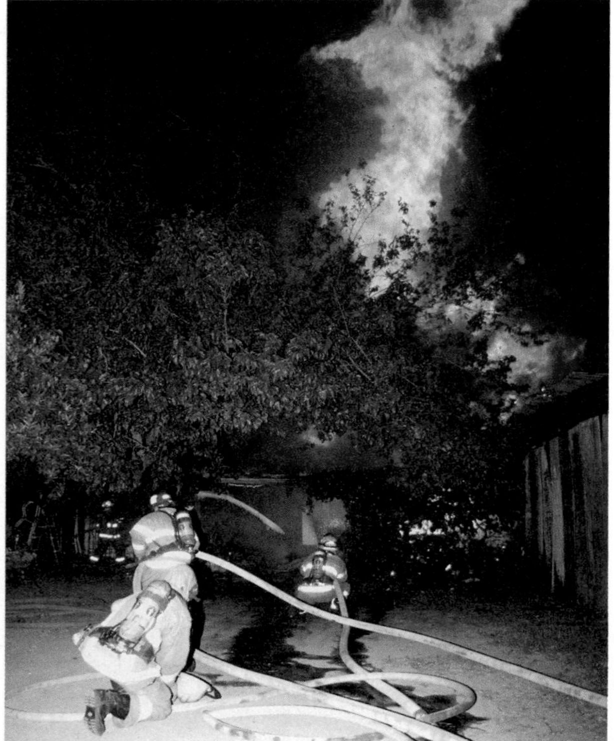

Courtesy of Edwina Davis

FIGURE 6-35 Firefighters advancing hoselines to attack a structure fire.

FIGURE 6-36
Firefighters deploying supply line from hose bed.

Delmar/Cengage Learning

distances between water supplies. A standard pumper should be able to carry around 2,000 feet or more of 2½-inch hose in its hose bed. With a three-pumper relay, one pumper at the water source, one in the middle of the hose lay, and the third pumper at the fire, it would be possible to supply water for over a mile. As more pumpers are added, it is theoretically possible to extend the hose lay indefinitely. The relay pumpers are needed to boost the pressure as a 1,000-foot hose lay at 250 gallons per minute would require approximately 125 pounds of pressure to overcome the friction of the water going through the hose. A modern compromise in this area is the three-inch hose with 2½-inch couplings. It can flow more water with less friction loss, without being too large to manage when charged. The 2½-inch couplings allow it to be used with standard 2½-inch hose (**Figure 6-37**).

In a metropolitan or industrial setting, the emphasis is on larger diameter hose (LDH) for supply lines, four and five inch, to better supply large volumes of water to large fires. This hose takes up more space per linear foot than two and half-inch hose, therefore reducing the total length of hose that can be carried on a standard pumper. With four-inch diameter hose, a thousand gallons of water per minute can be pumped a distance of 1,000 feet with only 20 pounds of pressure lost due to friction of the water against the inside of the hose. A hose five inches or more in diameter would have even less pressure lost to friction. This large diameter hose comes in 100-foot lengths and a length of it wet and rolled up may weigh over 100 pounds, making it very hard to hand up into the hose bed. You can imagine how much work it would be to pick up a 2,000-foot hose lay. These size lines are where synthetic hose is the most appreciated. It is rolled to remove the air and then unrolled as it is loaded back into the hose bed. There is no need to lift the full rolls into the hose bed for transport.

FIGURE 6-37
Hose laid in hose bed (left to right): 1¾-, 2½-, and 4-inch hose. The hose is laid in the hose bed in a "flat lay."

On engines that are used primarily for wildland firefighting, one- and 1½-inch hose are carried. The smaller diameter is used because the hose lays must often be put in by hand over rough terrain. The hose is carried in rolls or packs, allowing for easier carrying (**Figure 6-38**). Wildland firefighters must master the skill of extending a hose lay by adding lengths of hose while actively fighting the fire as they advance.

NOTE The smaller diameter is used because the hose lays must often be put in by hand over rough terrain.

Some pumpers are equipped with a hose reel that contains either ¾- or one-inch hard rubber line (**Figure 6-39**). This is used for controlling small fires and saves time because it can just be rerolled on the reel and is ready to go.

The hard suction hose is rubber or plastic with wire wrapping on the inside that makes the walls stiff (**Figure 6-40**). When a vacuum is created inside the hose, it will not collapse. When drafting water from a static source, an engine requires a hose that will stay rigid and let the water through. Hard suction hose is carried on apparatus for taking water from swimming pools, reservoirs and other static sources. It comes in 10-foot lengths

FIGURE 6-38 Rolled hose set up for quick deployment at wildland fires.

Delmar/Cengage Learning

FIGURE 6-39 Firefighter deploying hard rubber line from hose reel.

Delmar/Cengage Learning

and is typically 2½ or 4½ inches in diameter. For 1,500-gallon-per-minute pumpers it is required to be six inches in diameter. This size is rarely carried because it is extremely heavy and hard to use. A certain amount of volume is sacrificed for ease of operation and 4½-inch hose is often carried instead.

FIGURE 6-40
Hard suction hoses, 2 ½- and 4-inch diameters.

Delmar/Cengage Learning

Another method of taking water from static sources is the floating pump. It is a small gasoline-powered pump mounted on a floating housing that will pump a 1½-inch line at around 90 gallons per minute. Some apparatus also carry nonfloating portable pumps (see **Figure 6-57**) that are set up at the site and draft water, pumping it through a hose line into waiting vehicles or portable tanks.

Nozzles

After the water leaves the pump and travels through the hose it is applied to the fire through the use of nozzles. There are nozzles for most sizes of hose, ranging from garden hose to master streams. The nozzles used for wildland firefighting are for one-inch diameter hose. They turn on and off and adjust the stream by rotating the nozzle head in relation to the base. They flow approximately 23 gallons per minute and can be used in a straight stream or fog pattern (**Figure 6-41**). A quick shutoff can be added for conserving water. They are made of aluminum to save weight. Another hose used in wildland firefighting is 5/8 inch in diameter (garden hose sized) and is commonly referred to as

FIGURE 6-41
Firefighters operating nozzles showing straight stream and fog patterns.

Delmar/Cengage Learning

"pencil hose." It allows the distribution of small amounts of water while being light in weight (synthetic) and easily carried and moved. It is often used for mop up operations as it does not flow enough water for attacking flames of any great intensity. An added advantage is that it requires standard garden hose fittings and nozzles which are lightweight and inexpensive.

The nozzles used on 1½-, 1¾- and 2½-inch attack lines are made of chrome-plated brass or pyrolite. They are called combination nozzles if they have the capability of straight stream or fog patterns. Combination nozzles may have an adjustment ring on them for controlling the amount of water they will flow per minute. This adjustment would typically be from 60 to 125 gallons per minute on a 1½-inch nozzle and 125 to 250 gallons per minute on a 2½-inch nozzle. Combination nozzles have rubber around the head so they will not be damaged if bumped into things (**Figure 6-42**). Nozzles are not to be used to break out windows. This can embed glass in the rubber and cut your hands when you adjust the stream. A better way is to spray water on a hot window pane, causing the pane to break, or use the proper tool for the job, such as an axe or pike pole.

Others have straight bore tips that project a solid stream. To generate a broken stream of water from a straight tip nozzle, emulating a fog pattern, the firefighter would have to bounce the stream off a wall or ceiling or place their finger over the opening to break up the stream. Fire departments differ in whether they carry combination or straight stream nozzles on their attack lines. There is a bail handle on the top for opening and closing the nozzle. This is designed so that it is shut off when the bail is moved toward the front of the nozzle. This safety feature causes the nozzle to shut itself off if dropped. The nozzle is equipped with a swivel female coupling for connecting it to the fire hose.

FIGURE 6-42
Assortment of hand-held nozzles.

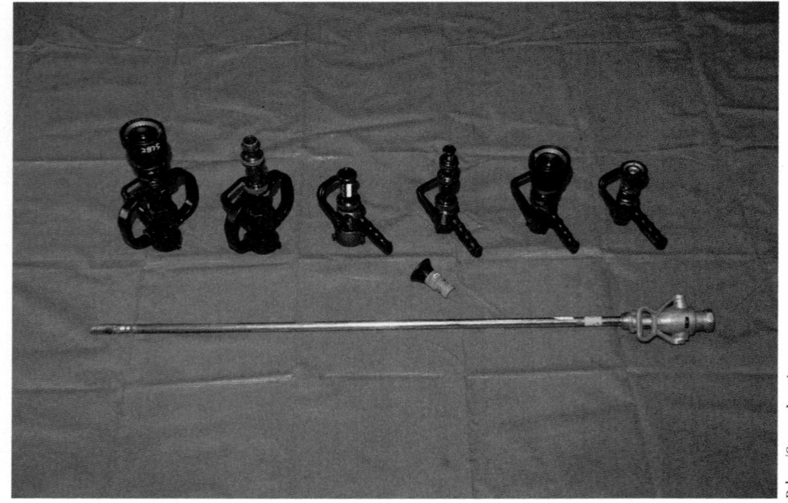

Delmar/Cengage Learning

SAFETY The nozzle is shut off when the bail is moved toward the front of the nozzle. This safety feature causes the nozzle to shut itself off if dropped.

Nozzles can vary in appearance and performance. Some handheld nozzles are equipped with a pistol grip handle. Some combination nozzles have a plastic ring on the tip that spins when the nozzle is set for the fog pattern. This spinning action breaks up the fingers of water into a finer fog. There is another class of nozzles called automatic nozzles. They are designed to give the same reach at different pressures. They depend on a set of springs inside that adjust the flow rate to the pressure.

A recent innovation is the low back pressure nozzle. As the water leaves the nozzle, the laws of physics dictate that there will be an equal and opposite reaction, felt as back pressure. This makes it hard to control even average amounts of flow at normal pressures. If you are flowing 125 gallons per minute at 100 pounds nozzle pressure, you can definitely feel it. These nozzles reduce the amount of back pressure to the point that a normal- to smaller-sized person can effectively control and maneuver a fire stream.

As the demand for fire flow, in gallons per minute, increases, so do the sizes of the nozzles. The larger nozzles are not designed for hand-held use and come in two configurations. There is the adjustable flow, adjustable stream type, and a set of straight tips of assorted sizes. The adjustable type commonly flows from 350 to 1,000 gallons per minute and are combination nozzles that can be used for straight or fog streams. The straight tips typically range in size from one to two inches in diameter. They are arranged in such a manner that if the smallest one on the end is unscrewed the next size is available and so on. The advantage to straight tips is their reach. At the same pressure and flow, a straight tip will far outreach a combination nozzle. Straight tip nozzles also give better penetration when directed into interior fires or used to knock out windows or ceiling panels. The general rule of thumb is that fire streams directed from the street are only effective to the third floor; for this reason these large flow nozzles are also mounted on aerial apparatus where they can be directed through windows, onto roofs, over walls, or used to cool convection columns.

NOTE As the demand for fire flow, in gallons per minute, increases, so do the sizes of the nozzles.

Pumper apparatus carry a **master stream appliance** that is built in and/or removable. If built in, the appliance is mounted on the top of the apparatus (**Figure 6-43**). Some of these are detachable and come with a mounting base that is attached for operation remote from the pumper. The advantage to having the nozzle built in is it can be used for a quick, massive attack on a fire. The problem lies in that it may be a one-shot deal. If supply lines are not laid and the pumper is working solely from its tank, a nozzle set at 500 gallons per minute will empty the tank, depending on size, in one to two minutes. This can leave you with a fire that is still out of control and no water. However, it may be able to stop a fire before it has a chance to spread.

Another creative use of this type of setup is to equip the top-mounted nozzle with a straight tip for attacking roadside grass and brush fires, or for situations with steep terrain or wind where a ground attack will not catch the fire. The master stream is used to try to snuff the head of the fire before it gets out of reach. The reasoning is that if you do not catch the fire now, it will grow beyond control in a few minutes. You might as well take your best shot.

master stream appliance Large-bore nozzle equipped with a base. Not designed for hand-held use.

FIGURE 6-43 Pumper-mounted master stream device in operation at structure fire.

Courtesy of Edwina Davis

A detachable, remotely operated master stream, called a monitor nozzle, is used when the pumper cannot gain access to the location where the nozzle is needed or it would be unsafe to locate the pumper and personnel close to the fire. The nozzle base is equipped with several 2½-inch inlets that are equipped with clapper valves. Clapper valves are one-way check valves that allow water into the base from the hose lines, but will close off any hose inlet that is not in use, so not all of the inlets need to be used. This allows the monitor nozzle to be placed into operation while only one hose line is attached and others are being laid. To operate in a remotely located situation the monitor nozzle is placed and supply lines are laid between the pumper and the nozzle. This operation is commonly used in oil refinery fires. In this situation the nozzle is placed and secured and the personnel withdraw to a safe location while fire control operations are performed.

NOTE If someone thought of a better way of applying water, a nozzle was developed to fill the need.

A type of nozzle that comes in all of the previously described sizes is the foam nozzle. This nozzle is designed to aerate the foam solution coming through the hose, giving the foam a light fluffy appearance that makes it easier to see, and in some situations, making the foam more effective (**Figure 6-44**).

All of the types of nozzles available to today's firefighter are too numerous to list. There are wall-piercing nozzles, cellar nozzles, distributor nozzles, and many others. Basically, if someone thought of a better way of applying water, a nozzle was developed to fill the need. As building construction methods and materials change, nozzles will too.

Another appliance that has found its way onto most fire pumpers is the foam eductor. This device is equipped with a female and male coupling and is inserted into the hose line. It has a suction tube that is inserted into a five-gallon container or 55-gallon drum of

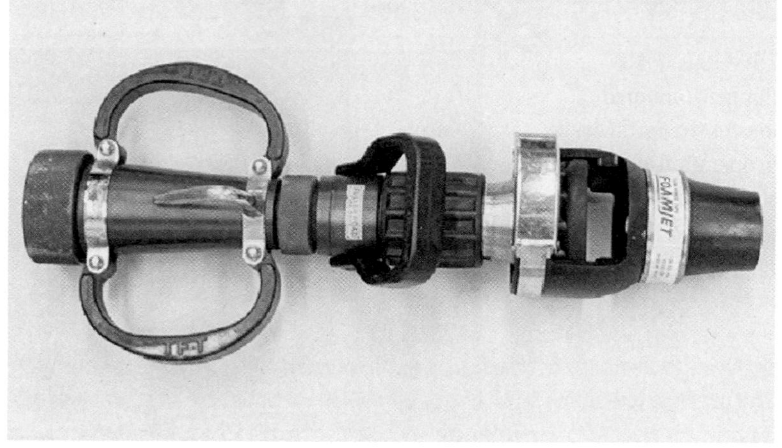

FIGURE 6-44
Foam nozzle attachment.

foam concentrate. Through the use of the venturi principle the foam concentrate is drawn up through the suction hose and enters the hose line to be discharged as foam solution at the nozzle. When using in-line foam eductors, it is extremely important to follow the manufacturer's requirements as to the pump pressure, flow rate set on the nozzle, length of line after the eductor, and elevation of the nozzle over the eductor. All of these factors influence the quality of the foam produced. A caution here is that an in-line educator may not work with an automatic nozzle as nozzle flow rate is variable due to hoseline pressure and not manually selected by the operator.

For more information on foam eductors, please refer to DVD clip *Fire Suppression, Fire Extinguishment—Ignitable Liquid Fire.*

When using in-line foam eductors, it is extremely important to follow the manufacturer's requirements as to the pump pressure, flow rate set on the nozzle, length of line after the eductor, and elevation of the nozzle over the eductor. A caution here is that an in-line educator may not work with an automatic nozzle as nozzle flow rate is variable due to hoseline pressure and not manually selected by the operator.

Fittings

To give firefighters versatility in constructing hose lays and accessing water supplies, pumpers carry a wide variety of fittings (**Figure 6-45**). There are double male and double female fittings in all of the hose sizes carried on the pumper. These allow the firefighter to connect two hose lays together that were laid in different directions. Otherwise one of the hose lays would have to be reversed. There are reducers and increasers. When describing fittings they are described by the female fitting first. If a fitting were adapted from 2½-inch female to 1½-inch male, it would be a reducer; and if it were a 1½-inch female to 2½-male, it would be an increaser. Some adapters are for changing the thread. If we wanted to extend a one-inch line off the end of a 1½-inch line, we may have to change from national standard thread to iron pipe thread as well as changing the coupling diameter. This fitting could be called a *reducer/adapter*. A very common adapter is used for attaching the four- or 4½-inch front mount suction hose on a pumper to a hydrant with a 2½-inch outlet. In areas where water is available from plumbing systems on wells and tanks, such as rural areas, a well-equipped pumper carries a set of adapters to attach its hose fittings to iron pipe thread fittings in the 1½-, two-, and three-inch pipe sizes.

FIGURE 6-45 Reducers and adapters for connecting various sizes of hose lines.

Delmar/Cengage Learning

In areas where **vacuum trucks** are prevalent, many pumpers carry fittings that allow the adaptation from cam lock fittings to the thread type carried by the department. This gives firefighters the capability of using the vacuum truck as a water tender.

Other fittings are used to divide and combine hose layouts (**Figure 6-46**). A wye is used to divide a hose line into two hose lines. The wye is equipped with a female fitting on the incoming side and male fittings on the discharge side. These can be the same size, as in one 2½-inch to two 2½-inch or, more commonly, one 2½-inch to two 1½-inch. If the wye is equipped with shut-offs, it is called a gated wye; without shut-offs, it is a straight wye.

The fitting used for combining hose lines is called a "siamese." The siamese is equipped with two female fittings on the incoming side and a male fitting on the discharge side. It can be equipped with clapper valves that close automatically under pressure. These clappers allow water to enter from one female end without leaking out of the other female–male end if only one line is attached or charged.

vacuum truck Tank truck equipped with a pump that evacuates the air from inside the tank causing it to draw a vacuum. Used for picking up liquids from spills or tanks, commonly used at crude oil production facilities.

Ladders

The NFPA *Standard 1901, Automotive Fire Apparatus* requires that pumpers carry a minimum of one straight ladder at least 14 feet in length with roof hooks. The standard also requires an extension ladder of at least 24 feet in length and a folding

FIGURE 6-46
Wyes and siamese for splitting and joining hose lines.

Delmar/Cengage Learning

FIGURE 6-47
Portable ladders (left to right): extension ladder, roof ladder and folding attic ladder. In this photo the attic ladder is in the unfolded (ready for use) position.

Delmar/Cengage Learning

ladder—commonly called an attic ladder—that is 10 feet in length. The attic ladder is used primarily for gaining access to the access hole to the attic in structures. The ladders carried on all apparatus should meet the required specifications for firefighting use (**Figure 6-47**).[10]

The ladders carried on all apparatus should meet the required specifications for firefighting use.

For more information on ladders, please refer to DVD clip *Ladders*.

Self-Contained Breathing Apparatus

SCBA are required equipment on pumpers and are designed for firefighting.[11] SCBA allow the firefighters to work safely in environments with inhalation hazards such as toxic smoke, elevated air temperature, and oxygen deficiency. The SCBA consists of a backpack, air bottle, face mask, and regulator. It is designed to operate in the **positive pressure mode** so if there is a face mask seal leak, the firefighter does not breathe any harmful products due to positive pressure from the air bottle flowing into the facemask. The backpack is designed to be donned quickly. It has adjustable shoulder and waist straps to fit different-sized users. The air bottle carries the compressed air that the firefighter will be breathing. It may be steel, carbon fiber, or fiberglass-wrapped aluminum. It is equipped with an air gauge that shows how much air it contains and a handwheel for turning it on and off. The mask covers the face with a clear plastic face piece for visibility. Straps on the mask hold it securely to the head. A new trend is to the built-in communication system that allows the person wearing the mask to communicate on the radio and/or through a speaking diaphragm. Another innovation is the in-mask heads up display that gives the wearer up-to-date information on the amount of air remaining in the air bottle. The regulator adjusts the pressure of the air from that in the tank to a pressure that can safely be breathed. It is equipped with a low-pressure warning bell or whistle to let firefighters know when they are close to running out of air and should leave the work area for a safe area.

positive pressure mode SCBA regulator function that keeps positive pressure in the mask face piece at all times.

For more information on self-contained breathing apparatus, please refer to DVD clip *Self-Contained Breathing Apparatus*.

SCBA regulators are designed to operate in the positive pressure mode so if there is a face mask seal leak, the firefighter does not breathe any harmful products.

Hand Tools

Firefighters are always looking for ways to do their jobs with greater speed, safety, and efficiency with fewer than the necessary people at scene to get the job done. Often lives and much property are at stake. Firefighters follow the Boy Scout motto of "Be prepared." A question one must keep in mind is, "If you are at scene and you need it immediately and did not bring the equipment with you, just exactly how are you going to get it?"

Firefighters carry just about every type of hand tool imaginable. As well as regular hand tools such as wrenches and screwdrivers, the fire engine is equipped with tools specially designed for firefighting needs. Hose wrenches that aid in the tightening and loosening of fittings are called spanners. Hydrant wrenches are carried for opening and closing fire hydrants (**Figure 6-48**).

NOTE Firefighters are always looking for ways to do their jobs with greater speed, safety, and efficiency with fewer than the necessary people at scene to get the job done.

For vehicle extrication and rescue, fire apparatus carry specially designed rescue tools. Designed to be placed in the gap between the car door and the body, these tools exert

FIGURE 6-48
Spanners and hydrant wrenchs.

Delmar/Cengage Learning

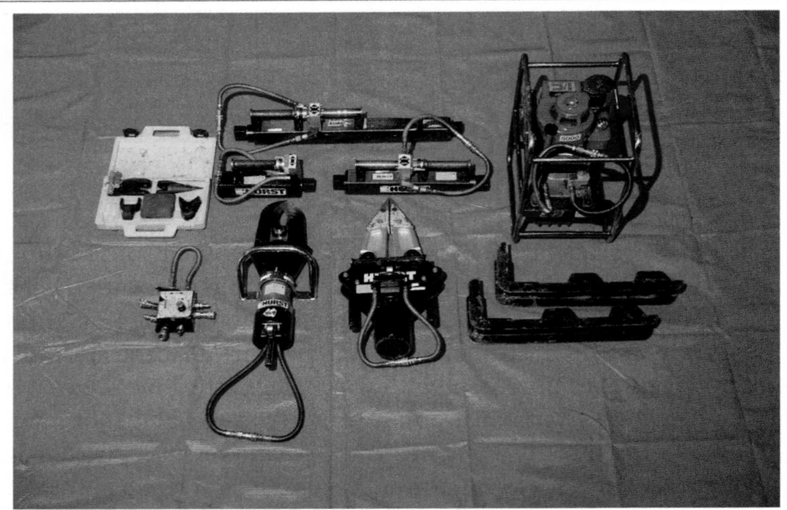

FIGURE 6-49 Hydraulic rescue tool with cutters and spreaders.

Delmar/Cengage Learning

up to 60,000 pounds of force to pop the door open. They can also be equipped with cutters to cut the pillars that attach the roof (**Figure 6-49**). Equipped with special high-strength chains, they can be used to pull the steering column up and away from the victim. Many departments are now carrying hydraulically powered rams, specially designed to be used with their rescue tools, to push car bodies apart to gain access to victims. Gas-powered circular saws with metal cutting blades used to be in common use for vehicle rescue operations. After several accidents involving flammable vapors and sparks produced by cutting operations, which injured and killed firefighters and victims, their use has been curtailed. For taking out the side windows on a car for victim access the spring-loaded hand-held punch works very well. It shatters the glass without spreading it all over the victims as striking the window with an axe would.

Another rescue tool is the air bag system (**Figure 6-50**). These bags are inflated through their own regulator from a self-contained breathing apparatus bottle and, depending on size, are capable of lifting from 12 to 70 tons. They are used for lifting vehicles and heavy objects off victims and for other lifting jobs. They can also be used to roll/lift the dash by running the chain across the top of them and then inflating. Their advantage is in their thinness, less than an inch when deflated, which allows them to be easily slipped into narrow spaces. They are effective in soft dirt where a rescue tool would dig into the ground and are not a source of ignition when flammable vapors are a danger to the operation.

For ventilating roofs several tools are available. For working on **composition roofing** the axe was a traditional choice. The next development in this area was the

composition roofing Tar paper and shingles or tar paper covered with roofing asphalt.

FIGURE 6-50
Air bags used for lifting heavy objects.

Courtesy of Edwina Davis

gas-powered circular saw with a wood-cutting blade. The gas-powered circular saw with a metal-cutting blade is effective on metal sheeting roofs. The tool of choice for today's firefighter is the carbide chain–equipped chain saw. Quick, lightweight, and easily used on different thicknesses of roofing, it is gaining in popularity. When clay or concrete tile roofing needs to be removed, to get at the wooden roof sheeting with a saw, a sledge hammer or pike axe is used to break the tiles. Once the cuts are made, an axe, rubbish hook, or pike pole can be used to remove the desired material. The ventilation crew should always take a pike pole with them to poke out the ceiling below the roof to complete ventilation to the outside.

A tool that has made great inroads into the firefighting field is the power fan. Traditional smoke ejectors are electric fans that draw the smoke from the building. They had to be hung in windows and doorways and supplied with electricity to operate. They tended to be inefficient and often got in the way. The new tool for this purpose is the power fan that is powered by a gasoline engine, electricity, or water, depending on type (**Figure 6-51**). When used, it is placed back from the doorway on the outside of the building so that its cone of forced air covers the opening. This placement leaves the doorway open for access by the firefighters. These fans move so much air that when used properly they can evacuate the smoke from a three-bedroom home in a few minutes. They can also be used to pressurize stairwells in high-rise buildings to keep them free of smoke when firefighting operations require doors onto fire floors to be opened. Their greatest advantage is that they help reduce the number of times that roof ventilation is required at structure fires. In today's lightweight construction, roofs are collapsing sooner and

FIGURE 6-51 Gasoline-powered ventilation fan.

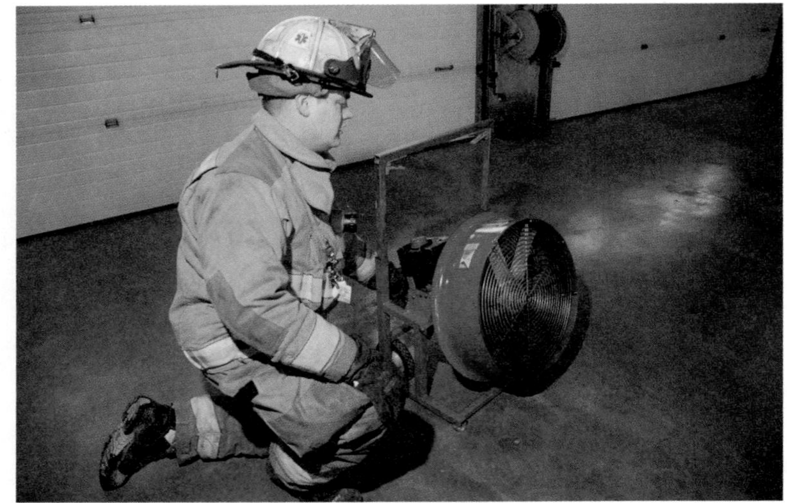

Delmar/Cengage Learning

firefighters are in great danger of falling through and being injured or killed while performing roof ventilation. A power fan can be placed and started by one person in a matter of minutes and then left to operate. A roof ventilation crew usually consisted of at least three people going above the fire. It took them a while to place their ladders, evaluate the roof, and effect ventilation.

In today's lightweight construction, roofs are collapsing sooner and firefighters are in great danger of falling through and being injured or killed while performing roof ventilation.

Salvage covers are tarps used by firefighters to protect a building's contents from water and falling debris damage (**Figure 6-52**). A salvage cover spread on the floor makes quick work of cleaning up a room after pulling the ceiling and spilling the insulation all over the floor. It can also be used to channel water down stairs and to make temporary catch basins. A salvage cover laid over ladders can be used to make a sump for drafting.

For more information on salvage covers, please refer to DVD clip *Salvage Covers*.

FIGURE 6-52
Results of (A) proper and (B) improper salvage operations.

Delmar/Cengage Learning

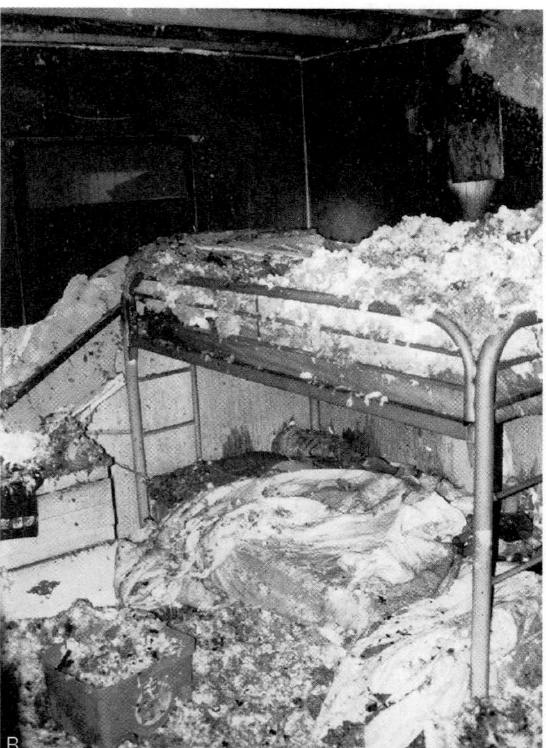

Delmar/Cengage Learning

Almost all fire apparatus carries a fire extinguisher of one type or another. Many fire departments have a person detailed to carry a 2½-gallon water extinguisher into structure fires. This amount of water is often sufficient to extinguish a small fire without pulling hoses inside.

The amount and complexity of the medical aid equipment carried on fire apparatus depends in large part on the level of life support provided. The two basic components of a life-support system are the resuscitator and the medical aid kit (**Figure 6-53**). The resuscitator allows the firefighters to administer oxygen to patients. It can also be used to ventilate victims during cardiopulmonary resuscitation and has a suction device for removing vomit and other material from the patient's mouth. The medical aid kit is a first aid kit containing an assortment of bandages and tools necessary to stop bleeding. The next step above basic life support (BLS) is the automatic external defibrillator (AED). It is used to convert the heart rhythm of a patient in fibrillation to an organized rhythm. In a department where advanced cardiac life support (ACLS) is provided, the equipment and the treatment available are more complicated.

Most fire engines are equipped with a gasoline-powered generator and detachable lights. These lights are used to illuminate the scene when the electricity has been turned off or there is none available. If the engine carries electrically operated tools, the generator should be capable of powering them also.

Thermal Imaging Cameras

Thermal imaging cameras (**Figures 6-54 and 6-55**) are used by the fire service for rescue and finding heat sources. They are battery operated and either hand held or helmet mounted. Looking through the viewfinder, the firefighter sees sources of higher heat as white or gray

FIGURE 6-53 Resuscitator and first aid kit being used by firefighters to treat victim. Firefighters are dressed in medical aid PPE.

FIGURE 6-54
Thermal imaging camera (TIC) in use.

Delmar/Cengage Learning

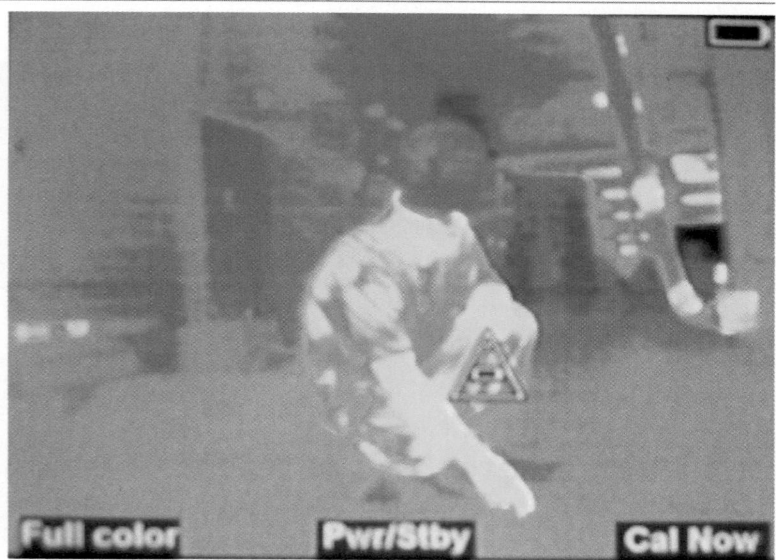

FIGURE 6-55
View of screen on thermal imaging camera (TIC).

Delmar/Cengage Learning

against a black background. The use of the camera allows the operator to determine heat sources in limited visibility situations. They are very effective at locating victims in smoky atmospheres or people ejected from vehicles in tall vegetation. They can also be used to find fire hidden in walls, hot electrical items, and other heat sources. Their use speeds up rescue and fire source location and minimizes damage in tearing open walls, floors, and ceilings to look for hidden fire. Firefighters must be properly trained and have experience in thermal imaging operation and use to ensure accuracy in identifying heat source images.

Wildland Firefighting Hand Tools

In areas where wildland fire fighting is a function of the department, specially designed tools will be carried (**Figure 6-56**). One of these tools is the McLeod, a tool that has a scraping blade on one side of the head and a rake on the other. The scraper is used to remove fuel, such as grass, down to mineral soil to form a fire break. The rake side of the McLeod is used to remove heavy **duff** and **forest litter**. The Pulaski consists of a grubbing blade on one side of the head and an axe on the other. The grubbing blade is used for digging out roots and removing sage and other types of brush and for loosening the ground in preparation for the scraping tools that follow. The axe side of the head is used for removing the lower limbs of trees ("lollipopping") and chopping heavier fuels. The Pulaski, due to its awkward balance, is not a very good chopping tool compared to an axe and great

duff Leaves, pine needles, and other dead forest material.

forest litter The components of duff including tree limbs.

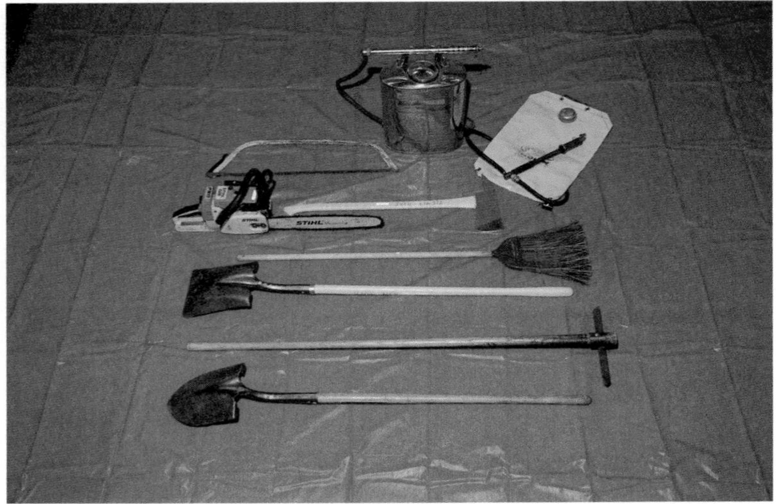

FIGURE 6-56 Wildland firefighting hand tools.

Delmar/Cengage Learning

care should be taken in its use. The shovel carried for wildland firefighting is shorter than a standard shovel and has a more pointed head. This makes it a better scraping tool. It can also be used for throwing dirt to cool down hot spots. When properly sharpened, it can also be used to lollipop tree limbs. Lollipopping prevents fire from climbing the dead limbs at the lower levels of trees to prevent **crown fires**. A new tool is the Combi, which has a swivel head much like a trenching shovel. It can be used as a shovel, grub hoe, and chopping tool. Chain saws are also used for wildland firefighting. They are used to fell trees, buck up limbs and logs, and to create fire breaks in brush. Wildland firefighters prefer to keep their tools very sharp and great care must be exercised when operating around or with these tools as they can cause very bad cuts.

Wildland firefighters prefer to keep their tools very sharp and great care must be exercised when operating around or with these tools as they can cause very bad cuts.

Portable tanks and portable pumps are often set up on forest fires (**Figure 6-57**). Through the use of the pumps, tanks, and hose lays, water supplies can be established in areas that are inaccessible to vehicles. The water is sometimes used to fill backpack

crown fire Fire in the tops of trees. These fires move very rapidly and defy control efforts.

FIGURE 6-57 Portable pumps (left to right): backpack pump and floatable pump.

Delmar/Cengage Learning

pumps. A backpack pump consists of a five-gallon water bag or can with shoulder straps and a hand-operated pump wand that forces the water out of a nozzle. Backpack pumps are primarily used for mop up due to their limited carrying capacity.

HEAVY EQUIPMENT

Fire departments that have wildland areas or special needs have heavy equipment in the form of bulldozers, commonly referred to as "dozers," to establish fire control lines in brush, timber, and grass. In the West, dozers are equipped with blades and discs. In the southeastern United States, they are often equipped with a plow for creating fire lines in heavy tangle undergrowth and boggy areas. Some of these dozers are equipped with built-in water tanks and pumps with hose reels (Pumper Cat). They are used to provide water in areas inaccessible to wheeled apparatus. To get the heavy equipment to the fire, the department has to have heavy transports, consisting of truck tractors and low bed trailers.

In most areas, dozers of the D-5 or D-6 size are used. Bigger dozers are so large and have such a wide blade that it is hard to transport them on narrow mountain roads. They are also so heavy that they tend to destroy old roads that were originally designed to carry stagecoaches. (See **Figure 2-3.**) The exception to this is areas with heavy concentrations of mature brush. The Los Angeles County Fire Department utilizes bulldozers in the D-9 category to push through dense brush, and these dozers can also cut fireline uphill at a faster rate of speed than the smaller dozers.

In areas where oil production and refining operations are a problem, a special foam unit is an asset. The foam unit consists of a large tank for the foam concentrate and a pump with a motor to add the concentrate to the fire stream. The foam unit may also carry special foam nozzles and appliances not carried on other apparatus. When met at the scene by several engines or a water tender, the foam unit has firefighting capability away from established water supplies.

PERSONAL PROTECTIVE EQUIPMENT (PPE)

To protect firefighters in various types of dangerous environments many types of specialized protective clothing are required. The layering of protection provides a greater margin of safety.

Station/Work Uniforms

Modern firefighter work uniforms are made primarily of fire-resistant fabrics,[12] which provide another layer of protection under the PPE worn for firefighting. The station/work uniform is not a substitute for wearing the proper PPE. For any kind of firefighting, cotton underwear should be worn. Any material made from nylon or related synthetic fibers has the possibility of melting to your skin. You should never get in a situation with this kind of heat, but fires can be unpredictable and accidents do happen. The more layers you have on, the better you are protected from external heat. Some departments are experimenting with long sleeve T-shirts under wildland fire shirts for added protection. There is discussion about whether even plastic watch bands should not be worn as they may melt under extreme conditions.

Any material made from nylon or related synthetic fibers has the possibility of melting to your skin.

Structure Fire PPE

The firefighter's first line of defense against the harmful effects of heat and flame is personal protective equipment. This equipment is designed as a system for your safety and all components should be properly worn when the situation dictates. For structural firefighting, the turnout uniform and SCBA are used (**Figure 6-58**). The turnout uniform consists of a helmet with a hood or other ear protection, turnout coat, turnout pants, gloves, and

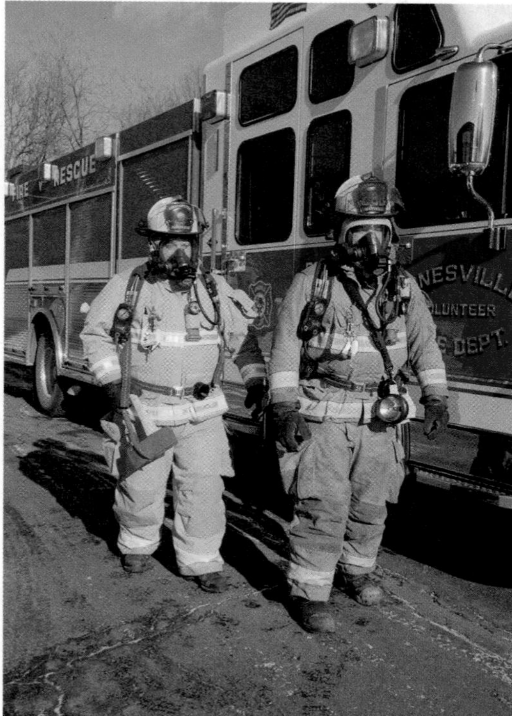

FIGURE 6-58
Firefighters in full structural-firefighting PPE.

Delmar/Cengage Learning

boots. Designed and tested to resist impacts, the helmet is equipped with an internal suspension system to help absorb impact forces.[13] The helmet can be constructed of one of several materials, with high-impact plastic becoming the norm. It should have reflective material on the exterior. It has a long brim on the back to keep hot water and melting roofing tar from going down the back of your neck. Eye protection in the form of a face shield or goggles should be included for use when an SCBA is not used. The turnout coat is constructed of an outer shell of a material that will not support combustion and has reflective stripes.[14] There are two layers of inner liner, one a vapor barrier to keep water and other liquids out and the other an insulating material to protect from heat. The coat should never be worn without the inner liner in place. The coat is equipped with either snaps and hooks or Velcro® and a zipper to secure the front flap, and a high collar that is raised and snapped in place. The pants are constructed in the same manner. Gloves are necessary to protect your hands from heat and debris. For a time, plastic-coated gloves were used: They kept the hands dry but tended to melt to the skin under high temperatures. Firefighting gloves are now made of leather with long knit cuffs, unless the cuffs are built into the jacket, to protect the wrists.[15] Remember, interior attack structural firefighting is performed in a hot, dark, smoky environment, often crawling around on your hands and knees, and all of your body needs protection. Firefighting boots are made of rubber or leather with tread soles and have steel toe and sole protection.[16] Several items that often come in handy and should be carried in your pockets are a knife, a piece of rope, and a flashlight.

For more information on structure fire PPE, please refer to DVD clip *Donning Personal Protective Clothing.*

NOTE The firefighter's first line of defense against the harmful effects of heat and flame is personal protective equipment.

Interior attack structural firefighting is performed in a hot, dark, smoky environment, often crawling around on your hands and knees, and all of your body needs protection.

Personal Alarm/Personal Alert Safety System

All firefighters are required to carry, either on their turnout coat or SCBA, a personal alarm device (PAL or PASS).[17] These devices emit a loud alarm signal when the person wearing them does not move for approximately 30 seconds. Their purpose is to help others find you if you become trapped or are rendered unconscious. It is your responsibility to make sure to turn it on before engaging in firefighting operations. Some new SCBA models have PASS that activates automatically when the air is turned on. With accounting for personnel being such an important part of firefighting safety, some departments are equipping members with removable helmet tags or other identifiers. These are gathered at a point at the incident to assist in keeping track of who is assigned where. In the future, each person may be equipped with a bar code, and a portable scanner would be used to record their passing a certain point at the scene.

Proximity Suits

The PPE worn primarily by aircraft firefighters is called a proximity suit.[18] It consists of a system that includes boots, pants, coat, gloves, and hood with internal helmet. The suit is made of an aluminized material to reflect heat and the facepiece is specially coated to reflect heat as well. The key point is that it is a proximity suit; it is not designed for walking through fire (see **Figure 6-32**).

Wildland PPE

The wildland firefighting uniform (**Figure 6-59**) consists of a hard hat–type helmet with goggles, and ear and neck protection in the form of a shroud attached inside the helmet. The helmet is plastic, so as not to attract lightning. A fire shirt of fire resistant material is worn to protect from heat and hot embers. The pants worn should be made of fire resistive material. Boots, at least eight inches in height with lug soles, should be worn.[19] Dry grass

FIGURE 6-59
Firefighters in full wildland-firefighting PPE.

Courtesy of Edwina Davis

is slippery, and ankle protection from hot material on the ground is a must. When stumps and roots burn out under the ground, they are not always evident. It is relatively easy to have the ground collapse and drop your foot into a bed of hot coals. You should always carry a compass and canteens of water with you.

Another critical part of the wildland PPE is the fire shelter (**Figure 6-60**). It is a small tent folded up in a container that you carry on your belt or on the outside of your line pack (backpack). It is made of an aluminized fabric and can be quickly unfolded. You should never leave the limited safety of your apparatus in a wildland fire situation without your fire shelter with you. Some firefighters have got into the habit of carrying the fire shelter in their backpack instead of having it readily available. The South Canyon incident of Storm King Mountain in Colorado in the summer of 1994 showed the importance of having the fire shelter ready and being able to deploy it quickly. At this fire, the situation changed in a matter of seconds and 14 firefighters were overrun and killed by the fire.[20]

In a wildland fire situation, you should never leave the limited safety of your apparatus without your fire shelter.

Delmar/Cengage Learning

FIGURE 6-60 Fire shelter deployed on road with fire approaching from left. Tools in the photo are modified shovel and fireline pack, which are to be left outside of shelter when deployed.

Emergency Medical PPE

For responding to medical aid incidents firefighters need to be protected as well.[21] A disposable, long-sleeved shirt of moisture-resistant material should be worn. Latex or vinyl gloves are worn on the hands. Eye protection and face mask protection is also a good idea. Use of these items prevents exposure to blood and other body fluids that may be splashed or dripped on you (see **Figure 6-53**).

AIRCRAFT

Aircraft are assuming a larger role in fire-fighting operations. There are basically two types of aircraft: fixed-wing (airplanes) and rotary wing (helicopters). Fixed-wing aircraft are used to transport crews to fires, sometimes across the country. Another method of crew delivery is the smoke jumper plane. These firefighters board the plane with their tools, supplies, and equipment and parachute into the area near the fire. When they are done with the fire, they either hike out to a pickup spot with their equipment or are picked up by helicopter.

The Thirtymile Fire incident on the Okanogan National Forest in Washington State in 2001 illustrated the importance of having the fire shelter ready and being able to deploy it quickly. Forest Service firefighters were cut off from escape and overrun by a wildland fire. Their proper use of fire shelters saved several lives, both firefighters and civilians.

Fixed-Wing

Fixed-wing aircraft are also used as air tankers (**Figure 6-61**). Tankers range in size from single engine airplanes to four-engine DC-7s and the DC-10. Depending on aircraft size they can deliver up from 350 to 30,000 gallons of fire retardant per load to be dropped on the fire. The largest used is a converted DC-10 that carries 30,000 gallons.[22] Their limitation is that they cannot operate at night, in high winds, or heavy smoke. In areas of steep, narrow canyons they may not be able to fly low enough to effectively drop on the fire. Their turnaround time to return to their base and refill slows them down. Their main advantage is their ability to cover long distances relatively quickly. If the only air tanker available is 300 miles away, it can still be over your fire in a fairly short period of time. A new concept in aircraft is the Canadair CL-215 air tanker. This aircraft can drop down over the surface of a large lake or the ocean and fill itself. It looks much like a sea plane with high-mounted engines, pontoons on the wings, and a boat hull–shaped fuselage.

FIGURE 6-61
Air tanker making retardant drop on wildland fire.

Delmar/Cengage Learning

Rotary Wing

Rotary wing aircraft (helicopters) are rapidly gaining in use by firefighting forces. They can transport crews to remote areas and drop them off at **helispots** that are nothing more than openings in the forest canopy, either meadows or ridge tops. Helitack crews are trained to **rappel** from the helicopter to the ground if no landing area is available. Helicopters can **sling load** supplies and equipment to and from the fire line. When dropping water or fire retardant, helicopters are capable of pinpoint accuracy on burning snags, buildings, and hot spots. They do not need an airport and runways to operate. Some are equipped with fixed tanks and others are equipped with buckets (**Figure 6-62**). The fixed tank craft either land and are refilled from a pumper or have a built-in pump with a tube hanging down that is dipped into a water source. The pump fills the tank while the helicopter hovers. These pumps are hydraulically driven and can fill the 2,000-gallon tank in a few minutes (**Figure 6-63**). They are capable of operating out of a water source only 18 inches deep. The bucket-equipped helicopter hovers over the water source and dangles the bucket into the water. Helicopter operations are hampered by high winds and heavy smoke, just like the air tankers. Several departments are now looking into the feasibility of using helicopters to attack fires at night.

Overall, the advantages of aircraft in direct firefighting operations are the delivery of fire retardant or water in areas that are inaccessible by pumpers or that are untenable

helispots Unimproved areas large enough to land a helicopter.

rappel To descend by means of a rope.

sling load Material transported by being placed in a net suspended underneath a helicopter.

FIGURE 6-62
Helicopter making water drop on wildland fire, supporting ground forces.

Courtesy of Edwina Davis

FIGURE 6-63
Helicopter filling belly tank with snorkel.

Delmar/Cengage Learning

on the ground. They can slow down a fast-moving fire enough to allow ground forces to gain the upper hand. They have also saved lives in their ability to place a drop when and where it is needed to keep ground forces from being overrun in a bad situation. Aircraft of both types are also used to fly reconnaissance missions. By flying over the fire area with **infrared sensing devices and GPS**, the personnel can determine where hot spots exist and map the fire edge through smoke and poor visibility.[23] They can also determine where fire control lines have been completed and where they need to be reinforced or extended.

NOTE The advantages of aircraft in direct fire fighting operations are the delivery of fire retardant or water in areas that are inaccessible by pumpers or that are untenable on the ground.

infrared sensing devices and GPS Devices that can detect heat energy through smoke and clouds. Used for aerial mapping of fire edges and locating hot spots.

SUMMARY

This chapter has only scratched the surface of the tools and equipment available to firefighters. New tools and equipment are constantly being developed or adapted to firefighting use to make the job safer, faster, and more efficient.

The facilities described in this chapter are not available and not needed at every fire department; they are only a representative sample of the facilities that do exist.

The apparatus and equipment covered varies due to department needs. Depending on the types of incidents encountered and requirements placed on the fire apparatus, it is adapted and modified. Fancy apparatus and equipment is not necessary to get the job done. Before the invention of all of the power tools mentioned, fires were put out and victims were rescued. The new tools certainly do make it easier and safer for firefighters to perform their jobs. It is important not to become so dependent on power tools that you become ineffective when they are not available or will not start.

The one resource not covered was the most important one, the firefighter. Without intelligent, trained, aggressive firefighters, the fanciest and most expensive facilities, apparatus, and equipment in the world are worth nothing. Always remember that skill and knowledge are what makes the whole system work. The best firefighters are capable not only of using all of the tools available to them, but also adapting those tools to new uses and performing in their absence.

REVIEW QUESTIONS

1. List several advantages of a fire department having its headquarters separate from a fire station.
2. List three structures that may be located at the training facility and their uses.
3. What is meant by an enhanced 9-1-1 system?
4. Why is the diesel motor chosen for most pumper apparatus?
5. Explain the difference between centrifugal and positive displacement pumps.
6. Why is a centrifugal pumper equipped with a positive displacement pump in conjunction with the main pump?
7. List three types of squad vehicles and their functions.
8. What is meant by a "twinned system" on Aircraft Rescue Fire Fighting (ARFF) apparatus?
9. What is the difference between attack and supply hose lines? Give examples of each.
10. List the four main components of a self-contained breathing apparatus (SCBA).
11. What are the uses of a resuscitator?
12. List three wildland firefighting tools and their uses.
13. What are the components of structural personal protective equipment (PPE)?
14. What are the components of wildland PPE?
15. What are the components of EMS PPE?
16. What fits the description of fixed wing fire fighting aircraft? Give two examples of their uses.
17. What fits the description of rotary wing fire fighting aircraft? Give two examples of their uses.

DISCUSSION QUESTIONS

1. Why is the centrifugal pump chosen as the main fire pump for most fire apparatus?
2. Why does the fire service need a variety of vehicle types to perform its function efficiently?
3. Why is it important to wear all of your PPE for a specific fire (structure, wildland, ARFF, etc.) every time?

NOTES

1. National Fire Protection Association, *Standard 1402, Building Fire Service Training Centers* (Quincy, MA: National Fire Protection Association, 2007).
2. National Fire Protection Association, *Standard 1403, Live Fire Training Evolutions* (Quincy, MA: National Fire Protection Association, 2007).

3. National Fire Protection Association, *Standard 1911, Standard for the Inspection, Maintenance, Testing, and Retirement of In-Service Automotive Fire Apparatus* (Quincy, MA: National Fire Protection Association, 2007).
4. National Fire Protection Association, *Standard 1221, Communications, Emergency Services* (Quincy, MA: National Fire Protection Association, 2007).
5. National Fire Protection Association, *Standard 1901, Automotive Fire Apparatus* (Quincy, MA: National Fire Protection Association, 2009).
6. National Fire Protection Association, *Standard 1911, Standard for the Inspection, Maintenance, Testing, and Retirement of In-Service Automotive Fire Apparatus* (Quincy, MA: National Fire Protection Association, 2007).
7. National Fire Protection Association, *Standard 414, Aircraft Rescue and Fire-Fighting Vehicles* (Quincy, MA: National Fire Protection Association, 2007).
8. National Fire Protection Association, *Standard 1961, Fire Hose* (Quincy, MA: National Fire Protection Association, 2007).
9. National Fire Protection Association, *Standard 1963, Fire Hose Connections* (Quincy, MA: National Fire Protection Association, 2009).
10. National Fire Protection Association, *Standard 1932, Use, Maintenance and Service Testing of In-Service Fire Department Ground Ladders* (Quincy, MA: National Fire Protection Association, 2004).
11. National Fire Protection Association, *Standard 1981, Open-Circuit Self-Contained Breathing Apparatus (SCBA) for Emergency Services* (Quincy, MA: National Fire Protection Association, 2007).
12. National Fire Protection Association, *Standard 1975, Station/Work Uniforms for Emergency Services* (Quincy, MA: National Fire Protection Association, 2009).
13. National Fire Protection Association, *Standard 1971, Protective Ensemble for Structural Fire Fighting and Proximity Fire Fighting* (Quincy, MA: National Fire Protection Association, 2007).
14. Ibid.
15. Ibid.
16. Ibid.
17. National Fire Protection Association, *Standard 1982, Personal Alert Safety Systems (PASS)* (Quincy, MA: National Fire Protection Association, 2007).
18. National Fire Protection Association, *Standard 1971, Protective Ensemble for Structural Fire Fighting and Proximity Fire Fighting* (Quincy, MA: National Fire Protection Association, 2007).
19. National Fire Protection Association, *Standard 1977, Protective Clothing and Equipment for Wildland Fire Fighting* (Quincy, MA: National Fire Protection Association, 2005).
20. *South Canyon Fire Investigation* (Washington, DC: U.S. Government Printing Office, 1994).
21. National Fire Protection Association, *Standard 1999, Protective Clothing for Medical Emergency Operations* (Quincy, MA: National Fire Protection Association, 2008).
22. National Fire Equipment System Course Manual, S-370 Intermediate Air Operations, Aircraft Types (Boise, ID: National Interagency Fire Center, 2002).
23. Ibid.

CHAPTER 7

Fire Department Administration

LEARNING OBJECTIVES

Upon completion of this chapter, you should be able to:

- Describe the six principles of command.
- List and describe the six components of the management cycle.
- Identify the four methods of communication.
- Describe fire department chain of command.
- Fill out a typical fire department organizational chart.
- Identify different fire department types.
- Identify the different ranks and their general responsibilities.
- Explain the terms customer service, the one department concept, team building, and incident effectiveness.

INTRODUCTION

One of the most important jobs in the fire department is that of administration. The fire chief must balance the needs of the community and the department with the resources available. It is not enough to have a group of personnel willing to take risks to save lives and property. These people must be properly led and supported in performing their functions. The chief and the fire department administrative staff perform their duties to ensure that the personnel of the department are trained, equipped, and supplied with the necessary support services.

NOTE The fire chief must balance the needs of the community and the department with the resources available.

To understand how the fire department functions, it is necessary to understand the management principles behind the organization. It is also necessary to understand the command structure and how it operates on a day-to-day basis and under emergency scene conditions.

This chapter is about administration, but some of the illustrations used are based on firefighting operations to illustrate the points more clearly.

PRINCIPLES OF COMMAND

The guiding principles for the organization of a fire department can be referred to as the Principles of Command.[1] These principles are not used in the same manner in every fire department, because a department should be organized to best serve the needs of the jurisdiction it protects. These principles are general guidelines. As you review these principles, you will see that they are used in both emergency and nonemergency organizations. The Principles of Command are divided into six areas in this text with Accountability and Responsibility grouped under Delegation of Authority. Some texts refer to them as eight principles because they single out Accountability and Responsibility and address them separately.

NOTE The guiding principles for the organization of a fire department can be referred to as the Principles of Command.

Unity of Command

The first command principle is that of unity of command. This means that each person in the organization has only one boss. This concept also requires that everyone in the organization have a clear understanding of whom they are supervised by and whom they supervise. It is a proven fact that personnel work best when they are supervised by only one person (**Figure 7-1**). When this principle is violated, the person receives orders, often conflicting, from several people at the same time. Obviously, you cannot satisfy all of those making requests of you. This leads to confusion, inefficiency, and sometimes nothing gets done or the wrong thing gets done. To address this issue in incident command situations the National Incident Management System (NIMS) in Chapter 13 is based on Unity of Command as its organizational structure.

This command principle is probably the most violated, especially at emergency scenes, when proper procedure is not followed. For example, when operating at the scene of an emergency the firefighter may be involved in performing a task for the company officer. The chief approaches and sends the firefighter to perform another assignment. The firefighter knows that the chief outranks the company officer and is placed in the position of not knowing exactly what to do. This question is commonly asked on oral interview examinations and requires judgment on the part of the prospective firefighter:

> "You are operating at a structure fire and your company officer sends you to the engine for a couple of salvage covers. While you are at the engine, the battalion chief approaches you and wants you to go and turn off the utilities to the fire building. What will you do? Justify your answer."

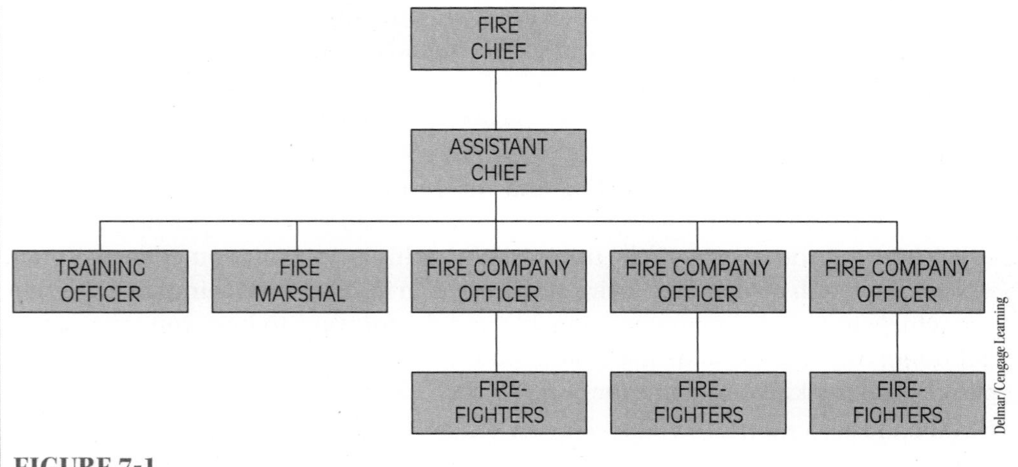

FIGURE 7-1
Fire department organization.

You already know that the chief outranks the company officer and has a very good reason for asking you to turn off the utilities. Possible answers are:

1. Do what the chief requests without question or comment.
2. Tell the chief you will get to it when you have time.
3. Ignore the chief's request and continue with your original assignment.
4. Return to your company officer's location and ask her which task you should perform first.
5. Tell the chief that you already have an assignment from your company officer and ask the chief to let the company officer know that you will be delayed.

Which one did you choose? The correct answer is the last one. The reasons why the answers are correct or incorrect are listed in order.

1. If you do as the chief requests without question, he may feel that you are performing as ordered. Your company officer is going to be understandably upset when you do not return. The situation may be such that the company officer becomes concerned about your whereabouts and leaves her position on the assignment to look for you. This leads to two persons being removed from their positions on the assignment, you and the company officer. You will almost certainly end up in a consultation with your company officer that will not be to your liking. The company officer is going to be upset and is much more likely to take it out on you than on the chief.
2. If you tell the chief that you will get to it when you have time, you are certainly not going to make any points with the chief. As a matter of fact, the chief is probably going to have a one-sided discussion with you about the rank structure of the fire department and where you fit into that structure. The chief is then most likely going to tell you to go and do as you were told. After the incident is completed, the chief is probably going to have a discussion with your company officer about your attitude and training needs. The company officer is already going to be unhappy with you because you did not show up with the salvage covers as requested. If you add the fact that the company officer had to come and check on your safety, it is just going to compound the situation.
3. If your choice is to ignore the chief and return to your company officer with the salvage covers, the chief is going to be very dissatisfied with your performance. The chief expected you to turn off the utilities. When this is not done someone may come into contact with an energized wire and receive an injury. When things like this happen, chiefs tend to become unreasonable and do not want to hear your excuses for not doing as you were told. The chief is going to come down on you and your company officer, making problems for both of you.
4. If you choose to return to your company officer's location and ask for clarification, you are going to be considered to be of somewhat less than average intelligence. You have not brought the salvage covers, nor have you done as the chief requested. No

one is going to be particularly happy with you at this time. You have now become so ineffective that neither task is accomplished.

5. You could tell the chief that you are already doing a job for your company officer. Request that he notify the company officer that you will be delayed or not return at all. The responsibility is then placed on the chief for changing your orders. The chief can then decide which assignment is more important. He will have to find someone else to turn off the utilities, or have you go ahead and do it, advising your company officer of his decision. This lets the company officer know that you are reassigned and not lying in a hole somewhere, so she does not have to come and check on you. If the chief forgets to notify the company officer, you have at least tried to do what you can in the situation. Another option you may have available is to contact your company officer on the radio and advise her that the chief has requested you turn off the utilities and that you will be delayed. This keeps the company officer on top of the situation and she may send someone else after the salvage covers.

The purpose of this was not to portray company officers and chiefs as unreasonable people with the poor firefighter caught in the middle. It was to illustrate how things can go wrong, and often do, when the principle of unity of command is violated. This most often happens in the heat of battle at emergency scenes and you must be prepared to deal with the situation should it arise. There are right and wrong ways to act in situations like the one just presented and it is up to you to perform appropriately.

Chain of Command

The second command principle is chain of command. The term chain is used because there is a linkage between the different levels of the organization. The levels do not operate independently of one another. The decisions made at the top affect the levels below to a greater or lesser degree. When the chief makes a decision, it is presented to the assistant chiefs, then to the battalion chiefs, and then to the company officers and firefighters.

From the viewpoint of the new firefighter, the chain of command pretty much ends with the company officer. The company officer is the company commander and the first-level supervisor. If the firefighter has a question or problem, the company officer is usually the one to deal with it. As the firefighter rises in rank, the chain of command dealt with extends further upward through the organization. The exception to this is when the firefighter is assigned to a special project, such as public education. He may then answer directly to a higher-level officer such as the chief of prevention.

The chain of command also represents the formal path of communication through the organization. Communication flows down from the top and up from the bottom through the chain. An example would be if the firefighter has a policy question, she should first ask the company officer. The company officer may not be able to answer the question. The firefighter is then asked to write the question on a memo and forward it to the battalion chief. If it is a question that the battalion chief does not have the answer for, it can be passed further up the chain of command.

This method of operation may seem rather cumbersome and inefficient at times, but it serves a very good purpose: It allows the command structure of the organization to maintain control. If firefighters went around their company officer when they wanted something and the battalion chief allowed it to happen, the company officer would have less authority over the firefighters. Every time the company officer wanted something to happen that the firefighters did not like, they would question the orders to the battalion chief and work would not get done in a timely fashion, if at all.

In every organization there is also an informal communication system called the "grapevine" or "rumor mill." The problem with the informal communication system is that it often disseminates information that is either wholly or partially untrue or is based on speculation. Many times it has turned out that the information has hurt someone's career or feelings and was totally unfounded.

Everyone should be cautious about spreading rumors about others or what is going to happen in the department. As has been previously mentioned, gossiping is not a trait that others are going to respect.

In an organization that has a clear chain of command, it is to be strictly adhered to. No one in the organization appreciates someone going around them or "over their head." When this happens the person bypassed usually feels slighted and may react accordingly. The only time the chain of command may be violated is in extreme circumstances or in emergency situations that require immediate action, such as a safety matter that requires immediate attention.

The chain of command also applies to the concept of unity of command. When viewing an organization chart, it is obvious who answers to whom in the organization. It also specifies the relationships of personnel to those above them, below them, and across from them on the chart. Do not think that the organization chart specifies all of the relationships between personnel in the organization. Not all personnel on any particular level are perceived as equals. The probationary firefighter may be on the same level of the organization chart as the 10-year firefighter, but they are not viewed the same by the command structure or their peers. When the senior firefighter tells you to do something, it is not the time to tell him that you are his equal and take orders only from the company officer. Keep this in mind, as every organization has an informal organization as well as an informal communications structure. If it becomes a problem for you, discuss it with your company officer.

The organization chart of a small department will be much less complicated than that of a large department. Large departments tend to have many different bureaus and activities not performed by the smaller departments, in which more activities are grouped under one person as they are less complex in nature. Fire department organizations tend to be pyramid shaped. The organization is narrow at the top, usually beginning with one person, called the chief, chief director, commissioner, or other similar title. Regardless of the title, the person at the top has the ultimate responsibility for what goes on in the organization. When a firefighter is injured or killed or a major fire occurs, this person has to answer to the public.

NOTE The chain of command provides for the transfer of authority at incidents.

The chain of command provides for the transfer of authority at incidents. When a higher-ranking officer arrives at scene, it is the policy in most departments that the higher-ranking officer has the option of taking over command of the incident. He will do this if he feels it is necessary. On a major fire the chief officer may take over command from the company officer. This is not done to infer that the company officer is incapable. It is done because the higher ranking officer has more experience and training in managing large incidents and diverse operations.

It also frees the company officer to return to managing his own company or a portion of the incident. The chief officer may very well use the company officer as an information source if the incident is in the company officer's first-in area. On a small incident, the chief officer may arrive at scene and just observe the operations being performed. There is no reason for him to take command from the company officer. He will be observing the actions of the at-scene crew to see that they are performing efficiently and safely. If either is seen as lacking, the company officer is going to receive some guidance as to the training needs of his crew.

This system works in reverse as well. When an incident has escalated to the point where the chief officer has taken charge, it will eventually be controlled. As the incident is brought under control and resources are released, the chief officer will return command of the incident to the company officer. The company officer retains the responsibility of seeing the incident to completion. The company officer also has the final responsibility of making sure that the fire is completely out or that the incident is properly terminated before all resources are released.

NOTE Under the chain of command concept, organizations are divided into line and staff functions.

Under the chain of command concept, organizations are divided into line and staff functions. Line functions are those directly related to the goals and objectives of the fire department.[2] Department goals and objectives include the preservation of life and property. The methods used to achieve these goals are prevention, suppression, and rescue.

Staff functions are all the other functions necessary to support the primary goals and objectives.[3] Functions include training, vehicle repair, station maintenance, supply, and personnel. These functions indirectly support the goals and objectives of the line personnel. Without a personnel section there would be no one to make sure that new firefighters were properly selected or that the currently employed personnel received their pay and benefits. Once hired, the new firefighters must be trained and outfitted. Without vehicle repair the equipment would not be able to leave the station to respond to the incident.

In many departments, the line and staff division becomes blurred in the area of fire prevention. The primary goal of the fire department is the preservation of lives and property. The best way to do this is to not have fires—thus prevention. Personnel working in the fire prevention section of the department may not respond to fires, so they are considered staff. They do contribute directly to the primary objective, so they could be considered line.

The fire department organization cannot survive without both line and staff functions. In times of tight budgets, there is often a demand on the part of the line personnel to cut the staff. Their reasoning is that without firefighters on the equipment, you do not have a fire department. In real life this is an oversimplification. Unless the firefighters in the stations are capable of performing all of the functions of the staff in addition to their other duties, the staff is necessary.

NOTE The fire department organization cannot survive without both line and staff functions.

Span of Control

The third management principle is that of span of control. Regardless of rank or education, a supervisor can effectively supervise only a certain number of personnel. Some supervise more than others, due to the nature or complexity of the work being performed. In firefighting, the effective span of control is considered to be three to seven personnel per supervisor, with five being optimal in most situations. This illustrates the need for intermediate levels of management between the chief and the firefighters. The chief of a large fire department may indirectly supervise more than 3,000 personnel. If you were to look at the organization chart, you would see that the chief has from three to seven personnel that confer directly with him. His decisions are dispersed through the chain of command through these key personnel and he receives input from them in return. If the same department were to have a chief and firefighters with no intermediate levels, the chief would be concerned with and receive input directly from the other 2,999 personnel in the department. As you can see from this extreme example, the chief could quickly suffer information overload. Using, for example, vacation requests, the chief would have to coordinate the requests of all of the personnel at one time. This would be quite impossible.

NOTE Regardless of rank or education, a supervisor can effectively supervise only a certain number of personnel.

The same concept applies to the management of emergencies. At emergency scenes, the situation is not only dynamic, but dangerous as well. As you will see in Chapter 13, the recommended span of control at incidents is three to seven with the optimum being five. This prevents the incident from being so subdivided that a coordinated attack cannot be made. It also prevents information overload on the part of the management personnel. An incident manager needs information as to what is happening overall but does not need to know that firefighter Johnson was sent to get more chain saw fuel. The object is for the subordinates to pass on the important and required information without supplying too much that is unnecessary.

Division of Labor

The fourth management principle is division of labor. The work to be performed needs to be divided into specific areas to prevent duplication of effort, to apply the most appropriate resources to the job, and to determine responsibility for completion of the assigned work. Division of labor is also assigned based on area, skill, and complexity.

The division of labor is clearly seen in the classification of functions into line and staff. Some positions have responsibility for both, especially in small- to medium-sized departments. Assigning certain functions to the staff, such as record keeping and research, allows the direct application of effort without the distractions incurred by line personnel responding to emergencies. It would be extremely difficult for line personnel to schedule and attend meetings if they were sent out on emergency incidents on a regular basis.

Assigning areas of responsibility avoids duplication of effort and promotes efficient use of resources. The fire prevention bureau performs plan checking and building approval whereas the arson unit investigates suspicious fires and assists in prosecution of arsonists. The appropriate resources are applied because the fire prevention personnel are trained in plan checking and inspection and the arson personnel are trained in investigation, collection of evidence, and arrest procedures.

NOTE Assigning areas of responsibility avoids duplication of effort and promotes efficient use of resources.

Fire stations are located so that they each have an area of direct responsibility.

They are required to perform fire prevention and prefire planning in their area along with other tasks. When needed, they can respond out of their area and assist adjoining companies in their areas of responsibility. This division of labor by area provides for a more efficient use of resources. The first-in company knows its area and its complexities better than any other company. The increased knowledge is a great advantage in structure or wildland fires. The adjoining companies coming in to assist can utilize the knowledge of the first-in company when attacking the fire. If all of the companies worked out of one central fire station in a large jurisdiction, little time would be spent in the outlying areas.

In other functions of the fire department, specialized skills become important. The hazardous materials team responds to hazardous materials incidents anywhere in the jurisdiction. This allows the department to train a few select personnel to a high level and maintain their skill level, saving the expense of training all of the department personnel to that level. Any one station may only be involved in one hazardous materials incident a year, whereas the hazardous materials team personnel respond on these incidents on a regular basis.

NOTE The complexity of certain fire department jobs requires the efforts of personnel assigned to that function alone.

The complexity of certain fire department jobs requires the efforts of personnel assigned to that function alone. Dispatching in a busy department requires specialized skills in using the dispatch center equipment and possibly **emergency medical dispatch** training. Examples of other specialized skills used on major wildland fires are those of weather experts and fire behavior specialists.

emergency medical dispatch A system where dispatchers are trained to give medical advice to the persons at the scene until emergency help arrives.

Delegation of Authority

The fifth management principle is that of delegation of authority. For work to get done, the manager must delegate authority to the subordinates. Managers that cannot do this spend so much of their time managing and reviewing what their subordinates are doing that they have no time for planning or other important functions. The fire department hiring process is selective enough that managers should be able to assume that their subordinates are capable of handling the jobs that they are trained to perform. It is the supervisor's responsibility to ensure that the personnel have the required training to perform their jobs. It is considered normal and good practice to check up on the job being done once in a while, but it is not necessary or desirable to closely monitor personnel when they are performing work for which they are qualified.

Delegation of authority requires that the manager give the subordinate the authority and responsibility, while holding them accountable, for taking action on a specific mission. It also requires that the mission be broken down into segments that are assignable. When the fire chief places the fire marshal in charge of prevention, the fire marshal is given the mission of reducing the loss of life and property due to fires by preventing ignition of hostile fires. The fire marshal is not capable of inspecting every business in the jurisdiction by himself. The fire marshal then empowers the inspectors to make inspections and implement the necessary changes to make the businesses more fire safe. The fire chief has delegated authority—along with responsibility and accountability—to the fire marshal, and the fire marshal has delegated it to the inspectors.

NOTE Delegation of authority requires that the manager give the subordinate the authority, along with responsibility and accountability, to take action on a specific mission.

The mission is broken down into assignable segments by the fire chief, specifying that the fire marshal prevent fires, and not have to suppress them as well. The fire marshal divides the workload among the various inspectors.

This concept is often used at an emergency scene. In the case of a train wreck with mass casualties, the incident commander (IC) has responsibility for the whole incident.[4] The IC determines the objectives to be achieved during the incident. These would include making sure that people are rescued, treated, and transported; that fires are suppressed; that any spilled diesel fuel is contained; and that the safety of firefighters and civilians is provided for during the operations. The IC then assigns an operations section chief (OSC), who implements the tactical operations to accomplish the objectives.[5] The OSC assigns units to rescue and remove people from the wreckage. The OSC makes sure those rescued are treated and transported, provides personnel to suppress any fires that may have occurred as a result of the incident, assigns others to control any diesel fuel leaks or spills, and will probably assign a qualified person to act as the safety officer to ensure that the operations are conducted safely.

All of the operations just listed are occurring at once, and it is easy to see that the IC cannot control all of them at the same time. There are too many different activities competing for his time and attention. Through delegation, he is releasing the authority of making the needed decisions to the lowest possible level. With competent and properly trained personnel this works very well. If the personnel at the scene were untrained, it would not work well and would be unsafe.

The alternative to delegation is that the IC requires that every decision be cleared through him. This would bring the operation to a standstill. Every individual rescue, every treatment decision, suppression action, and spill control operation would require consultation before action was taken.

One thing that must be kept in mind is that authority for decision making can be delegated, but overall responsibility cannot. The fire chief is ultimately responsible for the actions of the subordinates. The fire chief must answer to the public when things go wrong. As personnel rise in rank, they assume more responsibility for achieving the organization's objectives. The company officer is responsible for seeing that her subordinates are properly trained and prepared to perform emergency functions. She is also responsible for seeing that the program work assigned to her shift at the station is completed in a timely and satisfactory fashion. As a firefighter, you are on the bottom of the organizational ladder and may think that you do not have much in the way of responsibility, but this is incorrect. You have the responsibility of coming to work prepared to perform your duties. This includes showing up in the proper uniform, remaining physically fit, and training yourself to perform your job. You are the one visible to the public on a day-to-day basis. To the people you assist in your job and the public, you are the fire department.

NOTE Authority for decision making can be delegated, but overall responsibility cannot.

Program work, such as hydrant maintenance, is tedious and not nearly as exciting and rewarding as firefighting, but it is important to the overall mission of the fire department. When the company officer sends you out to service hydrants, you have been delegated the authority and responsibility to service the hydrants as you deem necessary. If what you deem necessary is sitting somewhere drinking a soda pop and filling out paperwork, without doing the required work, you are asking for trouble. It may not be a problem this year or even next year, but sooner or later there is going to be a fire in which that hydrant is needed for water supply. The engine crew tries to make a hookup and cannot remove the caps that you indicated you removed and lubricated just last month. As it turns out, they are rusted on. The firefighters return to the station, look in the hydrant maintenance book to see who serviced the hydrant last, and there is your name or initials. At this point accountability for the failure is going to be placed. At the least, you are going to get a severe verbal reprimand. If the fire cost major property damage, or even worse, loss of life, you are partially to blame. How will you feel about that? The company officer will be in trouble as well as she was responsible for ensuring that you were doing your work as assigned. If the press acquires the information that the fire escaped initial control due to a hydrant being unavailable because of poorly scheduled maintenance, the whole department takes a public beating.

As previously stated, authority can be delegated, but overall responsibility cannot—that lies with the fire chief. It is the obligation of the person assuming the authority to take responsibility of seeing that the function is performed correctly within department guidelines. You must hold yourself and the others around you accountable to do as they are directed.

Exception Principle

The sixth management principle is the exception principle. This principle states that the person delegating authority wants to be informed in situations of major importance. To put it another way, the supervisor could say, "I only like surprises on my birthday." The principle refers to the fact that certain situations are going to arise that the supervisor needs to be informed about, even if the subordinate handles them. These types of situations could include personnel matters, major incidents, and incidents involving major expense to the department.

There may be certain situations that the supervisor wants to be consulted on before the subordinate commits the department to a course of action. This is certainly the case when personnel matters are involved. If the company officer is forced to remove a subordinate from duty, the supervisor should be notified. If this process is done incorrectly, the company officer has opened the department up to a tremendous amount of liability for damages from the person relieved from duty.

Another situation that could easily be encountered is a citizen volunteering a piece of equipment at a fire, in this example, a water tender. The good citizen sees the need for the fire department to use his water tender and volunteers it to the company officer working at the fire. The citizen then finds out that the department is paying the local construction company $2,000 a day for its own water tender. The citizen then decides that he should also be paid the same amount for his water tender that he initially volunteered for free. After all, it is only fair to pay for both of the water tenders, isn't it? This then places the department in the position of having to decide what to do about the volunteered–water tender owner's claim. In most cases, the fire department is going to go ahead and pay the claim. The whole process could have been avoided if the principle of exception was applied and the company officer had checked with the supervisor before approving the use of the water tender in the first place.

A more drastic example of the exception principle is a scenario that happens in emergencies when someone is trapped under a heavy object, such as a car that has rolled over. The tow truck arrives and volunteers to hook up to the car and remove it from on top of the trapped person. In the haste to remove the car, the company officer approves the operation. In so doing, the tow truck operator has an equipment failure and the car falls back on top of the injured victim and kills her. When this goes to court, as it most likely will, the company officer will be asked why he did not wait for the proper rescue equipment to arrive at the scene to perform the operation. He will also be asked, "What is standard department procedure and was it followed?" His answer will most likely be that, "No, that is not standard procedure, but the situation required removing the car as soon as possible." The company officer is so close to the scene that his judgment may be affected by the situation in front of him. It is easy to get caught up in the heat of the moment and lose proper perspective. When in doubt as to the proper course of action to be taken in a situation, follow department policy. Had he contacted his supervisor and asked him what to do, the answer probably would have been to wait the extra time for the proper rescue equipment to arrive. In some instances, the principle of exception can be used as a system of checks and balances to either approve or disapprove a course of action.

NOTE It is easy to get caught up in the heat of the moment and lose proper perspective. When in doubt as to the proper course of action to be taken in a situation, follow department policy.

THE MANAGEMENT CYCLE

The management cycle is an organized thought process to achieve the desired goals of the organization.[6] There are six components of the management cycle: planning, organizing, staffing, directing, controlling, and evaluating (**Figure 7-2**). As previously stated, the

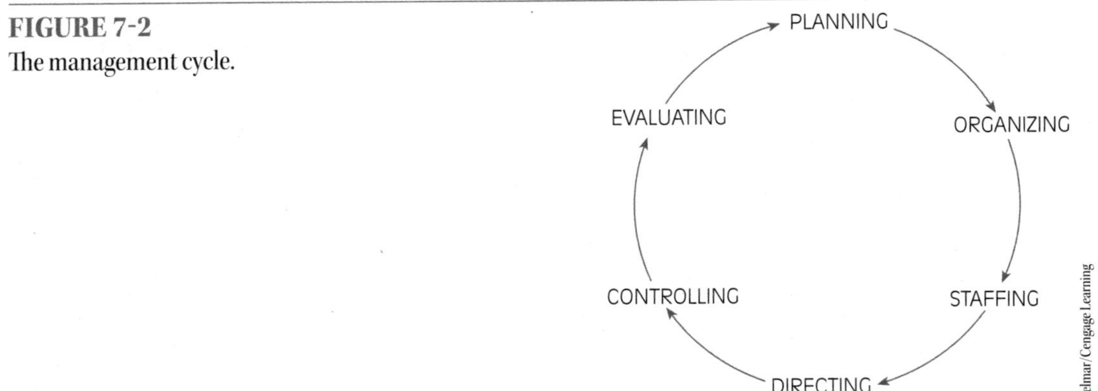

FIGURE 7-2
The management cycle.

department must have clearly defined goals. These are often described overall in a mission statement, such as the following:

> The goal of the fire department, through its members, is to provide effective life safety and emergency services in the most efficient and cost-effective manner possible to ensure public safety and minimize economic loss.[7]

The mission statement just quoted is not clear in that it does not spell out anything specific. It is conceptual in nature and the fire department administration is charged with setting the objectives to achieve the stated goal.

Goals

Goals tend to be nonspecific and unmeasurable in that they are too broad. The goal stated is to provide "effective life safety." Just exactly what is effective life safety? The goals must be supported with statements that specify just what exactly the department can expect to accomplish. Not all fire deaths can be avoided. Accidents are going to occur, no matter how effective the firefighting force and no matter how much the jurisdiction spends on fire protection. There could be a fire station on every corner in town and fires would still occur. People will die from medical and other emergencies. The department must determine where its resources can best be applied to provide the most good for the most people.

One method of doing this is for the chief to concentrate resources in the high-value districts of the jurisdiction. If there is a downtown area, that is where most of the resources will be stationed. In the outlying, even rural areas, the staffing can be reduced and the response areas for individual fire stations increased.

Objectives

Objectives are statements of measurable results to be achieved with the resources available. All objectives must be specific, measurable, attainable, realistic, and timely. The objectives must be oriented toward the stated goal. The fire department must have the resources available or arrange for their acquisition. The resources can be in the form of money, personnel, and equipment. An objective would be to reduce deaths from fires to so many per 1,000 of population. This may sound cold and calculating, but it is the only method against which progress can be measured.

The management cycle is used as a problem-solving process to accomplish the goals and objectives of the department within the context of the mission statement. Each step in the process is a unit unto itself, but it is related to the steps taken either before or after. The steps must be taken in the proper order or the process will fail.

Planning

The first step in the cycle is planning. Planning consists of determining the objectives and deciding which resources may be used in the most effective and economical manner to achieve those objectives.

Planning requires the use of policies. Policies are described as a definite course or method of action. Policies tend to be broad, encompassing different situations. The department's policies are to be written and clearly understood by all members of the fire department. In many organizations, policies were determined on an assignment and live on in the culture of the organization. An example of a policy that could have been initially developed this way—but that should be written down—would be that all personnel are to wear breathing apparatus on incidents when smoke is showing. By writing and maintaining clear and applicable policies, the management of the department will aid in the smooth running of the organization and avoid conflict.

The following is an example of a policy:

> To guard against incurring injuries during drills and the handling of basic fire fighting tools, the following rules will be adhered to:
>
> 1. Engine companies participating in drills shall wear the structure helmet, turnout coat, and gloves, when protective clothing would normally be worn, while performing the same evolutions in emergency situations.

Procedures are defined as a particular way of accomplishing something or acting in a specified situation. An example of a procedure would be that the oncoming shift is to make a face-to-face shift change with the off-going shift to discuss any problems with the equipment or other matters that need to be passed on. Procedures are more specific than policies. A specific description of how a form is to be filled out, such as a fire report or prevention inspection form, would be a procedure. The everyday operations of the department are mostly guided by procedures. Without the procedures, there would not be a standardized way of doing things and each shift and station would be like its own little fire department. A large part of studying, as a new firefighter and for promotion, is learning the policies and procedures applicable to the position in particular and the rules of the organization as a whole.

The following is an example of a procedure:

The procedure for purchasing an off-duty badge shall be as follows:

1. Any member wishing to purchase an off-duty badge should have the vendor submit a request for verification of employment to the Administrative Deputy Chief.
2. Upon verification of employment, the Administrative Deputy Chief will send the verification back to the vendor. A copy of the verification will be retained by the department.
3. The department is to be notified immediately should the badge be lost or stolen or in the event another badge is to be ordered.

Using objectives, policies, and procedures, the management sets the course of the organization to reach the desired goals. All of the items must remain within the confines of the mission statement. It may be nice for the chief to have his own helicopter to fly around in and observe operations, but it is not very cost effective.

Organizing

The second step is organizing. When organizing, the managers of the department bring together essential resources and incorporate them in a structured relationship. The framework of the organization is the chain of command and the table of organization of the department. Without a structured framework, the organization could not function because people would not know whom to go to for answers to questions or authorization to do what needs to be done.

Organization is done after planning because the manager needs to know what has to be done to structure the proper organization. The manager must also be able to determine whether the **fiscal** and personnel resources are available to fill the positions in the organization. It does no good to have a position of public education/information officer in the organization if there is no money to hire someone to perform those functions. If this is the case, the manager is going to have to restructure and place the responsibilities elsewhere, usually in addition to the duties of another position.

fiscal Financial.

Staffing

The third step is staffing. Staffing is the assignment of resources to the needs that have been identified in achieving the objectives. Staffing applies to both the line and staff functions.

One of the biggest debates in the fire service is the one over what constitutes adequate staffing on an engine company. This debate became so heated that in 1995, the International Association of Fire Fighters (IAFF) withdrew from the National Fire Protection Association (NFPA) for a period of time over this very issue.[8] Studies have been done over the effectiveness of a three-person company versus a four- or five-person company. Timed evolutions were performed to determine the results. It was found that a larger company could more quickly perform the basic evolutions, but there did come a point where the increase in personnel gave a diminishing return and did not justify the added expense.

In a 1995 study, several conclusions were drawn.[9]

1. The more firefighters available, the more rapidly the task is completed.
2. More complex tasks require more firefighters. With complex tasks, more people available allowed for some of the steps to be performed simultaneously instead of one at a time. One of the overriding factors was that complex tasks require careful coordination to be performed efficiently and safely.
3. When operations require working above or below grade (ground level), it complicates the operation. When resources, equipment, and personnel must be moved either up or down stairs or ladders, the time required to perform the operations is extended. This also has a tendency to require that personnel be rehabbed more often to maintain their level of performance. Carrying and climbing is much harder on personnel than just carrying.
4. The ambient temperature is a critical factor. Temperature extremes, either hot or cold, tend to use people up at a faster rate. With more personnel to share the workload, the operations can be performed with less individual exertion.
5. The quality of the leadership makes a difference, especially in complex operations. A group of firefighters making an uncoordinated attack will probably get the fire out eventually. A group of firefighters making a well-coordinated attack can perform with more efficiency and get the job done more quickly.
6. Level of training, physical fitness, and competence of the firefighters is important. Conditioned firefighters who know their jobs and equipment can perform the task with efficiency and safety with little time lost in deciding what to do and how to do it. This also reinforces the point that good leadership is a must.

NOTE Conditioned firefighters who know their jobs and equipment can perform the task with efficiency and safety with little time lost in deciding what to do and how to do it.

A study completed in 2010 by the National Institute of Standards and Technology drew the following conclusions when comparing two-person companies to three-, four-, and five-person companies.[10]

- *Overall Scene Time.* The four-person crews operating on a low-hazard (one-, two-, or three-family dwellings and some small businesses) structure fire completed all the tasks on the fireground (on average) seven minutes faster—nearly 30%—than the two-person crews. The four-person crews completed the same number of fireground tasks (on average) 5.1 minutes faster—nearly 25%—than the three-person crews. On the low-hazard residential structure fires, adding a fifth person to the crews did not decrease overall fireground task times.
- *Time to Water on Fire.* There was a 10% difference in the "water on fire" time between the two- and three-person crews. There was an additional 6% difference in the "water on fire" time between the three- and four-person crews (i.e., four-person crews put water on the fire 16% faster than two person crews). There was an additional 6% difference in the "water on fire" time between the four- and five-person crews (i.e. five-person crews put water on the fire 22% faster than two-person crews).
- *Ground Ladders and Ventilation.* The four-person crews operating on a low-hazard structure fire completed laddering and ventilation (for life safety and rescue) 30% faster than the two-person crews and 25% faster than the three-person crews.
- *Primary Search.* The three-person crews started and completed a primary search and rescue 25% faster than the two-person crews. The four- and five-person crews started and completed a primary search 6% faster than the three-person crews and 30% faster than the two-person crew. A 10% difference was equivalent to just over one minute.
- *Hose Stretch Time.* In comparing four-and five-person crews to two-and three-person crews collectively, the time difference to stretch a line was 76 seconds. In conducting more specific analysis comparing all crew sizes to the two-person crews, the differences are more distinct. Two-person crews took 57 seconds longer than three-person crews to stretch a line. Two-person crews took 87 seconds longer than four-person crews to complete the same tasks. Finally, the most notable comparison was between two-person crews and five-person crews—with more than 2 minutes (122 seconds) difference in task completion time.

Large fire departments, with numerous companies available, have implemented the practice of staging a tactical reserve, called rapid intervention teams (RIT; see in Chapter 14). The tactical reserve is used to stand by at the fire scene. They are in full PPE, equipped with the necessary tools and ready to go in, but uncommitted unless there is a firefighter emergency. What qualifies as a firefighter emergency is a firefighter(s) trapped, lost, low on air, or injured either due to structural collapse or other situations that endanger firefighters' lives. To many in smaller departments, such reserves seem like a luxury. Some firefighters may even question the need, feeling that everyone at the scene should be involved in controlling the fire. The added margin of safety in having the RIT Team available at a moment's notice is not to be discounted.

The staff positions must be adequately filled as well. If effective fire prevention is to take place, enough personnel must be assigned to that function. They are better trained in fire prevention and committed to it as a means of achieving the organizational goals. In addition, personnel must be assigned to training on a full-time basis to keep a department of larger size trained on the new developments and techniques.

Directing

Directing can be described as guiding and supervising the efforts of subordinates toward the attainment of specified objectives. Objectives come in a variety of forms. Directing is accomplished through rules, standard operating procedures, job descriptions, and assigned duties. Having many of the guidelines written relieves the administrator of having to give the subordinate detailed guidance in most situations, and lets the subordinate know what is expected of him.

Rules set the limits for behavior on and sometimes off the job. A standard rule would be, "No person shall absent themselves from duty without the notification and approval of their supervisor." This rule makes very clear what is considered acceptable behavior. Rules give guidance for what is required as well as a set of guidelines for determining when discipline is required. Within the framework of the rules there can be circumstances that provide for leeway. The supervisor may be empowered to allow the subordinate to be absent for up to an hour in certain circumstances. This would be a policy modifying the rule. The trouble is it is often not written down and could cause problems if discipline were applied against one person and not another.

NOTE Rules give guidance for what is required as well as a set of guidelines for determining when discipline is required.

Job descriptions are an important part of any organization. From the very beginning of the selection process for new firefighters, the job description is utilized (see Chapter 2). Through job descriptions, the department decides what is to be included in the testing and training processes for positions. They are also used to **validate** the testing process. The fire department usually does not ask management level questions of a firefighter applicant, because management of other personnel is not part of a firefighter's job description. Job descriptions let people know what is expected of them in the position they are assuming as well as defining the limits of authority and responsibility of the assigned position. Obviously a job description for battalion chief is much different than the one for firefighter.

validate To make sure that the items included are actual requirements of the job.

One phrase included in many job descriptions is "and other duties as assigned." Job descriptions are often used as a minimum standard of what is expected, not the maximum. A firefighter performing a line function would have much different duties than one assigned to staff.

NOTE Job descriptions are often used as a minimum standard of what is expected, not the maximum.

Controlling

Controlling is the process of determining if the organization is working toward its goals and objectives. When discrepancies are found, it is the personnel's responsibility to perform the necessary corrective actions to get back on course. Controlling is done in many ways.

One of the controls that has the largest effect on the fire department is the annual budget. In most departments, the budget period is a fiscal year, from July 1st to June 30th, The Federal budget cycle is October 1 to September 30th. During the time the budget covers, the chief is allowed only so much money to accomplish the fire department's goals. The chief has personnel working for him that assist in this function. The financial officer keeps the chief up to date as to where the department is within the year's approved budget. If the overtime budget is being depleted faster than is reasonable because of several large incidents, the chief must either shift funds from another budget area or seek more money from the jurisdiction.

NOTE One of the controls that has the largest effect on the fire department is the annual budget.

An area that all personnel can have an effect on is the use of utilities. Heating and cooling a fire station is quite expensive over the course of a year. As you may know, just turning off the lights when you leave a room is a way of conserving energy and reducing costs. Another area is the proper use and care of equipment. Many departments have the procedure that the oil is to be checked in fire apparatus on a daily basis. If this is not done, low oil could destroy a motor. To rebuild a large diesel, like those in fire apparatus, costs thousands of dollars. Often through proper maintenance and care, problems like this can be avoided. The money would be much better spent on new PPE or other tools to enhance the department's efficiency and effectiveness.

Evaluating

Evaluation is the process of determining whether the organization's goals and objectives are being met. The mission statement identifies part of the evaluation in the phrase, "in the most efficient and cost-effective manner possible." The method used to evaluate is to measure the performance of the fire department against its own goals and objectives. Are they being met in a timely, cost-effective manner? The objectives, as previously stated, must be measurable and attainable. Without the ability to measure progress toward the determined objectives, management cannot tell if they are being achieved.

NOTE Without the ability to measure progress toward the determined objectives, management cannot tell if they are being achieved.

The evaluation of the fire department is carried out both internally and externally. Internally, progress reports, program records, personnel evaluations, and apparatus and station inspections are used. Externally, the ultimate evaluator is the public. Is it satisfied with the performance of the fire department? Its opinions are expressed through direct contact and through elected officials. Proactive fire departments may go to the extent of sending questionnaires out to the public to determine if it is satisfied with the department's effectiveness and to identify areas where the department can improve.

The budget is also a valuable tool in evaluating the fire department. If the fire prevention budget was increased 50% and there is no reduction in the number of accidental fire starts, further study needs to take place. The study is performed to determine whether the program is effective, and if not, why not.

Evaluation must be carried out objectively. One fire in which a firefighter made a spectacular rescue does not mean that the whole department is doing its job well.

Evaluation is to be carried out in comparison with accepted standards. If fire department A has five persons per engine and fire department B has only three, the losses in comparable fires should be less for department A.

From the first time you fill out a job application for firefighter, you are being evaluated. All during your career, no matter what rank you attain, you will be evaluated. Those below you will evaluate you as a leader and those above you will evaluate you as a subordinate. Everyone answers to someone else. The firefighter is evaluated by the company officer, the company officer is evaluated by the battalion chief, and so on. Even the chief must answer to the fire commission or the directly elected officials of the jurisdiction.

NOTE From the first time you fill out a job application for firefighter, you are being evaluated.

Evaluation of employees is an ongoing process. It does not only occur when your annual evaluation form arrives at the station. The company officer and the remainder of the crew are also evaluating you for your strengths and weaknesses every time you are on duty. The ability to accept evaluation for what it is, a tool to aid in attaining the goals and objectives, will assist you in becoming a better firefighter and an asset to the organization.

Do not overlook the first phrase in the mission statement, "The fire department, through its members. . . ." Never forget that what makes a great fire department is not money; it is the people who work there. When these people are trained and motivated, they can do great things. The lack of the latest in equipment and facilities should not severely limit the department's abilities. Many a life has been saved and untold numbers of fires have been controlled without the latest equipment. What really get the job done are the desire, resourcefulness, and abilities of the personnel on the fire engine.

NOTE Never forget that what makes a great fire department is not money; it is the people who work there.

FIRE DEPARTMENT TYPES

The U.S. Fire Administration estimates that there are more than 30,000 fire departments across the United States involving 1.2 million firefighters, of which approximately 200,000 are career and one million are volunteer. There are numerous types of fire departments throughout the nation. Depending on the needs and resources of the jurisdiction, the departments vary in size. As departments increase in size and number of persons served, they increase in complexity.

Volunteer Fire Department

The first fire departments in the United States were volunteer. Their spirit lives on in the personnel who give their time and effort to suppress fires and perform other necessary functions without pay. Of all fire departments in the United States, 90% are composed entirely or mostly of volunteers. These departments protect 42% of the U.S. population.[11] Most of the volunteers (94%) are in departments that protect fewer than 25,000 people, and more than half are located in small, rural departments that protect fewer than 2,500 people.[12]

NOTE Of all fire departments in the United States, 90% are composed entirely or mostly of volunteers.

Volunteer fire protection service serves as the preliminary first-step fire service, to be followed, as the jurisdiction grows in size and population, by some form of paid service.[13]

The volunteers must often participate in fund-raising activities to buy the necessary equipment. Fire engines and equipment are not cheap. Often older engines can be purchased from paid departments at a nominal fee. Other times equipment is donated outright. Departments are becoming more reluctant to donate equipment as they will be named in the lawsuit if injury results from use of the equipment.

Just because a department is made up of volunteers is no reason for it to lack in effectiveness. There are numerous training programs available from the National Fire Academy and state fire-training organizations. One advantage volunteer departments have is their ability to spend most of their money on equipment. A paid department can spend upwards of 90% of its funding on personnel costs alone.[14]

NOTE Just because a department is made up of volunteers is no reason for it to lack in effectiveness.

Within the realm of volunteer fire departments, there are differences. Some volunteer departments are totally volunteer; these have no paid members. Others have several paid members supported by volunteers. A common paid position in a primarily volunteer department is that of driver (**Figure 7-3**). By having a paid driver on each shift, the department has someone who is responsible for maintaining and ensuring the readiness of the equipment. The paid person also maintains the fire station. In areas with long responses, the paid person can respond in the engine and the volunteers can meet him at the scene. This prevents the problem of all of the volunteers responding to the station, finding that the equipment has already left with the volunteers who got there first. The other problem scenario is that all of the volunteers respond directly to the scene and no one brings the engine.

In the past, the person who received the alarm would transmit it through a system of bells or sirens. The sequence of the siren sounding, a series of long and short blasts, notified the volunteers as to which part of town to respond. Now most volunteers have voice or alpha-numeric pagers and can be notified at home or work as to the exact address of the assignment.

The obvious limitation is that it is not predetermined how many personnel are going to arrive at the scene. When a fire officer has a regular crew that she works with, she knows their strengths and weaknesses. She also has a good idea of the next-in company's response time and capabilities.

FIGURE 7-3
Volunteer fire department organization with paid driver/operator position.

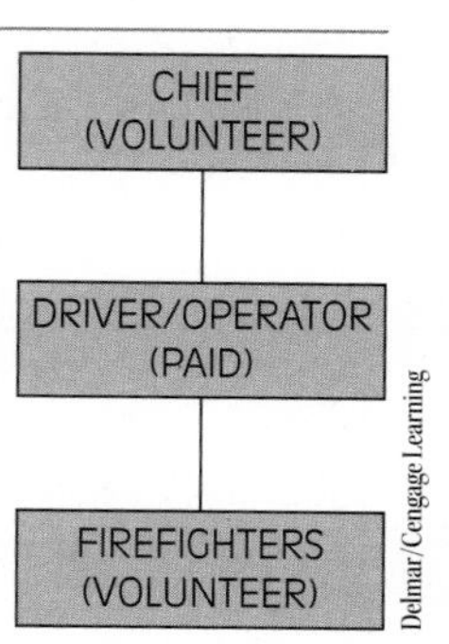

There are also variables among volunteers. Some volunteer firefighters are in good physical shape and spend lots of their free time becoming trained; others are members, but lack training and commitment.

The term "volunteer" is being replaced with "on call" or "paid call" in many states. Labor law court rulings have determined that volunteer firefighters are employees of the jurisdiction they serve and must be compensated for their work as well as be covered by Workmen's Compensation Insurance should they be injured in the line of duty.

Combination Fire Department

In combination departments, a large part of the staff is paid and the volunteers are used as supplemental personnel (**Figure 7-4**). They are either used to cover the station when the regular crew is on an assignment or they respond directly to the scene. This method of staffing saves the cost of paying a large staff of firefighters when they are often not needed. One advantage is that the community receives protection at a prescribed minimum level all of the time. Only when circumstances dictate are the volunteers paged to respond. This is much less disruptive to the volunteers' jobs and businesses. They do not need to respond to every small incident that can be handled by the paid personnel. When the inevitable large incident occurs, there is a reserve force of personnel to help in mitigating it.

Another advantage is that the paid firefighters are available to receive training and pass it on to the volunteers. They are also available to receive specialized training and perform fire prevention activities. In this type of organization, the position of chief is usually paid as the paid staff is increased to the point where a professional manager is required for planning and budgeting.

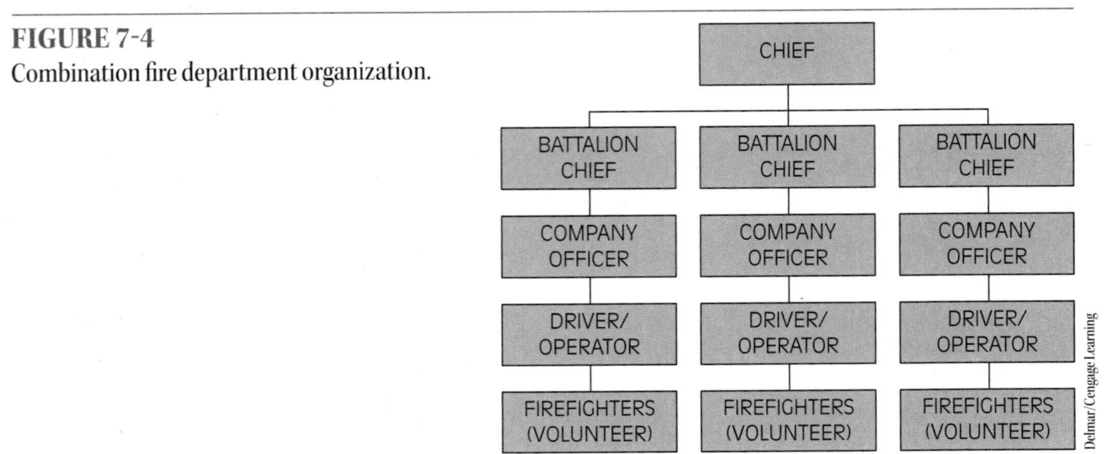

FIGURE 7-4
Combination fire department organization.

Many departments use this concept in the form of a force of reserves. The reserves are usually comprised of young people with a career in the fire service as their goal. This program serves both parties well as the reserves seek training and experience and the department gains a group of trained personnel at little or no cost.

Another variation on volunteers is paid-call firefighters, reserve personnel who are paid by the incident. If they are paged to respond to **cover** a station, an incident, or a drill, they are paid a certain amount. This program is often used by departments to supplement forces in small towns and outlying areas. The ability to pay the reserve force is an incentive for the reserve personnel to attend the necessary drills and covers.

A problem that needs to be addressed in a proactive manner is the training mandated by state and federal law. From a liability standpoint, the jurisdiction is required to train the personnel to at least minimum levels. Some of this training is very time consuming and is hard for volunteers to complete due to their regular job commitments. As time goes on, jurisdictions are going to have to find ways to address this issue or suffer the consequences when someone is hurt or killed.

cover To move resources into a fire station when the regular crew is assigned to an incident. In some departments this is called a *move up.*

Public Safety Department

In an effort to save money and better utilize personnel, some jurisdictions have set up public safety departments (**Figure 7-5**). Under this concept, the police and fire departments are under the same department head. They may go as far as to have the personnel cross trained so that they can go out on patrol as police officers and respond to fires as needed. The advantages are that the jurisdiction gets productive time out of their personnel most of the time that they are on duty. The disadvantage is that if there is a crime or major fire, most of the resources are going to be tied up on one or the other. Another disadvantage is that both police work and firefighting require a lot of time spent on specialized training and a person who is cross trained is not likely to have the training required to be very proficient at either job, or they will concentrate on one over the other.

Career Fire Departments

The basic premise of a paid fire department is that all of the personnel are paid a salary, or in the case of the U.S. Forest Service, Bureau of Land Management, and Park Service, paid by the hour. In the large cities, such as Boston, Los Angeles, Chicago, Nashville, and New York, the departments are of such a size that a fully paid department is a necessity. Their operations are too large and complex to be performed by volunteers.

One advantage of a paid fire department is that the jurisdiction has much more control over the fire personnel. The jurisdiction has the ability to hire and fire as necessary.

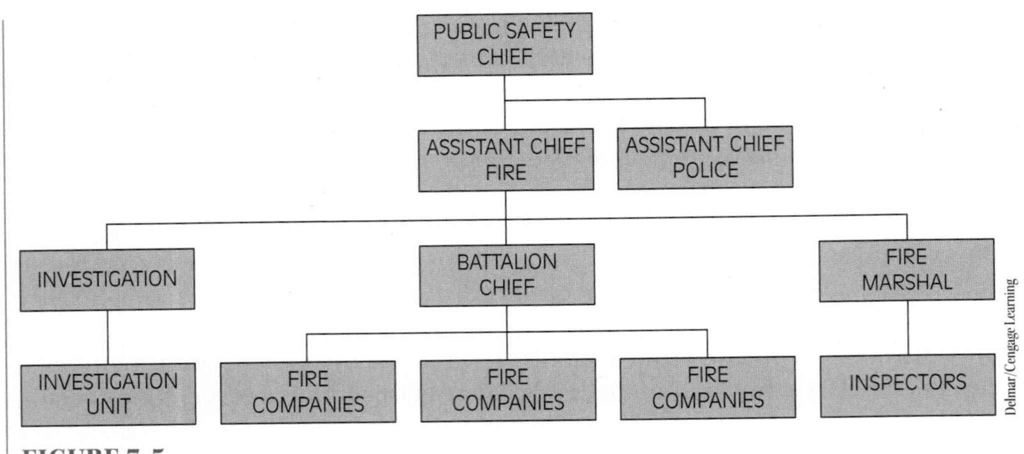

FIGURE 7-5
Public safety department organization (law enforcement organization not shown).

When people are hired they are offered a rewarding career with pay and benefits. The person is expected, in return, to abide by the rules, procedures, and policies of the department.

NOTE One advantage of a paid fire department is that the jurisdiction has much more control over the fire personnel.

Paid fire departments, especially the large ones, require expert management (**Figure 7-6**). It is becoming less common for the top officer to be from a purely firefighting background. The best firefighter in the department, from an operations standpoint, is not necessarily a good administrator. The chief of department, in the large departments, does not even respond to fires. The job of administrator requires his full attention. Administrators of large fire departments set the tone for the organization through policies and goals. Their work is accomplished by delegating authority to others to make sure that rules and regulations are followed and the mission of the department is accomplished. By referring back to the National Professional Development Model in Chapter 1 it is easy to see that the new generation fire administrator requires both training and education to perform the job effectively.

The people in the top jobs may have come to work as firefighters, but they have educated themselves in the areas of public or fire administration. An easy way to see what is required for the top spots is to look at the job offers for Chief in the back of magazines such as *Fire Engineering*, *Fire Chief*, or *Firehouse*. These job offerings require administrative backgrounds with an advanced education, usually a bachelor's degree minimum, with a Master's Degree preferred. Often successful completion of the Executive Fire Officer Program at the National Fire Academy is required.

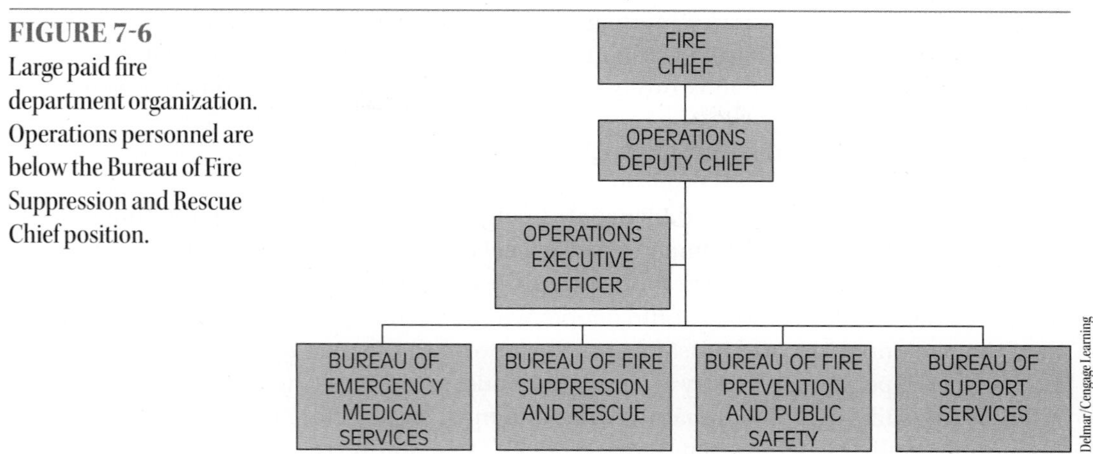

FIGURE 7-6 Large paid fire department organization. Operations personnel are below the Bureau of Fire Suppression and Rescue Chief position.

One of the most valuable attributes of a chief is the ability to communicate. When you are in a position where you must make it clear to others what you want done, without performing the function yourself, it is very important to be able to communicate what exactly it is you want. This means face-to-face, on paper, and over the phone. The chief also needs to be somewhat of a diplomat to present the needs of the fire department to the governing body and to interpret the wishes of the governing body and convert them into fire department action.

Some of the management concepts that are popular with chiefs are customer service, the one-department concept, team building, and incident effectiveness. Customer service refers to the fact that the public is the customer and, like any other business, the fire department must serve the customer's needs or be replaced. The department will not actually go out of business, but budgets can be cut and personnel replaced or other alternatives can be explored.

NOTE Customer service refers to the fact that the public is the customer.

The one-department concept is especially important in large departments with many stations. The fire department is already divided among the shifts. When this is compounded in large geographical areas with varying needs, it is difficult to have a cohesive department that works basically the same on all shifts at all stations. Standardization is sought as much as possible. In a department that covers municipal and rural areas, there will be some differences in operations and activities. Standardization is often established as far as practical so that a firefighter working overtime at another station does not have to learn a whole new set of procedures used at incidents. The one-department concept also applies to attitude. The personnel who work in a metropolitan area closer to headquarters or at a battalion headquarters station are liable to be held to a higher standard of dress and performance than those working in stations in outlying areas. You must remember that you are paid to act and look professional no matter what station you are assigned to.

NOTE Standardization is sought as much as possible.

Team building is performed by people being willing to work together within department guidelines. For team building to succeed each person must know his or her duties and responsibilities and where they fit in the overall scheme of things. They must also know the desired results of their actions. Some people will fill certain positions better than others. All of the personnel involved must be able to conform to the plan to work for the common good. All of the personnel on duty at the station must be willing to help each other out, even in simple tasks like equipment maintenance and meal preparation. Each shift must be able to support the other in seeing that things get done at the station regardless of who is on duty. If the other shift had a fire just before they went off duty and had to leave a bunch of dirty tools, the oncoming shift should take care of them without complaint. Sooner or later you will be leaving it for them when you have a fire. This is not just at the single-station level. All of the stations that respond to incidents together need to be able to work together as an integrated unit. This concept carries right on up to the top of the organization.

Incident effectiveness can be defined as the fire department's ability to function quickly and efficiently when called upon to act. Incident effectiveness is achieved through focus on the four guiding principles of drill, equipment maintenance, programs, and station maintenance. If everyone in the department performs his job well, the department will be effective. Performing drills raises the proficiency of individuals and the team as a whole. Equipment maintenance ensures that the apparatus will make it to the scene and that all of the tools are in working order. Programs make the personnel more aware of the problems in their response area and contribute to the accomplishment of the department's mission. Station maintenance leads to a more pleasant work environment. If one shift leaves the station a mess every time it goes off duty, it will create tension with the other shifts. Something as simple as emptying the trash cans and wiping out the sinks after use makes the station a more pleasant place for everyone.

NOTE Incident effectiveness can be defined as the fire department's ability to function quickly and efficiently when called upon to act.

In the modern fire department, more and more of the upper management positions are being filled by civilians. These people are chosen for their particular knowledge, skills and abilities. They are not directly responsible for firefighting activity, so do not need firefighting skills. The jurisdiction derives a cost savings by not having to pay these people a safety section retirement. The common complaint is that they are "bean counters" and do not understand firefighter needs and priorities. The downside of this is outweighed by the fact that these personnel are trained in administration and do their jobs very well. They are experts at budgeting and resource management. Their ability to squeeze the maximum benefit from every budget dollar positively affects the firefighter in an indirect way. Most departments have recognized the need to keep the line functions under the control of managers with firefighting experience and limit the hiring of management personnel with no firefighting experience to staff positions.

In adhering to the principle of span of control, depending on the size and complexity of the department, there are layers of management between the chief and the firefighter. Not all fire departments have the same name for the positions, but the duties and responsibilities of the positions are much the same.

Under the chief are the various assistant and deputy chiefs. These personnel are not required to perform the same functions as the chief but are required to have a high degree of ability as administrators as well. A deputy chief in charge of operations in a major department may have 2,500 employees under his direct and indirect control. His communication powers must be of the highest ability as well. This chief will not know all of the firefighters in the stations, but he is responsible for their safety and efficiency.

The person in charge of several stations is the battalion or district chief. This person usually works the same duty schedule as the firefighters and is responsible for the activities of one shift, whereas staff personnel usually work a 40-hour week (an 8–5 workday). The battalion chief responds to emergencies when necessary. The battalion chief probably does not respond to medical aids, car fires, and other minor incidents but does respond to any major incident, such as a structure fire, and other incidents that could become major or complex. It is his responsibility to ensure that the company officers and their crews are trained and performing to the expected level of proficiency. This is done through company officers' meetings and evaluating engine company proficiency by observing drills and performance on emergencies. The battalion chief evaluates the company officer on the performance evaluation (**Figure 7-7**) and assists the company officer in determining goals for the next rating period.

When the battalion chief performs a station inspection, he inspects all of the station's assigned programs, records and reports, maintenance of the station, and condition of the apparatus (**Figure 7-8**). The station personnel are inspected for the proper work uniform, PPE, conformity to the grooming standard, and possession of a valid driver's license. The station inspection is a perfect opportunity for the battalion chief to ask the crew questions about the apparatus and department policy to make sure that they have been studying and keeping up on policy and procedure changes.

At the station level, the company officer is the first-level supervisor. He is responsible for ensuring that his crew is trained and equipped to respond to emergencies. The company officer evaluates the driver/operator and the firefighter on his performance evaluation and assists him with determining his goals for the next rating period. In the case of a probationary firefighter there may be an evaluation every three or six months until probation is completed.

EMPLOYEE PERFORMANCE REPORT
PERSONNEL DEPARTMENT

DEPT. NO.	DEPARTMENT NAME	CLASSIFICATION	EMPLOYEE NAME	EMPLOYEE NO.

REASON FOR RATING:

IF 6 MONTH PROBATIONARY DO YOU RECOMMEND PERMANENT APPOINTMENT? YES ☐ NO ☐ ☐ SPECIAL ☐ SEPARATION

RATING PERIOD FROM TO

SECTION A - ITEMIZED CHECK LIST

EMPLOYEE'S IMMEDIATE SUPERVISOR SHOULD CHECK EACH ITEM IN THE APPROPRIATE COLUMN. REPORT MUST BE COMPLETED IN INK. ANY CHANGES MADE IN THE REPORT SUBSEQUENT TO THE EMPLOYEE'S SIGNING REQUIRE INITIALING BY THE EMPLOYEE AND PERSON MAKING THE CHANGES.

(READ INSTRUCTIONS ON BACK)

	DOES NOT APPLY	OUTSTANDING	ABOVE STANDARD	STANDARD	IMPROVEMENT NEEDED	UNSATISFACTORY
ALL EMPLOYEES:						
1 ATTENDANCE						
2 PUNCTUALITY						
3 PHYSICAL FITNESS						
4 SAFETY PRACTICES						
5 PERSONAL NEATNESS						
6 COMPLIANCE WITH RULES AND REGULATIONS						
7 COOPERATION						
8 ACCEPTANCE OF NEW IDEAS AND PROCEDURES						
9 APPLICATION OF EFFORT						
10 INTEREST IN JOB						
11 ACCURACY OF WORK						
12 QUALITY OF JUDGMENT						
13 PUBLIC RELATIONS						
14 WRITTEN EXPRESSION						
15 ORAL EXPRESSION						
16 EQUIPMENT OPERATION						
17 NEATNESS OF WORK						
18 PERFORMANCE WITH MINIMUM SUPERVISION						
19 PROMPTNESS IN COMPLETING WORK						
20 VOLUME OF WORK PRODUCED						
21 PERFORMANCE UNDER PRESSURE						
22 PERFORMANCE IN NEW WORK SITUATIONS						
EMPLOYEES WHO SUPERVISE:						
1 COORDINATING WORK WITH OTHERS						
2 ACCEPTANCE OF RESPONSIBILITY						
3 ESTABLISHMENT OF WORK STANDARDS						
4 TRAINING AND LEADING STAFF						
5 PLANNING AND ASSIGNING WORK						
6 FAIRNESS AND IMPARTIALITY TO STAFF						
7 CONTROL OF STAFF						
8 ADEQUACY OF INSTRUCTIONS						
ADDITIONAL ITEMS:						

SECTION B - OVERALL PERFORMANCE

CHECK OVER-ALL EVALUATION WHICH MUST BE CONSISTENT WITH THE FACTOR RATINGS. ALTHOUGH THERE IS NO PRESCRIBED FORMULA FOR COMPUTING THE OVERALL PERFORMANCE, SPECIFIC WRITTEN COMMENTS ARE REQUIRED TO JUSTIFY OUTSTANDING OR UNSATISFACTORY RATINGS.

OUTSTANDING | ABOVE STANDARD | STANDARD | IMPROVEMENT NEEDED | UNSATISFACTORY

COMMENTS:

EMPLOYEE'S CERTIFICATION: (CHECK ONE)

☐ I HEREBY CERTIFY I HAVE REVIEWED THIS REPORT. I UNDERSTAND MY SIGNATURE DOES NOT NECESSARILY MEAN I AGREE WITH ALL THE MARKINGS.

☐ I REQUEST AN APPOINTMENT TO DISCUSS THIS RATING WITH MY DEPARTMENT HEAD.

EMPLOYEE'S NAME ______________________ DATE ________

RATED BY ______________________ DATE ________

TITLE ______________________

REVIEWED BY ______________________ DATE ________

TITLE ______________________

REVIEW INSTRUCTIONS ON REVERSE SIDE

PERSONNEL DEPARTMENT COPY

PRINTED ON REGENESIS TM POST CONSUMER RECYCLED PAPER

Courtesy of Captain Steve Pendergrass, Kern County Fire Department

FIGURE 7-7
Personnel evaluation form.

STATION INSPECTION RECORD

STATION ____________ SHIFT ____________ DATE ____________

CAPTAIN ____________ BATTALION CHIEF ____________ DIVISION CHIEF ____________

✓ - O.K
X - SEE REMARKS

1. PERSONNEL

A. Uniform
B. Protective Clothing
C. Grooming & Cleanliness
D. Driver's License
E. I.D. Card
F. E. C. Card

5. SAFETY

A. Overhead Storage
B. Caution Signs
C. Safety Bulletin Board
D. Tool Storage
E. Combustibles Storage
F. Insecticide Chemical Stor.
G. SCBA Tests
H. Extinguisher (s)
I. Smoke Detector (s)
J. RADEF Equipment

6. EXTERIOR

A. Paint
B. Doors
C. Sidewalks
D. Ramps
E. Parking Area
F. Yard
G. Hose Tower
H. Gas Pump
I. Flag Pole

2. OFFICE

A. Files
B. Maps
C. Log Book
D. Shift Change Log
E. Desk
F. Bulletin Board
G. Tourist Information

3. ENGINE HOUSE

A. Floor
B. Work Bench
C. Hose Rack
D. Storage
E. Turnouts
F. Apparatus

4. PROGRAMS

A. Hazard Reduction
B. In Service Inspections
C. Pre Fire Plans
D. Water Systems

7. INTERIOR

	Office	Kitchen	Day Room	Dormitory	Rest Room	Heater Room		
A. General Appearance								
B. Cleanliness								
C. Floors								
D. Windows								
E. Lights								
F. Furniture								
G. Storage Area (s)								
H. Woodwork								
I. Walls								
J. Ceiling								

REMARKS ____________

Fire 580 2415 016 Catalog # 9020 (Rev. 2/90)

Courtesy of Captain Steve Pendergrass, Kern County Fire Department

FIGURE 7-8
Station inspection form.

NOTE At the station level, the company officer is the first-level supervisor.

As a training tool, the company officer may delegate authority to complete some of the station level programs to the driver/operator and firefighter. This requires the crew members to be more familiar with the program and teaches them management skills. It also helps to prepare the crew members to take promotional examinations. The driver/operator is responsible for the equipment and knowing how to operate it. Most driver/operators will take the time to train the firefighters in the use and operation of all of the equipment.

The firefighter position's responsibilities were covered in Chapter 1.

Industrial Fire Brigades

Many businesses have organized private fire brigades. These may be organized to protect a manufacturing plant, oil refinery, or other location. The brigades are made up of personnel hired by the company. Often the personnel have other jobs and the fire brigade is an additional duty that is performed when necessary. They receive equipment from their employer. The training they receive may be provided by the employer, a contracted service, the local fire department, or a combination of these. They are often organized into teams. They may respond to alarms in all parts of the property, or each geographical or functional area may have a separate fire brigade organization according to the needs of the property. The organization is such that a fire brigade is on duty on each working shift and at periods when the plant is shut down or idle.[15]

Contract Fire Protection Service

There are private sector companies that provide fire prevention and suppression services. These companies arose in the sprawling suburban and rural communities of the United States and are involved in every aspect of the fire prevention and suppression business. They provide services in national parks and forests, airports, nuclear plants, commercial businesses and industrial firms, and rural and residential neighborhoods.

The private fire protection companies provide fire protection service either by contract or subscription. Contract service is offered to local governments or special fire districts; subscription service is offered to residents or property owners through the payment of a subscription fee.

In the case of contract fire protection service, the officially designated representatives of a local jurisdiction (usually a town, city, or special tax district) award a private fire protection company the right to service that jurisdiction for a specified time period. The company is paid a fixed and contractually agreed upon sum, either through the jurisdiction's general tax revenues or through a special fire tax levied by the jurisdiction.

In the case of subscription fire protection service, individual property owners or residential associations contract directly with a private company for fire protection service. When a fire occurs, the people who have a subscription on their property are not charged. If they do not have the subscription coverage, they are charged suppression costs.

The private fire protection companies are able to generate cost savings in a number of ways:

- *By reducing personnel costs.* These companies rely heavily on trained volunteers and reservists. They often pay salaries and benefits commensurate with market rates, which are typically less than those earned by public sector fire department employees.
- *By making productive use of otherwise idle time.* When not actually fighting fires, private firefighters are engaged in a number of other endeavors, including building and refurbishing fire apparatus, providing combination fire and security patrols, training industrial fire brigades, operating alarm monitoring and installation services, and sponsoring fire prevention campaigns and other educational activities.
- *By using innovative strategies and technologies to prevent and combat fires.* As an example, some of the private sector companies are now actively encouraging the use of sprinkler systems in residential buildings.[16]

COMMUNICATIONS

There are four basic methods of communication in common use. These are face-to-fac
radio/telephone, written, and electronic. The best method is face-to-face. Mess
passed in this way are the most likely to be understood. Through this method the
and receiver can see how one another react as clues to how and if the message w
erly understood. You can often tell by the look on someone's face and body lan
they did not really understand what you told them even if they are saying t
lows the asking of questions and clarification of misunderstanding. This i
choice for transferring command at incidents.[17]

Radio communications are essential because personnel and eq
out at an emergency scene and the personnel are too busy or the
important for them to meet. A problem is that personnel may be
clarification of an order because they will feel stupid. It is bet
than to misunderstand the order and do something that end
cipline must be maintained when using radios to commun
are not CBs, and there are specified procedures as to how
is that many people have scanners, especially the news
ment frequencies. Radios should not be used to con
firefighters have been in dangerous situations an
someone was tying up the air with nonessential

common use is the cellular phone. It is good for transmitting and receiving information in that you can have a more or less private conversation. Unlike hard-wire phones, persons with the proper equipment can listen in on cellular phone conversations. A disadvantage of radios is that they are hard to use when wearing a SCBA mask, unless it is equipped with a built in radio interface, and may have poor transmission and reception when working in remote or hilly areas. They also do not work in some buildings. In high-rise buildings, the method of choice is the internal telephone system. Telephones are a method of communication that, much like radios, depersonalizes the transmission and receipt of messages. Their advantage is that they provide a good connection without static or other interruption.

NOTE Radios should not be used to converse on unimportant matters.

Written communications are used when time is not a critical factor. They are used to record the policies and procedures of the department. They are also used to maintain a hard-copy record of communications for future reference and to reduce the possibility of error in interpretation. The problem is that written messages either need to be handled by courier or mailed, which slows their delivery.

With the invention of fax machines and electronic mail, the transfer of written messages has been greatly speeded up. Through the use of modems and networks, computers can e linked and messages can be quickly transmitted, with a hard copy produced when desired.

ARY

...r we have looked at the im- ...inistration at all levels in ...mission of the fire depart- ...tion is made up of peo- ... to widely accepted concepts and standards. Depending on the size of the department and its needs, the organization of the administration can vary widely. The most important point to remember is that without the support of the administrative staff, the line would not be able to function efficiently and effectively.

REVIEW

1. Summar... command.
2. List the six ... nagement cycle.
3. Why is it nec... ent cycle to be a continuing process?
4. List the four fo... nd their strengths and weaknesses.
5. Which form of c... effective? Why?
6. In the chain of co... fighter answer to?
7. Draw and fill in the ... tion chart with at least five levels of rank represented.

8. What are the general responsibilities of the firefighter?
9. What are the general administrative responsibilities of the company officer?
10. Who is the first-level supervisor in the fire department?
11. If you had a question about policy or procedure, who should you ask?
12. What is the difference between line and staff functions?
13. Give an example of division of labor in the structure of the fire department.

DISCUSSION QUESTIONS

1. What is meant by the term "incident effectiveness"?
2. Why is customer service such an important concept in the modern fire service?
3. When issued conflicting orders by supervisors, what should you do?

NOTES

1. National Fire Education System Course, *I-430 Operations Chief* (Boise, ID: National Interagency Fire Center, 2002).
2. David B. Gratz, *Fire Department Management: Scope and Method* (Beverly Hills, CA: Glencoe Press, 1974).
3. Ibid.
4. Fire Protection Publications, *Incident Command System* (Stillwater, OK: Oklahoma State University, 1983).
5. Ibid.
6. National Fire Education System Course, *I-430 Operations Chief* (Boise, ID: National Interagency Fire Center, 2002).
7. Kern County Fire Department, *Manual of Operations: Administration* (Bakersfield, CA: Kern County Fire Department, 2009).
8. International Association of Firefighters, 1750 New York Ave. NW, Washington, DC 20006.
9. J. D. Cortez Lawrence, "Company Staffing, The Proof Is in Your Numbers." *Fire Engineering*, April 1995.
10. J. D. Averill, A. Barowy, L. Moore-Merrell, K. Notarianni, R. D. Peacock, R. Santos & D. Wissoker, *Report on Residential Fireground Field Experiments* (Washington, DC: National Institute of Standards and Technology, April, 2010). http://www.nist.gov/manuscript-publication-search.cfm?pub_id=904607
11. John R. Guardino, David Haarmeyer, & Robert W. Poole, Jr., *Fire Protection Privatization: A Cost Effective Approach to Public Safety* (Los Angeles, CA: Reason Foundation, 1993).
12. National Fire Protection Association, *Fire Protection Handbook*, Twentieth Edition (Quincy, MA: National Fire Protection Association, 2008).
13. John R. Guardino, David Haarmeyer, & Robert W. Poole, Jr., *Fire Protection Privatization: A Cost Effective Approach to Public Safety* (Los Angeles, CA: Reason Foundation, 1993).
14. International City Management Association, *Managing Fire Services* (Washington, DC: International City Management Association, 1979).
15. National Fire Protection Association, *Fire Protection Handbook*, Twentieth Edition (Quincy, MA: National Fire Protection Association, 2008).
16. John R. Guardino, David Haarmeyer, & Robert W. Poole, Jr., *Fire Protection Privatization: A Cost Effective Approach to Public Safety* (Los Angeles, CA: Reason Foundation, 1993).
17. National Fire Education System Course, *I-430 Operations Chief* (Boise, ID: National Interagency Fire Center, 2002).

CHAPTER 8

Support Functions

LEARNING OBJECTIVES

Upon completion of this chapter, you should be able to:

- Identify the support functions required by the fire department.
- Describe the duties and responsibilities of the support functions.
- Explain the need for the support functions.
- Explain the difference between a managerial support function and a technical support function.

INTRODUCTION

Fire department operations can be divided into several areas, which can be further divided into those that are incident-focused and those that are not. Not all of the personnel who work for the fire department respond to incidents; many work in supporting roles that aid the incident responders in performing their jobs. The modern fire department needs personnel in the fire stations to respond to incidents; but it also needs support and assistance in other areas. Without this support, the personnel in the fire stations would not have the equipment, training, and facilities they need to perform their jobs. For the purposes of this chapter, those personnel not directly engaged in fighting fires and responding to incidents are considered support personnel.

DISPATCH

The fire department requires the capability of receiving requests for emergency service and sending its units to the scene of the incident. Even in a one-station department there needs to be someone assigned to answer the phone and alert the station crew that their services are required. Small departments may rely on the police department dispatch system to alert their personnel. Volunteer fire departments issue their personnel pagers so they can be alerted at their homes or jobs. In larger fire departments, the dispatch section is usually a purely fire department function. The personnel assigned may be firefighters or civilians, depending on the department's preference. Assigning firefighters to dispatch positions has its positives and negatives. On the positive side, the personnel assigned are familiar with departmental operations and can often anticipate the needs of the units assigned to the incident. During intense activity, a trained firefighter acting as a dispatcher can be of great help to the incident commander. When a situation is developing rapidly, the incident commander is trying to keep track of resources assigned and responding. She is also trying to formulate a plan and give assignments to the equipment that has already arrived or is about to arrive. It is difficult to think of everything when the incident is gaining the upper hand. It takes a while for everyone to arrive and to amass the resources needed to control the incident. Having someone who understands the situation assisting with the tracking of resources is a help in that the incident commander is more able to focus on what is happening and about to happen at the scene.

NOTE Even in a one-station department there needs to be someone assigned to answer the phone and alert the station crew that their services are required.

On the negative side, having firefighters as dispatchers is much more expensive in pay and benefits than having civilians in the positions. Another negative is the feeling of helplessness felt by firefighters when they are sitting at a dispatch console listening to the radio traffic as other firefighters engage in a tough fight. Firefighters tend to be, and are trained to be, action-oriented people. When acting as dispatchers they are forced to sit there thinking that their friends and co-workers need help and there is no direct action they can take. There is no release for the adrenaline they build up while listening to what is going on. When watching sports on TV, they can at least jump up and down and cheer. When dispatching, they have to sit there and pay close attention in case someone needs assistance. In too many incidents, firefighters have gotten themselves in bad situations and no one heard them call for help. In some cases, they ended up dying because no one came to their rescue. Dispatchers cannot leave the console and go do something else to relieve their frustration. Until they get used to being in the dispatch position, they want to go out, jump in their vehicle, and respond to the scene. Many firefighters never do adjust to the dispatch position and return to an operations assignment at the first opportunity.

Expanded Dispatch

In certain situations, incidents reach a point that they are beyond the immediate capabilities of the dispatch center. One way that departments handle the communications for incidents that become large or complex is to set up an expanded dispatch, sometimes called the command center or emergency operations center (EOC), to take some of the load off the dispatchers performing their normal duties. When a large incident occurs, there is an increased volume of radio traffic between the units at the scene and between the command units and the dispatch center. The rest of the department does not shut down when there is a large incident in the jurisdiction. In a jurisdiction with a large geographical area there may be a major wildland fire occurring. During operations at the wildland fire, several structure fires need to be dispatched as well as a hazmat (hazardous material) incident, medical aid incidents, and normal traffic, such as equipment going in and out of service. It does not take very long for the dispatchers to become overloaded.

The expanded dispatch is set up at a location outside the normal dispatch center. Some departments have a mobile command post and others move into a conference room at headquarters or a classroom at the training center. This allows the personnel assigned to the expanded dispatch to perform their functions without disturbing the regular dispatchers. When the department has several different radio channels available, all of the radio traffic for the large incident is switched to a designated channel. In this way, the regular dispatchers turn that channel off and are not disturbed by all of the incident traffic. At the expanded dispatch center, the orders for resources are handled by dispatch recorder personnel. Other personnel are assigned to making the necessary phone calls to get the resources ordered and to gather necessary information, such as estimated time of arrival.

An example of an expanded dispatch center is the Geographic Area Coordination Center (GACC) concept used by Federal wildland firefighting agencies and some states. In California, the stateis divided into two regions represented by the South Operations (South Ops) dispatch center in Riverside, California, and the North Operations Center (North Ops) in Redding. South Ops is staffed by U.S. Forest Service and Cal Fire personnel. When a major wildland fire occurs in the Southern Region and the responsible fire department has to go outside of its jurisdiction for resources, it contacts South Ops. South Ops then dispatches Federal and state wildland firefighting resources as necessary to the incident. This allows for the response of numerous resources from various agencies quickly and efficiently. When coordination is required on a larger scale, resource requests are sent to the National Interagency Fire Center (NIFC) in Boise, Idaho. It is not uncommon for Hot Shot crews from the Angeles National Forest, in Southern California, to be unavailable because they are assigned to fires in Montana. When a fire starts in Southern California, a crew may have to be flown in from the East Coast to take their place. This may seem inefficient, but you must fight the fire that is burning. When the fire in Montana starts first, that is where the closest available resources are sent.

South Ops and the National Interagency Fire Center track the daily availability of air tankers, helicopters, hand crews, engines, water tenders, shower units, catering services, bulldozers, and just about everything else that could be used for suppressing fires or supporting incident operations. By having one central order processing center, the closest available resource can be dispatched when requested. In a region where numerous large-scale incidents occur at the same time on a regular basis, this system is necessary. When South Ops receives a resource request, it checks status information and sends a resource order to the jurisdiction nearest to the incident that can fill the request. There have been several incidents in the last few years where the requests were for hundreds of engines for structure protection for one incident. An order of this size requires the request of engines from fire departments, both volunteer and paid, from all over the state and sometimes out of state. Air resources have even come into the country from Canada to attack fires. At times like this, a well-organized and coordinated system is necessary to track who is available, who is responding, from where, and when they will arrive. All resources ordered must be given order and request numbers, **resource designators**, travel routes, communications frequencies, a rendezvous point, and a reporting location.

resource designator An identification system using numbers and letters to identify resources by agency and type.

TRANSMISSION OF ALARMS

There are several methods in which alarms are transmitted to the fire department. The 9-1-1 system has taken the forefront (**Figure 8-1**). Before 9-1-1, the fire alarm box was very common in cities. Attached to a street light or power pole, the alarm box was always

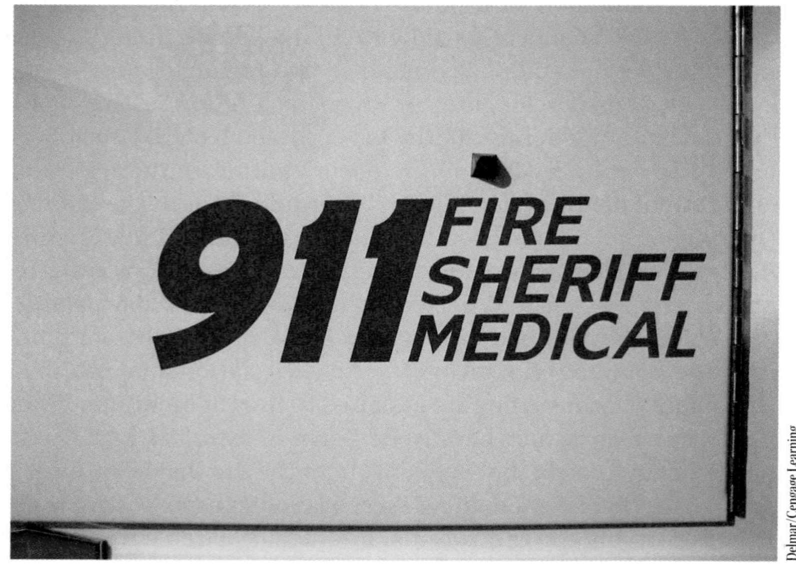

FIGURE 8-1
9-1-1 sticker on apparatus compartment door.

Delmar/Cengage Learning

available should a fire need reporting. Alarm boxes had several weaknesses as a system. There were frequent false alarms in some districts.

The boxes were sometimes vandalized or stolen. Another fault was that they were a one-way communication to the fire department. The dispatcher receiving the alarm had no information from the reporting party as to what the problem was. They were required to send a "full box" to alarms received. This could require the response of a ladder truck, a couple of engines, and a chief to a minor traffic accident. An argument in favor of the alarm box on the corner is that it is easy to understand and use. With the installation of telephones in almost every residence and business, the day of the alarm box is coming to an end. As an added benefit, dispatchers can now ask the reporting party what the problem is and respond with the appropriate resources.

NOTE As the fire department has moved into the area of emergency medical services (EMS), two-way communication has gained in importance.

As the fire department has moved into the area of emergency medical services (EMS), two-way communication has gained in importance. Now the dispatcher can ask the reporting party just exactly what the problem is. In some jurisdictions, the level of EMS response is determined by this conversation. A heart attack will get a paramedic response and a cut finger will get an EMT response. Many departments are training their dispatchers in emergency medical dispatch. Using this system, a trained dispatcher asks questions as to what is wrong with the victim and can then turn to a page in a pamphlet to give instructions to the person at the scene. The instructions include CPR, artificial respiration, control of bleeding, imminent childbirth, and the treatment of other medical emergencies. The system is a proven lifesaver. In a medical emergency, when seconds count, the instructions given to the persons at the scene can help preserve the life of the patient until qualified medical help can arrive. This system requires one dispatcher to gather the information and type it on a computer screen. Another dispatcher actually dispatches the call and initiates the response of emergency units. The problem occurs when the dispatch center is busy with more than one incident. In critical cases, at least one of the dispatchers must stay on the line with the persons at the scene until units arrive. This ties up a dispatcher and prevents him from answering incoming phone calls and dispatching incidents to the stations.

In many areas, there are people in the community who do not speak English. When these people call in for emergency services, there may not be anyone on duty in the dispatch center who can translate for them. Translation services are now accessible in the form of a conference call initiated by the dispatch center. When the department has an enhanced 9-1-1 system that shows the reporting party's address, it helps. Another problem is that many recent immigrants do not understand the 9-1-1 system. It did not exist in their native country. They do not speak English well enough to pick up on their own that the system is available and how to use it. They cannot read English, or are unable to read, and are not able to get the information from a pay phone or the phone book. When there is an emergency, they will drive around looking for a policeman or fire station seeking help. By the time they find help, it is quite possible that someone who could have been helped has bled to death or died of some other injury that could have easily been handled through quick medical intervention.

There has been an increase in private alarm companies that monitor homes and businesses for burglary and fire. When they receive a fire alarm, they contact the fire department and a response is started. In many cases, the alarms turn out to be false. Police departments across the country require alarm permits to be purchased on an annual basis and charge extra after one or two false alarms. Fire departments are sure to follow as the number of false alarms continues to rise. When a burglar alarm is received, one or two policemen respond in their cars. When a structure fire alarm is received, the fire department responds with several pieces of heavy apparatus and numerous personnel. The difference in cost is evident. In areas with volunteer departments, the volunteers leave their jobs and places of business to respond, placing an unnecessary financial burden on them. The risks associated with responding with red lights and siren are a factor that must be taken into account. The initiation of a structure response may leave a first-in area uncovered while the personnel are responding. While they are gone, there may be a legitimate need for their services in their first-in area. All these issues have to be balanced with the need for quick response to a structure fire. Very few firefighters will call off a response or lessen the amount of equipment responding because it might be a false alarm. The one time that it is done, the first-in unit may arrive and report flames through the roof with rescue necessary.

Lookouts

A position that has been used for a long time to report fires in forest areas is the lookout tower. The fire lookout person reports to his or her position in the tower early in the morning and does not come down, except for short periods, until dusk. This can be quite an exciting position in one of the older towers. The tower is approximately 75 feet high and stands on metal legs. The lookout person sits in a room about eight feet square. The chair he sits in has glass insulators on the ends of the legs. When lightning storms occur, the whole tower can become energized and the lookout person cannot touch anything for fear of being electrocuted. The lookout person is forced to remain in the chair until the storm is over, no matter how much the tower sways in the wind or how many times it is struck by lightning. A modern lookout tower is much lower to the ground and includes the person's living quarters. These types of lookouts depend on their location on top of a ridge or knob to give them visibility of the surrounding area.

In the center of the room is a table with a fire finder. This is a sighting device mounted on top of a revolving base plate. When a fire or smoke is sighted, the fire finder is rotated until the sighting device lines up with the smoke. The degrees of the compass heading are read from the table. When two or three lookouts are able to give a fix on the same smoke a very accurate location can be determined.

GIS/MAPS

Departments in areas of any great size need updated maps on a regular basis. As new streets are added and new subdivisions are built, the maps in all of the stations that respond into the affected area need to be updated. There is also a need for parcel maps to show property ownership. These are used for hazard reduction work, to contact the owners of vacant lots, and to determine ownership of acreage burned in wildland fires.

With the introduction of geographic information systems (GIS) (**Figure 8-2**), the job of the person responsible for the department's maps has changed. The GIS uses maps entered into a computer as a base. The master map can then be overlaid with other information. The map coordinates are tied to different databases that contain the information required. There can be overlays of place names. These include information such as business names, property owner names, and location names in rural areas. In town it is helpful for the dispatchers to be able to access the address of the local supermarket without having to look it up in the phone book. In rural areas the name may be used to determine the location when a call comes in for a tractor accident at "Yurosek's farm." In wildland areas the location given may be "the head of Haypress Canyon." An aircraft sighting given off a global positioning system (GPS) is given in longitude and latitude. These coordinates are not on most maps. The GIS has this information and the ground location, and its relationship to roads can be quickly established. This information is also useful when dispatching medical helicopters or firefighting aircraft to an incident scene.

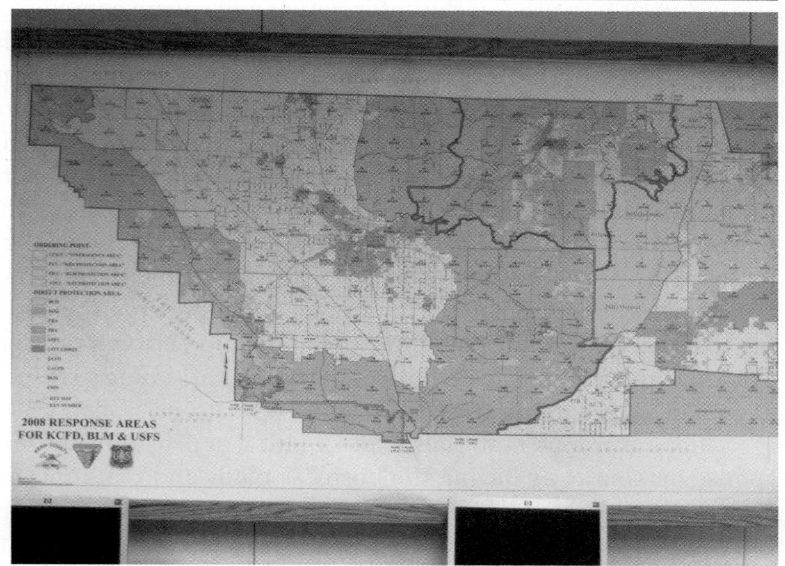

FIGURE 8-2
GIS-created map.

Delmar/Cengage Learning

Pilots are much better able to plot a course and identify a location by longitude and latitude than by ground-based references such as the intersection of Tucker Road and Round Mountain Road. Using the GIS, the dispatcher can quickly find with a few keystrokes any location, if the information is in the database, and determine in whose first-in area it is located. Use of this information speeds up the dispatch of equipment and thereby reduces response times.

Using information from a GPS, a fire can be mapped fairly easily. A ground or air unit, using a GPS, travels the perimeter and records the coordinate points. The information is then fed into the GIS program and a detailed map of the perimeter can be created as a map layer on a map with other layers showing property ownership, roads, and any other information that is entered into the database that can then be printed out. In wildland areas, where a fire may cover several jurisdictions, this information is very important when it comes to determining each jurisdiction's share of suppression costs. It is also necessary for statistical purposes when justifying budget requirements based on acres burned.

HAZARDOUS MATERIALS CONTROL UNIT

In recent years, an area that has received much public attention is that of hazardous materials. Several major incidents have occurred that have led the public to demand closer control of hazardous materials stored and used in their towns and cities. States have passed *Right to Know* laws. These laws require the producer or user of materials considered hazardous to file a business plan and inventory of the hazardous materials they are storing on site. To process this information and ensure compliance with legal and code requirements, the agency responsible sets up a hazardous materials unit. This unit is not to be confused with a hazardous materials response unit, which responds to spills and leaks.

Many fire departments wanted the responsibility of hazardous materials control because they were the ones who would respond to incidents involving hazardous materials. The fire department already had the expertise of inspection technique from fire prevention activities. In other jurisdictions, the responsibility lies with the health department or environmental health department. In some jurisdictions, the responsibility is shared between the fire department and the health department. Often in these situations, the fire department and health department personnel undergo their hazardous materials training together. This aids in building a relationship between the departments from the bottom up. The personnel from both departments are going to respond to incidents together and need to be familiar with each other's operations.

The hazardous materials control unit can provide vital services to the incident response personnel by having the business plans and materials inventories available. The information can be accessed by modem and printed out on a computer printer or a copy can be sent to the field by fax machine. Having this information available can save lives and prevent environmental damage at incident scenes. It is preferable to know what is in a chemical warehouse when it is on fire so the right decisions can be made as to how to handle the incident. There may be times that the chemicals are such that a let-burn policy is the safest choice.

NOTE The hazardous materials control unit can provide vital services to the incident response personnel by having the business plans and materials inventories available.

The hazardous materials control unit also has personnel with technical expertise to assist in decision making. There are very few firefighters with a strong background in chemistry. From an inspection standpoint, the control unit personnel are better prepared than firefighters to identify a hazardous situation. They would more likely recognize the storage of a powerful oxidizer next to a fuel than the average firefighter would. Identifying these types of hazards may help prevent an incident from occurring.

INVESTIGATION UNIT

One of the most heinous crimes committed is that of arson. When a hostile fire is ignited, there is no telling what will burn and who will be injured. Arson fires are started for many different reasons including spite/revenge, profit, sexual gratification, and to cover up other crimes. The arsonist may be targeting one specific individual, but when gasoline is poured in the stairwell of an apartment building and ignited, often numerous innocent persons are injured or killed. In areas with dry wildland fuels, one arson fire can destroy hundreds of homes and result in losses in the millions of dollars. When these fires are started, the lives of firefighters and the general public are put at great risk. On January 5, 1995, four firefighters died and five others were injured fighting an arson fire in Seattle, Washington.[1] There is no doubt that arson will continue and so will the tremendous costs in lives and property. The direct costs are those of the buildings and contents destroyed in the fire. Indirect costs are the cost of the suppression effort and the loss of natural resources. Another indirect cost is that when units are fighting an arson-caused fire, their first-in districts may be left uncovered because of lack of manpower. If another fire starts while they are out, it has a chance to gain headway, resulting in higher losses.

On January 5, 1995, four firefighters died and five others were injured fighting an arson-caused fire in Seattle, Washington. The fire was deliberately set and the firefighters were killed when the floor collapsed and they plunged into a burning basement that they were unaware of. As a result of this fire tragedy the Seattle Fire Department has made it policy that crews responding to a fire be alerted to arson threats and any other known potential hazards. Today, crews around the city routinely conduct inspections on buildings to become familiar with the layouts and potential hazards.

The crime of arson can be determined in many cases, but the conviction of the person responsible is difficult. It is the very nature of fire to destroy evidence of the crime. Much of the material is consumed in the ensuing fire. Fingerprints are destroyed, and structural collapse and firefighters with hose lines disturb the scene. In wildland situations, the **incendiary device** may be thrown from a passing auto and the criminal is miles away before the fire is reported. Arson investigators dream of catching the person with the match in their hand, but this is very rare. Often, long hours of investigation and gathering of facts lead to conviction of the person(s) involved for defrauding the insurance company because the identity of the actual arsonist cannot be proven.

incendiary device An incendiary device is a device used to light a fire. This can be as simple as a lit cigarette folded into a matchbook or more complicated, involving chemical mixtures and a timer or cell phone to activate.

Arson Bureau

Under the fire prevention chief or in its own unit is the arson bureau. The personnel in the arson bureau are responsible for determining the cause of suspicious fires (**Figure 8-3**). They are not needed at every fire scene. In many cases the company officer and crew at the scene can determine the cause of the fire and no further investigation is necessary. In

FIGURE 8-3
Arson investigator searching for fire cause and origin.

Delmar/Cengage Learning

cases where a crime is believed to have been committed or the cause is unclear, the arson investigator is called on. Arson investigators should also be called any time there is serious injury or death due to a fire. Due to their investigative abilities, the arson investigators may also be utilized as an internal affairs bureau for the fire department, investigating harassment and other complaints, either brought by one employee against another or a member of the public against an employee.

Arson Investigators

Arson investigators should be specially trained in the gathering and preserving of evidence.[2] If the evidence is not properly preserved, the case is thrown out of court and the criminal goes free to threaten lives and property again. Arson investigators are trained in giving testimony as witnesses in a court of law and are trained in the laws of the state regarding arson. They are also trained in the art of interviewing suspects and witnesses. In a typical investigation, they interview the firefighters at the scene, law enforcement officers, building or area occupants, property owners, witnesses, and bystanders.

NOTE Arson investigators should be specially trained in gathering and preserving evidence.

They must exert authority at fire scenes and over suspects. A background in fire-fighting is an asset. When they are on the witness stand, they are often asked for their opinion. Their opinion carries more weight when they come across as trained, experienced professionals.

The qualifications for arson investigators are that they must be knowledgeable in the chemistry and physics of fire. They must have a good understanding of how a fire burns to interpret the scene. They must be systematic and careful in gathering evidence and building the case. The legal system requires the investigator to follow set procedures in order for the evidence collected to be admissible in court. The investigator must have a knowledge of the laws that pertain to arson. An investigator must be able to gather the facts and put together a complete picture of what happened and when. She must also be able to present this information in a professional and understandable manner from the witness stand. When called to testify, the arson investigator is explaining the facts to a jury, not other firefighters.

The arson unit of the fire department maintains a close relationship with the law enforcement community. By working closely together their effectiveness is greatly enhanced. In many jurisdictions, the fire and police departments have officers working together on single cases as well as on arson task forces. Arson is sometimes used in an attempt to cover up other crimes and a united fire and law enforcement effort is required to convict the persons responsible.

The arson investigators also work closely with the district attorney's office in preparation for presentation of the case. As previously stated, the evidence must be properly handled and preserved to be admissible in a court of law.

Cost Recovery

An effort that is becoming more common is that of cost recovery. Through the use of civil suits, the fire department is now starting to pursue the recovery of costs from persons who cause fires. It is not uncommon for the suppression cost of a major wildland fire to run into the millions of dollars. This amount is not totally recoverable in most cases. The recoverable amount is usually limited to the amount of insurance carried by the person who caused the fire. This can be thousands of dollars and is better than nothing. Once the word gets out that cost recovery will happen, it will tend to make people more careful.

In cases where property is burned for profit or other motive, the fire department may be able to collect as well. The amount expended putting out a car fire may be in the hundreds of dollars and the recovery of these costs frees up department monies for other activities such as prevention and equipment.

Some jurisdictions have passed extraordinary hazard ordinances, which are used to recover money spent on the mitigation of hazardous materials incidents. Other areas where cost recovery is pursued are criminal negligence and criminal acts. One scenario is where a person has caused a vehicle accident while driving under the influence of drugs or alcohol. Once someone is convicted for driving under the influence of alcohol or drugs, the department can pursue recovery of the costs incurred under the applicable penal code section.[3]

PERSONNEL

The fire department personnel clerk is involved with matters regarding the employees of the department. In a department with few employees, the city personnel clerk takes care of personnel matters. Larger departments may have their own personnel clerk assigned.

When personnel are to be hired or promoted, the personnel clerk fills out the necessary paperwork to requisition the position from the agency personnel department. The personnel department sends the required number of names from the list of eligibles for interview. Once the person is notified that they have been hired, the personnel clerk schedules the medical exam and drug screen. Once cleared, the newly hired employee sees the clerk to complete the required paperwork for setting up payroll deductions including income tax withholding, and, if the employee chooses, union dues. The personnel clerk handles the enrollment forms for health insurance and other benefits. When personnel are terminated or retired, the personnel clerk ensures that all of the proper forms are completed.

The personnel office issues personnel evaluations and retains them, when they are completed, for record-keeping purposes. When an employee is on work-related injury leave, the personnel clerk makes sure that all of the proper forms are completed and sent to the state and other agencies that are involved in worker's compensation. If there is personal property damage to an employee's property, the personnel clerk handles the claim forms. In some departments, the personnel clerk completes the estimate of salary and benefit costs for preparation of the budget. This office also maintains a personnel database to record the home addresses and phone numbers of fire department personnel. The personnel clerk works with the payroll clerk to make sure that personnel receive their base salary, overtime, pay raises, and incentives and are paid the correct amount.

Outside the fire department is the jurisdiction-wide personnel department. This would be countywide in a county fire department and citywide in a city fire department. The name becoming more common for these departments is human resources. Their job is to advertise for applicants when the fire department needs to hire new personnel. They receive and evaluate the applications for employment.

The personnel department administers the entrance and promotional written tests and assists with the oral examinations. The personnel department is given this function to avoid the filing of charges of favoritism. In departments with civil service, the personnel department makes sure that all the rules and laws regarding personnel are followed. It is not uncommon for the personnel department to assign one or more personnel to handle all matters regarding the fire department. In this type of system, the fire department budget is charged for personnel department services.

The personnel department ensures that all federal and state guidelines regarding affirmative action and equal opportunity employment are followed. It also has personnel who are trained in the Americans with Disability Act employment requirements. Were the fire department to want to change any part of an approved entrance or promotional testing process, personnel department staff would be involved to ensure fairness and compliance with the law.

A personnel analyst may be assigned to assist in writing the job requirements for the different job classifications, such as firefighter and company officer. As the jobs have changed over the years in terms of responsibilities and requirements, these classifications need to be updated. Not only do they affect the entrance requirements, they also justify pay increases as personnel are required to operate at a higher level of responsibility, skill, and training. These requirements are represented in the job announcement and the testing procedure. An example of this is when federal guidelines regarding driver's licenses were instituted: All job descriptions for firefighters required to operate heavy vehicles had to be updated to reflect the changes.[4]

A large department may have its own personnel department that allows it to be heavily involved before new personnel are hired. The personnel department is involved with the active recruitment of prospective firefighters. There are pre-academy tutorial programs to assist persons in preparing to take the firefighter test. Some of these programs involve only mental skills and others involve weight training and other activities to help recruits pass the physical agility test. It has an affirmative action section that ensures that there are no discriminatory practices and that the targeted groups are reached. Through the use of these types of activities, the department does not have to adjust its standards to meet its recruitment goals.

INFORMATION SYSTEMS

The fire department always seems to have more information than it can manage. There is information coming from all sides. All expenditures already made and expected need to be recorded for budgeting purposes. Specifications for the new equipment must be generated and the bids sent out to the manufacturers and suppliers. Statistical information from fire and medical aid reports needs to be put into an understandable form. Information that pertains to contracts and agreements must be gathered and kept track of.

Training information on new techniques and procedures must be reviewed and applied if advantageous. New laws, codes, and ordinances must be reviewed for their effect on the department and its operations. Personnel and payroll information has to be gathered and recorded. Personnel are on vacation, sick leave, or injury, and others are working overtime in their place. These are just some examples of the type of information that has to be managed by the fire department.

NOTE The fire department always seems to have more information than it can manage.

To manage all of this information, the information systems bureau has been developed. In the smaller departments, much of the information will be managed from outside the department. In larger departments, the information systems bureau can range in size as to the number of personnel assigned. Some of the most important people in this bureau are the repair technicians. They make sure that the computer system is up and functioning. The people working in the headquarters need to have access to information and to update it as necessary. Even in a small headquarters, with few personnel, when the **local area network** goes down, the work starts to slow down and become backed up. The problems are compounded when the fire stations are linked through remote terminals to the headquarters. When decentralized, more personnel are required as the technicians have to make site visits outside the headquarters to make repairs.

In the information systems bureau are the personnel who actually handle the information as it comes in. These personnel are assigned to data entry. They may be assigned to entering information from the fire reports received at headquarters or entry of other information into the department's database. An emerging trend is to connect all of the stations to headquarters with wide area networks. The fire reports and other communications can then be entered from remote terminals instead of being written on hard copy and transferred to headquarters through the department mail.

Data management personnel control the flow of information and make sure that it is directed appropriately. They may suggest improvements to the system and to how data is generated. They manipulate the information available to present as requested by the department staff in the form of reports. These are used to collect money, justify contract amounts, justify budgets, and track expenditures.

The systems analyst maintains the computer operating system for the department. Depending on how computerized the department is, this person may have a wide array of responsibilities. The analyst helps users with their applications, assisting them in the use of their software programs and automating functions commonly used to save time

local area network A system linking the computer terminals of a department on a local scale—for example, at the different desks at headquarters. A wide area network links the computers of several different geographic locations, such as between headquarters and the fire stations.

and increase productivity. This is done through determining user needs and configuring hardware and software. The analyst specs out new equipment for the department so it purchases what it needs.

BUSINESS MANAGER

An important part of the fire department management team is the business manager. This person can be compared to a chief financial officer in a private enterprise. This is usually a civilian position because the person is hired for her financial expertise, not her firefighting skills. The business manager and the department staff prepare the final budget document to be presented to the jurisdiction for approval. They also participate in the research necessary to justify the budget request. The business manager assists with budget preparation through revenue prediction activities. The department management cannot plan ahead if it does not have any idea how much money will be available. The main budget is a combination of all of the subordinate budgets: the budgets for training, prevention, vehicles, and equipment. When there are separate staff officers in charge of all of the subordinate functions, they submit their budget requests to the business manager, who determines if there is enough money to cover all of the proposals. If not, which is usually the case, negotiation among the staff officers takes place.

NOTE An important part of the fire department management team is the business manager.

The budget is then presented to the jurisdiction's budget analyst for review. The budget analyst works with the business manager and the fire chief on the final proposal that is submitted to the elected officials of the jurisdiction. The elected officials want to make sure their money is well spent and may challenge the fire department's requests in several areas.

A part of the business manager's responsibilities is to keep track of all expenditures. Monthly reports are submitted to the department administration regarding the status of the budget to make sure that the department is staying within its budget and does not run out of money three months before the new budget is approved. If one area of expenditure is too high, the other areas will suffer as there is only so much money to cover all of the department's expenses. Special studies are also conducted when a specific project or problem is identified, for example, the cost of maintaining the fleet of department vehicles. This information is used to identify which pieces of apparatus are becoming too expensive to keep in service and need to be replaced.

The business manager must be well versed in all of the federal, state, and local laws that govern the actions of the fire department when it spends or receives money. There are ordinances that govern all aspects of how an agency functions. When equipment is donated, the offer often has to be brought to a vote before the elected officials before being officially accepted. Every penny has to be accounted for. At any time the jurisdiction can step in and audit the department.

The business manager's staff does all of the day-to-day accounting for the department. They pay the bills and cash the incoming checks. Many departments contract for services from other agencies, such as the personnel department, and budget transfers must be made. When other agencies, in or outside of the jurisdiction, contract with the fire department for services, they must be billed. An example of this is when there is a major fire and the department responded out of jurisdiction on an assistance-for-hire basis. A final bill for the apparatus, personnel, and support services must be prepared and sent. In some states, if the bill is not sent in a timely fashion, it is dismissed. One area in which the department can recover costs is on motor fuel taxes. When apparatus is operated off road, idled, or pumped for long periods of time, fuel is consumed. The taxes collected for the fuel when it was bought can be recovered from the state if this consumption can be proved. A fire department with many apparatus consumes enough fuel under these circumstances to make it well worth assigning someone to spend the time to recover these funds. In areas with outlying stations, the fire department may be the only department that maintains fuel pumps. Other governmental agencies may access this fuel and they need to be charged for the amounts used because the fire department is paying the fuel vendor for the deliveries.

The business manager's staff does all of the ordering for the department. The staff keeps track of what has been delivered and when to pay the bill. It may also have responsibility for the warehouse/central stores inventory as a double check to prevent pilferage or misuse of departmental resources.

FIRE BUSINESS MANAGEMENT

All of the agreements and the system to activate resources are set up beforehand. Just the work involved in designing, updating, and administrating the resource ordering system requires thousands of personnel hours annually.

Prior to any privately owned resources being used at an incident, there needs to be a contract signed between the owner and the requesting agency. There are contracts to be written and signed with vendors for every item that may be used on the incident. Persons with equipment, such as helicopters, air tankers, and water tenders need their equipment inspected on an annual basis before the contract is allowed. At the incident itself, there are equipment time recorders and check-in personnel who keep track of who is at the incident and their status. When equipment is used, the time must be recorded so the owner can be paid for its use. When supplies are delivered, they have to be accepted and signed for. Personnel time must be kept so that crews and other personnel are paid.[5] The personnel filling these positions may never even see the incident they are assigned to. Management teams are brought in on large incidents just for this purpose. The people assigned to them have been trained in keeping records and preparing the necessary forms so the information can be properly processed.

TECHNICAL SUPPORT

The requirements placed on the modern fire department are such that technical assistance is often required in many areas. One of these areas is legal services. The office of the county counsel or the city attorney reviews contracts and agreements that involve the fire department. Fire districts contract with private attorneys for these services. The fire department may enter into agreements involving **mutual aid** or **automatic aid** with adjoining agencies. When this is done, there are legal questions that must be answered about liability and worker's compensation coverage for employees working outside their jurisdictions. If the department decides to lease fire apparatus instead of buying, the department's attorney should review the lease agreement. In many areas, county fire departments contract with cities inside their boundaries for providing fire protection. Another possible contract is a county contracting with the state for protection of state land within the county. The county counsel is consulted to make sure that all of the necessary language is in the contract and that it is legal under the county charter or state law to enter such a contract. Counsel is also necessary any time the department runs into legal problems regarding its relationship with an employee or the public.

Investigative technical support can be provided in an agreement in which the police crime lab assists the fire department arson personnel. Very few fire departments are able to afford, or even need, a full crime lab. When the arson personnel have gathered evidence, it is sent to the crime lab for interpretation. The crime lab can determine if there are traces of flammable liquids or other accelerant in the samples submitted. They can also research fingerprints and other evidence collected at the scene. The police agency is usually the one that administers lie detector tests of suspects, when required.

For incidents where weather can be a major factor, such as hazardous materials incidents and wildland fires, the National Weather Service can provide assistance. At hazardous materials scenes the wind direction and relative humidity can affect the size and direction of the evacuation or shelter in place ordered. One of the factors which drives wildland fire behavior is the weather. There have been numerous wildland firefighters killed or injured because unexpected winds changed a fire's direction or fanned a dormant fire to life.[6] General weather forecasts are available for large areas, much like the ones used by the television news. So-called spot forecasts are available to firefighters on a local scale. These give a much clearer picture as to what is likely to happen at the incident. Information is gathered at the fire's perimeter and faxed to the National Weather Service. They can then make a local forecast and fax it back to the requesting agency. On long-term

mutual aid When departments draw up an agreement that they will assist each other upon request.

automatic aid Under this system departments assist each other without regard to jurisdictional boundaries. Often used in areas where there are county islands within city limits or in interagency areas where a fire starting in one agency's area is a direct threat to another's jurisdiction.

incidents (those that last for more than a couple of days), a weather observer can be requested for the incident. He sets up his trailer at fire camp to take local observations. He uses his own observations, those from the personnel on the fire line, and information from the computers of the National Weather Service. By using this information he can make weather predictions specific to the incident area. Weather information is also used at the scene of hazardous materials incidents to predict the speed and movement direction of vapor clouds.

There have been numerous wildland firefighters killed or injured because unexpected winds changed a fire's direction or fanned a dormant fire to life.

In the area of hazardous materials, technical assistance is a must. Computer databases are available to determine the hazards of various materials. Research resources in the form of books are also available. With approximately 2,000 new chemicals developed each year, books soon become outdated.[7] By subscribing to a database service, the information is constantly updated. The health department can be of great assistance in these types of incidents. They have personnel trained in the field of environmental health. These personnel usually have chemistry backgrounds and are trained in the determination of an unknown chemical's properties. It is often necessary to determine at the scene whether a spilled white powder is poisonous, corrosive, or harmless and whether it may require collection and disposal as a hazardous material or just be washed off the highway. Until this determination is made, no real action can be taken as the hazardous properties are unknown. When a major highway has to be shut down at rush hour, a quick and accurate determination is appreciated by everyone involved, especially the public backed up in traffic.

Other technical sections may be set up in cases in which the jurisdiction contains oil refineries or other special hazards. The Los Angeles City Fire Department has a technical section devoted to the Metro Rail. They have already had a major fire in a tunnel under construction and recognized the need for such a section of technical expertise. Many of the requirements in model fire codes came from fire department technical sections recognizing the need for protection systems or engineering, and drawing up ordinances to be adopted by the local jurisdiction.

When departments are involved in emergency medical services on the higher levels, they may have a technical section for this function. As laws and protocol change, the fire department must stay abreast of the changes. The costs in public image and liability are too high not to. Smaller departments can rely on the county EMS department for their needs.

WAREHOUSE/CENTRAL STORES

The warehouse/central stores facility is where the items needed on a regular basis by the fire stations and offices are kept (**Figure 8-4**). This is also the site where much of the department's non-motorized equipment, such as fire hose, is repaired. The warehouse manager and stock clerks receive, sort, and store items that are supplied to the stations. The warehouse also acts as a central receiving facility for items ordered by the department, making it easier to ascertain whether orders were received than if they were sent directly to the fire stations. When the stations send in supply requisitions, the requested materials are gathered and either delivered by the warehouse personnel or picked up by the station personnel.

The warehouse may also be capable of making repairs to equipment. Hose that has been damaged can often be recoupled and placed back in service. When the damage is near the end, the coupling is removed, the bad section cut away, and the coupling reattached. The hose is no longer the standard 50 feet in length, but it is long enough to be reused. If the break was toward the center, the resultant hose will be too short. It can be reused by cutting it into several shorter hoses to be used as soft suction hoses or tank fillers. Every so often nozzles, gated wyes, and other appliances need repair. They are sent to the warehouse and the personnel either repair them or replace them as needed.

Many departments have a self-contained breathing apparatus repair technician. This person makes the necessary repairs to SCBA. When new SCBA are received, the technician tests them to make sure that they are working satisfactorily before they are placed in service. By having this capability in-house, the department can reduce the cost of going outside for service and provide for shorter down time for SCBA. The fire department SCBA

FIGURE 8-4
Fire department warehouse.

Delmar/Cengage Learning

will be the technician's number one priority. The technician's shop may also be equipped with a breathing air compressor for the filling of SCBA bottles. A stock of filled bottles is kept on hand for exchange when a station needs filled bottles. By keeping filled bottles on hand, the stations can get their equipment back into service more quickly after a major incident that has depleted their supply of filled air bottles. The SCBA for other governmental departments can be repaired here and their bottles filled as time permits. When the department provides emergency medical services, the technician may also have a facility for refilling medical oxygen bottles.

The U.S. Forest Service and Bureau of Land Management have much the same facilities. The U.S. Forest Service has a warehouse for the entire forest and then a smaller one located at the headquarters of each ranger district in the forest. Their warehouses supply the needs of all of the personnel, not just fire. These facilities supply materials for suppression, prevention, recreation, timber, and other needs. It is not uncommon for the fire crew assigned to the district headquarters to repaint picnic tables, trail signs, and fire prevention signs in the spring. Usually located at the district headquarters station is a 50- to 200-person cache of fire tools that must be maintained. Once fire season is over, the firefighting staff that stays on is used to maintain equipment, plant trees, and work on other projects. When not fighting fires or training, the fire crews always have plenty of work to do.

The smoke jumpers spend the winter and any other down time making helicopter rappelling harnesses and other equipment. They are well known for their design and construction of many items of wildland firefighting equipment.

REPAIR GARAGE

The fire department may have its own repair garage. The repair garage services the fire apparatus. The mechanics at this facility are civilian employees who work a 40-hour week directly for the fire department. When a fire department is large enough to support its own repair garage, it should have one. Some reasons for this are that the mechanics assigned have specialized knowledge of fire pumps and related equipment and that fire equipment will have the top priority at a fire department facility. People in the community may not notice when a fire engine is out of service, but they will definitely notice if a garbage truck is unavailable and their garbage is not picked up. Mechanics do not want garbage trucks sitting around the lot stinking up the place either, which may lead to them receiving priority.

RADIO SHOP

Fire departments need communications capability. Communication is carried out through the use of radios installed in the dispatch center, the headquarters, the stations, the vehicles, and with hand-held units. After installation, all of these units will need maintenance. When the department covers a large and topographically diverse area, a system of repeaters is usually installed. These boost the radio signal and provide the capability for any unit

Delmar/Cengage Learning

FIGURE 8-5 Radio repair technician's work bench.

in the department to talk to any other. The radio technicians install and service all of the radio equipment, including pagers, used by the department (**Figure 8-5**).

ADJUTANT/AIDE

A position that can be considered support is that of adjutant or aide. The firefighter assigned to this position assists the chief officer in his duties. The job includes running errands, typing memos, and researching information for projects. The aide also acts as a dispatch recorder and message runner at incidents. It is important that the chief keep track of all of the units at the scene, the ones still responding, and their time of arrival. The aide can record this information, freeing the chief to make strategic and tactical decisions for the incident. The aide drives the chief around so he can conduct business on his cellular phone and review written documents or material on his personal computer. When responding to a preplanned incident, the chief can review the preplan as the aide drives.

NOTE This is an opportunity for firefighters to learn more about the department and how it works.

Basically the aide does the things requested of him by the chief. This is an opportunity for firefighters to learn more about the department and how it works. It is also an opportunity to learn about emergency and nonemergency management. Spending time as a chief's aide can help you prepare for promotion.

SUMMARY

A modern fire department requires many personnel working in the area of support functions to operate. These functions are widely varied in nature and require personnel with expertise in each of the different areas.

For a fire department to operate, it must be able to receive calls for service and dispatch units to the scene. This requires a staff of trained professional dispatchers. Dispatchers need a system that will help them do their jobs in a timely fashion without mistakes. As the public's expectation of level of service increases, the fire department must keep up. Technology has developed to the point where there are new systems available to aid in the dispatch function. One of the greatest needs of any at-scene unit is information. This information can be gathered from GIS systems through the use of databases.

To control hazardous materials in the jurisdiction, the hazardous materials control unit is set up. It inspects businesses to ensure compliance with codes and ordinances regarding storage and reporting of inventories. It also provides information to incident responders in the field when the need arises.

It is the responsibility of the fire department to investigate all fires that occur in its jurisdiction and determine their cause. With the help of specially trained arson investigators, the cause can be determined and criminals can be convicted when necessary. These personnel can also aid in cost recovery efforts when fires are found to be caused by illegal activities, or maliciously set.

The fire department needs to hire, retain, and provide benefits to its personnel. Personnel takes care of these needs. As the department grows in size, the needs for personnel staff increase. A small department may use the jurisdictions' personnel department whereas a large fire department may have its own.

Support in the area of provision of supplies and repair of vehicles is also a necessity. The personnel in the fire stations need to have operable equipment for response and the necessary items supplied to ensure their readiness.

To operate in a complicated legal and political environment, the modern fire department needs a staff of personnel capable of managing information. The information is generated from within the department and from the outside. Without management of this information, the department cannot determine how it is doing and where it is going. The information is in the form of money, statistics, and personnel.

A big part of the job in any fire department is the management of its budget. There are numerous federal, state, and local regulations and ordinances that specify how a department must operate. Today's budgets, for even small departments, are in the millions of dollars. These large sums must be professionally managed to ensure that they are spent wisely and within legal constraints. When a large multijurisdictional fire disaster does occur, a whole system of business management must be set up to service the needs of the forces operating at the incident. A large incident almost takes on a life of its own and vast resources are mobilized. All of the supplies and personnel must be accounted for and the bills generated by the incident must be paid.

Along with all of the other areas of support, the fire department requires specialized technical support, in the form of weather reporting or other services.

REVIEW QUESTIONS

1. List three reasons why a fire department requires a dispatch center.
2. What is meant by the term "expanded dispatch"?
3. List several ways that alarms can be transmitted to the dispatch center.
4. What valuable information can the hazardous materials control unit provide to operations personnel at the scene of an incident involving hazardous materials?
5. Other than investigating fire cause, what are the functions of the arson unit?
6. Why is it necessary to assign a personnel clerk to the fire department?
7. What functions does the business manager perform when dealing with the department's budget?
8. Why is a business management (finance) section set up on major incidents?
9. Why is a weather person important on major incidents?
10. What are the advantages of a fire department having its own repair garage?
11. If you were offered the position of chief's aide, what do you see as your responsibilities?
12. Your fire engine has a noise in the transmission and the radio is out of service. Whom do you contact for service?
13. Can the fire department exist without the support services? Justify your answer.

DISCUSSION QUESTIONS

1. Why is there a need for expanded dispatch on major incidents?
2. What are the advantages of the GIS system?
3. As fire departments become more computer dependent, how will the role of information systems expand?

NOTES

1. E. Nalder, D. Wilson, The Seattle Times, "*The Pang Fire: What Went Wrong—A Disaster Marked By Bad Preparation, Poor Communication And Many Other Mistakes*," retrieved from http://community.seattletimes.nwsource.com/archive/?date=19950611&slug=2125846

2. National Fire Protection Association, *Standard 1033, Fire Investigator Professional Qualifications* (Quincy, MA: National Fire Protection Association, 2009).
3. State Penal and Health and Safety Codes.
4. U.S. Government, *Commercial Motor Vehicle Safety Act of 1986* (Washington, DC: U.S. Government Printing Office, 1986). These regulations should also be included in your state vehicle code and/or Commercial Vehicle Operator's Handbook.
5. National Fire Equipment Systems Course, *S-260 Fire Business Management Principles* (Boise, ID: National Interagency Fire Center, 2005).
6. Carl C. Wilson and James C. Sorenson, *Some Common Denominators of Fire Behavior on Tragedy and Near-miss Forest Fires* (Broomall, PA: USDA Forest Service, 1978).
7. Frank L. Fire, *The Common Sense Approach to Hazardous Materials* (New York, NY: Fire Engineering, 2004).

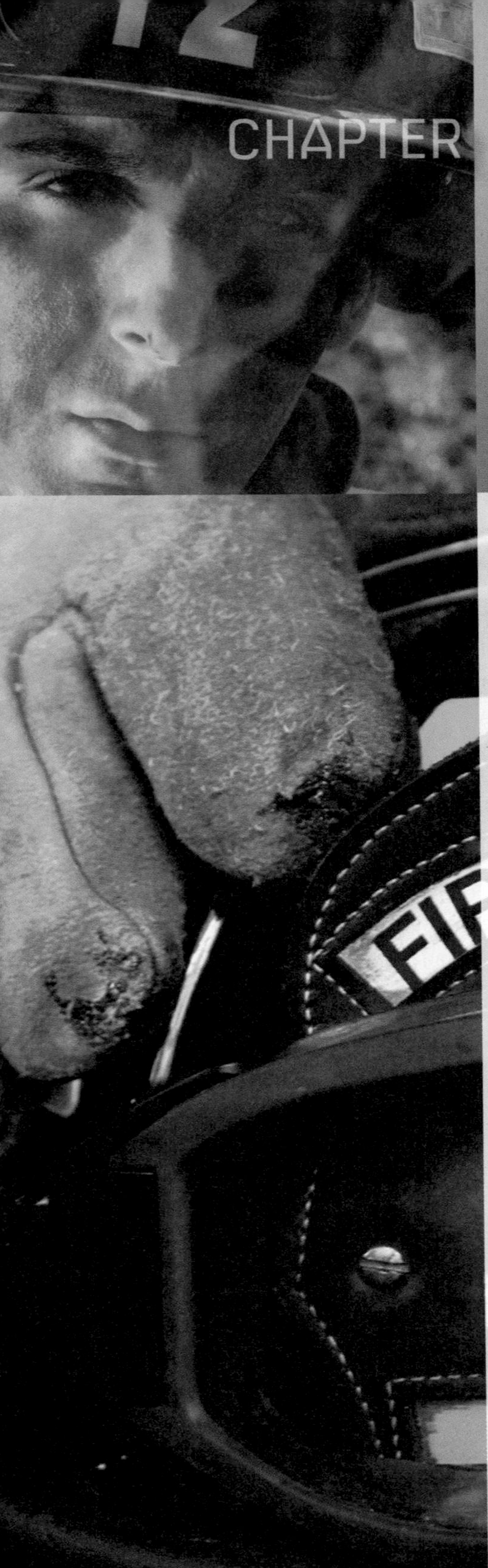

CHAPTER 9

Training

LEARNING OBJECTIVES

Upon completion of this chapter, you should be able to:

- Identify the personnel and positions that make up a training bureau.
- Describe the need for training in the fire service.
- Explain the difference between technical and manipulative training.
- Describe how an adequate level of training is determined.
- Describe how performance standards are determined.
- Explain how skills are developed.
- Explain the importance of skills maintenance.
- Explain how training level applies to incident effectiveness.
- List areas in which firefighters require training.

INTRODUCTION

It is dark, smoky, and hot. You are alone inside a burning structure and running out of air in your SCBA. You cannot even see your hand in front of your face and you are totally disoriented. The heat is starting to increase tremendously. The linoleum under your knees is starting to melt. Your ears feel as if they are on fire. How did you get in this situation? What are you going to do? What should you do? Do you have enough time to make a carefully considered decision? Should you stand up and run? Should you do the low crawl and try to find a doorway? What has happened to the rest of your crew? As panic starts to set in, you have one thing to rely on at that moment and that is your training. You must not let panic overcome you and reduce you to helplessness. You need some idea of where to go and how to get there. You should be able to conserve your air and activate your personal alarm. You should be able to make the decisions to save your own life. Will you be able to?

The same type of scenario can happen on a wildland fire. The fire has swept around below your position on the hillside. It is now starting to make a run in your direction. When do you deploy your fire shelter and when do you run? If you run, should it be uphill, cross hill, or downhill? Where is your escape route? Where is the safety zone? Can you run through the flames to safety? Are you wearing your proper protective equipment so you have a chance of surviving in your shelter? Have you ever deployed a shelter in wind as strong as it is right now? Where is a good deployment spot? What do you do with your pack and the chain saw you are carrying? Do you surrender them to the fire before you run or try to save them by carrying them with you? Any of these decisions made correctly can save your life. Made incorrectly they can cost your life or result in serious injury. You only have seconds to formulate your plan of action and carry it out.

Are you prepared to make these decisions?

The foregoing scenarios could happen to you sometime in your career as a firefighter. Will you be prepared? Are you going to pay attention in classes and drills? Are you going to practice and maintain your skills? Only you can answer these questions.

After firefighter fatalities in the line of duty, investigations are made and reports written. These reports are made available on the Internet. Many fatalities, in no matter what type of firefighting, have common denominators that have been recognized in previous fatality situations. The statement "those who do not study history are doomed to repeat it" is definitely true. An example of this is the four firefighter fatalities on the Thirty Mile Fire on the Okanogan-Wenatchee National Forest in the State of Washington in 2001. There were violations cited of the "Ten Standard Firefighting Orders" and the "18 Situations That Shout Watch Out" (see Chapter 14).[1] We must gather and examine information available to us so we do not become a case history ourselves. We must also pay strict attention to the material already available. Most of the material was developed after firefighters lost their lives.

Many fatalities, in no matter what type of firefighting, have common denominators that have been recognized in previous fatality situations.

In the information-rich environment that we operate in today, innovations are developed on an almost daily basis. These innovations, such as the positive pressure ventilation fan (power fan), can make us more efficient and provide a wider margin of safety in a dangerous profession. If firefighters do not avail themselves of this information, they are the ones at fault. There is no one else to blame. Training is a career-long commitment that all emergency service personnel must make.

TRAINING BUREAU

The personnel who devote their time and efforts preparing firefighters to do their job safely and efficiently both on the everyday incidents and under extraordinary circumstances are the personnel of the training bureau. Their job requires them to stay abreast of the latest developments in firefighting technique and safety; they then plan, prepare, and present this information to the personnel in operations.

NOTE The personnel who devote their time and efforts to preparing firefighters to do their job safely and efficiently both on the everyday incidents and under extraordinary circumstances are the personnel of the training bureau.

Small departments may have only one person in charge of training for the entire department. Larger departments require a complete training staff. Regardless of the size and capabilities of the department's training staff, firefighters are ultimately responsible for their own safety and education. You will be the firefighter you choose to be. No one can make you better or worse. You have to realize that it is in your own best interest to be the best educated, most physically fit, and best trained firefighter you can be. It may just save your own life and/or the lives of other firefighters.

Staff Function

Some departments look on the training bureau as a staff function. The personnel work a five-day-a-week schedule. They have offices, usually at headquarters or at the training facility.

The advantage of working the five-day schedule is that they can get more productive work done in a week than if they were on the operations schedule and working out of a fire station. However, the personnel in operations often see an imaginary dividing line between themselves and the training staff. Like any other staff position, the training personnel are there to support the function of the line (operations) personnel. As previously stated, without the training staff, the operations personnel may not have the required training to perform safely, efficiently, and in compliance with the law. Instead of viewing training as a group of people who just give them more work to do, operations personnel should view the training staff as a valuable asset to their personal welfare and ability to perform their jobs.

Operations Function

In other departments, the training staff is under the command of the operations section. Many feel strongly that this is where it belongs. With training officers responding to incidents as safety officers and in other capacities, this only stands to reason. Some departments have battalion training officers. These are usually company officers who have expressed an interest and aptitude for training. An advantage to having training personnel show up on incidents is they can then view the operations taking place and assess training and equipment needs and overall performance.

In large departments with stations spread over large geographic areas and differing greatly in the types of incidents they handle, it is only natural that different areas have different strengths and weaknesses. The needs of each area differ as well. A station in a suburban area performs operations differently than one in the downtown area with high-rises and manufacturing occupancies. The training personnel must develop programs that will maintain the proficiency of the outlying companies so they will be able to operate efficiently when called to a fire in the downtown area. The personnel assigned to training also assist in the preparation of the downtown companies to respond into the hills when a wildland fire is threatening homes.

Personnel

The overall requirements for a training officer are varied.[2] They can be divided into aptitude and attitude. Aptitude will be discussed first. The training officer must be able to plan effectively. The planning must be flexible enough that a company attending a training session can respond if it is needed to handle an emergency and have the training session resumed when it returns. The planning also requires the coordination of a wide variety of resources. A wildland class may include instructors from various agencies, such as the local fire department, U.S. Forest Service (USFS), and Bureau of Land Management (BLM). A structural firefighting class could include experts from the building department and the local utility provider. The training officer must make sure that all of the personnel are available and understand what portion of the training each is required to present. Nothing looks worse and destroys the credibility of a training session than two of the instructors engaging in a heated discussion as to a difference in strategy or tactics in firefighting operations in front of the class. Planning and conducting a drill with numerous agencies and resources, including hand crews, helicopters, dozers, and engines, is understandably difficult. When there is a breakdown in the planning process, the resources are not properly utilized and the whole drill may be more or less a waste of time for all involved. Training officers should go through the training to be certified to the levels specified in NFPA *Standard 1041, Fire Service Instructor Professional Qualifications.*

Training officers must have the capacity to research and develop instructional outlines (lesson plans) for personal presentation or presentation by others; this requires well-developed communication skills. The training officer must be able to communicate clearly and concisely, orally and in writing. Enough information must be included to get the point across and answer the common questions, but not be overly technical or boring. The material must be well organized and build knowledge in a logical sequence.

In addition to preparing material to be presented orally, the training officer must be able to develop instructional packages for individual or group self-study. The ability to take a video presentation from the concept stage to the completed video is a valuable attribute. It used to be that most of the visual aids used in training were in the form of slides, films, or chalkboard presentations. With today's availability of video technology, closed circuit TV, televisions, digital cameras, electronic slide programs, computers, computer projectors, and VCR/DVDs, training posted to the department's web site and video are viable alternatives. Not only do these tools provide for general productions, they also allow trainers to address issues specific to the needs of certain personnel in a department—for example, a video about the inspection of the brakes on fire engines operated by the department or a virtual tour of occupancies within the jurisdiction.

The training officer must be able to present concepts and ideas to groups in a teaching situation (**Figure 9-1**). Nothing requires you to know more about a subject than to have to present it to others. When you study the material and prepare your lesson plan to make that presentation, you had better know the material in depth. You must not only be

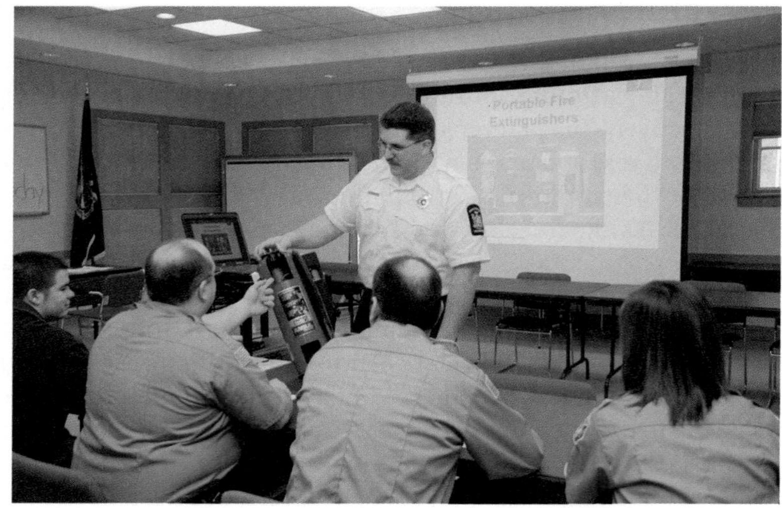

FIGURE 9-1 Training officer instructing class on fire extinguishers.

Delmar/Cengage Learning

able to present it in a logical manner, but you must also be prepared to answer questions. No instructor wants to keep saying "I do not know. I will have to look it up and get back to you on that." Be sure that you do get back to the student in a timely manner. Training others in concepts is probably the hardest material to present. You cannot hold a concept in your hand and point out its features and parts to the class. Taking the concept of the fire tetrahedron for example, you cannot see the production of free radicals. You just have to take it on faith that they are there. When instructing a class on the parts of a ladder, you can point them out and show their relation and function to the students. It is much easier to grasp the idea of an item in front of you than to understand a concept or theory.

NOTE The position of training officer requires that the person have a positive attitude concerning the importance of training.

The position of training officer requires that the person have a positive attitude concerning the importance of training. The officer must understand the importance of training and its relationship to the organization as a whole. A training officer should have experience in as many of the department's areas of activity as possible. Many of the people who apply for these positions have one or more areas as a specialty, such as hazmat, but have only a general knowledge of the rest of the department's functions. There is no substitute for experience when it comes to having credibility in the eyes of one's peers. The training officer must have demonstrated initiative in past assignments and be a self-starter. The position requires that the person be able to work with minimum supervision on complicated projects. In departments in which the position is considered a specialty position and not awarded on seniority alone, credit may be given in the selection process for having shown an interest in training in the past and having taken courses to better prepare to function in the position.

As a whole, the members of the training bureau, with respect to their rank, have the greatest effect on the policies of the department. The training officer has the ability to affect change in the overall direction of the department on a regular basis. When a company officer trains his crew to do things a certain way, the only personnel affected are those under his direct control. When the training bureau writes a new policy or procedure, it can change operations in every fire station in the department.

Along with the training officer's capability to implement change is the exposure to scrutiny by the whole department. When the training bureau develops a course or program, everyone in the department sees the results. The company officer's work is seen only by the immediate subordinates and supervisor. If the company officer makes a mistake, very few will ever know about it. When the training officer makes a mistake, most everyone is aware of it. This leads to a greater responsibility to not only perform top quality work, but also an opportunity to assist in setting the trends for the future and providing for the safety and efficiency of everyone in the department.

Training Chief

The overall responsibility for the training bureau is usually given to a person of chief's rank. The responsibility and importance of this function warrant a person of this position in the organization. Having the head of the training bureau hold a high-ranking position also lends more authority to her decisions when it comes to requiring personnel to assist or participate in training programs. Many times the training bureau is charged with introducing new procedures and methods to the organization. For them to be accepted requires careful decision making as to their usefulness and method of introduction. One person is often in overall charge of training. This will help establish a coordinated effort to have all of the shifts trained to the same level on the same material and in the same procedures. There needs to be a person who sets the tone for the bureau's activities and leads the planning for a comprehensive program that addresses the department's needs. There also needs to be a central contact person for the higher ranks who can prioritize the requests that come from them and adjust the workloads of the training staff accordingly. Situations will come up that require dropping the normally scheduled program and that demand immediate attention. These are often safety-related problems that must be addressed immediately.

In situations where the training bureau is closely tied with the local college, the training bureau administrator may serve as the **liaison officer**. The college's fire science representative will most likely be a person with an extensive education background and a master's degree or higher. The fire department person needs to be able to operate comfortably on this level.

liaison officer A contact person for outside agencies.

Company Officers

Departments that have a chief officer in charge of training usually have company officers as the primary training officers. These officers schedule the training and arrange the programs. They also perform much of the actual instruction and needs assessment. This is beneficial for the company officers involved as they are exposed to working closely with chief officers on a daily basis. They improve their planning and evaluation skills through the exercise of their duties. Many company officers who aspire to the chief's rank spend some time working in the training bureau to improve their communication skills and to aid in developing themselves for the staff part of the job that all chiefs perform. In many departments, the training staff prepares and revises the department's **operational** and **administrative procedures**. These will surely be included on promotional tests and there is no better way to prepare yourself for testing on them than to be intimately involved in their preparation and introduction to the line personnel. It is considered a given that the best way to retain the material learned in a subject area is to have to present it to others.[3] Not only does the position of training officer aid in preparing oneself for promotion, but also, in many departments, proven effectiveness in a staff position is a prerequisite for promotion. At the very least, it does not hurt as the interviewers on the promotional oral panel you will face are more than likely training or prevention officers.

On the downside, the training officer does not often have the same measure of control as an operations company officer. The station officer can pretty much plan most of the activities of the crew and schedule program work and other functions as he sees fit, as long as they fall within department guidelines. The training officer works on long-term projects when there is time between assisting instructors and handling last-minute changes to schedules. This can be compared to the station officer having the scheduled work interrupted due to responding to emergencies. Another factor is that the station functions in the company officer's absence. The program work needs to be completed even if the company officer responsible for that shift is on vacation. When the training officer is at a class or on vacation, the work has a tendency to pile up and be waiting when he returns. In many departments the training bureau must borrow the apparatus it needs from operations to put on a class, such as driver/operator training, or a firefighter academy. This can lead to problems when the operations personnel do not deliver the equipment on time or other problems crop up. After the equipment is used, it had better be returned to the operations people clean and in good working order. If not, the training officer may receive a phone call or memo from an irate station company officer asking why the chain saw was returned out of fuel and dirty from the drill.

operational procedures Written procedures for performing operational functions.

administrative procedures Written procedures for performing staff-related functions such as reports and other paperwork.

Instructors

The instructors are another very important part of any successful training program. Instructors can be from the same department, other departments, or from related agencies, such as ambulance companies. It is a proven fact that good instructors are made, not born. They can attend courses and seminars on presenting instruction. They can learn from observing and speaking with other instructors. They can access information on training topics and instructional methodology in publications and the Internet. Some people seem to have an easier time addressing a group in a training situation. If you were to ask them, they would tell you that it does not come naturally. They studied, prepared, and practiced to become comfortable with the material and to develop their presentation style.

NOTE The instructors are another very important part of any successful training program.

In a busy department, the personnel of the training bureau do not have the time or the subject matter expertise to teach every class. A part of the training officer's job is to recruit and assist instructors in their presentations. Fire personnel may be the best at presenting course material at the level and in the language that other fire personnel understand. However, some areas of expertise are hard to find among the fire personnel available to a department, so the use of civilians is required. One of these areas is weather as it relates to wildland fires. Few firefighters have the depth of weather knowledge to cover this subject thoroughly. However, the local television weatherman is likely to go over the students' heads immediately and not present the material effectively. There must be careful coordination with the instructor as to the expectations for the course and the subject matter to be covered.

The professional qualifications for the different levels of instructor and what is required at each level is contained in NFPA 1041, *Standard for Fire Service Professional Qualifications*. In this document instructors are divided into three levels with Level I being the lowest and Level III being the highest. In summary, Level I instructors are able to utilize prepared material to set up and deliver courses. Level II instructors are able to develop lesson plans and evaluate and coordinate the activities of other instructors. Level III instructors are able to develop comprehensive training curriculum and programs, conduct organizational needs analysis, and develop training goals and strategies.[4]

In departments that are tied closely with the local college, credit is given for the courses. The college can collect money from the students or the state and use the money to pay the instructors. The National Fire Academy (NFA) uses federal funds to pay its instructors. The people who instruct for the NFA are from all over the country, firefighters and civilians included, and are required to apply for acceptance into an instructor course for the subject they wish to teach. On successful completion of the instructor course, usually held at the NFA in Maryland, the instructors are certified. The instructors are then notified as to where and when the courses are to be held and submit a bid for the amount they wish to receive to instruct the course. The instructors are not allowed to bid courses taught in their home state to give the courses a more national perspective.[5]

Electronic Media Technician

With the increased use of video and other electronic media in today's training environment, an electronic media technician is a valuable resource for any training bureau. One well-planned, 10-minute video, DVD, podcast, on-line course, or other method of electronic transmission can be distributed to every station in the department and instruct all of the shifts very economically. To have a single person perform this training in person in a 50-station, three-shift department spread over a large geographic area would be just about impossible. Just about the time the instructor got set up, the personnel to be instructed would receive an emergency assignment and leave. With electronic media the students can turn the video off when they leave and resume when they return. Members who were absent when the rest of the crew saw the video can view it upon their return.

NOTE With the increased use of electronic media in today's training environment, an electronic media technician is a valuable resource for any training bureau.

The purchase of electronic media equipment is expensive, but it can return its initial cost in a short time. The return on investment can be justified by the cost in mileage, fuel, and fire loss experienced by having uncovered first-in areas when having companies report to a central location for training purposes. There are numerous sources of fire- and medical-related videos on the market. These videos can be used by themselves or supplemented with department-specific information.

The acquisition of electronic media staff and equipment gives the fire department the opportunity to perform this function for other intra- and interagency departments and related groups. In addition to buying goodwill, the department can arrange for reimbursement through exchange. This avoids one department having to transfer funds to another. Preparing a short video for another department may get you the instructor you wanted for a particular course, at no charge. It also gets the name of your department in front of others in and out of the firefighting profession. When done properly, through the development of professional materials, your department's image can be enhanced.

Light Duty

As a way to gain some usefulness from people who are restricted from active fire duty because of injury, numerous departments are requiring affected firefighters to go on what is often called light duty and assist the training and other bureaus. They may be given data entry, equipment transfer, or other assignments. Light duty gives the department a way to recover some of the expense of paying these people while they are on injury leave. It also gives the personnel assigned to light duty the benefit of exposure as to what is required of the training and other staff positions.

INTERAGENCY

One of the finest examples of training cooperation is when the fire department jointly trains with other agencies. These agencies can be federal, state, or local. Joint training creates an atmosphere of mutual respect among all concerned. When training with the USFS and the BLM, the fire department gets a national perspective on its problems with wildland fires. Through the National Wildfire Coordinating Group (NWCG), the federal personnel have training resources not normally available on a local level. The NWCG is an association of agencies that develop training and reference materials for wildland firefighting. They also have a qualification system that can prepare local department firefighters to participate in firefighting on a national scale.[6] The USFS and BLM personnel, in turn, gain a new respect for the problems faced by local firefighters in the areas of structural firefighting and hazmat. By sharing their expertise, all of the involved agencies gain knowledge and increase effectiveness.

On the local level, fire departments that fight fires together should train together. Through the use of multiagency drills they become familiar with each other's procedures and capabilities. When all of downtown is going up in flames is no time to find out that your SCBA bottles do not fit their breathing apparatus. History has proven that when things are at their worst, we rely on our neighbors the most to help bail us out of a bad situation.

NOTE Fire departments that fight fires together should train together.

Some departments have taken this concept to the point that they have joint recruit training academies. When people go through the academy together, they get to know each other from the very beginning. They are well versed in the procedures of all of the departments involved. It also helps to avoid the bad feelings that sometimes arise between departments operating in close proximity to one another. If there is a move to consolidation in the future, the joint training of personnel from their first day will pay dividends in the long run.

Joint training also occurs between the fire department and the private sector. Industrial fire brigades train with the fire department regularly. This gives the fire brigade the benefit of the firefighting expertise of the fire department. The fire department gains information on the processes protected by the private fire brigade. In some cases, private industry sponsors fire department personnel's attendance at special schools; otherwise many fire departments do not have the training budget to send their personnel. This partnership can extend as far as the private sector purchasing equipment and donating it to the local fire department. The fire department could not afford it and industry does not have to provide personnel time to train in using, operating, and maintaining the equipment.

Ambulance Companies

A group that sometimes is overlooked when joint training takes place is the local ambulance company (**Figure 9-2**). Firefighters who are EMTs are not allowed to start intravenous lines (IVs) but should at least be prepared to set the equipment up so it can be started by the on-scene paramedic. Ambulance personnel require access to the patient when fire personnel are performing **extrication**. They need to be taught the dangers of being unprotected from flammable liquids, broken glass, and sharp metal edges at an accident scene. How many times have you seen a picture in the paper of firefighters in full turnouts working a vehicle accident and right in there with them are ambulance company personnel in shirt sleeves without helmets? Everyone is trying to do his job to the best of his ability to save a life, but personnel safety is the number one priority.

Another example of the training cooperation between fire departments and private ambulance companies is firefighters going through paramedic school. They then act as ambulance company personnel to gain the hours required to finish their paramedic clinical training. One way to gain advanced medical training and to exchange information and ideas with the local paramedics is to attend base meetings. These meetings are conducted at the local hospitals that serve as the informational contact bases for the paramedics when treating and transporting patients. Base meetings are conducted because paramedics are required to keep their certification current by completing a prescribed number of continuing education units (CEUs) between recertification testings. These classes are presented regularly. The presentations are made by doctors or specialists in the subject

extrication The act of removing trapped victims. Term usually used in reference to vehicle accidents.

FIGURE 9-2
Firefighters training with ambulance personnel.

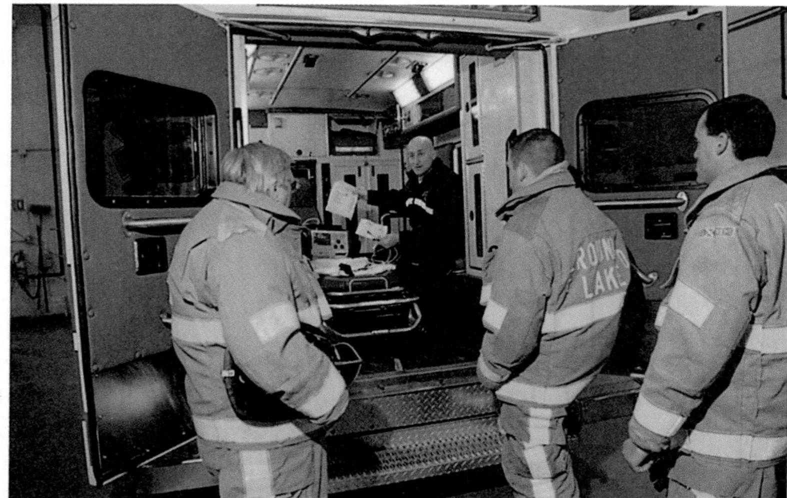

Delmar/Cengage Learning

area. The subject areas can include burn treatment, snake bite, head injury, and others. Because medical providers are required to work closely with the fire department at accident scenes, it benefits the fire department to present material on their operations and incident command system at least annually.

An area that requires special training and coordination is working around medical helicopters, commonly referred to as air ambulances. These craft have come into common use in many areas. Firefighters not used to working around helicopters are at great risk of injury without the required training. When the local department does not have its own helicopters, it is necessary to coordinate with the air ambulance provider to receive the necessary training. With the high cost of insurance and helicopter parts, the provider will be anxious to assist you in training your personnel.

TRAINING FACILITIES

Training can take place anywhere: The fire department does not require expensive props or facilities to train its personnel. In a search for realism departments may choose to contact local building owners when buildings are scheduled for demolition to perform ventilation, firefighter rescue, and other drill scenarios. The most important thing is that the drill be organized and conducted in a safe manner. Realism is good, but being too realistic, such as setting a building on fire with personnel on an upper floor, is unnecessarily dangerous. There should always be a qualified safety officer designated and a plan of operations in place prior to any training that may result in injuries being experienced. It is safer and more efficient if training has been conducted before an emergency situation arises. It is often not possible for firefighters to be trained in every type of situation they will face in their careers. They must take their training and past experience and relate it to new situations as they arise. Firefighters must be able to learn from their mistakes. Once you burn your hand by forgetting to wear your gloves, you will be more careful in the future. If you are not, you may need to reevaluate your ability to receive training.

There should always be a safety officer designated and a plan of operations in place prior to any training that may result in injuries being experienced.

NOTE Training can take place anywhere.

Classrooms are a necessity for many types of training. When training a large number of personnel or dealing with technical subjects, a classroom is a must. When people are comfortable and distractions are limited, they learn better. When dealing with technical material, dry erase boards or other means of display are often necessary. Many of today's training programs are available on video and the required equipment is necessary for presenting the material.

Another valuable asset to the fire department is a training or drill tower, for ladder training, both ground and truck mounted. Drill towers also allow practice for operations in multistory structures (see **Figure 6-4**). When the department does not have a drill tower, a multilevel parking structure can serve the same purpose.

A burn building is nice to have when actual live fire training in structures to be demolished is not available (see **Figure 6-5**). The burn building provides a controlled environment where operations can be performed at little risk to personnel. Through the use of smoke machines and small fires, the burn building can last for years and provide much valuable training.

A drafting pit is necessary for pumper testing and can double as a training prop for pump operator skills testing and certification (see **Figure 6-7**). By varying the level of the water in the pit, the difficulty of the drafting procedure can be adjusted. By assigning a person to oversee the annual pump testing and sending the operators in with their equipment, the operation can serve as a refresher course on drafting as well as completing the pumper tests.

As the fire department moves into new areas of responsibility, the training facilities must be included. Hazardous materials props are expensive to construct and take up large amounts of room. These types of facilities are usually provided on a regional basis. The required props include leaking pipes, flanges, and chlorine cylinders. Railroad tank cars are another necessary prop. The props are plumbed with compressed air and water to simulate pressure leaks. The props need to be as realistic as possible while providing a measure of safety for personnel (**Figure 9-3**). It is extremely difficult to do anything in a

FIGURE 9-3 Hazardous materials training.

Delmar/Cengage Learning

fully encapsulated suit, especially climbing ladders and working on top of tank cars. It is a long way to the ground if you slip.

The inclusion of an electronic media studio allows editing of videos taken during training sessions. Once edited and prepared, the videos can be reproduced and sent out to the stations so that every engine company can gain at least some value from the training, even if they could not attend the actual drill (refer to **Figure 6-10**).

fully encapsulated suit A suit that includes total body protection. When worn with gloves, it gives head to toe protection from certain chemicals. The interior is sealed from the outside air.

Off-Site Training

In just about any area, off-site training can be conducted in numerous locations. Every time you complete a fire prevention inspection, you are performing training. When the alarm comes in at 2 A.M. and you enter the office you inspected last week, you will stand a better chance of finding your way around in the dark. Even the simple operation of driving around in your first-in area serves as training. As you become more familiar with the streets and traffic patterns in your area, your response times can be lowered.

Wildland fire training must be done as safely as possible with provisions made to keep the fire within the drill site; and adequate personnel, equipment, and water supply must be available.

A prime example of off-site use is wildland fire training. By going out in the wildland area and burning off areas during drills, several purposes can be served. The most important one is training personnel in wildland firefighting. This does not mean just lighting off a couple of acres and watching them burn. That may be informative as to fire behavior but gives little training value. The suppression of training fires in a wildland setting can serve to prepare personnel for the upcoming summer fire season. As always, this must be done as safely as possible with provisions made to keep the fire within the drill site and adequate personnel, equipment, and water supply must be available. The site must be surveyed for safety hazards and prepared before any fires are lit. Topographical features that pose a hazard should be identified and brought to the attention of all participants. In areas with oil field operations, the grassy area should be searched for rocks, pipes, holes, and pumping unit foundations that may not be readily visible. These areas may contain overhead electrical wires and, because electricity can arc through smoke, they should be avoided. All power poles should have the grass removed around the bases so the fire does not spread to them (**Figure 9-4**). Fence lines with wooden posts must be protected to

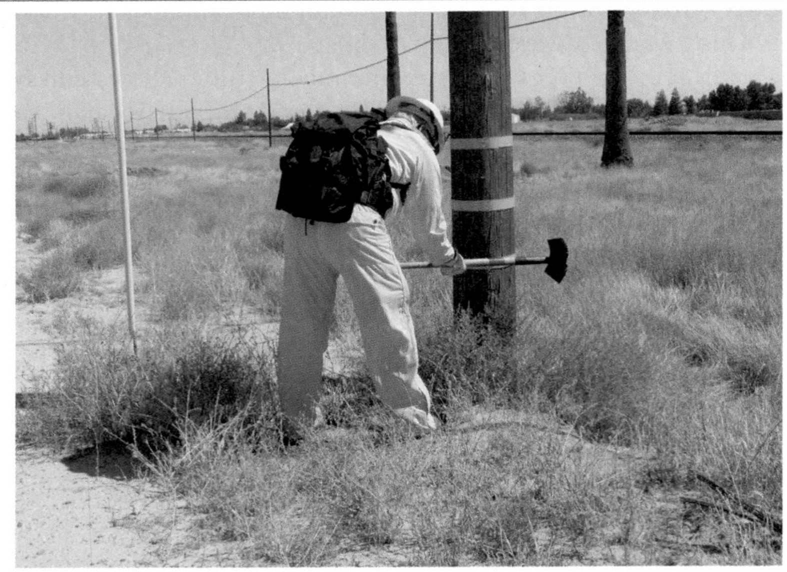

FIGURE 9-4 Removing fuel from base of power pole prior to vegetation-fire hot drill to prevent fire from damaging the pole.

prevent damage. High heat will weaken barbed wire fences and fuels should be reduced or removed underneath them. Basically anything that you do not want to burn should be adequately cleared around it before any fires are lit. Piles of tumbleweeds and other high heat producing fuels must be considered in planning the drill.

NOTE The purpose of drills is to train personnel, not to punish them.

Another advantage to wildland fire drills is the reduction of existing hazards. An example is areas where certain roadsides suffer numerous fires during the dry season. They may be able to be burned off to provide a buffer zone so that vehicle fires do not spread to the wildland or grazing lands.

Bear in mind when performing grass firefighting operations in agricultural areas that the aerial and ground spraying of pesticides may have coated a certain amount of the grass. Vegetation firefighting is often done without breathing apparatus and the presence of pesticides can be an unseen hazard.

In areas with high-rise structures under construction, it may be possible to secure permission to drill in the building on a weekend or holiday. Before the floor and wall finishes and furnishings are installed, hoses can be dragged around with little chance of damage to the structure. The drill can also acquaint engine and truck company personnel with the construction methods used in the building.

As with any other type of drill, proper preplanning is necessary to provide maximum training during the time allotted and to provide for personnel safety. The drill area should be checked for electrical and other hazards. Vertical shafts must be secured, including any hole large enough for someone to fall into. Construction workers fall into open elevator shafts on a regular basis. There is no reason to think that it could not happen to a firefighter on a drill.

NOTE Proper preplanning is necessary to provide maximum training during the time allotted and to provide for personnel safety.

An opportunity to engage in training is during your preplan inspections of the high-hazard occupancies in your district. These walk-throughs usually require the participation of several engine companies and serve as an opportunity to put some hose on the ground. With several companies at the drill site, one engine can remove its hose load and all hands can help it get back into service. During the preplan, building construction, hazards, strategy, and tactics can all be discussed and reviewed (**Figure 9-5**). This is also a good time to test communications from inside the building to the outside. In some structures, hand-held radios are unable to transmit or receive. It is better to know this before the incident happens and plan accordingly.

With the use of a smoke generator, search and rescue drills can be conducted in different types of buildings. When the proper type of smoke machine is used, there is no residue and the drill can be very realistic from a visibility standpoint. The only thing lacking is the heat. This type of drill can also be performed without smoke by placing a piece of waxed paper over the mask of the SCBA to simulate low visibility. The advantage of this method is that the instructors can easily observe the actions of the personnel performing the drill.

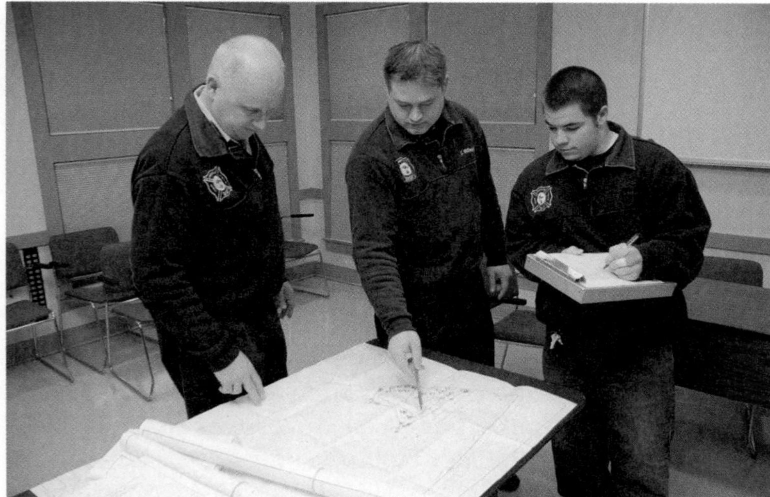

FIGURE 9-5
Incident preplanning for shopping center using plot plan.

In areas with harbors, ships are a hazard that must be addressed. Shipboard firefighting has many hazards not present in any other type of structure.[7] With the possibility of a fire below deck, the ship turns into a multilevel basement fire that is particularly hard to ventilate. There is always the possibility of falling down one of the ladders that serve as stairs between the decks. The drill should also include a close inspection of the various types of ships one is likely to encounter. Although locations will vary, somewhere on the ship there should be a fire control plan. With many ships having non-English-speaking crews, communications can be a definite problem. The metal structure of the hull tends to prevent communication by hand-held radio as well.

Aircraft firefighting has special techniques and hazards. The equipment used at major airports is specially designed for the suppression of fires in aircraft[8] (**Figure 9-6**). Personnel have to be trained in attack, rescue, extinguishment, and overhaul in an airport setting.[9] Just learning to operate the radios and how to deal with the control tower to receive clearance to cross active runways requires extensive training. If the operator of the crash rescue truck were to get excited and pull out in front of a landing jetliner, the results would be catastrophic. The personnel must be trained to operate safely around both propeller and jet aircraft. They must learn how to approach with the engines running. In a

FIGURE 9-6
Aircraft rescue firefighter (ARFF) training.

Courtesy of Jeff Riechmann

military setting, the aircraft may well be loaded with different types of ordnance that pose a great hazard to firefighting personnel.

Training in oil firefighting is a requirement for firefighters working in areas that have oil refining or production. Due to the large volumes of smoke produced by these types of live fire training facilities, they must be placed far from any residential areas.

PURPOSE AND IMPORTANCE OF TRAINING

The primary purpose of instruction is to change behavior.[10] The first-day-on-the-job firefighter may not know one end of a fire engine from the other. They certainly do not know how to effectively use the wide variety of tools and equipment available to them on the apparatus. It would be a poor idea to expect this person to respond to a fire and take action without training in what to do, how to do it, and when to do it. Just telling the person "we will show you what to do when we get there" is not effective training either. One of the most important jobs in any department is the thorough training of personnel. The personnel have the right to demand good training and the department has the obligation to provide it.

NOTE The primary purpose of instruction is to change behavior.

NOTE One of the most important aspects of any training program is safety.

One of the most important aspects of any training program is safety. It is not enough to just show someone how to perform an operation; they must be instructed in how to perform it safely. If you were just told to advance the hose line at every fire and "put the wet stuff on the red stuff" you could soon find yourself in a life-threatening situation without a clue as to what to do. As has already been discussed, firefighting is a carefully coordinated operation in which a great many people play a part. When safety is not stressed from the very first training that you receive, when can you be expected to learn it? There are many ways to perform most operations, but only a very few safe ways. Safety must be included and stressed in every training session. It is extremely important that firefighters develop a safety attitude as it helps to protect them and the other firefighters working with them.

There are many ways to perform most operations, but only a very few safe ways.

Some parts of the firefighter's job are practiced to the point that they become habit. The act of donning a SCBA over and over until you perform all of the steps to department specification is one of these. The idea is that people will do as they are trained, even under times of extreme stress. Being confronted with a situation that requires a rescue is no time to forget how to don your SCBA. It should be so familiar that you can do it almost without thinking. Your mind will be occupied with the thoughts of how to rescue the trapped person, not whether you remembered to test your face mask seal when donning the SCBA.

One of the areas that must become so ingrained that it is habit is "size-up." When faced with devastation it is important that you are able to size up the situation and determine the course of action to take that will have a positive effect on the situation and not make you one of the victims. Rushing blindly into a bad situation is just going to make it worse for all involved. As you move up in the rank structure and are expected to come up with a plan and assign resources, this training becomes even more important because it is not only your life that you are endangering by your decisions.

Along with the development of habits in the simple tasks comes the development of the thought processes that are required to perform under extreme pressure. By training yourself from the very beginning to think in a logical and orderly fashion in emergency situations you develop a **command presence**. Persons with command presence seem very cool and collected no matter how bad things have gotten. This does not mean that they do not feel the stress and pressure of the situation. They just know how to deal with it. They do not yell and use profanity on the fireground. They keep their wits about them and perform in an orderly and decisive fashion. One method of dealing with extremely bad situations is to turn your back on the incident for a moment, while you talk on the radio. This does not mean that you should ignore what is going on for long periods. It means you should turn around and collect your thoughts so you can present a calm demeanor when speaking, in the face of destruction and devastation. People who are calm and collected tend to have that same effect on those around them. Conversely, the opposite is true; if you lose your cool, it is likely that others will too.

As you become better trained and experienced in the performance of your job, you will develop the necessary knowledge to make changes to your plan if it is not working. The knowledge of what other methods and tools are available to handle the situation is critical. At first the knowledge level that using a sledge hammer instead of an axe is more effective on a tile or slate roof is where you will be. As you progress you should be capable of determining different strategies and tactics to handle situations. At that point you will be able to determine whether an offensive or defensive attack is the best strategy for the fire you are in charge of. You should also develop the ability to determine which areas of the plan need to be changed and how to implement the changes. To become a well-rounded firefighter and incident commander does not happen overnight. It takes long hours of study and years of experience.

command presence The ability to maintain composure in situations that are stressful.

TECHNICAL TRAINING

Not all fire-related training is manipulative (**Figure 9-7**). The world is changing at a rapid pace and so are the hazards and situations that firefighters must face in the line of duty, so firefighters must be trained to meet the new challenges.

NOTE The world is changing at a rapid pace and so are the hazards and situations that firefighters must face in the line of duty, so firefighters must be trained to meet the new challenges.

To perform fire prevention functions, firefighters must be able to read and interpret codes and ordinances. They must be able to speak and write well. They are required to come across as professional and knowledgeable when dealing with the public. The reputation of the department as a whole rests on your shoulders when you make contacts with business people in the community. If you come across as well trained and competent, that is how the department will be perceived. If you do not, you may be the subject of discussion at the business association meetings. When this is the case, the business owner may possibly even call the chief or local politician to report your shortcomings.

Much hazardous materials training is technical. Learning chemistry, even at a basic level, is difficult for many people. It requires study and application of effort. Learning how to select the correct PPE is not only technical, but also your life could depend on it. Hazmat team personnel must be able to perform tests on samples to categorize chemicals (**Figure 9-8**). This requires observing the results of the tests and making decisions. A careless or incompetent person may be placed in a dangerous situation when he handles the spilled material incorrectly.

FIGURE 9-7 Firefighter performing technical training.

Delmar/Cengage Learning

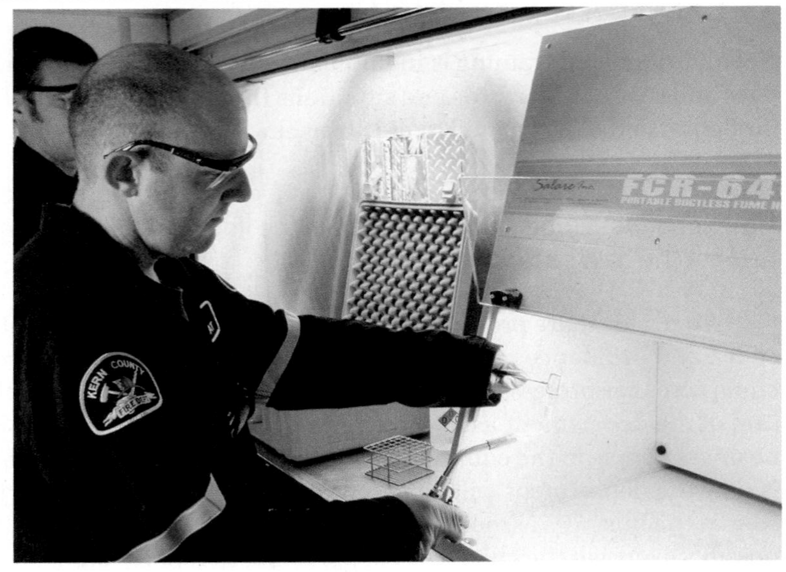

FIGURE 9-8 Hazardous materials technicians testing unknown substance to determine hazard category (haz cat).

In the provision of EMS, much of the learning is technical. You must learn about the human body and its systems. Paramedics learn how to read and interpret EKGs and what information to pass on to the base hospital so the proper drugs can be administered in the field. On the manipulative side, you will be required to learn how to immobilize spinal injuries and properly package victims for removal from the accident scene. The information must be remembered because you may only deliver one baby in your entire career, but the penalty for assisting incorrectly can cost the newborn its life (refer to **Figure 9-2**).

Building construction is one of the areas of technical instruction. Seeing photos and visiting construction sites can provide you with the required knowledge about the relative strengths, weaknesses, and problems encountered with roof construction of varying types. It is not necessary to have spent years working in the construction trades to understand truss roof construction and its reaction to fire. This information is essential when determining whether it is safe get on the roof to perform ventilation and where to place the hole.

Extinguishing agents are based on both chemical and physical reactions. It is not enough to know to open the nozzle and spray water on the fire. You should understand what is happening and why. As extinguishing agents, such as foam, become more complex, you must be able to choose the right agent for the type of fire and determine whether there is enough on hand to start the attack or if you should wait for more to arrive. The situation could easily arise in which you use up the foam on site and do not extinguish the fire. By the time more foam arrives, the fire is back to full intensity and you still do not have enough foam for extinguishment. If you had waited a few minutes for the reinforcements to arrive, the first try would have been successful. These types of decisions are directly connected to technical training in application rates and coverage areas.[11]

When inspecting or supporting extinguishing systems, it is necessary to know how they are designed. It is not enough to know that pipe A connects to pipe B and then to pipe C. You must understand how the valve system works and be able to figure out from how many sprinkler heads are open whether you can adequately support the system (**Figure 9-9**). If your water source is inadequate, you must be capable of working through the command structure to request more units to assist you.

A thorough knowledge of the incident command system used by your department is critical. You not only need to know how to initiate the system as the initial attack incident commander, but also whom to contact to get the things that you need.

NOTE A thorough knowledge of the incident command system used by your department is critical.

The most important area of technical training is safety. You must be knowledgeable about the design limitations of all of your equipment. This includes PPE, SCBA, and everything else. Your knowledge of flashover and back draft can save your life. A common reason for the failure of aerial ladders is operator error in exceeding the limitations of the equipment. This information is technical. Yes, the ladder was extended in the proper sequence with all of the steps followed. The locks were set and the controls were locked. What went wrong was that it was placed at such a low angle while fully extended, resulting in exceeding the design limitations, and the ladder failed.

FIGURE 9-9 Sprinkler system riser cutaways showing internal parts of valves.

You must be knowledgeable about the design limitations of all your equipment.

All apparatus has limitations as to what it can do. A good operator knows the design limitations of the apparatus thoroughly. When you try to climb too steep a hill and exceed the apparatus's climbing ability, you have made a technical error in judgment. When you are the pumper operator and the company officer calls for another line to be stretched to attack the fire, you must be able to determine whether you have a sufficient water source to supply the additional line.

Fire equipment is furnished with all manner of tools. The use and application of these tools is an important part of your training. When dealing with forcible entry tools, the study of locks, their construction, and their removal methods are technical information. It is difficult to gain access to locks of all of the different types that you may encounter in your career. When it comes to lifting a rolled-over car off of a person, you must know how much weight your rescue air bags will lift. If they are insufficient, you will need to call for more resources to accomplish the task. The car is crushing the life out of the person and your knowledge of your equipment and quick decision making may just save him.

NOTE A thorough knowledge of the communication system is necessary to perform efficiently.

A thorough knowledge of the communication system is necessary to perform efficiently. In large departments, numerous channels may be used. It is necessary to know what channel to switch to when performing on an emergency. It is very frustrating to have equipment assigned to the same incident on different channels and to have to hunt them down when you are trying to coordinate resources. In departments with large geographical areas, the radio system may have **repeaters**. It may be necessary to turn one repeater off and turn another on to have clear communications when operating in certain areas. Clear communications are vital to a coordinated attack and personnel safety at an incident. Your ability to use the system correctly and effectively can have a definite effect on the outcome of incident operations.

Another area of communications that requires technical training is written communications within the department. Most departments have forms made up for different purposes. Supplies from central stores or an oil change for the engine often require different forms. One of these is the fire report form. If it is not filled out correctly, the

repeaters Serve the purpose of receiving radio transmissions and boosting the signal. Used in areas where topography or tall buildings interrupt clear communications.

information needed will not be available for planning purposes and for developing statistical information.

MANIPULATIVE TRAINING

The area of training with which you will become intimately familiar as a new firefighter is manipulative training. Every time you advance a hose line or don an SCBA during the initial training, you are undergoing manipulative training. Manipulative training differs from technical training in that you are performing hands-on operations of equipment and tools (**Figure 9-10**). When you learn how much air is contained in your SCBA cylinder and at which pressure the alarm goes off, that is technical training.

NOTE The area of training with which you will become intimately familiar as a new firefighter is manipulative training.

One of the rules of manipulative training is that perfect practice leads to perfect performance. You will be required to perform simple operations over and over under the

FIGURE 9-10
Firefighter performing forcible entry.

Courtesy of Edwina Davis

watchful eye of your instructor according to department guidelines as to the steps to be followed. Learning by trial and error is not conducive to good performance. You will be watched closely and corrected at any misstep to make sure that you do things correctly.

NOTE Perfect practice leads to perfect performance.

Performing drills with hose and other equipment is called *performing evolutions*. They start as the apparatus comes to a stop and you exit your seat. At first they will be very simple and involve only performing one task, such as pulling off the supply hose. As you progress, they will include combinations of tasks, for example, exiting the apparatus, donning your SCBA, and advancing a hose line to the door of the burn building. As you become proficient at the steps of the evolution, they will become more and more complex. By the end of your initial training and assignment to operations, you should be able to show up in the right location with the right equipment, as specified by your company officer, and be ready to perform.

Much of your time will be spent learning to operate the wide variety of tools carried on your assigned apparatus. You must become proficient to the point that you can operate these tools in the dark with limited visibility under stressful conditions. You should also be capable of minor repairs as equipment has a way of malfunctioning when it is most needed. One way you are taught to do this is by properly cleaning the equipment after it is used. There is hardly a better way to become intimately familiar with the workings of a chain saw than to clean it and return it to service after each use.

When you become an engine operator, you must be able to drive safely and perform all operations required at the pump panel. These operations can include hydrant hookups, drafting, relay pumping, supplying attack lines, and supplying master streams. You must be able to perform all of these operations under normal circumstances and troubleshoot when something malfunctions. This position requires a wide array of manipulative and technical knowledge. You must not only be able to hook everything up correctly, but also be able to pump at the proper pressure. Ladder truck operators have much the same requirements. They must be able to place the apparatus properly so the ladder is kept within design specifications when it is used. Operators of wildland firefighting equipment must be able to operate in four-wheel drive and read terrain so they do not get on a side hill and roll the apparatus. They must be able to traverse rough terrain without tearing up the equipment or getting stuck. Having your apparatus stuck in a hole with the fire rapidly advancing in your direction is no place to be.

One thing that is sought during manipulative training is realistic conditions. Drills should not be conducted only during good weather and under optimum circumstances. Drilling at night is performed to acquaint firefighters with the actual conditions that they will commonly work under. Performing drills in darkness stresses the need to be intimately familiar with your equipment. When crews only drill during the day, they tend to forget about setting up adequate lighting during night operations. When operating at a fire scene at night and you put out the fire, there goes your light. Other advantages are that some of the locations, such as commercial areas, that would not be available for practice during the day are available when the stores are closed and traffic is at a minimum.

NOTE Drills should not be conducted only during good weather and under optimum circumstances.

When dealing with people who have already mastered the basics, a method of practice is to perform stress drills. When performing stress drills, the drill site is selected to be as close as possible to realistic conditions. The purpose of stress drills is to practice skills under extreme conditions, not to get anyone hurt.

The purpose of stress drills is to practice skills under extreme conditions, not to get anyone hurt.

A type of stress drill is the SCBA drill in a pitch black, smoky room, or maze. It is amazing how realistic this can be, even when you know that there is no fire and you are just on a drill using a smoke machine or a covering over the face piece of the SCBA. The claustrophobia that sets in when confronted with total darkness with a SCBA mask on your face and you are basically lost can be very frightening. The way to keep yourself from panicking when faced with this situation is to practice. The advantage to performing this type of drill under controlled conditions is that if you do panic, tear off your mask, and stand up to run for the door you will not be harmed by deadly fire gases and heat produced in a fire. Familiarity with this type of situation may just save your life when confronted with it for real. Another situation faced by firefighters that is very claustrophobic is performing rescue in a **confined space**. This type of situation must be practiced when wearing SCBA as well (**Figure 9-11**). The firefighter must get used to the limited mobility of working in tight spaces.

These spaces may include, but are not limited to, underground vaults, tanks, storage bins, pits and diked areas, vessels, sewers, and silos.

confined space (As defined by OSHA).

- Has limited or restricted means of entry or exit
- Is large enough for an employee to enter and perform assigned work, and
- Is not designed for continuous occupancy by the employee.

FIGURE 9-11 Confined space rescue training.

Courtesy of Edwina Davis

To further complicate the situation, the instructor may tell one of you that your SCBA has just failed and you must perform **buddy breathing** with one of the other members of the crew. This skill should be learned and well rehearsed before trying to perform it under stress drill conditions. You should know how to do it before you are tested.

When performing SCBA drills, a common session is to have trainees search for and rescue a downed firefighter. Dragging someone in full turnouts with a SCBA on is very hard to do. The victim must wear full PPE to avoid injuring them as you drag them out.

Trying to find someone in smoke and darkness should reinforce that you need to activate your personal alarm device whenever entering a building. You may think that you can find the other firefighters by the sound of their breathing through their SCBA. The problem is your SCBA and your partner's are both making the same noise. When the person you are searching for has his personal alarm activated, you may just be able to locate him, disentangle him from whatever he is caught in, and drag him out before either you or he runs out of breathing air.

Another type of stress drill is to have crews perform wildland fire line construction and tell them that they are about to be overrun by the fire. When this happens, they must proceed to the safety zone or deployment site and deploy their fire shelters. This drill should only be performed without fire present. The drill must include shedding tools,

buddy breathing A technique in which two people share the same SCBA air supply to avoid breathing smoke or toxic fumes.

packs, fuel bottles, and **fusees**. The personnel then deploy and take up position in their shelters. The hard run to the deployment site aids in the realism and the instructors yelling at them to hurry adds to the stress. When this drill is performed in a location with high winds that simulate an actual overrun situation, the realism is increased. When winds are light or nonexistent, several power fans can provide sufficient wind to simulate real conditions. The drill should be performed in as realistic conditions as possible. Performing the drill on the station lawn does not adequately prepare the personnel for realistic conditions. The practice of having the crew deploy and lighting a fire that actually overruns their position is dangerous and unnecessary.

Your level of physical fitness will become evident during your performance of stress-type drills. When making the run uphill to the safety zone or dragging a firefighter around, your overall fitness is tested. It is much better to find out beforehand that you need to work out more than to find out under life-threatening conditions.

NOTE Your level of physical fitness will become evident during your performance of stress-type drills.

A point that must be emphasized when talking about stress-type drills is that safety is still the number one priority. To perform search and rescue drills, you do not need to set the building on fire for realism. The skills to be practiced are search and rescue, not putting out the fire. It would be extremely dangerous to place someone in a burning building, leave him alone, and tell him not to come out until he is found. That would violate just about every safety rule that there is. Always remember the purpose of the training, and determine the safest way to perform the drill without unnecessarily endangering the safety of the attendees. It is better to focus on one skill at a time than to construct a situation in which things can get out of hand.

NOTE The first criterion for determining the adequate level of training for any job is whether it is being performed safely.

fusees Another term for road flares.

DETERMINING ADEQUATE LEVELS OF TRAINING

When determining the adequate level of training for any job, several criteria must be considered. The first criterion for determining the adequate level of training for any job is whether it is being performed safely. All personnel who perform the job must be able to perform it in the safest manner possible. If the safe way to perform the job requires it to be accomplished while wearing gloves, then you must be able to perform with your gloves on. This is not as easy as it sounds. The design of many pieces of equipment adapted from other disciplines did not take into account a firefighter wearing thick gloves.

The complexity of the job has an effect on the adequate level of training. A simple job may require only minimal training, whereas a complex job with many steps, such as raising an aerial ladder or drafting from a static water source, will require more extensive training. Not only do these jobs have many steps that must be performed in the proper sequence to accomplish them, but there are also safety concerns as well. If certain steps are left out, the results could be catastrophic.

The third criterion is with what frequency the job is to be performed. Some jobs are performed on a regular basis, such as donning SCBA and advancing attack lines. In the initial training period, these will be focused on heavily. When you are to the point that you can perform them almost automatically, within department guidelines and in a safe manner, the focus will shift to more complex operations.

As you gain in experience and skill, you will be required to learn jobs that are performed less frequently. Referring back to the shelter deployment drill, many firefighters are never required to perform this operation under real conditions.

That is no excuse not to master the skill. Firefighting has never been a routine job; every incident presents a different situation. When you become complacent and think that nothing is ever going to go wrong, something surely will. When it comes to any job that affects your safety and survival, being caught up in the situation is no time to realize that you should have prepared yourself better to deal with it. Remember, the job you perform the least may just be the one you need to be the most familiar with when the situation arises and you are in trouble.

NOTE Department policies, and in many cases the law, require that personnel be trained to a minimum prescribed standard.

Department policies, and in many cases the law, require that personnel be trained to a minimum prescribed standard. To achieve maximum effectiveness, personnel should be trained beyond the minimum level for efficiency and safety reasons. The minimum acceptable level for wildland firefighters is 32 hours of training.[12] This is enough to put them out on the line in relatively nonhazardous locations. It is surely not enough to place them on the head of an active fire that is exhibiting erratic behavior.

PERFORMANCE STANDARDS

The basic underlying factor that determines the performance standard for any job is whether it was performed correctly, which means that it was performed within department guidelines as to time and safety and with a minimum of errors.

One of the most commonly used criteria for setting performance standards is time. The department specifies that evolutions that require a single skill or combination of skills are to be performed within a time limit. An example of this is the almost universal standard that you'll be able to don your breathing apparatus in under one minute. Time is one of the hardest criteria to justify. It is better to use the criterion that the operation be performed smoothly with no major mistakes and zero safety violations. When too much focus is placed on time, competition develops between crew members and companies, and safety is sacrificed for speed. An area that makes this very evident is ladder training. One of the primary points of raising ladders is to look up before raising the ladder. When personnel compete against each other at the drill tower for time, with no overhead wires, the emphasis becomes one of speed, not the proper steps. When these people go to work in the field, they may very well be in a hurry and raise their ladder without looking up first. The one time that the ladder comes into contact with overhead electrical wires is probably the last time that person will ever raise a ladder, or do anything else.

There is much discussion among training officers as to what percentage of errors is acceptable. Some will be satisfied with a passing grade of 70%; others think it should be 80%. Realistically, the percentage will vary with the importance of the material being presented. Knowing the numbers of all of the forms used in the department may have an acceptable rate of error of 30%. Forms are used in the office and have titles on the top. They are not usually used in emergency situations where speed and accuracy are of great importance. If you make an error on a fire report, it is not going to cost anyone his life or property.

When it comes to safety, the acceptable error level is 0%, which is referred to as "zero tolerance." If the acceptable level of error were 20%, that would mean that on the average, for every 10 responses, personnel could be expected to be injured, injure someone else, or damage equipment two times. Morally and financially this rate is unacceptable. Part of the professional attitude of modern firefighters is that they approach their jobs with due consideration for the consequences of their actions. When you do not secure the axe you are using on the roof and it falls on someone below it is your fault, not theirs, even if the reason he ended up in the hospital was he was not wearing his helmet.

When it comes to safety, the acceptable error level is 0%.

SKILLS DEVELOPMENT

The definition of skill is the ability to use one's knowledge effectively and readily in execution or performance; a developed aptitude or ability.[13] When you come to work for the fire department, you are not expected to be skilled. You are expected to be able to become skilled through instruction, study, and practice. You are also expected to be physically fit to the point that you will be able to perform the skills you are taught.

A master is defined as a person whose work serves as a model or ideal.[14] Your goal as a student of your job should be to achieve mastery of the tasks expected of you when performing the required jobs. In training a person to the level of mastery, each individual task is required to be performed to a set standard. The scores achieved on the individual parts of the testing are not averaged. In this type of training, the emphasis is on performing the tasks correctly, not on time.

NOTE Your goal as a student of your job should be to achieve mastery of the tasks expected of you when performing the required jobs.

As a firefighter, you are expected to be able to perform at a consistently high level after a very short time. The job requires that you exhibit mastery of at least the basic tasks of properly donning your PPE and performing suppression actions. Your supervisor does not have the time at an emergency scene to watch you closely. In a normal fire situation, everyone is very busy and the price paid if you make a mistake may be very high.

As you master the basic sequence of skills, you will soon be expected to be able to deal with more complicated situations. This requires the application of both manipulative and technical skills. Firefighting is a dynamic process and the situation is subject to change rapidly. You will be expected to adapt your operations accordingly. Some examples of this are a wildland fire spreading to the roof of a structure, a fire with burn victims, or a vehicle accident with hazardous materials and rescue of trapped persons required. In these situations, you cannot limit yourself to one aspect of the situation. You must be able to tie together all of your training and experience and act appropriately. All of the mentioned scenarios require decision making. The major burden of the decisions falls on the supervisor, but they are not always standing next to you or available. You must become proficient to the point where you can at least begin to take the appropriate action while they are being notified. At the very least, you should be capable of recognizing the hazards present and not unnecessarily endanger your safety or that of others at the scene.

SKILLS MAINTENANCE

After you have mastered your basic skills and moved on to more complicated operations, you must maintain your proficiency at every skill, even the basics. As you are learning to operate the pumper, you must still be able to don your SCBA in the required time. The only way to maintain skill level is to practice. Once you have mastered something, the tendency is to take it for granted that you are still capable of performing at a high level. This does not hold true for most people. Only through constant review and practice can you maintain your skill level.

NOTE The only way to maintain skill level is to practice.

SKILLS ASSESSMENT

Once skills are learned, the ability of the student to perform them must be assessed. This is done through testing performed on both technical and manipulative skills. Many different types of testing are used. One of these is the pencil and paper test of technical information; another way is through verbal testing. An exam about ladders could require you to name all of the parts when given a drawing or you could be asked to name all of the parts when looking at one lying on the ground. At the station level, a form of technical testing is the company officer observing how proficient you are at filling out reports and doing other required paperwork.

In manipulative testing, you will be expected to perform operations and evolutions in front of an evaluator. This goes on all of the time at the station level. When the company officer has the crew perform a drill, she is evaluating the performance of the personnel and assessing training needs. If the crew is slow or hesitant in performing the required operations, more time should be spent practicing to increase proficiency. The company officer has the added incentive that when the chief wants to see a drill performed, she is going to receive guidance if the crew is inefficient or sloppy and especially if it is unsafe.

At the end of your initial training, you may receive a comprehensive manipulative and technical examination. The manipulative test requires you to show a high level of proficiency in the skills you are expected to perform. The technical test is to determine if you learned the policies, procedures, and rules and regulations of the department. There may also be specialized testing in the areas of hazardous materials and emergency medicine.

In departments that have structured post-academy training programs for new firefighters, you can expect a manipulative and technical test at the end of your probationary period. In many cases, the successful completion of this test is required to complete probation and retain your position with the fire department.

A reason for getting companies together and performing testing is the "one department concept." When companies are brought together from different areas of the department's jurisdiction, they are expected to be able to perform together as they would with companies that they drill with on a regular basis. With the standardization of apparatus, any firefighter should be able to approach any station's apparatus and find the location of an SCBA or other tool on the first try. With standardized hose loads, any combination of personnel should be able to advance hose lines off any of the apparatus. At a typical drill, several engine companies are brought together at one location and told to perform a coordinated attack. As they do this, the drill requires that they support each other's operations with personnel and equipment. When there are vast differences in the way operations are performed, it is immediately evident in the inefficiency and time it takes to complete the required operations. If the companies spend a lot of time talking over what they are going to do before they do it, it is evident that they are not trained to perform the operations the same way.

NOTE Standard operating procedures are written procedures that specify what companies will do when they arrive at the emergency scene.

STANDARD OPERATING PROCEDURES

The way to achieve unity and coordination is through the use of standard operating procedures or guidelines (SOPs/SOGs). Standard operating procedures are written procedures that specify what companies will do when they arrive at the emergency scene. SOPs are developed to apply to the areas of command, communications, safety, tactical priorities, and utilization of companies.[15] With the inclusion of SOPs, the department can function more smoothly, as the incident commander knows what to expect from the arriving companies. It also reduces the need for fireground communications as the incident commander does not have to spell out everything required of the companies at scene. When the first-in truck takes care of turning off the utilities upon their arrival at scene, the incident commander does not have to remember to request it or specify what is to be done. The following is an example of a simple SOP. It can also be used as a standard drill to build and test company proficiency.

Standard Operating Procedure Horizontal Ventilation With Power Fan

Purpose: To adequately ventilate a structure in a rapid manner with due regard to firefighter safety the following procedure will be utilized whenever possible and applicable.

The first engine at scene will give a size-up and determine the attack mode appropriate to the situation. When offensive mode interior attack is chosen, ventilation will be performed in the following manner:

1. Personnel with SCBA, charged attack lines, and forcible entry tools will advance to the entry point and prepare for operation.
2. A firefighter will be detailed to determine the location of the seat of the fire and create a ventilation exit hole, without entering the structure. When this is done, the firefighter will advise the entry team.
3. The power fan will be placed into operation and the entry point opened.
4. As the smoke is cleared from the path of attack, the entry team will enter and proceed to the seat of the fire.

TRAINING RECORDS

A necessary step in the training process is maintaining training records (**Figure 9-12**). Records are necessary to document that training was received by the personnel involved and how they performed. Over a period of time these records can be reviewed and areas of greater need can be assessed. The review of these records can also identify areas in which training was lacking or missed all together. When some operation goes especially right or especially wrong, the training records can be accessed to help determine the reason. Also numerous government regulations and laws require that firefighters receive training in specified areas. Training records are necessary to verify that the required subjects were covered and to what extent.

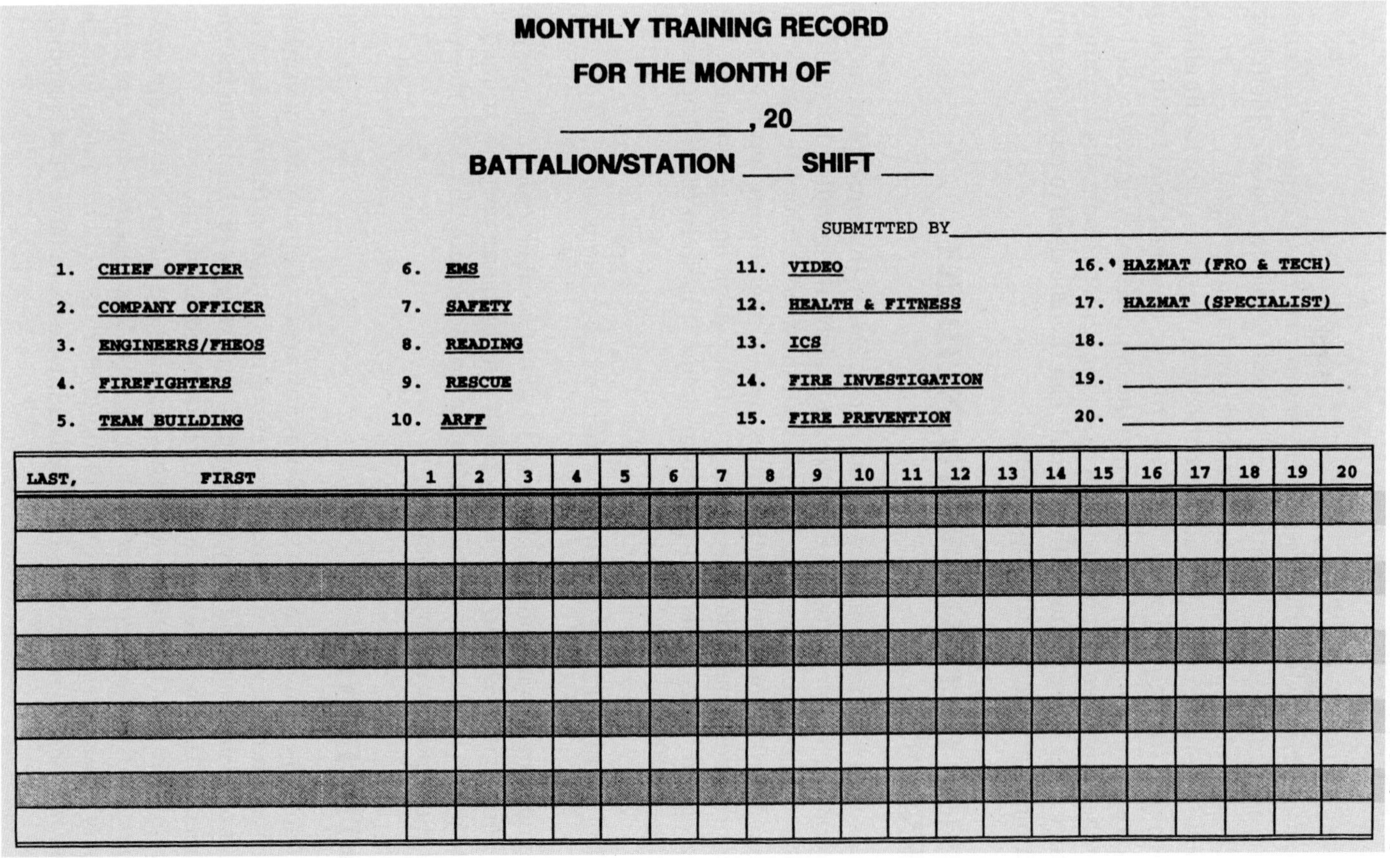

MONTHLY TRAINING RECORD

FOR THE MONTH OF

______________, 20___

BATTALION/STATION ___ SHIFT ___

SUBMITTED BY____________________

1. CHIEF OFFICER	6. EMS	11. VIDEO	16.• HAZMAT (FRO & TECH)
2. COMPANY OFFICER	7. SAFETY	12. HEALTH & FITNESS	17. HAZMAT (SPECIALIST)
3. ENGINEERS/FHEOS	8. READING	13. ICS	18. ________
4. FIREFIGHTERS	9. RESCUE	14. FIRE INVESTIGATION	19. ________
5. TEAM BUILDING	10. ARFF	15. FIRE PREVENTION	20. ________

LAST, FIRST	1	2	3	4	5	6	7	8	9	10	11	12	13	14	15	16	17	18	19	20

Courtesy of Kern County Fire Department

FIGURE 9-12
Monthly training record.

NOTE A necessary step in the training process is maintaining training records.

When a situation arises where the department has to go to court to justify its actions in a case, the training records of the personnel involved will be requested as evidence. It is imperative that the records be factual, up to date, and easy to understand.

Some governmental programs, such as apprenticeship programs, provide funding to reimburse departments for training costs. These can only be accessed through the maintenance of records of hours spent on training. When the department is linked to the local college and the college receives funding from the state for student hours, the training records are necessary to receive the funds. These funds can then be used to pay instructors and purchase training packages for the department.

RELATIONSHIP OF TRAINING TO INCIDENT EFFECTIVENESS

The overall purpose of training is incident effectiveness. Under the concept of incident effectiveness, operations are performed efficiently and safely. The incident is controlled with minimum loss and everyone goes back to his station without injury. In some areas, staffing has been reduced to the point where a three-person engine company is expected to perform the work traditionally done by a four- or five-person engine company. Firefighters have been given better tools and procedures; now it is up to the firefighters to learn how to use them effectively and efficiently.

NOTE The overall purpose of training is incident effectiveness.

The best trained engine company in the world will not be able to put out every fire upon its arrival, nor will it be able to save every life. It will, by being properly trained, be able to perform to a high level and make as much of an impact as possible with the available resources at the incidents they respond to.

REQUIRED TRAINING

The modern firefighter is required to be skilled in many areas of emergency operations. The public expects them to be able to handle just about any situation that arises. When people dial 9-1-1, they want help and they want it fast. An example of this is the training requirements placed on the fire department by the Federal government.

Federal requirements for hazardous materials training are referred to in the *Superfund Amendments and Reauthorization Act* (SARA) of 1986. The requirements are found in OSHA 29 *Code of Federal Regulations* (CFR) 1910.120. Part 1910.120 requires firefighters and other first responders to hazardous materials incidents to have First Responder

Awareness (FRA) training. This training is required for personnel who are likely to witness or discover a hazardous substance release and who have been trained to initiate an emergency response sequence by notifying the proper authorities of the release. Their only action would be to notify the proper authorities of the release. They are required to have:

- An understanding of what hazardous materials are and the risks associated with them in an incident
- An understanding of the potential outcomes associated with an emergency created when a hazardous substance is released
- The ability to recognize the presence of hazardous substances in an emergency
- The ability to identify the hazardous substances, if possible
- An understanding of the role of the first responder awareness individual in the employer's emergency response plan
- The ability to use the DOT *Emergency Response Guidebook*
- The ability to realize the need for additional resources and to make appropriate notifications to the communication center

First responder operations-level (FRO) personnel are those who respond to releases or potential releases as part of the initial response to the site for the purpose of protecting persons, property, or the environment. They are trained to act in the defensive mode without actually trying to stop the release. They are to contain the release from a safe distance, keep it from spreading, and prevent exposures (basically to isolate, identify, and deny entry). In addition to the requirements of the awareness level, they are required to know the basic hazard and risk assessment techniques; know how to select and use proper PPE provided to the FRO level; have an understanding of basic hazardous materials terms; know how to perform control, containment, and/or confinement operations and how to rescue injured or contaminated persons within the capabilities of the resources and PPE available with their unit; know how to implement basic equipment, victim, and rescue personnel decontamination procedures; and understand the relevant operating procedures and termination procedures.

29 CFR then goes on to specify the training requirements for the more highly trained positions of hazardous materials technician, specialist, and incident commander/on-scene manager. The average firefighter should be trained to at least the FRO level, as they are expected to take action when they respond to the incident.

The Federal Aviation Administration specifies the training requirements for aircraft firefighting personnel in the Federal Aviation Regulations Part 139. The personnel involved in this type of operation are required to be initially and recurrently trained in the following:

- Airport familiarization, aircraft familiarization, rescue and firefighting personnel safety, and emergency communications systems on the airport including fire alarms
- Equipment use including use of the fire hoses, nozzles, turrets, and other appliances required for compliance with Part 139

- Application of the types of extinguishing agents required for compliance with Part 139
- Emergency aircraft evacuation assistance
- Firefighting operations
- Adapting and using structural rescue and firefighting equipment for aircraft rescue and firefighting
- Familiarization with firefighter's duties under the airport emergency plan

All rescue and firefighting personnel must participate in at least one live fire drill every 12 months. At least one of the personnel on duty must be trained and current in basic emergency medical care. The training must include at least 40 hours covering bleeding, cardiopulmonary resuscitation, shock, primary patient survey, injuries to the skull, spine, chest and extremities, internal injuries, moving patients, burns, and triage.

In the area of emergency medical services, the requirements are set on the state or local level. As the level of medical expertise increases from basic first aid through emergency medical technician (EMT) and on through paramedic, the training hours and requirements to maintain proficiency increase.

There is no set time to be spent on firefighting skills maintenance. This can be considered to include all of the manipulative and technical training and practice the department requires. A standard rule of thumb in many departments is two hours of drill per day. This is not two hours of standing around talking about it, but actual application of skills. The skills to be practiced are hose evolutions, use of PPE, tools, and apparatus. A standard combination drill would be to don PPE, including SCBA; advance an attack line and forcible entry tools to a doorway; enter; and conduct a primary search. All of this is to be done within department SOP adhering to safety guidelines and within a reasonable time. The drill would also be evaluated for its smoothness, coordination of personnel, and that all of the required tools and equipment were used in the proper manner.

A valuable part of your training will be that of preparing to perform the duties of the position above you. This requirement is specified in many job descriptions. In many departments the firefighter is able to act as the driver/operator and the driver/operator as the company officer. In this way, if one of the personnel is missing for any reason, the others can fill in and perform the required operations.

It would be an ineffective crew that could not charge their hose lines at the scene because the driver/operator had fallen and twisted his ankle.

TRAINING SAFETY

In training, as in any other operation, safety is of the utmost importance. Training accidents have occurred, and when instructors and participants are not constantly vigilant, they will continue to occur. Training injuries can be grouped in several main areas:

- Falls—typically from ladders or roofs.
- Being struck—by apparatus, heavy equipment, and falling objects.

- Overexertion—leading to heart attack, thermal stress, and dehydration. Strains and sprains are often the result of overexertion.
- Burns—during live fire training.
- Carbon monoxide (CO) poisoning.

Injuries from training accidents have ranged from minor to fatal. Through the proper use of PPE, planning, and vigilance, most if not all of these types of injuries should be avoidable.[16]

SUMMARY

As a new firefighter most of your time will be spent learning the skills necessary to perform your job. You will be instructed in the various tools and equipment you will be required to use and how to use them. As your skill level increases, you will approach mastery of the basics and not only know how to use tools, but also where and when to use them without having to be told.

As you progress in your career, the need for learning never stops. There will always be new hazards to be overcome and new techniques to do so. There is a constant flow of information and new equipment on the market to help you perform your job better. If you fall behind in this learning, you may never be able to catch up.

With promotion comes the added responsibility of training those under your supervision. As a manager, making sure that your subordinates are trained to the highest level possible is a big part of your responsibilities. Their safety as well as yours depends on the training of the whole crew. Firefighting is a team effort and teams are only as strong as their weakest member.

One of the most important aspects of your training is in the area of safety. You must be able to size up a situation and decide how to approach it without endangering your life or health any more than absolutely necessary. Part of this comes from developing a safety attitude and having a zero tolerance for safety violations. When you exhibit this safety attitude and perform on this level, you set the example for those around you. There are those in the emergency service who do not use good safety practices and have been lucky, so far. One of these days their luck will run out. Do not let it happen to you. If you train carelessly, without regard to proper procedure, that is how you will perform at emergency scenes. Sooner or later it will catch up to you. By paying attention to detail and learning all you can about your job, you can be the best firefighter you are capable of being.

One training practice that cannot be emphasized enough is that of personnel who respond together training together. This practice builds mutual respect and increases the efficiency at emergency scenes, when time is of the essence. It is also a good way to discover incompatibilities in equipment and operations. By discovering them before the emergency occurs, they can be addressed and contingency plans can be made.

REVIEW QUESTIONS

1. Why is the training officer position given to a high-ranking officer?
2. Why do the members of the training bureau have such an effect on the direction of the fire department as a whole?
3. Which level of government sets the requirements for first responder operations-level training?
4. Which level of government sets the requirements for emergency medical service training?
5. Why is time alone not a good indicator of the performance level of an engine company?
6. Why is it important for firefighters to train with the local ambulance company?
7. When there are multiple jurisdictions in the area, what are the benefits of joint training?
8. What are the two basic types of training a firefighter receives?
9. Which type of training is based on the operation of various tools?
10. Which type of training is learning firefighting-related chemistry?
11. What is the importance of maintaining training records?
12. List two factors in the determining of an adequate level of training.

DISCUSSION QUESTIONS

1. Is the training bureau more of an operations or staff function?
2. What is the one most important factor in any training?
3. Why are stress drills performed?
4. What is the importance of standard operating procedures?
5. You are the fire chief and have been asked to cut the training budget. How will you defend not making the cut?

NOTES

1. *Fire Engineering*, "Tragedy on Storm King Mountain," January 1995.
2. National Fire Protection Association, *Standard 1041, Fire Service Instructor Professional Qualifications* (Quincy, MA: National Fire Protection Association, 2007).
3. International Fire Service Training Association, *Fire and Emergency Services Instructor*, 7th ed. (Stillwater, OK: Fire Protection Publications, 2006).
4. National Fire Protection Association, *Standard 1041, Fire Service Instructor Professional Qualifications* (Quincy, MA: National Fire Protection Association, 2007).
5. Federal Emergency Management Agency, Procurement Office, 16825 S. Seton Ave., Emmitsburg, MD 21727.
6. National Wildfire Coordinating Group, *310-1: Wildland Fire Qualifications* (Boise, ID: USDA Forest Service, 2002).
7. *Fire Engineering Magazine*, "Firefighting in Our Waters," October 1994.

8. National Fire Protection Association, *Standard 414, Aircraft Rescue and Fire Fighting Vehicles* (Quincy, MA: National Fire Protection Association, 2007).
9. National Fire Protection Association, *Standard 402, Aircraft Rescue and Fire Fighting Operations* (Quincy, MA: National Fire Protection Association, 2008).
10. International Fire Service Training Association, *Fire and Emergency Services Instructor*, 7th ed. (Stillwater, OK: Fire Service Publications, 2006).
11. National Fire Protection Association, *Standard 11, Low-, Medium-, and High-Expansion Foam* (Quincy, MA: National Fire Protection Association, 2005).
12. National Fire Education System, *S-130, Firefighter Training* and *S-190, Introduction to Fire Behavior* (Boise, ID: National Interagency Fire Center, 2002).
13. Webster's New Collegiate Dictionary (Springfield, MA: G. & C. Merriam Co., 1994).
14. Ibid.
15. Alan V. Brunacini, *Fire Command* (Quincy, MA: National Fire Protection Association, 1985).
16. David Dodson, *Safety in Training for the Fire Instructor* (Albany, NY: Delmar Publishers, 2005).

CHAPTER 10

Fire Prevention

LEARNING OBJECTIVES

Upon completion of this chapter, you should be able to:

- Describe the importance of fire prevention.
- Describe the activities performed by a fire prevention bureau.
- List methods of public education as it relates to fire prevention.
- Explain how the authority to enforce fire prevention regulations is derived.
- Describe a typical fire prevention bureau organization.
- Describe the importance of fire information reporting and the National Fire Information Reporting System.
- List the uses of fire-related statistics.

INTRODUCTION

The United States has one of the highest fire death rates per capita in the world.[1] To prevent these deaths, we must reduce the number of hostile fires that start. Research conducted by the Department of Homeland Security and other groups has shown that for every one dollar spent on prevention, four to seven dollars can be saved in responding to incidents.[2]

One of the most important and least recognized jobs the fire department performs is that of fire prevention. Prevention does not make the headlines when it is successful. When it is unsuccessful, however, the community suffers fire-related deaths and property losses. Not all fire-related deaths and property loss can be prevented, but through prevention efforts they can be reduced.

NOTE One of the most important and least recognized jobs the fire department performs is that of fire prevention.

One of the true measures of a fire department's effectiveness is the amount of loss experienced in the community or jurisdiction. If hazards and unsafe acts can be reduced, there will be a resultant reduction in the area's fire experience. Through proper record keeping this can be proved over time. When firefighters spend their time sitting around the station waiting for the next emergency, they are not fulfilling their duties as charged by the mission statement to "protect life and preserve property." In order to reduce the losses due to fires, effective, focused fire prevention effort must take place.

FIRE PREVENTION BUREAU

Fire prevention is performed by people. In technical or high-risk areas, these people are often fire department employees assigned to the fire prevention bureau. They inspect occupancies using technical processes, and those of high life-loss risk, such as factories, schools, and hospitals. They are professionals with advanced training in fire prevention. Many are specialists in the field due to the amount of specialized knowledge required to perform their jobs. It is the responsibility of every firefighter to have at least a minimum amount of knowledge in the field of fire prevention and to act to correct hazards when they are encountered. A number of various fire prevention activities are performed by a fire department. As our environment and surroundings become more complicated, many of the fire prevention activities also become more complicated. Typically, the activities that require a higher level of experience or training are handled by the fire prevention bureau.

Staff Function

A fire inspector is specially trained in the science of fire, fire prevention inspection, and enforcement. This person is responsible for performing inspections to identify and reduce hazards and risks. A fire inspector is also responsible for maintaining written records and reports on the inspections made and the findings of those inspections. The inspector is required to have enough knowledge of fire chemistry, electricity, building construction, safety practices, and codes and ordinances that he can recognize a hazard when he sees it. It is his responsibility not only to identify the hazards and risks but also to be able to work with the property owner to mitigate them. This may require education of the owner as to what is wrong, suggestions on how to correct the problem and, if necessary, forcing compliance through legal action. The position of fire inspector requires a certain amount of salesmanship to gain the property owner's cooperation to reduce the hazards and risks. It is counterproductive and time consuming to have to gain compliance through legal action. When performing fire prevention work, public relations are very important. In many cases, the fire inspector is the only contact the owner will ever have with the fire department. The professional demeanor and capabilities of the inspector are under scrutiny any time an inspection is made. The fire inspector must know his job and be professional in appearance and presentation style.

Operations Function

As a firefighter assigned to an engine or truck company, you may very well be called on to perform fire prevention inspections of businesses in your first-in area. Your knowledge of the related codes and ordinances and prevention techniques will be tested in these situations. If you require an owner to purchase an expensive system or make a building change that is not required by law, you may find yourself in serious trouble when the owner learns the truth. If you overlook a hazard and there is a resultant fire, it reflects poorly on your ability and knowledge and in some cases may result in legal action being taken against the department.

Personnel

Staffing of the fire prevention bureau differs depending on the size and resources of the department. In large departments, the prevention bureau is managed by the chief of prevention, often called the fire marshal. In smaller departments, it may be run by a person of a lower rank. Whatever their rank, the personnel assigned to the prevention bureau must understand the importance of the bureau in achieving the overall mission of the fire department.

Fire prevention bureaus are often staffed with a mixture of uniformed (firefighters) and civilian personnel. The uniformed personnel are necessary because they are charged with enforcing the fire prevention codes and ordinances. They are empowered to write citations and to stop unsafe operations when necessary. In some areas, civilian personnel are given this authority as well.

Personnel are selected to work in the prevention bureau based on aptitude, attitude, and experience. The personnel must be able to understand codes and ordinances and to make a reasonable interpretation and enforcement application of them. They must possess a positive attitude about the role fire prevention plays in the fire service. They must be able to maintain their composure in stressful situations. Last, but not least, they must be able to sell fire prevention. Their experience should include active participation in and a thorough understanding of fire prevention programs.

The assignment of personnel on light duty and firefighters retired for medical reasons to these positions does not always work out so well. Although they are not fit for active duty, they usually miss their friends at the station and tend to spend a lot of time trying to find out what is happening in operations. The light duty personnel are assigned to the prevention unit only until they are well enough to return to active duty and often lack the training and commitment to performing well as prevention personnel.

Fire Prevention Chief

On the state level, the head of fire prevention is typically the state fire marshal, so called because the job requires more than just fire prevention work. The state fire marshal office often includes research, arson investigation, training, and fire prevention divisions. The Office of State Fire Marshal is often charged with responsibility for facilities deemed above the level of local jurisdiction such as prisons, hospitals, and pipelines.

The local department level fire prevention bureau is usually headed up by a person of chief officer's rank. They may have the title of Fire Marshal as well. Due to the complex nature of the bureau and its wide range of responsibilities, a person of this rank is required. In large fire departments, prevention bureaus have numerous employees. The fire marshal must often come up with recommendations to the fire chief as to issues that have a wide-ranging effect on the businesses and residences of the jurisdiction. On this level, costs and politics play a large part in making changes to existing policies and requirements. It is a position of major responsibility and visibility. People in this position must be politically astute as well as able to present themselves well in a public forum.

Inspection Officers

Usually working directly for the prevention chief are inspectors who are equivalent in rank to company officers. They are in charge of the major subdivisions of the prevention bureau. These personnel may be in charge of subfunctions such as

petrochemical, public assembly, plans review, public education, institutional occupancies, or other areas of special expertise. The personnel in these positions make decisions that can be very costly to businesses in the jurisdiction. They need to be well trained so that they do not make mistakes. They also need to be well spoken as they are trying to sell their recommendations to the business owner.

petrochemical Dealing with oil refining and production facilities.

Inspectors

When officers are in charge of divisions of the fire prevention bureau, firefighters may act as the inspectors. These personnel have received more training than the average firefighter on the engine company. Their job is to make the actual inspections. If you were to look at the requirements for a fire prevention inspector, it would be obvious that an assignment to the fire prevention bureau is a good career move to prepare yourself for promotion.

Civilians

In the fire prevention bureau there may be technical specialists, called water engineers, plans check specialists, and other titles. These personnel are hired for their ability to perform these jobs, not for their firefighting skills. An advantage to hiring civilian personnel is that they are selected for their expertise in this one area. Their job descriptions usually contain requirements for advanced education in the field for which they were hired. They are typically not paid as much as firefighters, nor do they get the more expensive pension benefits paid to firefighters; therefore it is easier for the department to get a highly qualified person for less money. From an attitude standpoint these personnel want to do what they are doing.

PROFESSIONAL STANDARDS

The training requirements for a skilled fire prevention officer are contained in NFPA 1031, *Standard for Professional Qualifications for Fire Inspector and Plans Examiner.*[3] An inspector of any level may seek training in fire prevention from numerous sources. State fire training programs are available in many states. Courses are available from colleges of all levels that have fire science–related programs. In most cases, courses in fire prevention are required for an associate's degree in fire science. Each of the agencies that publish model building and fire codes provide training and certification to accompany the codes. The National Fire Academy presents fire prevention courses on a national level. The NFA courses include Fire Inspection Principles, Fire Prevention Specialist, Code Management, Plans Review for Inspectors, and Fire/Arson Investigation.

Another way to receive training in inspection technique is to accompany fire prevention inspectors from the local department on an inspection. Observing how they perform the inspection and interact with the public can be of great benefit. Knowledge of the code and how to enforce it is not enough. The ability to prioritize hazards, develop the plan of correction, and write reports are necessary qualifications for the inspector. Public education is an area of fire prevention in which courses are available. The NFA has a course titled "Developing Fire and Life Safety Strategies." This course, coupled with others on course development and instructional methodology, can help you become an effective communicator in fire safety. Other courses are available from state level organizations in some areas. The courses in instructional technique are of value because they give you the techniques to assess needs and tailor the program to your audience. You must be able to prepare and present instruction that is clear and understandable yet motivational as well. The purpose of the program is to get people to change their behavior and become more fire conscious and therefore more fire safe.

A way to find out the latest techniques and methods is by joining fire prevention organizations and subscribing to fire prevention publications. The major fire-related magazines, such as *Fire Engineering* and *Fire Journal*, regularly have articles on code enforcement and other fire prevention–related subjects. In some areas, conferences and forums are held for people working in fire prevention.

These gatherings focus around presentations made by those preeminent in their field of expertise. They also give the person attending the opportunity to share their concerns and network with others in the field. It is not necessary to reinvent the wheel every time you run into a problem. By attending these gatherings, you can become acquainted with others in your field and a phone call may be all it takes to find the solution to your problem. Web sites (Appendix H) can be a source of help by allowing you to access the ideas and methods of other people working in prevention.

PURPOSE OF FIRE PREVENTION ACTIVITIES

The purpose of fire prevention activities is to prevent the loss of life and property due to fire. This simple statement sums up a complicated system. As we have seen previously, the loss of just one major facility can have a severe negative financial impact on a community or area. When a business that is a major employer burns, many people are out of work and the jobs may never be available again. Even worse, there are numerous documented cases of fires in structures that have led to major loss of life due to correctable hazards, such as locked or inadequate exits.

When the fire is a major forest fire, large amounts of natural resources are destroyed. This has a direct negative effect on the local timber industry as well as tourism.

When Yellowstone National Park burned in 1988, the park was severely impacted. From the tourism standpoint, the businesses in the affected area lost revenue because few visitors wanted to see vast amounts of scorched acreage. Much of the timber that burned was in the range of 150 to 300 years of age. The purpose of fire prevention is clear: to prevent hostile fires from starting; to provide for life safety in case of fire; and to prevent the spread of fire from one area to another. This broad statement can be broken down into the various activities performed by fire prevention bureaus and engine companies.

NOTE The purpose of fire prevention is clear: to prevent hostile fires from starting; to provide for life safety in case of fire; and to prevent the spread of fire from one area to another.

FIRE PREVENTION ACTIVITIES

Fire prevention activities can be divided into four areas: engineering, education, enforcement, and fire cause determination. The areas are interrelated and are all necessary parts of a fire prevention system. Through the use of all four, in an integrated manner, the fire department's objectives of fire and life safety can be accomplished.

The activities performed include (1) the design of fire-safe assemblies and systems; (2) review of plans prior to buildings being built or remodeled; (3) inspection of fire safety equipment and devices once installed; (4) inspections to ensure that the devices are kept in working order and not prevented from proper operation by occupant error, intention, or modification; (5) enforcement of codes and ordinances related to fire prevention; (6) public education in the methods and benefits of fire prevention and fire safety; (7) education of the legislative body in the need for the enactment of fire safety-related legislation; (8) investigation to determine fire cause and prosecution of arson when applicable.

Fire prevention does not take place only in urban settings. In rural and wildland areas, fire prevention is conducted as well. In these areas the focus is not so much on devices and installations as it is on clearance between combustible vegetation and improvements, such as structures. In wildland areas, public education is extremely important as many of the people in any given area are campers and others who are not permanent residents.

The overall goal of fire prevention activities is to keep people and property fire safe. It is not to issue citations and fines. The most effective and successful fire prevention programs are those that gain compliance through voluntary means. If the program is adversarial, the public is likely to go right back to doing what they were doing before you arrived as soon as you leave.

NOTE The most effective and successful fire prevention programs are those that gain compliance through voluntary means.

FIRE PREVENTION TERMS

The term fire prevention encompasses the use of inspections, engineering, and education to reduce or eliminate the causes of fires. A fire prevention inspection is the act of making a systematic and thorough examination of a premises or process to ensure compliance with fire codes and ordinances.

In terms of fire prevention, anything that can burn is a potential hazard. Hazards can also be defined as anything that can cause harm to people or equipment. Risks are the activities undertaken in relation to the hazard. A person washing an automotive part in soap and water is not exposed to much of a hazard and is taking relatively little risk. However, that same person washing parts in an open pan of gasoline by a gas water heater is creating a hazard and operating in a risky manner.

Hazards also occur in the design of buildings. When we allow structures to be built or modified in ways that prevent the orderly and safe exit of the occupants in the event of a fire, that presents a hazard. When persons enter that structure in such numbers that the exiting capacities are exceeded, their life safety is potentially at risk.

An occupancy is defined as the use or intended use of a building, floor, or other part of a building. Occupancies are divided into general types. One type is called a place of public assembly, such as restaurants, night clubs, or churches. Another is educational, which includes schools and day-care centers. Hazardous occupancies are those that pose more than an ordinary hazard, such as vehicle repair facilities with welding or spray painting, aircraft hangars, and cabinet shops. Institutional occupancies include facilities such as jails, hospitals, and group homes where occupants are hindered or incapable of self-preservation.

The activities performed in a structure determine its occupancy classification. A restaurant would be one kind of occupancy and an aircraft repair facility would be another. Fire prevention revolves around occupancy classifications in many settings. Different occupancies present different hazards and are treated differently in the code and in inspection procedures.

METHODS OF FIRE PREVENTION

Fire prevention does not begin only after a hazard is in place. Fire prevention activities start before the building is even built. Zoning regulations determine what kind of occupancies can be built and where. A high-hazard occupancy, such as an oil refinery, would not be allowed to be built next to a school and vice versa. In planning for the construction of a building, there are required setbacks from property lines. Setbacks are not totally due to fire safety concerns, but they are a consideration. Another consideration is the accessibility for placement of fire apparatus. If ladder trucks are going to be necessary for rescue and suppression, many jurisdictions require special pads to be poured for this purpose. The fire lane required around shopping malls and other facilities provides for fire department access (**Figure 10-1**).

FIGURE 10-1 Fire lane sign with ordinance identified.

Delmar/Cengage Learning

Typically, when building plans are submitted, the required fire flow is determined to ensure that adequate water is available to fight a fire if one does start. When subdivisions are constructed, one of the first considerations is a hydrant system. It is important that the system be installed and in service before the first foundation is poured, to provide a water supply in case of fire when the structures are in the framing or later stages of construction. In rural areas, water supply is one of the most important considerations from a firefighting standpoint. Hydrant systems are not often available and fire water tanks may be substituted (see **Figure 12-12**).

NOTE In the design and construction of buildings, one of the first considerations is fire protection.

In the design and construction of buildings, one of the first considerations is fire protection. These concerns are addressed through engineering. Engineering activities are evident in the design and installation of fire protection and suppression devices and fire-rated assemblies. These devices include fire sprinkler systems, standpipes, detection and notification systems, and fire-rated construction used in the design of the building itself. Through adopted codes, the fire and building departments require the architect to design in certain features that can prevent the spread of a fire if one does start. By enclosing vertical shafts, such as laundry and mail chutes, in fire-resistive construction, the upward spread of fire through the structure can be prevented. Areas where hazardous functions are performed are separated in much the same manner. In high-rise buildings, the structural steel is covered with fire-resistive materials, and holes for cables and pipes that "poke through" floors are sealed with material to resist the spread of smoke and heat through such penetrations. Something as simple as an intact ceiling assembly can prevent the spread of fire to upper floors by concentrating heat that allows for the proper operation of sprinklers or blocking fire spread.

One type of construction that does not take fire into consideration and causes much concern and danger to firefighters is lightweight construction.

One type of construction that does not take fire into consideration and causes much concern and danger to firefighters is lightweight construction. In this form of construction, the floor and roof supporting systems are made from the least amount of material possible (**Figure 10-2**). As with any system, the assembly is only as strong as its weakest member. When attacked by fire, these systems have a tendency to fail and fail early in the fire. If you are inside or on the roof, it is very likely that you could become trapped or injured due to this failure. Extreme caution must be exercised when operating at incidents where this type of construction is used. The simplest way to identify this type of construction is to observe the construction of the building. However, you will come across buildings that have been built and occupied for several years. In these cases, a thorough inspection can also reveal lightweight construction. As always,

FIGURE 10-2
Truss roof construction.

Delmar/Cengage Learning

a thorough knowledge of the hazards and where they are located in your response area is essential to your safety.

Devices

Some activities are hazardous to the point where all fuel and ignition sources cannot be controlled. A simple example is a commercial deep fat fryer. In these types of installations, devices and systems are installed to either control the fire or prevent its spread (**Figure 10-3**). These devices could include automatic and/or manual fire extinguishing systems, interlocks to shut off the gas or electric supply to the fryer, interlocks to control the ventilation system in the exhaust hood, and portable fire extinguishers. Fire prevention devices can be as simple as a lid that automatically closes on a parts washer when there is a fire. They can be as complicated as a smoke removal system that increases the likelihood that a safe exit route for occupants is available in case of fire in a high-rise building.

The most common types of devices you will encounter are portable fire extinguishers and automatic fire sprinkler systems. In every case, the type and number of devices installed will depend on the hazard to be protected against.

Assemblies

Assemblies come in many forms. A fire-resistive door with fire-resistive walls is considered an assembly. When assemblies are properly designed and installed, they can prevent the spread of fire from one part of the structure to another. They must be inspected on a regular basis as owners and occupants often negate their effectiveness by removing part of the assembly, block their operation, or bypass them altogether. Examples of this would be a self-closing door blocked or wired open or holes cut in a fire wall.

FIGURE 10-3 Class K fire-extinguishing equipment installed in range hood.

Delmar/Cengage Learning

HAZARD EVALUATION AND CONTROL

The purpose of hazard evaluation is to identify possible accidents and estimate their frequency and consequences. In hazard evaluation, an accident is defined as a specific unplanned sequence of events that has an undesirable consequence. The first event of the sequence is the initiating event. There are usually one or more events between the initiating event and the consequence. The intermediate events are the result of the response of the system and the operators to the initiating event. Different responses to the initiating event often lead to different accident consequences. In theory, it is possible to reduce the consequences of the accident by reacting appropriately to each of the events between the initiating event and the ultimate consequence.

There are two basic methods of hazard evaluation and control in use: (1) adherence to good practice and (2) predictive hazard evaluation.

Adherence to good practice includes observing rules and regulations, meeting the requirements of the accepted standards, and following the practices that have proved best from years of experience with the same processes, the same plant designs and requirements, and the same operating and maintenance procedures. Hazard evaluation procedures such as checklists and safety reviews are used to identify deviations from accepted standards and good practices.

Good practices evolve from experience. Experience has been documented in the form of standards or recommended procedures. The standards summarize today's accepted good practice. The NFPA has numerous standards directed toward establishing proper safeguards against loss of life and property by fire. Many of these standards and good practices are the result of lessons learned from analysis of accidents and near accidents. Two of the methods used to identify deviation from good practice are checklists and safety reviews. These methods are used to ensure that design specifications are met, that previously recognized hazards can be identified, and that operating and maintenance procedures conform to principles and practices that have evolved from experience. From a firefighter's standpoint, a safety review can be equated with a fire prevention inspection. When you conduct a plans check, preapproval for certificate of occupancy, or after-occupancy inspection, you are conducting a safety review. This is usually done using some sort of checklist. The checklist can be as simple as a fire prevention inspection form that lists common hazards, or may be as complicated as the checklists often used by plans checkers to assure compliance with code, design, and installation requirements.

Predictive hazard evaluation procedures have been developed for analysis of processes, procedures, systems, and operations that are sufficiently different from previous experience that adhering to good practice may not be adequate. They may also be used when evaluating a very low probability accident with very high consequences for which there is little or no experience. The concept addresses both the probability of an accident and the magnitude and type of the undesirable consequences of that accident.

Once a hazard is identified, it is necessary to evaluate it in terms of the risk it presents to the employees, the public, and the property involved. It is necessary to identify the initiating event, the intermediate events, and the consequences of each accident involving the hazard.

The first step is to identify the hazards that are an inherent feature of the process and/or plant and then focus the evaluation on events that could be associated with the hazards. Several recognized procedures are used to identify the hazards, including the "What-If" method, hazard and operability studies, failure modes, effects and criticality analysis, and human error analysis. To identify intermediate events, procedures such as fault tree analysis, event tree analysis and cause-consequence analysis can be utilized.[4] A complete description of these processes and their uses is beyond the scope of this text. Further information is available in texts on the subject.

PUBLIC EDUCATION

It is not possible to have a fire inspector on duty in every hazardous location at all times that the occupancy is in use. It is also not possible or good public relations to prosecute every case of violation of the fire codes. It is more cost-effective and more productive—while bettering public relations—to convince people ahead of time of the importance of fire prevention. When you visit a business, the people on-site are concerned with good practices and fire prevention measures while you are inspecting. By educating people as to the importance of fire prevention, they will be concerned about it even when you are not there.

NOTE It is more cost-effective and more productive to convince people ahead of time of the importance of fire prevention.

One of the most effective forms of public education is that carried out in the school system. By presenting a fire safety message to an assembly at a school, a large audience can be reached in a relatively short time. The children targeted must be old enough to understand what you are talking about. In numerous cases, very young children have learned to dial 9-1-1 and have done so in emergencies. Another program that has worked well with even very young children is STOP, DROP, AND ROLL. It is simple in concept and easy to understand. If the children are old enough, they can be instructed in more advanced concepts, such as electrical safety, and given fire prevention check sheets to take home and inspect their own houses. Not only are you reaching the children at school, but they are also old enough to go home and talk to their parents about the importance of having a smoke detector and a home exit plan. It is very important that you target your message to your intended audience (refer to **Figure 10-5**).

NOTE Civic groups present an opportunity to spread the fire prevention message as well as serving as a potential source of funds for purchasing fire safety education materials.

An opportunity to reach children with safety responsibilities is through programs such as the YWCA Super Sitter. This program brings together children who will be baby-sitting with representatives of the fire and police departments in an educational setting.

Civic groups present an opportunity to spread the fire prevention message as well as serving as a potential source of funds for purchasing fire safety education materials. Often fire departments do not have the funds to purchase hand-out materials and fire safety education videos and props. By making fire safety presentations and appeals for monetary assistance to civic groups, often these funds can be secured. Some groups have contributed enough money to purchase such expensive props as fire safety trailers. These cost approximately $25,000 and are very effective, but beyond many departments' budgets (**Figure 10-4**).

Another source of funding is industry. Many companies can use discretionary funds from their advertising budget to purchase equipment for the fire department to use in presenting the fire safety message to the community. This has the twofold effect of helping out the fire department and promoting the public image of the company involved. A fire safety trailer set up at a shopping center is a rolling billboard for the company that has its name painted on the side.

NOTE Another way to get the fire prevention message out is to have fire station tours.

Another way to get the fire prevention message out is to have fire station tours. Groups from the community, such as the Boy and Girl Scouts and school groups, visit

FIGURE 10-4 Fire safety education trailer.

Delmar/Cengage Learning

the station. During the visit they are shown the station and the equipment. During these tours and other programs, it is more effective to introduce yourself as "Firefighter John" than "Firefighter Jones." It puts the children more at ease and they are more likely to ask questions and have a valuable experience. Young children are more interested in the size of the apparatus and a demonstration of the siren and red lights than they are in a compartment-by-compartment description of the tools and equipment carried and their uses. This is also a good time to cover the proper procedure for children to follow when they are crossing the street or riding their bicycles when a vehicle is approaching with red lights and siren in operation. Another activity that children enjoy is being allowed to spray water from the hose. This must be done with great care as there has been a case where a hose ruptured—knocking a school teacher to the ground—and the fire department was subsequently sued.

One thing that is traditionally done is to have the children try on articles of turnout gear. Should you choose to do this, make sure the gear is free from contamination and that the children are old enough to support the helmet without injuring their necks. During this demonstration, you can discuss how well firefighters are protected before entering a burning building, and that the children should never reenter to find a pet or toys, because this has cost children their lives. It is possible to contract a case of head lice from a child putting on your helmet, so the station should have a helmet without a liner for demonstrations. The helmet will be too big for the children anyway and they do not notice the lack of the liner. It is not an acceptable alternative to have the children try on the equipment of the personnel who are off duty that day. Possibly the best alternative is to have a lightweight plastic helmet to use for this purpose.

NOTE In a fire, children should not hide under beds or in closets as this makes them extremely hard to find.

An effective demonstration is to have a firefighter don full turnout gear and SCBA. Having the firefighter crawl into the area as if they are searching gives the children a feel for how firefighters would be looking for them in an emergency. This is done to show the children that it is only a person in the suit. The appearance of a firefighter in full turnout gear and the sound of the breathing apparatus in action are frightening to small children and they may be afraid to call out to you when you are searching for them in smoky conditions. At this time, you can cover the point that, in a fire, children should not hide under beds or in closets as this makes them extremely hard to find. An aid for teaching young children is robots (**Figure 10-5**). Their size is close to a child's and they tend to hold a child's attention while the safety message is presented.

Public education activities that can reach a large number of people in the community in a short time are public service announcements (PSAs). With the help of the local television and radio stations, PSAs can be produced to spread the fire prevention message. Television spots are especially effective in that they include both audio and

Delmar/Cengage Learning

FIGURE 10-5 Teaching preschoolers fire safety.

visual information. Just hearing about something is not nearly as effective as seeing it at the same time. The television station may be willing to use some of their previously shot raw footage and edit it for you to do a **voice-over** of your message. A 30-second spot with three seconds of each image you want to be presented can show 10 fires or other incidents. When done correctly, this message can have a tremendous impact.

Television and radio can also be used effectively through the news conference format. When there is an issue deemed newsworthy, such as the opening of brush fire season or the high incidence of fires caused by heating equipment, the news departments are often more than willing to give you free air time.[5] This is mutually beneficial because you get your message across and they are always looking for a good story.

The media always seems to be interested in doing a fire department story at Christmas. An effective way to handle this is to have the media attend a news conference where a dried out Christmas tree is set up with packages around the bottom. The tree is ignited and the rapid spread of the fire is evident. A setup like this has great visual impact and instills the need to be especially careful when celebrating the holidays with a live tree. Public education needs to continue year-round, with Fire Prevention Week being an especially good time to get the media involved. Every year there is a new theme for the week. Some used in the past have been Operation EDITH (Exit Drills In The Home) and "I'm alarmed; smoke detectors save lives." These phrases give the media a good hook from

voice-over A presentation style where video is shown and a person's voice is heard without seeing the person. Commonly used in documentaries.

which to build their story. Combined with the pertinent statistics and a fire officer in dress uniform or full PPE, the message can be presented very effectively.

One method of building a good working relationship with the media is setting up a "Media Day." On Media Day, representatives from television, print, and radio are invited to the training center. They are equipped with turnout gear and equipment and, accompanied by regular firefighters, they are allowed to put out small fires in a burn building or other scenario. This gives them the firsthand experience of heat and smoke as well as the weight and difficulty of working in full turnout gear. If done properly and well planned, the experience is enjoyable for all involved and gives the media personnel a new respect for the job firefighters do and the conditions under which they are required to perform. That night you can just about be guaranteed that every station is going to run footage of the activities.

The positive public relations reaped from this are worth the small amount of money spent to present the activity.

Another way of spreading the fire prevention message is to post signs (**Figure 10-6**). These can come in the form of billboards, roadside signs, and bumper stickers. Some billboard owners will donate space for fire prevention messages when the sign is not otherwise used.

NOTE Another way of spreading the fire prevention message is to post signs.

One of the most important groups that needs to be reached with the fire prevention message is elected officials. These are the people who enact the laws. The only way the fire service can enforce fire safety requirements and regulations is through these people

FIGURE 10-6 Fire safety message on roadside sign.

Delmar/Cengage Learning

FIGURE 10-7
Wood shake roof after fire.

making them law. If the required measures are not law, firefighters can only seek voluntary compliance. An example of this is the controversy over wood shake roofs (**Figure 10-7**).

In areas with dry vegetation, wildland fires often spread to structures by flying brands: particles of burning material carried aloft in the convection column that land on dry wood shake roofs. The roof catches fire and spreads its own fire brands as the fire increases in intensity. When the roof of one house catches on fire, the fire is spread through convection of burning material, radiated heat, and direct flame impingement to adjacent structures. If the wind is blowing, the spread can rapidly exceed control efforts and capabilities. This leads to a conflagration. It is not uncommon for large numbers of homes to be lost in these types of fires.

For years fire department spokesmen have tried to get wood shake roofs outlawed, at least in wildland areas. But the wood shake manufacturers have lobbied against restrictive legislation. Fire retardant treatments have been developed, but until recently these products have only lasted for a few years and then become ineffective. A wood shake roof normally lasts for 25 years. New technology and improvements have now created fire retardant treatment for wood shakes that will last throughout the life of the roof. The ability to install aesthetically pleasing wood shakes which are also fire resistant will eventually have a tremendous impact on the spread of fire from rooftop to rooftop. Fireworks are another hotly debated issue in many jurisdictions. The fire department would like them restricted or outlawed due to the number of fires and injuries that they cause annually. However, community groups and businesses raise money by having fireworks booths around the Fourth of July, making it politically difficult to outlaw fireworks in many areas. Part of the

FIGURE 10-8 Fireworks sales facility located near state line.

Delmar/Cengage Learning

problem is that one jurisdiction may outlaw them, for example an incorporated city, while being surrounded by a county in which they are legal. Another example is fireworks being outlawed in one state and legal in an adjoining state, with fireworks sales very close to the state line, making them readily accessible to areas where they are outlawed (**Figure 10-8**). In some states, the state fire marshal's office limits what can be sold to so-called "safe and sane" fireworks. They do not explode like firecrackers, but they do emit showers of sparks. The idea is that they will be used safely, but this does not stop people from throwing them on shake roofs or into dry grass.

Another area with this kind of competing interest is sprinkler legislation. In most cases, if the sprinklers were not required when the building was built, they are not required to be retroactively installed. These buildings are called an existing nonconforming use. The best that many fire departments have been able to do is get legislation passed stating that if the building is remodeled to a certain extent, it is required to be brought up to current requirements. Many designers and developers will also fight against sprinkler installation in new buildings due to the additional construction cost. This opposing view of sprinklers is usually a result of misinformation about the operation of sprinkler systems. Many people believe that when one sprinkler operates, they all operate. This is a myth that has been created by the movie industry, because that is how sprinklers are portrayed in the movies and on television. A strong public education campaign is needed to overcome this belief.

Smoke detectors are an area in which many jurisdictions have been successful getting legislation passed because of their proven life-saving effectiveness. They are required in rental property, motels, and hotels and may be required to be installed in existing private homes.

The only way that the fire department is going to get the needed codes and ordinances enacted is to continue to gather information on the causes of fires and bring these statistics to the attention of the local lawmakers. Allies in this fight are the insurance companies that underwrite fire insurance policies. They, too, have an interest in keeping fire losses to a minimum and can help in lobbying the legislators and providing technical assistance.

National organizations that can provide materials for fire education include the U.S. Fire Administration (USFA), National Fire Protection Association (NFPA), International Association of Fire Chiefs (IAFC), International Society of Fire Service Instructors (ISFSI), National Bureau of Standards, National Safety Council, U.S. Forest Service (USFS), Bureau of Land Management (BLM), Underwriter's Laboratories (UL), and others.

State and local level organizations that can be sources of public fire education materials are the office of state fire marshal, state forester, state fire chiefs' association, safety film producers, electric and gas utility companies, hospital burn units, state firefighter associations, and fire departments with video production capabilities.

COMPANY LEVEL FIRE PREVENTION ACTIVITIES

Fire prevention inspections on the company level serve several purposes. They are useful in that the company members walk through the businesses in their district, which acquaints them with the business owners, the layout and occupancy of the building, and any special hazards. This is also a good time to take a look in the closets, basement, and attic, checking these areas for hidden hazards and construction features. During the visit the location of electrical, water, and gas shutoff controls should be noted. The business owner can be consulted about where fire would do the most damage and salvage is a priority, usually in areas where records are stored. By having firsthand knowledge of a building and its contents, firefighting is safer and more efficient.

By having firsthand knowledge of a building and its contents, firefighting is safer and more efficient.

Performing these inspections gives the company a chance to carry out some public relations work as well. Forming a working relationship with the business owners in your district will encourage them to notify you of any major changes they make in their building or occupancy and encourage them to consult you as to the fire-related implications of the changes. The public likes to see firefighters out taking proactive measures to ensure its safety.

NOTE The public likes to see firefighters out taking proactive measures to ensure its safety.

A company-level inspection performed in departments with natural hazards includes weed abatement/hazard reduction. During these inspections the company tours its area and notes dry brush and grass that may threaten improvements (structures). By doing so, the company reduces hazards and makes the public more aware of the dangers. Contact is made either in person or by mail with the parties that need to remove combustible vegetation from around their structures. These inspections are an opportunity for the firefighters to emphasize the need for persons to take care with ignition sources in dry grass. Just the sight of the fire engine cruising the area and inspecting tends to make people more aware of the fire danger.

The intent of this program is not to get people to ruin their natural surroundings by removing every blade of grass and bit of vegetation on their property. It is to break the fuel ladder by preventing a fire from spreading from ground fuels to the structure itself. The usual requirements are a 30-foot break to mineral soil around all structures and within 10 feet of LPG (liquid petroleum gas) tanks, fences, and wood piles; removal of limbs 10 feet from stovepipes and chimneys; and removal of combustible materials like pine needles and leaves from roofs (**Figure 10-9**). These requirements are often

FIGURE 10-9
Defensible space, an area surrounding a structure cleared of highly flammable vegetation.

contained in adopted ordinances. Other ways to comply with the intent of these regulations are to landscape with plant species with low combustibility around the structure and to keep them irrigated.

Another form of hazard reduction inspection is checking the area for accumulations of tires. In the last few years there have been fires with thousands of discarded tires involved. These fires can lead to tremendous control problems for the fire department. They burn with acrid black smoke and are usually very time consuming and expensive to extinguish. The water used to extinguish the fire must be contained to keep the runoff from polluting streams. Another type of business that has a large accumulation of combustibles is wooden pallet manufacturers and recyclers. These facilities can occur just about anywhere and once a fire gets started it is usually beyond control in a short time. The problem of stacks of combustibles and their arrangement combined with the design of a pallet make them highly combustible once ignited. In installations with large accumulations of combustible materials, the best the fire department can do is keep a hostile fire from spreading to surrounding structures.

NOTE In the last few years there have been fires with thousands of discarded tires involved.

LEGAL AUTHORITY

The fire department must have the legal authority to enforce fire-related codes and ordinances. This issue is discussed in Chapter 11.

FIRE PREVENTION INSPECTION

To properly conduct a fire prevention inspection of a business or facility, preliminary work must be done. Previous inspections should be reviewed to see what violations and corrections were made in the past. If the occupancy or facility is unusual, some preliminary research may be necessary to make sure that you are capable of noticing a violation when you see it. It is good to know which fire protection features are required as this will make you appear more professional because you are familiar with the code. When in doubt it is better not to give wrong information. You should advise the owner that you will conduct the necessary research in the code and get back to him.

The inspector needs to be equipped with the tools of the trade (**Figure 10-10**). These will aid in making the inspection and documentation more thorough as well as making the inspector appear more professional. One essential is a proper uniform or other form of identification to let the business owner know who you are. Protective clothing, such as a hard hat and coveralls, may be required when inspecting dirty or hazardous locations. Writing materials and clipboard, including graph paper, notebook, inspection forms (**Figure 10-11**), and violation notices are a necessity. A measuring device for

FIGURE 10-10 Fire prevention inspector performing check of building plans.

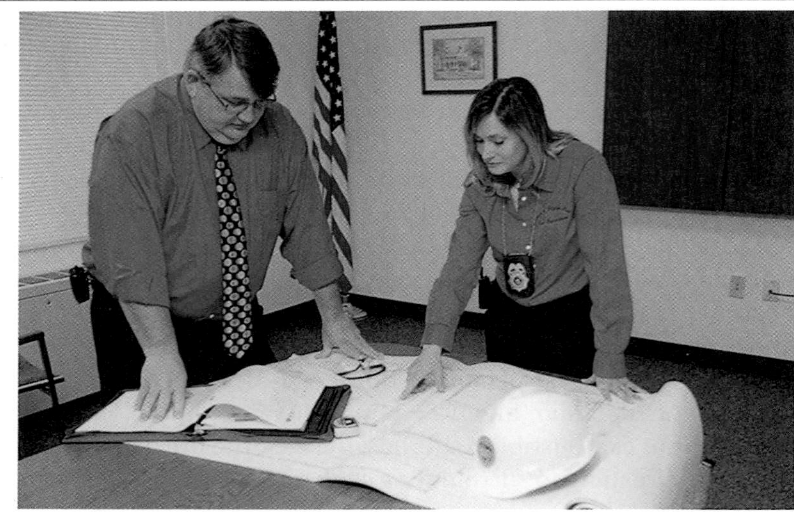

Delmar/Cengage Learning

determining distances and clearances is important; a 25-foot measuring tape is usually sufficient. A camera with flash to document hazards and photograph evidence can come in handy. Code books for easy reference during the inspection can be used when you are confronted with a situation that you are not sure about, when the code section needs to be shown to the owner, or when you are writing a citation or inspection report. A citation book and red tags for taking equipment out of service should also be available.

Fire codes typically allow entry for routine inspections or when the fire official has reason to believe an unsafe condition exists. Entry cannot usually be made by personnel without compliance with the following conditions: You must show proper credentials, request entry, and make a reasonable effort to locate the owner of a vacant building prior to entry. When an inspector is refused entry, an inspection warrant or administrative warrant may be obtained from a judge under civil code procedures. An inspection warrant is not the same as a criminal search warrant. They are not as specific as a search warrant and are designed to allow you to investigate any possible fire code violations. Legal precedent allows you to obtain an inspection warrant to enter the property if you can prove that a logical sequence of inspections is occurring in the geographical area.[6] The inspection warrant is used for the period of inspection but typically cannot exceed 14 days. An inspection warrant normally does not allow "forced entry" into the building, and it must be served during normal business hours. In some jurisdictions, part of the licensing procedure for businesses is that the owner agrees to periodic fire prevention inspections. You should know if this is the case in your area before you attempt to make the inspection.

FIRE DEPARTMENT
PREPLAN / INSPECTION REPORT

FILE NO. MISC. NO.

X New Inspection
X Purge
H = HAZ/MAT
E = EXEMPT
D = DELETE
TARGET HAZARD
P = PRIMARY
S = SECONDARY
D = DELETE

GENERAL USE

D.B.A.

LOCATION CITY OCCUPANCY CLASS PROPERTY CLASS

BUSINESS MANAGER BUSINESS PHONE NO. OF BLDGS. NO. OF STORIES BSMT. (B) JURIS-DICTION

RESPONSIBLE SHIFT

BUSINESS OWNER HOME PHONE EMERGENCY PHONE DATE OF LAST INSP.

Disposition Sym. 1 - Corrected 2 - Will Correct 3 - V.N. Issued. Call Back Necessary 4 - Citation Issued 5 - Referred

	APPROVED YES	NA		NO	DISP. SYM
BUILDING					
1. Exit Doors-Hardware Exit Signs-Lighted			1		
2. Exit Corridors			2		
3. Aisle Spacing-Seating			3		
4. Occupant Load Signs			4		
5. Vertical Openings			5		
COMMON HAZARDS					
6. Electrical			6		
7. Furnace/Boiler Rooms-Heating Equipment			7		
8. Cooking Equipment			8		
9. Decorations-Curtains-Drapes			9		
10. Housekeeping-Trash-Weeds			10		
FIRE PROTECTION EQUIPMENT					
11. Fire Extinguishers			11		
12. Automatic Extinguishing Systems			12		
13. Wet-Dry Standpipes			13		
14. Alarm Sysytem			14		
15. Fire Assemblies-Fire Walls			15		
16. Fire Sprinkler Records			16		
OTHER HAZARDS					
17. Grease Hoods & Ducts			17		
18. Warning Signs			18		
19. Compressed Gas			19		
20. Other (see Remarks)			20		
HAZARDOUS MATERIALS					
21. Training Records			21		
22. MSDS'S Available			22		
23. Chemical Inventory			23		
24. Labeling			24		

REMARKS:

REGULAR INSP.
RE-INSPECTION
MISC. INSP. (PREVENTION ONLY)
CLEARANCE
GRANTED
DENIED

TOTAL TIME TO COMPLETE INSPECTION

MISC. INSPECTION (PREVENTION ONLY)

SCHEDULED DATE OF RE-INSPECTION

OWNER / MANAGER

INSPECTOR SHIFT DATE

MONTH INSPECTION DUE

A = ANNUAL
O = ODD YEAR
E = EVEN YEAR

X - PREPLAN UPDATED

KCFD #97 CATALOG #9215

OFFICE

Courtesy of Kern County Fire Department

FIGURE 10-11
Fire prevention inspection form.

In most situations, when you approach the business owner and explain your visit as the standard fire prevention inspection, they usually invite you in and offer to show you around. Some businesses may request that you call and make arrangements before your visit. This is not always done to hide violations but to make sure that someone is able to assist you with your inspection. Remember that the goal of fire prevention is to educate and gain voluntary compliance. It is not always convenient for a business owner to drop what she is doing to show you around. Some businesses have peak business hours during the day and it is usually best to avoid these times. For example, a restaurant owner may be reluctant to accompany you on an inspection during the lunch hour but may be happy to join you in mid-afternoon. In seasonal businesses, such as agricultural packing facilities, it is best for you to visit when they are in operation, but not best for the owner as they are very busy. In these occupancies, an appointment is highly recommended. As you perform the inspection, approach it in a consistent manner. After you have introduced yourself and received permission to proceed, start around the outside and work your way inside. While outside you can check for accumulations of combustibles, location of utilities shutoffs, presence of fire hydrants and automatic extinguishing systems, posted and visible address, and other important data.

When you reenter the business, let them know that you are ready to proceed with the interior inspection. When performing the inspection, it is not critical whether you start at the top floor and work down or start at the bottom floor and work up. However, it is critical that you inspect the entire facility. You will develop your own style after you have completed several inspections. In a production or assembly facility, it is usually a good idea to start at the beginning of the process, with the raw materials, and go all the way through to the final product. As you tour the premises, take note and document any violations and ask that they be corrected. It may be necessary to explain to the owner or his representative exactly what you want and why as part of the education process. If you encounter a situation that you are unsure of, it is better to take the time to look it up, or at least document it so you can research it later, rather than just make a mental note and possibly forget or ignore it because you are not sure. Do not attempt to bluff the business owner regarding an issue that you are unsure of. There is nothing wrong with letting the business owner know that you will need to conduct some research about the topic and then get back together with an answer. It is much better to conduct yourself in this fashion and you will give the business owner the impression that you are taking an interest in his business and are trying to work with him. Do not be bashful about asking to look in closets and behind closed doors. Often combustibles are stored in closets right next to water heaters. Now is a good time to talk to the owner about any special concerns she may have about what she is doing or planning to do. Also determine if any remodeling has been done that may affect fire control systems or firefighting operations. When hazards and violations are found, they need to be prioritized as to their importance. The most important are those that jeopardize life safety and have the potential for excessive property damage. As you seek correction of these violations, some are going to be more time-critical than others. A blocked fire exit needs to be corrected in your presence as it is a direct threat to the life safety of the building's occupants.

A fire extinguisher overdue for maintenance or inspection can be given a reasonable abatement period. All hazards and violations should be documented, even if they were corrected in your presence.

This will provide a record of your findings. It can also provide a listing of the hazards detected at the facility for your records. The written documentation serves notice to the owner that the hazards need correction. In many instances, you will not be talking to the owner of the business or the building. You will be speaking with a manager. By providing this person with written documentation, she can present it to her boss and get the violations corrected.

NOTE The most important hazards are those that jeopardize life safety and have the potential for excessive property damage.

A blocked fire exit needs to be corrected in your presence as it is a direct threat to life safety of the building's occupants.

NOTE Always remember that the business owner is your customer and treat her as such.

Upon completion of the inspection, be sure to thank the business representative for her time and assistance. A little effort on your part goes a long way toward developing a working relationship between the fire department and local business owners. It also makes it easier for the person who inspects the business next time. Always remember that the business owner is your customer and treat her as such.

In some situations you may be required to develop a plan of correction. This situation occurs when there is a long list of violations or chronic hazard situations. By prioritizing the hazards and working with the business owner, you can develop a plan of periodic reinspections to assure progress. Through this process you can avoid having the business shut down while corrections are made. The plan of correction can be as simple as advising the owner, in writing, that you will be back in 14 days to reinspect the premises and if she is not in compliance at that time, a citation will be issued. In other circumstances, the reinspection and progress check process may take much longer, depending on the amount of work to be done.

When you reinspect and find the occupancy is still not in compliance, several courses of action are available. One is to provide a verbal warning by reminding the occupants of their legal obligation and that civil and criminal action can be taken against them. The other is to write a citation and send the matter to court or before a civil enforcement board. Depending on the immunity laws in your state, a failure to follow up on violations found can open you and the jurisdiction to a lawsuit. When you fail to act, you could be assuming a certain amount of liability. Long-term noncompliance generates poor attitudes on everyone's part. Inspectors tend to lose interest because of frustration. The fire department's credibility can be lost. Once the business owner sees that he can ignore your requests, he can develop the attitude that nothing you require is of any importance. When there is a disagreement between the fire prevention bureau and the owner, several courses of action are available. A board of appeals may hear an appeal from the owner or rule on the application of alternate materials and methods to correct a situation that is called for in many fire and building codes. Many fire codes also provide that the fire official may modify provisions of the code upon written application by the building owner when it is impractical to carry out the strict letter of the code.

DETERMINATION OF FIRE CAUSE

Typically, it is the responsibility of the fire official to determine the cause of every fire that occurs in his jurisdiction. As with other responsibilities, the determination may be delegated to the other members of the fire department. Scene responsibility lies with the highest ranking officer at the fire. The person who usually performs the primary investigation is the company officer. It is the responsibility of every member of the fire department at the scene to be aware that something had to cause the fire and to protect the area of origin as much as possible.

Cause determination is important from a prevention standpoint. If we do not know what is causing the fires, we will have a hard time targeting our prevention efforts. If the cause of numerous fires is children playing with matches, we may need to be more aggressive with the public education program in the schools. Cause determination has also uncovered problems with equipment and processes that can be handled through design changes or product recalls. Fire cause and fire statistics are often used to modify existing codes and standards. These measures can greatly reduce the fire loss in the community.

NOTE If we do not know what is causing the fires, we will have a hard time targeting our prevention efforts.

The first people inside the burning structure are the firefighters on the nozzle. Their actions are critically important to the determination of the cause. A cause investigation starts with the observations made by the personnel at the scene. Many indicators lead to the site of the fire's origin. Some of the questions you should be prepared to answer at a structure fire are:

- Was the door locked?
- What were the positions of interior doors?
- Were the windows broken when you arrived?

- Was the fire particularly hard to extinguish?
- Was the electricity on when you arrived?
- Did you notice anything out of the ordinary?
- What portion of the building was involved in fire when you arrived?

When the fire is extinguished, it is important not to disturb the suspected site of origin any more than necessary. Some portion of the overhaul may have to be delayed until cause determination is made. Taking a shovel and removing all burned material from the structure may be the usual procedure, but it can destroy evidence if done in the area of origin. One of the worst sins is to find a gas can and run to your officer with it in your hand, shouting "Look what I found." You have in essence just compromised that piece of evidence. This ranks right up there with telling the news media that the fire was started by electricity, then finding out that it was arson and the electricity was off. The defense attorney may very well show the film clip of your statement in court, during the trial, to discredit your testimony and credibility in the eyes of the jury.

NOTE When the fire is extinguished, it is important not to disturb the suspected site of origin any more than necessary.

For more information on cause determination at structure fires, please refer to DVD clip *Protecting Evidence for Fire Cause*.

At all fire scenes, the area of origin must be protected. In a wildland fire scenario, it is important to take note of the general area where the fire was burning on your arrival. As the fire burns, it can spread in all directions, making it hard to pinpoint the area of origin. Your initial observation can narrow this area considerably. At wildland fires it is important to protect the area of origin by keeping vehicles and foot traffic out. Causes have been determined from items as small as a single match or piece of carbon from an exhaust. One person walking through the origin area could easily destroy this fragile evidence. Another recommendation in remote areas is to take note of any vehicles leaving the area. Write down a description and license number whenever possible. The next step in the cause determination is the reconstruction. By working back from the furthest points of fire extension (i.e., area of least damage), the trained investigator can use numerous clues to narrow the way to the point of origin. Some clues are less obvious than others. Training in the science of cause determination is important to all firefighters so they do not inadvertently destroy evidence. Once in the area of origin, the tedious work begins. Material is removed a little at a time until the cause of the fire is uncovered. A part of the reconstruction is witness statements. Determining what people at the scene saw and an examination of the structure can create a chronology of the fire's development.

After the facts are determined from the first steps of the investigation, they are evaluated. Numerous factors can influence the behavior of a fire. The investigator should never approach the fire with any preconceived notions as to what happened. The facts should be evaluated as they are. If you are sure that it was electrical, you may overlook the gasoline burn pattern on the floor.

NOTE When the fire is extinguished, it is important not to disturb the suspected site of origin any more than necessary.

Once the facts have all been examined and evaluated, a conclusion may be drawn from the determination of the fuel source, the heat source, and the act or omission that caused the two to come together. Not all fires require a long drawn out investigation to determine their cause. In some cases, the cause is obvious. When the homeowner says the pan of grease on the stove caught fire and set the kitchen cabinets ablaze, the cause may be evident. In other cases, the investigation may take weeks to conduct. In any case, once the fire department releases the scene and all firefighters leave, any evidence found after that may be worthless in court.

The types and complexity of investigations can be divided into three basic categories. The first and most commonly conducted is the basic investigation. In this investigation, only general information such as time, place, and type of material burned is necessary. This is the minimum information needed for the fire report. When someone is working on his fuel system and accidentally sets his car on fire, a basic investigation is usually conducted.

A more involved investigation is the technical. These investigations are more in-depth to determine a more complicated cause of origin. When a certain process or device is suspected to be the fire cause, the investigation can get more technical. In these cases, evidence must often be collected and sent to the crime lab for further investigation to determine its chemical makeup. In some cases, personnel from the Bureau of Alcohol, Tobacco, Firearms and Explosives or other specialists are called in to assist with cause determination.

The third type of investigation is called for when the fire is **incendiary** or suspicious. In these investigations, the arson investigation experts are called in to assist with gathering and processing evidence. The cause of the fire may have already been determined by engine company personnel. A strong smell of gasoline in a dwelling after a fire is a good indication that a crime was committed. In most jurisdictions where there has been a death related to the fire, arson investigators are automatically called for cause determination. The cause may be as simple as smoking in bed, but no chances are taken with an incorrect determination in these cases.

incendiary A fire that is deliberately set.

In cases where the ranking officer is unsure or unable to determine the fire cause, the arson investigators may be called for. It is better to err on the side of caution than to leave the scene and then find out the cause was arson.

FIRE INFORMATION REPORTING

Fire reports are generated for several reasons. The statistical data generated from the reports can be used for budget justification, trend analysis, future planning, needs assessment, the identification of faulty equipment, code changes, enhanced safety devices, and legislation. Fire reports are used to generate statistical data on a local, statewide, and national scale. By keeping track of what types of fires are experienced, where they are occurring, why they are occurring, who is having them, and how they are starting, the fire service can identify focus areas for fire prevention efforts. Any statistics collected are only as good as the input. If the input or fact gathering that generates the statistics is faulty, the conclusions drawn from them will be faulty as well. For example, if a high-dollar fire loss occurs and the amount is not properly recorded in the fire reporting system, the dollar loss for the whole year will be incorrect.

In the management of any program, there must be a focus. There must be goals and objectives. To set the goals and objectives we must first determine the problem to be addressed. Without a plan of action, personnel can work for years and never really get anywhere in reducing the fire problem. Many personnel hours expended on fire prevention without a reduction in the fire loss should lead the department to question whether they are addressing the root problem. If most of the fires are started by faulty wiring and prevention efforts are only talking to people about combustibles stored by water heaters, the prevention job is not getting done.

The information that is required on a standard fire report and the coding for the information it contains are specified in NFPA 901, *Standard Classifications for Incident Reporting and Fire Protection Data*.[7] This guideline indicates how to tabulate the data collected so it can be used to compile statistics on a regional and nationwide basis. Most states require reporting of all fire-related civilian and firefighter casualties. The information is provided through the National Fire Incident Reporting System. Reports are transmitted electronically to the states and then to the National Data Center.

SUMMARY

It may be more glamorous and exciting to attack fires that have started, but the fire service's sworn duty is to "reduce life and property loss due to fire," and one way to do this is through aggressive fire prevention. Understanding the importance of fire prevention is necessary for every member of the fire department. It is not possible to place an engine company on every corner and have it available for immediate action whenever a fire starts, and even if it is, damage will already have been done prior to an effective attack being conducted.

Fire prevention is an important part of every firefighter's duties. Without it the fire department is always playing catch up and trying to stop the spread of hostile fires that have already started. A more cost-effective and successful method is to prevent these fires before they start.

Fire prevention consists of the four areas of engineering, education, enforcement, and cause determination. All four of these are necessary to have an effective fire prevention program. By being proactive and reducing not only the numbers of fires that start but also reducing the threat they pose through engineering and enforcement, the goal of reduced fire losses can be accomplished.

The fire prevention system consists of persons of all ranks and duties in the fire department. On the company level, the firefighter assists with public education and inspections. The fire prevention bureau inspector performs inspections and plan reviews. The staff evaluates data gathered from the fire reports generated after fires are extinguished, using them to plan for current and future needs. The data is also used to better engineer and design systems that will reduce fire starts and spread.

NOTE It may be more glamorous and exciting to attack fires that have started, but our sworn duty is to "reduce life and property loss due to fire," and one way to do this is through aggressive fire prevention.

REVIEW QUESTIONS

1. List the four areas of fire prevention and give an example of each.
2. What is meant by a fire assembly?
3. Give an example of a fire prevention device.
4. List two methods of presenting fire prevention education to the public.
5. You have been given the responsibility of preparing a fire prevention presentation for a third grade class. What are some of the things you can do to get your point across?
6. List two advantages to a fire company performing fire prevention inspections in their first-in district.
7. Why does the fire prevention bureau inspect the more complicated occupancies?
8. Explain the difference between a hazard and a risk.
9. What is the reason for weed abatement in an area with a dry climate?
10. What hazard do wood shake roofs pose from a fire standpoint?
11. Illustrate the chain of authority that allows you to require compliance with the fire code.
12. Which fire code is adopted in your area?
13. Illustrate the organization of a standard fire prevention bureau.
14. Where may a firefighter seek fire prevention training?

15. When faced with an owner who fails to comply with the required corrections, what actions should you take?
16. List three reasons why fire statistics are collected.

DISCUSSION QUESTIONS

1. How valuable is fire prevention in providing comprehensive fire protection to a community?
2. Why can't the fire protection problems in a community be solved solely by building more fire stations and hiring more emergency response personnel?
3. You are the fire chief and have been asked to cut the fire prevention budget. How will you defend not making the cut?

NOTES

1. *Facts About Fire in the U.S.* (Washington, DC: United States Fire Administration, 2008).
2. International City/County Management Association, *Fire Prevention Saves Lives and Budgets* (ICMA Web Site, 2009), http://icma.org
3. National Fire Protection Association, *Standard 1031, Professional Qualifications for Fire Inspector and Plans Examiner* (Quincy, MA: National Fire Protection Association, 2009).
4. The Center for Chemical Process Safety, *Guidelines for Hazard Evaluation Procedures* (New York, NY: American Institute of Chemical Engineers, 1985).
5. *Facts About Fire in the U.S.* (Washington, DC: United States Fire Administration, 2003).
6. Charles W. Bahme, AB, JD, *Fire Service and the Law* (Quincy, MA: National Fire Protection Association, 1976).
7. National Fire Protection Association, *Standard 901, Standard Classifications for Incident Reporting and Fire Protection Data* (Quincy, MA: National Fire Protection Association, 2006).

CHAPTER 11

Codes and Ordinances

LEARNING OBJECTIVES

Upon completion of this chapter, you should be able to:

- Explain the relationship between federal, state, and local regulations.
- Explain who is responsible for enforcing codes and ordinances at the different levels.
- Explain why codes and ordinances are created.
- Describe how codes and ordinances are adopted.
- Describe how codes and ordinances are affected by court decisions.
- Explain the relationship of codes and standards.
- Give the definition of legal terms as they apply to codes and ordinances.
- List several codes and ordinances that directly relate to firefighters regarding safety and operations.

INTRODUCTION

Codes and ordinances fall under the broad description of laws. Laws are written and adopted on all three of the levels of government: federal, state, and local. There are clearly laid-out relationships among the different levels of government and their influence on each other.

DEFINITION OF LAWS

Laws are pieces of enacted legislation. There are different types of laws in the United States. The supreme law, and that with which all others must not conflict, is the U.S. Constitution. All legal authority for governmental action comes from the Constitution of the United States. The Congress of the United States passes legislation within the confines of the Constitution.

Statutory laws are those that are adopted by Congress (federal statutes) and laws that have been passed by state legislatures (state statutes). Statutory law also includes local laws, usually called ordinances.

Codes are systematically arranged, comprehensive collections of laws. Regulations are rules designed to implement a statute based on an agency's interpretation of that statute. Such rules provide procedural requirements for program operations. Typically, they are published through an official process that allows for public comment. Regulations have the same effect as law and must be complied with once they are published in final form.

The federal government organizes its statutes applying to a certain subject by placing them in the code of federal regulations (CFR). States do much the same thing. The laws regarding public safety may be in the health and safety code. Criminal laws are organized into the penal code. Examples of codes that are adopted by local government ordinance are the building and fire codes. The hierarchy of legal enforcement is designed so that the state laws can augment or increase provisions in federal law, but they cannot weaken the federal law. The same relationship exists between local ordinances and state or federal laws. The types of codes are not to be confused. CFRs and state legislation are not the same as the adoption, by ordinance, of a building or fire code.

The constitutionality of laws is determined by the judicial system. When a law or the enforcement of a law is questioned, it may be taken before the courts. Once these issues are decided, they are referred to as precedents. These precedents are used as the basis for future interpretation and enforcement of the laws.

Laws are only applicable as long as they have been properly adopted and not been reversed by a court decision. Through court decisions, laws are changed and in some cases suspended until a final decision is reached. In modern times initiatives passed by voters may create a law whose enforcement is held up in the court system for years. Appeals may be made at successive levels until the Supreme Court agrees to hear the suit or refuses to hear it, thereby leaving the decision of the next lower level of court standing.

When the law is not specific in a matter, the laws and previous court decisions are reviewed to determine what is most closely the intent of the law. Laws are not created overnight and many new processes and situations occur that have not been addressed specifically in the law or in a code that has been adopted. In some states, in lieu of an existing law, a legislative counsel or attorney general's opinion governs conduct, action, or procedure.

It is important to have a good idea of what the applicable laws are in any situation. Firefighters are charged with the authority and the responsibility to only enforce laws that exist, not just what they think is a good idea. When you, acting under the authority of your position, tell someone to do something, they have two choices: to go ahead and do it or to challenge you. They can say that they want you to cite the provision of the code that requires them to have fire extinguishers serviced and maintained periodically. If you are not familiar with the code, and you are wrong, they have every reason to contact your supervisor and report your inadequacies. When you know the fire extinguisher regulations, you can produce the code section that requires compliance with your orders. A third alternative is for them to let you go ahead and issue the citation and take the matter before a judge. What they will ultimately have to do depends on the decision rendered by the court.

In some circumstances, the word of the fire department employee is not the final word, no matter what the law says. A business owner may be granted a variance to avoid complying fully with the letter of the law. A variance can be granted for a number of reasons. The code allows a variance provided that the intent of the code provisions is met. A variance allows the business owner an alternative to meeting the strict letter of the code as long as life safety and fire safety are provided. At other times, the variance may be granted for political or economic reasons.

LAWSUITS

We live in a society that increasingly likes to sue. Many people take no personal responsibility for their own actions and will sue anyone. It does not matter that you responded to the fire they caused and did the best you could. When their house burns down, they may sue the fire department for not saving it.

There are more attorneys than firefighters in this country. One of the ways that attorneys make money is to file lawsuits. Even if the matter never goes before a judge, the jurisdiction may settle out of court. This is often cheaper than defending itself. When this happens, the persons suing and their attorney both make money. In some jurisdictions, suits brought or negotiated down to a specified amount are not contested due to the cost, no matter how ridiculous the claim is.

It used to be that the fire department was so highly regarded it seemed above being sued, but that has changed. If the modern experience of police departments is any indication, we will see a sharp increase in the number of suits filed against fire departments.

The way to avoid lawsuits is to do your job correctly every time. You must also take the time to document, through the use of reports, what you did. When lawyers approach the jurisdiction to start a suit, the evidence of a job done correctly and properly documented can often stop the proceeding in its tracks.

NOTE The way to avoid lawsuits is to do your job correctly every time.

When you are sued, it is because of a tort. A tort is defined as a wrongful act. The term tort is used in civil, not criminal actions. Civil actions are brought with the intent of seeking monetary compensation. Torts can result from either nonfeasance, misfeasance, or malfeasance. Nonfeasance is defined as a failure to act. If you were to respond to a medical aid incident and failed to splint a broken leg before the person was moved, thereby further injuring them, that is a failure to act. Misfeasance is doing something wrong that you are lawfully allowed to do. Responding with red lights and siren is legal. It releases you from obeying the traffic laws to a certain point. It does not allow you to run stop signs at a high rate of speed without due regard to public safety, as is specified in many departments' Standard Operating Procedures and guidelines. Malfeasance is wrongdoing or misconduct. Driving the engine to a fire while intoxicated is an example of malfeasance. Yes, the engine needs to get to the fire, but when you are under the influence of alcohol you are violating the laws against driving under the influence. Using the red lights and siren does not release you from your obligation to obey the drunk driving laws. The greater public good is served by having a short delay while another person shows up to drive the engine than for you to try to operate it yourself.

The defense that you were just doing the best you could under the circumstances is often not good enough. When things do go wrong, it is our responsibility to find out what went wrong and take measures to prevent a recurrence. The ability to perform properly at incidents comes from experience and training. In many cases, you may lack the experience. If you paid attention to the training and practiced, you should be able to perform as expected. If not, your performance will be below that which is expected and you will have a hard time defending your actions in a court of law. The training must also be documented. Judges and juries are suspicious when documentation of actions taken is not available when requested.

One of the ways to address these issues is through policies. Policies clarify or provide direction for specific situations. The policies must be agency specific and regularly reviewed to ensure that they are still valid. The policies must be written and delivered to the members of the department so that they are understood. The department must be able to verify that they have been received and all personnel have not only seen them, but understand them as well. The only way to verify understanding is through regular testing.

All personnel in a supervisory capacity must follow the policies and set the example if they expect others to do the same.

There are several simple policies that you can use to limit your liability and help to protect yourself against a lawsuit:

- Develop a reminder process that will send up a notice when it is time to go back to a business to reinspect for previous problems.
- Conduct as thorough an inspection as you are capable of. In other words, do your best.
- Maintain accurate records of your inspections, variances, complaints, and so forth. These records can be used to justify the reason why you approached a business to conduct an inspection.
- In court, testify to those things you actually know.
- Have someone with you when making your inspections, preferably an employee or owner of the business. However, this action will not eliminate the possibility that someone could accuse you of misconduct.
- If you are refused entry, obtain an inspection warrant. Do not force your way in.
- Finally, treat all people fairly and honestly to eliminate the potential of a business owner claiming that she was being discriminated against or picked on.

The overriding issue is that the public pays you to do a job and do it well. It is not just done to avoid lawsuits. It is the responsibility of every firefighter to perform to the best of his or her ability and perform correctly. To you, it is may be just another incident; to the people involved it may be the most important event in their lives.

NOTE It is the responsibility of every firefighter to perform to the best of his ability and perform correctly.

PERSONNEL COMPLAINTS

Most departments have a standard procedure for complaints brought against members by the public or other members of the department. An example of a standard procedure is as follows:

1. The person wishing to file the complaint should speak to the accused's supervisor or the fire chief.
2. The complaint is discussed with the appropriate officer. The officer is to explain the options available to the complainant. Should they decide to pursue the matter, the personnel complaint form should be filled out.
3. The form is forwarded to the fire chief, or a designated representative, and a decision is made as to the correct investigative procedure based on the nature of the complaint.

4. At the conclusion of the investigation, a determination is made as to the disposition of the complaint and what action, if any, is warranted against the employee.
5. The person complaining is notified by mail, in writing, of the results of the investigation, as is the accused.

The person making the complaint should be assured that the complaint will be properly investigated. Also, the department should make sure that no adverse consequences occur to any person or witness as a result of having brought a truthful complaint or provided truthful information in any investigation of a complaint.

It is standard policy that any employee who attempts to discourage, delay, or cover up any complaint received by a citizen shall incur disciplinary action.

It is the complainant's responsibility to be truthful and as accurate as possible in presenting information that he or she believes should be investigated. The complainant should understand that malicious or false complaints against department personnel could subject him or her to possible criminal and/or civil action. The department may assist in pursuing such action. The complaint form should state that the person signing the complaint does affirm under penalty of **perjury** that the facts contained therein are truthful.

perjury Making false statements in a sworn document or testimony.

HARASSMENT-FREE WORKPLACE

The federal government and the courts hold management responsible for harassment in the workplace. Consequently, a clear understanding of what constitutes harassment is essential to the development of a harassment-free work environment.

Harassment is defined as coercive or repeated, unsolicited, and unwelcome verbal comments, gestures, or physical contacts, including retaliation for confrontation or reporting harassment. Harassment includes:

- *Physical conduct.* Unwelcome touching; standing too close; inappropriate or threatening staring or glaring; obscene, threatening, or offensive gestures.
- *Verbal or written conduct.* Inappropriate references to body parts, derogatory or demeaning comments, jokes, or personal questions; sexual innuendoes; offensive remarks about race, gender, religion, age, ethnicity, sexual orientation, political beliefs, marital status, or disability; obscene letters or telephone calls, catcalls, or whistles; sexually suggestive sounds; loud, aggressive, inappropriate comments, or other vocal abuse.
- *Visual or symbolic conduct.* Display of nude pictures, scantily clad or offensively clad people; display of intimidating or offensive religious, political, or other symbols; display of offensive, threatening, demeaning, or derogatory drawings, cartoons, or other graphics; offensive T-shirts, coffee mugs, bumper stickers, calendars, or other articles.

- *Work environment.* Any area where employees work or where work-related activities occur, including field sites, fire stations, buildings, and facilities. Also included are vehicles or other conveyances used for travel.
- *Responsibility.* Managers, supervisors, and employees, as well as contractors, cooperating agency personnel, and volunteers are responsible for creating and sustaining a harassment-free environment by their individual conduct, through job supervision, coaching, training, and other behavior and means. All employees, contractor personnel, and visitors must take personal responsibility for maintaining conduct that is professional and supportive of this environment. Employees who witness harassment are instructed to report it to the proper authority.

Individuals who believe they are being harassed or retaliated against should exercise any one or more of the following options as soon as possible:

- Tell the harasser to stop the offensive conduct.
- Tell an officer or supervisor about the conduct.
- Contact the fire chief, or any other individual who would take action, for example, a union representative or the agency equal employment opportunity representative.

THE COURT SYSTEM

Before looking at the court system and its organization, it is necessary to understand the concept of jurisdiction. The cases that a court can hear are considered to be within its jurisdiction. When a case can be first heard in a certain court, it is considered to have original jurisdiction. When the case must first be heard in a lower court and then the lower court's decision is appealed to the higher court, that is an appellate jurisdiction. In the case of fire departments, the meaning of jurisdiction is the limits of territory within which its authority may be exercised. An example is that the city fire chief cannot go into a county fire department station and give orders to the firefighters because he is outside his jurisdiction. Jurisdiction is a very important concept to understand whenever talking about who can legally do what, where, and when. If you are acting outside your jurisdiction, you are not legally taking action, no matter how good your intentions.

NOTE In the case of fire departments, the meaning of jurisdiction is the limits of territory within which its authority may be exercised.

The court system is divided into different levels. As you go up the levels, the matters heard tend to be more important. On the federal level, the highest court is the U.S. Supreme Court. Most cases heard before this court are questions about the constitutionality of a law. The next lower level is the circuit court of appeals, which hears appeals from the federal district courts. The federal district courts deal with matters of federal law. When a crime—such as an illegal campfire—is committed on federal property, the person must appear in federal district court. The illegal act occurring on federally managed land gives the federal court jurisdiction.

On the state level, the highest court is typically the state supreme court. This court usually hears appeals from the district courts of appeal. The district courts of appeal typically hear appeals from the district or superior courts. The district or superior courts are where most of the state trials are heard. When someone is charged with arson, the case is first heard in the district or superior court. In certain cities and counties, there are municipal or county courts where **misdemeanor** offenses are heard, with the exception of those committed by juveniles, which are heard in district or superior court. If you wrote someone a citation for illegal open burning of trash, she would be required to appear in municipal court. If someone were arrested for illegal fireworks, a **felony** in some states, he would appear in superior court. Both situations may start with the issuing of a citation, but the offenses are very different in the eyes of the law.

misdemeanor A crime punishable by up to one year in a county jail or by a fine usually not to exceed one thousand dollars or both.

felony A serious crime such as murder, arson, or rape for which the punishment is either imprisonment in a state prison for more than one year, or death.

RELATIONSHIP OF FEDERAL, STATE, AND LOCAL REGULATIONS

When discussing the relationship of federal, state and local regulations, the subject of jurisdiction comes into play. The jurisdiction is the limits of territory in which authority is exercised. Territory is not strictly geographical; it can also apply to situations.

When several levels of laws are encountered, usually the most stringent law is enforced. This is not always the case, especially when different levels of government jurisdiction are involved. A fire prevention person for the forest service would enforce federal regulations on federal lands. She would not perform a prevention inspection in a business in an incorporated city. Conversely, a local fire department would not have jurisdiction in a federally owned post office or other federal building. This should not prevent the local fire department from performing prefire planning at federal buildings. If there is a fire, it is the responsibility of the local fire department to respond.

The jurisdictional lines get somewhat blurred in that the local fire department does have jurisdiction over a contract post office operated from a privately owned building, common in suburban or rural areas. Another example is a locally protected housing area, on private land, in the middle of a national forest. The fire department and the forest agency should work together on hazard reduction to prevent a structure fire from spreading to the forest and a forest fire from burning structures. Through cooperation, the fire prevention mission can be accomplished within jurisdictional boundaries to satisfy the needs of both parties.

The state fire marshal usually has jurisdiction in state-owned buildings, such as prisons and government office buildings. The state fire marshal may make an agreement that allows the local fire department to enforce state regulations in certain occupancies.

Other governmental agencies are involved as well. The local zoning commission regulates what types of occupancies are allowed and where. It is uncommon for a high-hazard occupancy to be allowed in a residential area. This may be as simple as stopping someone from having a woodworking business in his garage. The local building department is responsible for the enforcement of building codes, plan checks, and the inspection of buildings during construction, whether new construction or remodeling. The building department also determines occupancy types allowed in buildings, dependent on construction type. A building that was originally a restaurant may not be allowed to be used as a cabinet shop without major modification. In the average jurisdiction, the building department may only be aware of remodeling when a building permit is requested, whereas the fire department makes periodic inspections of businesses and is likely to discover evidence of major work done or a change of occupancy. When the fire department discovers alterations or changes in occupancy, it should notify the building department and work with that staff to ensure that the facility complies with construction, fire, and life safety regulations. There may be times when the business owner is reluctant or hesitant to comply with the correction notice issued by the fire department. At that point law enforcement personnel may be needed to serve an inspection or administrative warrant or to arrest persons who willfully disregard action required to correct fire code violations.

Some problems are going to be beyond the legal authority of the fire department alone to handle. It is important for fire department personnel to know where their jurisdictional authority begins and ends. The violations found may not be of the fire code, but the building code or other ordinance. In these circumstances, personnel must know how and to which department to make a referral. An example is an abandoned house. It is a health hazard if left open and may be a life hazard if used as a sleeping area by transients. The provisions of the fire code may not provide jurisdiction for fire department action. A referral can be made to the building department and the hazard abated. Another common scenario is that you are out performing weed abatement and notice an accumulation of garbage behind a house or business. If it is not an accumulation of combustibles, it is not within your jurisdiction to correct it, but by giving a referral to the health department the nuisance can be cleaned up.

NOTE Some problems are going to be beyond the legal authority of the fire department alone to handle.

It is good public relations to have a working knowledge of the responsibilities and jurisdictions of the local public agencies. That way, when someone calls to complain about the neighbor's accumulation of trash in their yard, you can assist in resolving the problem.

In cases where the accumulation is an eyesore and not a fire hazard, no direct action can be taken by the fire department. If you can make a referral that handles the problem, the person who complained is going to remember that you, the fire department, were of assistance. You did not just say that there was nothing you could do about it because it was not your jurisdiction. By working together, the fire department and the other public agencies can assist each other and perform their functions to the public's satisfaction.

NOTE By working together, the fire department and the other public agencies can assist each other and perform their functions to the public's satisfaction.

FIRE PREVENTION

The fire prevention bureau has the legal responsibility and authority to enforce fire-related codes and ordinances. This authority does not just come about by the fire department deciding what they want to do. The fire department and its personnel cannot demand that action be taken by a property owner unless there is authority to inspect.

The most often cited case in relation to fire prevention is *See vs. City of Seattle* 387 vl. 541, 87 S. Ct 1737. In this case the U.S. Supreme Court has held that administrative entry, without consent, of the portions of commercial premises that are not open to the public, may be compelled only through prosecution or physical force within the framework of a warrant procedure.

The U.S. Supreme Court has also set forth guidelines for inspection agencies.

The following recommendations are given to assist fire inspectors in operating within these guidelines:

1. Inspectors must be adequately identified. It is recommended that inspectors wear some type of uniform. Commonly recommended is a blazer, which looks professional but is not as threatening as wearing a badge.
2. Inspectors must state the reason for the inspection. Many times people at the site will joke that you are there to "shut them down." This is not always taken as funny and you must be careful about what you say.
3. Inspectors must request permission for the inspection. There are times that are inconvenient for the occupants to undergo an inspection. This does not always mean that they are trying to hide something. It might just mean that they are extremely busy. Just make arrangements to come back at a more convenient time.
4. Inspectors should invite the person in charge to walk along during the inspection. When the manager or his designee accompanies you, it is a good chance to point out problems, ask questions pertinent to the inspection, and discuss voluntary compliance with the required changes. You must remain professional at all times and complete the inspection; the persons at the business have work to accomplish as well.

5. Inspectors should carry and follow a written inspection procedure, making it less likely to overlook something important. If you are called away or otherwise interrupted during the inspection, it also is easier to pick up where you left off.
6. Inspectors should request an inspection or administrative warrant if entry is denied. This will rarely be necessary. If all other methods of gaining voluntary submission to an inspection fail, then an inspection or administrative warrant will be necessary.
7. Inspectors may issue stop orders for extremely hazardous conditions, even if entry is denied, while warrants are being issued.
8. Inspectors should develop a reliable record-keeping system of inspections. By keeping records of past inspections, the inspector can update information as to phone numbers and contact persons, owner and business name information, and record of past violations. If the system of voluntary compliance is used, it is important to know when reinspecting whether the last violations were corrected. If they were not, stronger measures may be required to ensure future compliance.
9. Inspectors should have guidelines available that define conditions whereby they may stop operations without a search warrant or obtaining permission to enter. These guidelines are necessary in that if you shut down an operation, you are costing the business money and it may very well end up in a lawsuit. It is important that these procedures be applied uniformly and impartially.
10. Inspectors should be sure that all licenses and permits indicate that compliance inspections can be made throughout the duration of the permit or license. Before any inspection is made, the inspector must be sure that he has the right to inspect the premises. After you have required expensive modifications or shut down a portion of a business is not the time to find out that you were acting outside of your authority.
11. Inspectors must be trained in fire hazard recognition and in applicable laws and ordinances. When you enter a business, it is your responsibility to make a thorough, professional inspection. Ignorance of the law is no excuse for either party, you or the owner. When an inspection is made, you may be held liable for fire, injury, or death if there is a fire resulting from something you overlooked or failed to enforce. It is just about a sure bet that you will be named in the lawsuit. For this reason, personnel assigned to the fire prevention bureau receive special training and inspect the more complex businesses and processes. Personnel on engine companies are not as highly trained and, in most cases, perform the less complex inspections. As an inspector, you need to possess the technical expertise to identify "physical illegalities" such as substandard electrical installations, excessive amounts of flammable liquids stored, and lack of or inoperative fire extinguishers. "Procedural illegalities" include failure to obtain required permits for hazardous operations, unsafe welding activities, improper smoking areas, and unrestrained flammable gas cylinders that may tip over. Structural deficiencies include inadequate or breached fire walls, improper fire exit design, and improper installation of fire doors.

The state fire prevention laws and codes are based primarily on model national codes. In some states the codes are divided by type, such as a building code, health and safety code, welfare and institutions code, and administrative code. In most states, the state fire marshal interprets and recommends state-level legislation as it relates to fire and life safety. The state fire marshal may also be charged with the responsibility of enforcing regulations and codes in state buildings and in areas with no established fire prevention bureau, such as rural areas. The state fire marshal may delegate authority to the local jurisdiction when there is an established fire prevention bureau, usually in county, city, or district fire departments. The state fire marshal can delegate the authority to perform these inspections but still retains the responsibility to see that they are completed.

On the local level are the local fire departments and local codes and ordinances. Most jurisdictions adopt model codes through ordinance. These codes may be adopted in part or in whole, as the local jurisdiction wishes and finds politically acceptable. Adoption of the fire prevention code usually designates the fire chief as the primary enforcement authority. He then delegates that authority to his fire marshal and inspectors.

Adopting a model code does not mean that the jurisdiction cannot establish ordinances that are not addressed in the model code to deal with particular local problems. An example of this would be a weed ordinance in areas where dry vegetation is a summertime fire hazard. Another example would be regulations regarding high-rise buildings as there may not be language in the adopted model fire regulations that refer to this type of building. One area that is gaining in acceptance is a Residential Fire Sprinkler Ordinance. Another common local ordinance is one that specifies hydrant spacing in the jurisdiction. It is the responsibility of the local governing body to enact the required legislation, through ordinance, to make these requirements law.

MODEL FIRE CODES

The Building Officials and Code Administrators (BOCA), the International Conference of Building Officials (ICBO), and the Southern Building Code Congress International (SBCCI) have all developed model fire prevention codes. ICBO publishes the *Uniform Fire Code*,[1] BOCA publishes the *Basic Fire Prevention Code*,[2] and SBCCI publishes the *Standard Fire Prevention Code*.[3] Additionally, the National Fire Protection Association publishes a series of codes and standards known as the National Fire Codes; one of these is the NFPA 1, *Uniform Fire Code*.[4]

The three organizations, ICBO, BOCA, and SBCCI, recently teamed together to create a nationwide fire code together with a nationwide building code to go hand-in-hand with the fire code. This task was undertaken in an effort to create uniformity and ease of use of the fire codes across the nation. These two documents will replace the individual building and fire codes that each of the three model code organizations previously published. The benefit of creating a set of national codes is that building designers and architects can use the same set of regulations in Tennessee, Alaska, or California. In the year 2000, the first editions of the *International Fire Code* and the *International Building Code* were published by the newly formed International Code Council. This new set of codes is quickly gaining in popularity and has been adopted in all or part of 46 states across the country.

The use of a nationally recognized model fire prevention code is usually more desirable than a locally written code because the model codes have been nationally developed and represent a broad spectrum of fire prevention experience. The nationally recognized codes also give building experts, such as architects and engineers, a familiar base from which they can design a structure's built-in fire protection features. The national codes also offer a means to gain formal interpretations of the code's intent if a local inspector does not clearly understand a particular code requirement. Additionally, nationally recognized codes undergo a constant review process with a new, updated edition published every three years, which makes them more in line with current fire protection theories and technology.

NOTE The use of a nationally recognized model fire prevention code is usually more desirable than a locally written code.

Perhaps one of the single most important factors in adopting one of the model codes is that they are companion codes to the publishing organization's building codes. This fact is very important in that it minimizes the likelihood that there will be conflicting code requirements.

The model fire prevention codes are typically divided into sections or chapters that deal with certain topics of fire protection. Typically, the first chapters deal with administrative items, such as a model ordinance that officials can use to adopt the code. The codes typically then define the authority of the fire official and define the responsibilities of property owners to maintain their premises in a firesafe condition. One of the more important code sections usually follows next, and this section defines the fire protection terminology used in the code. The following chapters typically deal with the proper installation and maintenance of fire protection features, and lastly they offer specific requirements for the use and protection of equipment, processes, occupancies, and hazardous materials.

One important point is the recognition of the National Fire Protection Association's National Fire Codes. (A list of these is contained in Appendixes C and D). Many jurisdictions have, by ordinance, adopted the whole set of codes and standards, which has the positive effect of having almost any fire protection problem covered by a code or standard. It has the disadvantage of making the inspector responsible for the knowledge and research capabilities to understand many separate fire protection regulations.

At this point, it is important that you understand the difference between a code and a standard. A code is written in language that mandates what should be done and is written such that it can be enforced as a law. Standards, on the other hand, express how something should be accomplished and are typically written as recommendations. In other words, the codes stipulate when to install a fire sprinkler system in a particular building; the standards address how to design and install the sprinkler system. When standards are adopted as part of an ordinance, these recommendations have the force of law and may be enforced as such.

When a certain model code is adopted, it is imperative that a certain edition (year) be adopted. It is illegal for the ordinance to simply state, "Adopt the most recently published edition" of a code, as this denies the public its right to due process. The specific edition must be identified in the ordinance. This will give the public the opportunity to comment on, or protest, the enforcement requirements that are identified.

Occupancy Classification

When the building or fire code is to be applied, the first thing that must be determined is the occupancy classification of the building. It is important to select the classification that most accurately fits the use of the building. Most requirements of the code come from this classification. Several examples of occupancies are:

Occupancy	Letter Designation
Assembly	A
Business	B
Educational	E
Institutional	I
Mercantile	M
Residential	R
Storage	S

NOTE When the building or fire code is to be applied, the first thing that must be determined is the occupancy classification of the building.

Many of the occupancy examples have subcategories. These subcategories vary from code to code. An assembly occupancy may be subdivided based on the size of the occupant load or type of use, such as restaurant or theater. Educational occupancies may be subdivided into regular schools and day-care centers. Institutional occupancies may be subdivided into restrained or nonrestrained occupants. Residential occupancies may be subdivided based on the number of units or the type of residential setting (i.e., dormitory, board and care, hotel/motel, apartments). Finally, storage occupancies may be subdivided based on the combustibility of contents (fire load) stored in the building. The classification of the occupancy is important for many reasons. The building code requires limits as to the height and area of a building depending on the occupancy classification and the types of construction materials. Generally, the greater the fire resistiveness of the structure, the larger it is permitted to be. Additional area may also be added if a building is protected throughout by automatic fire sprinklers. It is also important, from a building code standpoint, to determine the occupancy of a building so that the mixed occupancy fire separations can be determined. For example, a model building code may require only a **one-hour fire-rated separation** between a business (such as an office) and a mercantile (such as a clothing store) use, whereas it may require a three-hour fire-rated separation between a moderate-hazard storage occupancy and a business occupancy.

one-hour fire-rated separation A fire-rated assembly that should resist breakthrough for a period of one hour. An example of this type of construction is the use of 5/8-inch-thick fire-rated gypsum wallboard or a combination of wallboard and plaster. All of the electrical boxes must be metal and not plastic. Any penetrations through the assembly must be properly protected to prevent the spread of fire.

Construction Types

Both model building codes and NFPA 220, *Standard on Types of Building Construction*,[5] can be used to determine the type of construction used in a building. This type of construction is typically denoted by a shorthand notation such as Type I, II, III, IV, and V. It may also be followed by a number or letters such as Type IV 2 HR, or Type IV Unprotected (**Figures 11-1** and **11-2**). These notations and numbers refer to what the building is constructed of and what the hourly ratings of its structural members are (i.e., exterior and interior bearing walls, columns, beams and trusses, floors and roofs).

The model building codes determine how close buildings of a certain construction type can be from another building or property line. The model building codes and NFPA 101, *Life Safety Code*,[6] determine other important factors, such as the occupant load of a structure, the size of the means of egress, the number of exits, and the travel distance to those

Delmar/Cengage Learning

FIGURE 11-1
Unprotected steel construction which may fail quickly when exposed to fire.

FIGURE 11-2
Unprotected steel construction after fire.

Delmar/Cengage Learning

exits. For multistory buildings, the codes require certain hourly ratings around stairwells and building shafts such as trash chutes and elevator shafts. The codes at times mandate the installation of fire sprinkler systems, fire alarm systems, or standpipe systems. They can also be used to determine the need for emergency lighting and exit signs. Additionally, these codes require the adherence to requirements on the types of interior wall and ceiling finish that is permitted in areas of the structure. These items are only a brief overview of complex codes that can contain up to 1,000 pages of codes and interpretations.

CODE DEVELOPMENT

Codes are most often developed in response to a disaster. After a major disaster that could have been prevented, there is a public outcry as to why it was not prevented. The fire codes requiring that exits open outward and be kept unlocked were developed after numerous large life-loss fires. Building codes requiring enclosing shafts in **fire-resistive construction** were the result of rapid fire spread in buildings caused by unprotected shafts. Examples of twentieth-century high–life-loss incidents are listed by occupancy type in **Table 11-1**.[7]

Structural steel is commonly covered with a fire-resistive coating or materials such as gypsum board.

NOTE After a major disaster that could have been prevented, there is a public outcry as to why it was not prevented.

fire-resistive construction Construction that has been designed to resist the effects of heat from fire.

TABLE 11-1 Major U.S. fires causing loss of life (more than 25 deaths).

Date	Location and Occupancy	Lives Lost
Dec. 5, 1876	New York, NY—Theater	300
June 30, 1900	Hoboken, NJ—Steamship piers	326
Jan. 12, 1903	Boyerstown, PA—Opera house	170
Dec. 30, 1903	Chicago, IL—Theater	602
June 15, 1904	East River, NY—Steamship	1,030
March 4, 1908	Collinwood, OH—Grammar school	175
March 25, 1911	New York, NY—Clothing factory	145
*April 10, 1917	Eddystone, PA—Ammunition company	133
May 15, 1929	Cleveland, OH—Clinic	125
April 21, 1930	Columbus, OH—State penitentiary	320
Sept. 8, 1934	New Jersey coast—S.S. Morro Castle	125
*March 18, 1937	New London, TX—School	294
April 23, 1940	Natchez, MS—Nightclub	207
Nov. 28, 1942	Boston, MA—Nightclub	492
July 6, 1944	Hartford, CT—Circus tent	168
*July 17, 1944	Port Chicago, CA—Munitions depot	300
*Oct. 20, 1944	Cleveland, OH—Gas company	130
June 5, 1946	Chicago, IL—Hotel	61
Dec. 7, 1946	Atlanta, GA—Hotel	119
March 25, 1947	Centralia, IL—Coal company	111
*April 16, 1947	Texas City, TX—S.S. Grand Camp	468
Dec. 21, 1951	W. Frankfort, IL—Coal company	119
Dec. 1, 1958	Chicago, IL—Grade school	95
Oct. 10, 1963	Indianapolis, IN—Fairgrounds	74
Nov. 23, 1963	Fitchville Township, OH—Nursing home	63
Feb. 7, 1967	Montgomery, AL—Restaurant	25
*April 5, 1968	Richmond, IN—Sporting goods store	41
Jan. 9, 1970	Marietta, OH—Convalescent home	32
Dec. 20, 1970	Tucson, AZ—Hotel	28
*Dec. 30, 1970	Hyden, KY—Coal mine	38
Feb. 3, 1971	Woodbine, GA—Chemical plant	25
Feb. 10, 1973	Staten Island, NY—Gas storage tank	40
June 24, 1973	New Orleans, LA—Nightclub	32

(continues)

TABLE 11-1 Continued

DATE	LOCATION AND OCCUPANCY	LIVES LOST
Nov. 15, 1973	Los Angeles, CA—Apartment house	25
June 30, 1974	Port Chester, NY—Apartment house	34
*March 9, 11, 1976	Oven Fork, KY—Coal mine	26
Oct. 24, 1976	Bronx, NY—Social club	25
May 28, 1977	Southgate, KY—Supper club	165
June 26, 1977	Columbia, TN—County jail	42
Dec. 22, 1977	Westwego, LA—Grain elevator	36
April 2, 1979	Farmington, MO—Boarding home	25
Nov. 21, 1980	Las Vegas, NV—Hotel	85
†Apr. 19, 1995	Oklahoma City, OK—Federal office building	168
†Sept. 11, 2001	Arlington Co., VA—The Pentagon	189
†Sept. 11, 2001	New York, NY—World Trade Center	3,000 +
Feb. 20, 2003	Providence, RI—Nightclub	100

*Indicates explosion.

†Indicates terrorist attack.

Codes are developed by recognizing the need for increased public safety. Legislation is sponsored and passed that sets the groundwork for the process to begin. Public input is sought and ordinances are written in answer to the needs and the input. An example of this process is public "Right to Know" laws. Right to Know laws were enacted after it was discovered that many businesses were storing large amounts of hazardous materials on their premises. Many times these businesses were operating in close proximity to residential areas and schools. After a few major incidents, the public came to realize the need for disclosure of just what was being stored, manufactured, and used in their neighborhoods. These laws require businesses to disclose, through a business or management plan and hazardous materials inventory, the materials they have on-site, their use, method of storage, and location.

Committees were formed of business personnel, technical experts, and public safety agency personnel to create model legislation. The legislation was designed so that the businesses were required to disclose their materials without requiring them to give away any trade secrets.

Along with the model legislation, model regulations were developed stipulating the storage and handling of the hazardous materials. These regulations were then added to all of the model codes and later adopted by local ordinance across the nation. The development of codes and ordinances is an ongoing process that has tried to become proactive instead of reactive as needs change.

RELATIONSHIP OF CODES TO STANDARDS

Codes are written as bodies of regulations that can be adopted in whole or in part by ordinance. Standards are recommendations on how things should be designed or done. They are usually adopted by policy or as part of a memorandum of understanding. The NFPA has many standards that are used in the design of fire apparatus and other equipment. Most fire departments specify that the equipment they are ordering from the manufacturer meets the current applicable standard requirements. Manufacturers in turn, design their equipment to meet the latest standard and mention this fact when they advertise in magazines and other literature.

When a standard is adopted as part of a memorandum of understanding between a jurisdiction and its employees, the items referred to must meet the standard. An example of this is the adoption of NFPA 1500, *Fire Department Occupational Safety and Health Program.*[8] When adopted in whole it means that all equipment bought by the department will meet the NFPA standard. It may go as far as to state that all equipment that does not meet the standard will be replaced by a specified date.

Standards are usually adopted as a matter of policy instead of by ordinance. This adoption method recognizes their use without their having the force of law. In the case of many model codes, a body of standards accompanies them that illustrates the points of the code.

OPERATION OF EMERGENCY VEHICLES

In 1988 the federal government passed legislation that required all operators of vehicles over 26,001 pounds gross vehicle weight (GVW) or towing trailers over 10,000 pounds to have a Class B driver's license.[9] A normal automobile operator's license is a Class C. This legislation affected fire departments in that many fire vehicles weigh over 26,001 pounds. The Class B license requires a valid medical card that must be renewed every two years. A medical card requires a physical examination by a medical doctor. The cost of having to send all personnel for a medical exam every two years was seen as being prohibitively expensive for fire departments. After public agency lobbying of state legislators, an exception was granted in many states. The exception relieves firefighters of having to see a doctor every two years for the driver's license medical exam. In some states, firefighters can complete a medical questionnaire and turn it in instead. Other states exempt firefighters from all Class B requirements.

When members of the fire department are operating emergency vehicles on a nonemergency basis, they are subject to all of the same traffic laws as the general public. When the vehicles are being operated as authorized emergency vehicles and are responding to emergencies, the laws specify certain exemptions. Usually, to be considered an authorized emergency vehicle, the vehicle must be equipped with a minimum of a siren and one steady burning red light.

Most state vehicle codes state that upon the approach of an authorized emergency vehicle displaying the required emergency warning devices, the driver of every other vehicle shall yield the right-of-way and drive to the right-hand edge or curb of the highway clear of any intersection and shall stop and remain stopped until the authorized emergency vehicle has passed. When approaching an emergency vehicle parked at the side of the road with emergency lights flashing, drivers must slow down and move over to the outside lane, proceeding with caution until safely past the emergency scene. All pedestrians shall proceed to the nearest curb or place of safety and remain there until the authorized emergency vehicle has passed.

One important fact is that the vehicle must be an "authorized emergency vehicle." The fact that the operator is a firefighter does not create the assumption that the vehicle is "authorized." This comes into question when volunteer firefighters respond to the station or the incident scene from home or work in their personal vehicles. They are only authorized to respond as emergency vehicles when properly equipped and it is allowed under the state vehicle code.

No state's vehicle laws relieve persons operating authorized emergency vehicles from the duty to drive with due regard for the safety of all persons and property. In effect, no matter how many red lights you display and how loud your siren, you are still limited, by law, to drive carefully and defensively. In many departments, the policy is that you never exceed the speed limit by more than five miles an hour and that you come to a complete stop before proceeding for all red lights and stop signs. Many times at red lights, the car in the curb lane will stop, but the car in the second lane cannot see you and keeps proceeding into the intersection. When caution is not exercised by the firefighter, they may collide.

NOTE No state's vehicle laws relieve persons operating authorized emergency vehicles from the duty to drive with due regard for the safety of all persons and property.

Some firefighters get the idea that the display of red lights and siren gives them the right to drive as fast as they want at any time they please. This leads to what is called "sirencide." If there is a traffic accident, they are going to have to prove that they were operating the vehicle in a safe manner and that they were responding to a true emergency. Responding to get a cat out of a tree is not considered an emergency.

INFECTIOUS DISEASE

An issue that is becoming more prominent in the workplace is HIV/AIDS and other infectious diseases. This issue involves the people with whom firefighters deal on emergency incidents and firefighters themselves. The Federal Rehabilitation Act of 1973 prohibits discrimination because of a handicap. The act applies to federal agencies and any organization receiving federal contracts in employment or the provision of services. In addition to covering people who are actually disabled, the act prohibits discrimination against those who have a history of a handicap and those perceived as having a handicap (*School Board of Nassau County v. Arline*).

A series of cases have held that persons with HIV are protected by the act. In 1987 the United States Supreme Court held that persons with contagious diseases were handicapped within the meaning of federal physical handicap statutes. In 1988, the Ninth Circuit Court of Appeals (*Chalk v. U.S. District Court*), found AIDS to be a contagious disease within the meaning of *Arline* and therefore a handicap within the meaning of the Federal Rehabilitation Act.

Many states prohibit discrimination in employment because of race, religion, color, national origin, ancestry, physical handicap, some medical conditions, marital status, or sex of any person.

Many agencies have issued opinions that emergency medical care personnel have a responsibility to provide emergency medical care to a victim when responding to emergency and rescue incidents. Emergency care personnel cannot wait for mechanical breathing devices to start CPR, as this may further endanger the life of the victim; therefore many departments have issued each firefighter a small shield device to be used when performing mouth-to-mouth resuscitation on victims. These devices need to be carried at all times while you are on duty.

When you are in the grocery store, in your uniform, buying food for the shift and someone has a heart attack is no time to realize that you are in violation of department policy by not having your shield with you.

In general, if you feel you have been exposed to HIV, you can report it to the receiving hospital and it will notify the health officer. One problem lies in the fact that the person you were treating does not have to undergo a test for HIV/AIDS if she or he does not want to. Should the person choose to be tested under confidentiality laws, the hospital may not be able to advise you of the results unless the patient agrees.

In many cases it is against the law to disclose to others the information that someone is HIV positive or has AIDS. If you went on a medical aid incident and the person told you she was HIV positive, that information is to be kept confidential. Writing the information on the board at the fire station or in any way advising the other shifts is ill advised. The absolute best way to avoid these problems is to wear your PPE every time you deal with patients, with no exceptions. Assume that every patient contact is a possible exposure and protect yourself appropriately.

Being able to get test results back from a patient from whom you suffered an exposure is no consolation. If they are HIV positive or have AIDS, finding out is probably too late to do you much good.

The absolute best way to avoid these problems is to wear your PPE every time you deal with patients, with no exceptions.

Local jurisdictions have enacted AIDS-specific discrimination laws. These laws vary in their content and scope and it is important that you become aware of them.

GOOD SAMARITAN LAWS

Numerous states have so-called Good Samaritan laws, which state that a person who voluntarily assists an injured person is not chargeable with responsibility for any errors or omissions in the care provided. This only applies if you are acting within the scope of your training. If you were to perform an emergency appendectomy on someone and you are only trained to the level of an EMT, you are way outside the scope of your training and could be held liable for injury to the victim. The line becomes more finely drawn if you were to give CPR to a patient and you did not clear her airway properly before you started. As you tried to ventilate her, you forced a piece of foreign material down into her throat that prevented her from receiving any air. This would render the CPR ineffective and prove you did not perform it properly. You were acting within the scope of your training by trying to perform CPR but did not perform correctly. You could be held liable to some extent.

You could also run into problems if you ask a person stopped at the scene of the traffic accident to assist you. When you do this without ascertaining his level of medical training, you should be very careful about what you ask him to do. If he were to take some action that injured the patient, he was acting at your request. If he injures himself while assisting you, he is acting at your request as well. It pays to be careful and think before you act.

NOTE It pays to be careful and think before you act.

NOTE 29 CFR Part 1910 requires that firefighters involved in interior structural firefighting operations be provided with and use SCBA.

PERSONNEL SAFETY

29 CFR Part 1910: Operating in IDLH Atmospheres

On May 1, 1995, federal OSHA issued compliance instructions, for states recognizing federal OSHA standards, to ensure uniform enforcement standards for all workers operating in atmospheres that are immediately dangerous to life and health (IDLH). This instruction further addressed the regulations requiring respiratory protection first issued in 1971. 29 CFR Part 1910 requires that firefighters involved in interior structural firefighting operations be provided with and use SCBA. The regulation goes on to specify that all firefighters involved in interior firefighting and utilizing SCBA shall operate in a buddy system with two or more personnel. The firefighters operating in the buddy system shall be in direct voice or visual contact or tethered with a signal line. Radios and other means of electronic contact shall not be substituted for direct visual contact for employees within the individual team in the danger area. Identically equipped and trained firefighters are required to be present outside the hazard area prior to a team entering and during the team's work in the hazard area in order to account for and be available to assist or rescue members of the team working in the hazard area.

All firefighters involved in interior firefighting and utilizing SCBA shall operate in a buddy system with two or more personnel.

A minimum of four individuals is required: two individuals working as a team in the hazard area and two individuals present outside the hazard area for assistance or rescue at emergency operations where entry into the danger area is required. One of these outside firefighters may be assigned to other duties such as incident command; the other can have no responsibilities other than to account for the team inside.

What this means to you as a firefighter is that you are not advised to make an interior attack on a structure fire with only two or three trained firefighters at scene. Every reasonable attempt must be made to provide a minimum of four personnel at the scene. The two who remain outside must be identically equipped, this means with full PPE and SCBA. One of those remaining outside must stay in direct contact with the entry team. The other may have other duties but must stay available to assist if necessary. The total number of personnel required at scene increases to five when one is assigned to operate the pumper, one is assigned to incident command, one is the contact person and two are inside. This is not a problem for personnel working for larger departments with four-person engine companies and a command officer in close proximity. In rural areas or on volunteer fire companies, it is entirely possible to have two or three personnel at scene and your assistance is 15 minutes or more away.

The regulation does allow entry with less than four personnel at scene if a rescue is needed. Another situation would be that the fire is small and you feel that you could extinguish it quickly with an interior attack. In these situations, you are faced with a dilemma. Do you obey the regulation as written and make an indirect attack or do you go in and attack the seat of the fire? The only way that you can literally stay within the regulation and still do the job that needs to be done is to change the atmosphere prior to entry. Your first operation may be to open the window to the fire area, make an indirect attack, and knock down the body of the fire. Then ventilate the structure to remove the "atmosphere immediately hazardous to life and health." This could be done by placing a power fan in the doorway and proceeding as soon as the smoke clears.

In many instances laws are enacted that do not consider every possible scenario. This regulation was not issued with the intent to limit firefighting operations, but it still directly affects firefighters.

23 CFR Rule 634: Firefighter High-Visibility Safety Apparel

In an effort to protect the safety of firefighters, emergency medical responders, and other workers who work in the right-of-way of federally funded highways, the Federal government has initiated 23 Code of Federal Regulations Rule 634. The rule states that:

> "All workers within the right-of-way of a Federal-aid highway who are exposed either to traffic (vehicles using the highway for purposes of travel) or to construction equipment within the work area shall wear high-visibility safety apparel. Firefighters or other emergency responders working within the right-of-way of a Federal-aid highway and engaged in emergency operations that directly expose them to flame, fire, heat, and/or hazardous materials may wear retro-reflective turn-out gear that is specified and regulated by other organizations, such as the National Fire Protection Association. Firefighters or other emergency responders working within the right-of-way of a Federal-aid highway and engaged in any other types of operations shall wear high-visibility safety apparel."

Definitions (634.2) within Part 634 cover what is meant by "workers" and "high-visibility safety apparel" (**Figure 11-3**). Any exceptions for emergency responders are incorporated in the definition of "workers":

"Workers means people on foot whose duties place them within the right-of-way of a Federal-aid highway, such as highway construction and maintenance forces; survey crews; utility crews; responders to incidents within the highway right-of-way; firefighters and other emergency responders when they are not directly exposed to flame, fire, heat, and/or hazardous materials." Structural firefighting PPE is not considered to be "highly visible safety apparel" as it does not meet the standard as specified in ANSI/ISEA 207, *Standard for High-Visibility Public Safety Vests.*

Safety vests should not be worn over firefighter PPE when there is a danger of flame, fire, heat, and/or hazardous materials contact.

The 2009 National Fire Protection Association's *NFPA 1901, Standard for Automotive Fire Apparatus* applies to all fire apparatus "contracted for on or after January 1, 2009." The standard requires "one traffic safety vest for each seating position, each vest to comply with ANSI/ISEA 207, *Standard for High-Visibility Public Safety Vests,* and have a five-point breakaway feature that includes two at the shoulders, two at the sides and one at the front."

Vests must be equipped with a five-point break-away feature to prevent firefighters becoming entangled in passing vehicles or other hazards.

Vehicle Markings

NFPA 1901 requires the marking of the rear of apparatus with **retro-reflective** lettering and striping. (Refer to **Figure 6-21**, the engine on the left has the required retro-reflective striping.). The striping is to be displayed in a chevron pattern with 6-inch-wide stripes, sloping downward from the center point and covering at least 50% of the rear of the vehicle. The colors allowed for the striping must be contrasting and are allowed to be red, fluorescent lime, fluorescent yellow, or yellow in color. Additional retro-reflective material may be used on the tail flap of the hosebed cover to enhance the visibility of the vehicle.

retro-reflective A surface, material, or device (retroreflector) that reflects light or other radiation back to its source; reflective.

FIGURE 11-3 Firefighter wearing reflective vest over firefighting PPE to increase visibility when working roadside incidents.

Delmar/Cengage Learning

SCENE MANAGEMENT

In many jurisdictions the agency in charge of the emergency scene is determined by law. Some states recognize the public agency with primary investigative authority as the scene manager. In the absence of a representative of the agency with primary investigative authority, the highest ranking member of the public safety agency at scene will be the scene manager. Once the agency with primary investigative authority arrives, the individual from that agency assumes scene management authority.

NOTE Some states recognize the public agency with primary investigative authority as the scene manager.

In the case of traffic accidents in a city, the agency with primary investigative authority is the local police department. On state and federal highways in rural areas this may be the state highway patrol. When the fire department arrives before law enforcement, they assume scene management responsibility. Responsibility for scene management would then be transferred to the agency with primary investigative authority when a member of that agency arrives on scene. As a matter of practicality, when the incident is primarily based on functions provided by the fire department, such as a vehicle fire, haz mat, or vehicle rescue, the law enforcement agency may relinquish command of the incident to the fire department. Another option is for the fire and law enforcement persons in charge (incident commanders) to participate in Unified Command of the incident as outlined in Chapter 13.

NOTE It is important to know and establish who has scene management authority as that agency is the lead agency ultimately responsible for the incident.

It is important to know and establish who has scene management authority as that agency is the lead agency and ultimately responsible for the incident.

In the case of fires, when the fire department has an arson unit, it is clearly the agency with primary investigative authority.

HEALTH INSURANCE PORTABILITY AND ACCOUNTABILITY ACT (HIPAA)

HIPAA affects firefighters in their jobs due to their response to rescues and medical aid incidents. HIPAA basically states that health information regarding a patient can only be given to someone directly involved in the treatment of the patient. In a practical sense, if you are at the scene of a medical emergency, you are only allowed to provide information regarding the patient's status to the ambulance personnel or other medical care provider directly involved. This precludes you from providing medical information to the news media, law enforcement, or other personnel at scene not directly involved with the treatment of the patient. It also covers any patient care report or station logbook information that you may gather; these records must be kept confidential.[10]

SUMMARY

The purpose of this chapter was to introduce you to the law and how it works. A law is interpreted in different levels of the court system. These court decisions have a direct affect on the intent if not the letter of the law. Every jurisdiction has laws that differ slightly from others. The laws that apply to everyone are the laws made by the federal government. The states can make their own laws as long as they are not in conflict with the federal laws or the U.S. Constitution. Local level laws are adopted through the use of ordinances. Ordinances can be used to adopt model codes or the creation of jurisdiction-specific law.

As a firefighter, it is important for you to have an idea of the laws that apply to the performance of your duties. You must know when you are legally required to act and to what extent you legally can act. If you exceed your authority, you may have acted illegally and assume liability for your actions. An area that requires specific knowledge of the law is fire prevention. It would be quite easy for you to exceed your authority and require a business owner to do something that he is not legally required to do. Great care must be exercised in this area. No one wants to become the victim of a lawsuit.

Numerous laws specify what is required of you in certain situations, one of these being the vehicle code. You may be exempt from some traffic laws but not exempt from displaying common sense. It is one thing to be right. It is quite another to be dead right.

As a firefighter, you have a moral obligation to perform your duties to the best of your ability within the scope of your training. As long as you keep this in mind and act accordingly, you should be exempt from liability in almost every situation.

REVIEW QUESTIONS

1. The supreme law of the United States with which no other laws must conflict is the:
2. May a state law be different than a federal law?
3. If a postal vehicle, fire engine, and private auto arrived at an intersection at the same time, who would have the right of way? Use the relationship of the levels of law as your guide.
4. A failure to act is considered which type of feasance?

5. What court case determines the responsibility of a fire prevention bureau to enter premises?
6. How are model codes developed?
7. Who may suggest changes to model codes?
8. How are model codes adopted at the local level?
9. What type of occupancy is a school?
10. What is meant by the term "one-hour fire-rated separation"?
11. Why is the location of a building in relation to the property line important from the standpoint of stopping fire spread?
12. What is meant by a Group A occupancy?
13. What are your responsibilities when responding to an emergency with red lights and siren activated?
14. Is it legal to refuse treatment to a person with HIV who is bleeding when you are not equipped with full medical PPE? Justify your answer.
15. If you act within the scope of your training at an accident scene while you are off duty and the patient dies, can you be held liable? Why, or why not?
16. Under federal OSHA standards, how many persons must be at scene before interior firefighting can be considered? What factors affect this decision?
17. What are the PPE requirements for operating on Federal-aid highways?
18. What are the exceptions to wearing "high-visibility safety apparel" at highway incident scenes?
19. According to the 2009 edition of the NFPA *Standard 1901, Motor Fire Apparatus*, what percentage of the rear of a vehicle must be covered with retro-reflective material?
20. At the scene of a motor vehicle accident, who usually has scene management authority and responsibility?

DISCUSSION QUESTIONS

1. Why are codes often not enacted until after a major incident?
2. Should firefighters be exempt from liability?
3. Do you think that the wearing of high-visibility vests and the installation of retro-reflective striping on the rear of emergency vehicles will greatly enhance responder safety on the highways?

NOTES

1. *Uniform Fire Code* (Whittier, CA: International Conference of Building Officials, 2000).
2. *Basic Fire Prevention Code* (Country Club Hills, IL: Building Officials and Code Administrators, 2000).

3. *Standard Fire Prevention Code* (Birmingham, AL: Southern Building Code Congress).
4. National Fire Protection Association, *Standard 1, Fire Code* (Quincy, MA: National Fire Protection Association, 2009).
5. National Fire Protection Association, *Standard 220, Standard on Types of Building Construction* (Quincy, MA: National Fire Protection Association, 2009).
6. National Fire Protection Association, *Standard 101, Life Safety Code* (Quincy, MA: National Fire Protection Association, 2009).
7. National Fire Protection Association, *Fire Protection Handbook* (Quincy, MA: National Fire Protection Association, 2003).
8. National Fire Protection Association, *Standard 1500, Fire Department Occupational Safety and Health Program* (Quincy, MA: National Fire Protection Association, 2002).
9. United States Congress, *Commercial Motor Vehicle Safety Act of 1986* (Washington, DC: U.S. Government Printing Office, 1986).
10. United States Department of Health and Human Services, *Health Insurance Portability and Accountability Act of 1996* (Washington, DC: 2003).

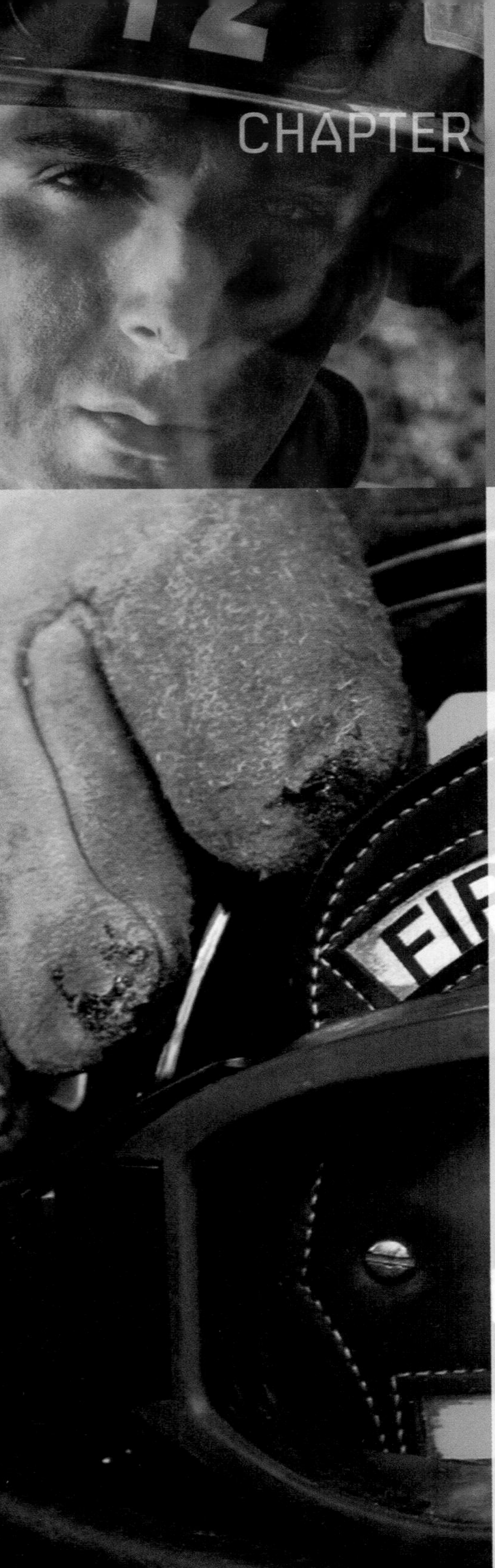

CHAPTER 12

Fire Protection Systems and Equipment

LEARNING OBJECTIVES

Upon completion of this chapter, you should be able to:

- Describe the components of a water supply system.
- Explain the importance of a dependable water supply system.
- Describe the components and importance of a fire department water supply program.
- Describe fire detection systems and their components.
- Describe different types of extinguishing systems and their components.
- Describe the different types of extinguishing agents.
- Explain how the various types of extinguishing agents work.

INTRODUCTION

Water is the most common extinguishing agent used for combating fires. Over the years water systems have been developed to the point where they have become dependable making water readily available at many fire scenes. Automatic fire fighting devices have been developed to aid in the application of water and other fire fighting agents. Additives have been devised for water to make it as effective as possible. Water is not the only extinguishing agent available to or used by modern firefighters.

NOTE Water is the most common extinguishing agent used for combating fires.

In occupancies or applications where water may cause damage or be ineffective, other extinguishing agents have been developed. These agents come in many different forms and firefighters must be acquainted with their uses and applications.

PUBLIC WATER COMPANIES

Water is so fundamental to firefighting that a good water supply is one of the most important single factors in municipal fire protection.[1] Close cooperation and communication are necessary between the fire department and the water company to ensure that an adequate water supply is available for firefighting operations. It must keep in mind that the primary reason for a water company to exist is to provide for the everyday needs of its customers.

Water companies are set up in several ways. One way is that the water company is set up under public utility laws, allowing it to act as a monopoly. It would be extremely rare for two competing water companies to provide service to an area under parallel systems. The water company usually has an elected board of directors with a president, board members, and secretary. These people are from the area the water company serves, much like a school board. The water company is allowed to charge for the water provided at rates set by the Public Utilities Commission. The money collected is used to administrate, maintain, and improve the system. An alternative is a water system owned and operated by the local government.

When new structures, such as large buildings or subdivisions, are in the planning stage, the fire department is often involved in specifying the water system requirements from a fire protection standpoint. This is done in cooperation with the builder, the building department, and the water company. By addressing concerns for fire water supply during the planning phase, serious problems with lack of **fire flow** can be avoided later.

fire flow Amount of water supply required for fire extinguishment expressed in gallons per minute (gpm).

In some smaller systems, a large fire can severely tax the capabilities of the whole water system. By having a good working relationship with the water company personnel and knowledge of the water system, it may be possible to have the pressure in the system boosted to provide for more fire flow. Likewise it is good practice to let the water company personnel know when the fire is under control so pumps and other equipment are not run needlessly. The water company should also be notified when any testing or flushing of the water system is to take place. Testing and flushing often stir up sediment in the water system and may cause customer complaints to the water company office. Testing and flushing can also cause the pressure to drop and demand on the water system to increase, the same as flowing large amounts at a fire would.

The agreement with the water company should also include that it let the fire department know when the system is undergoing major repairs, when water company personnel are flushing the system, or when individual hydrants are out of service. By knowing and respecting each other's needs, the water company and fire department can maintain a good working relationship.

PRIVATE WATER COMPANIES

Private water companies also exist, usually in industrial and commercial complexes.

They may store their own water in reservoirs or tanks or receive it from the public water company. The private water system maintains all of its own distribution and storage equipment. The fire department should check on the system periodically to ensure that it is in operating condition.

WATER SUPPLY SYSTEMS

All water supply systems must first have a storage capability. The size of the storage capacity and adequacy of the system are determined by several factors. The frequency and duration of droughts is a major factor. In the late 1980s and early 1990s, the West Coast experienced drought conditions for seven years in a row. Water supplies were so depleted that residents in some cities were prohibited from watering their lawns and washing their cars. Many fire departments conducted their firefighting drills without charging hose lines to conserve water.

A second factor is the danger to the system from natural disaster. Earthquakes can sever supply lines from reservoirs to pumping plants and individual mains, leaving whole areas of cities without water supply for days at a time. This happened during the earthquakes in San Francisco in 1906 and again in 1989. Fortunately, the fire department was equipped with fire boats that could pump water from the bay to engines on shore. Floods can affect water systems by making pumping plants inoperative. Flood waters can ruin the electric motors on the pumps, destroy electrical distribution systems, or wash out reservoirs and water distribution systems. Tornadoes and other natural disasters that disrupt power supplies and destroy the buildings housing the water company equipment can have the same effect.

Water systems take their supplies in a variety of ways. In the gravity system, the water source is at a higher elevation than the city. The water is collected in reservoirs and gravity-fed through pipes into the system (**Figure 12-1**). This gravity-forced concept is used in applications where water towers are used to keep a constant pressure on the system.

Direct pumping systems are used where a reservoir or river is a water source. The water is pumped straight from the source into the system. Automatic pressure controls are built into the pumps to maintain a consistent pressure on the system.

Combination systems are used where areas of need are removed from the water source, perhaps because of a higher elevation or remote location. The pumps in the system provide water directly to the system as well as pumping water into storage tanks. By filling the tanks at times of low use, the pumps do not have to provide the total supply to the system at times of high use (**Figure 12-2**).

In many systems, the storage is underground in water-bearing strata called aquifers. In a wet year, with lots of rain, the aquifer will have a high water level under the ground, called the water table. In dry years, the water table will recede as water is pumped out. Not all wells will be able to access the full range of the aquifer at any one time. The shallow wells will run dry as the water table recedes. Over the years wells tend to silt up and become plugged, at the least reducing their efficiency. In a system with only one or two wells, it is easy to see that an equipment failure could cause total loss of water supply to the system.

FIGURE 12-1 Municipal water pumping facility with vertical turbine pumps.

FIGURE 12-2
Water storage tower.

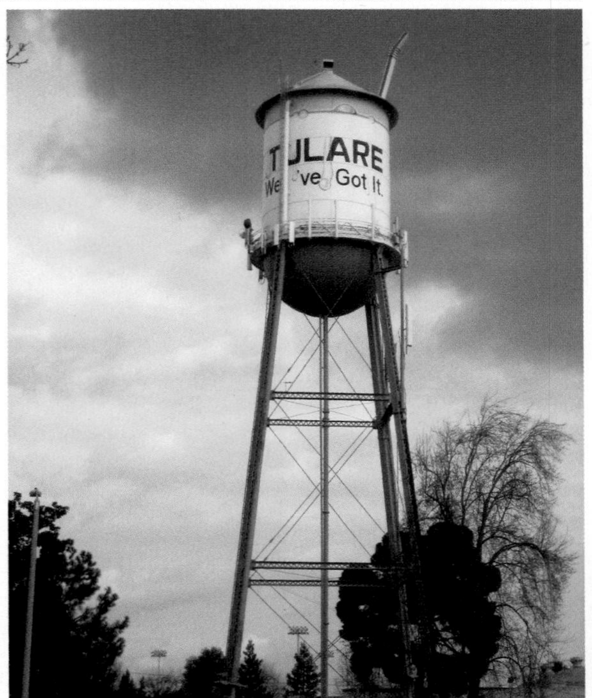

Delmar/Cengage Learning

To prevent loss of water supply, most systems of any size are equipped with duplication of wells, pumps, tanks, and sources (**Figure 12-3**). This way, if one part of the system is out of service, the other components can come online and take up the slack. Other safeguards used are gasoline- or diesel-powered pumps or generators to back up electric equipment. Electrical supply lines are laid underground to protect them from weather and vehicle accidents damaging power poles and other equipment.

In countries where terrorism and political unrest are a problem, two of the first things to be attacked are the electrical and water supply systems. In the United States, there have been very few instances of this type of activity and the systems are not protected. A power transformer can be taken offline by a single rifle shot. Surrounded by nothing but chain-link fences, these installations are vulnerable to attack, even by vandals.

The adequacy of a water system is gauged by its ability to meet several criteria. The first is the average daily consumption, figured over the last 12 months. The second is the maximum daily consumption, which is the highest demand in a 24-hour period over the last three years. If there has not been a major fire in the last three years, the system may be adequate for domestic, commercial, and industrial use, but inadequate in the case of a major fire. The peak hourly consumption is the maximum amount of water used in any given hour of a day. The maximum daily consumption is normally about 1.5 times the average daily

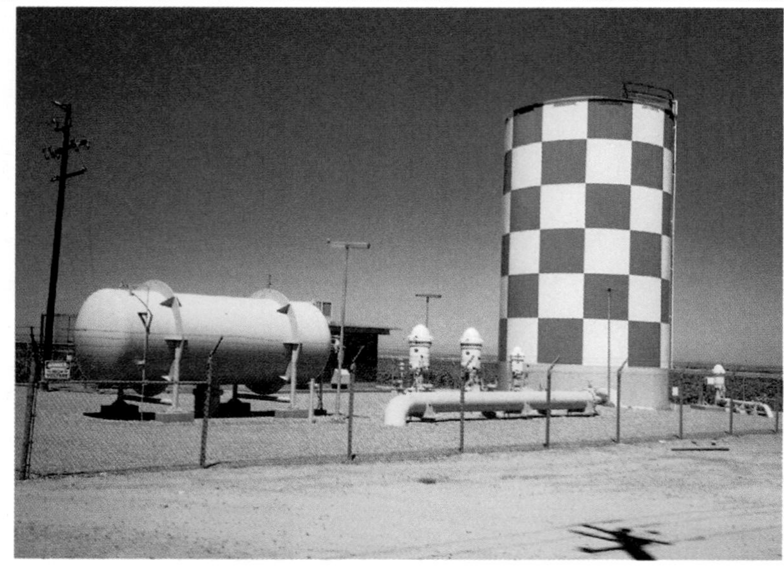

FIGURE 12-3
Well pumps with water storage tank.

Delmar/Cengage Learning

consumption. The peak hourly rate varies from two to four times a normal hourly rate. Both maximum daily consumption and peak hourly consumption should be considered to ensure that water supplies and pressure do not reach dangerously low levels during these periods and that adequate water will be available if there is a fire.[2]

For firefighting purposes, the minimum recognized water system is 250 gallons per minute for two hours.[3] This is not much in firefighting terms. A structure fire of any great size would require a much higher flow rate than this and quite possibly for a longer time. It is not uncommon to have the fire department applying 4,000 gallons per minute of water on a large fire in a warehouse or manufacturing facility.

Distribution System

Once the water is removed from the storage area, it goes through the treatment plant to make it fit for drinking. When it leaves the treatment facility, the water enters the distribution system. The distribution system is made up of underground piping of various sizes. These pipes are called water mains. The largest of these are the primary feeders. The primary feeders are widely spaced and carry the water to the various areas to be distributed by smaller mains. Like the other parts of the system, duplication is important. The primary feeders are often looped or cross connected so that water enters the system from at least two directions, preventing dead ends and pressure drops throughout the grid. Gridding becomes extremely important during times of high demand, such as large fires. If two high-flow pumpers are operating from the same main and water only came in from

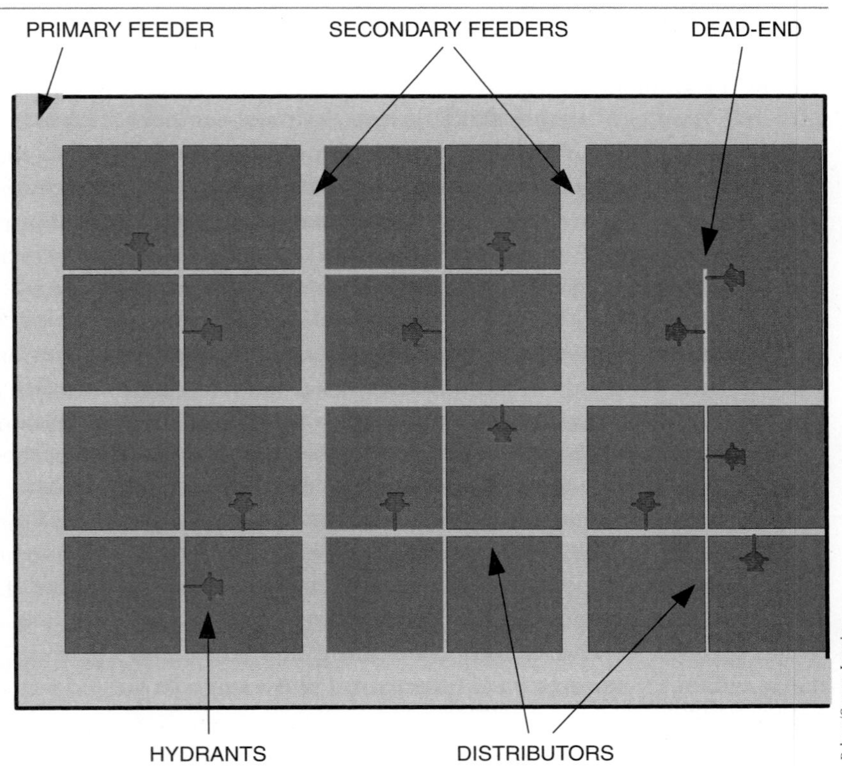

FIGURE 12-4 Gridded water main system.

one way, the pumper closer to the source would rob most of the water from the main. With gridding, the water flows in from two directions and both pumpers can be supplied to the limits of the system (**Figure 12-4**).

The water from the primary feeders then flows into intermediate size pipes called secondary feeders. The secondary feeders reinforce the grid within the loops of the primary feeders and concentrate the water supply in high-demand areas.

The piping that serves individual hydrants and blocks of consumers are the distributors. These mains may also be installed in a grid system, which allows them to reinforce each other in times of high demand.

The main sizes most commonly in use in newer systems are 8, 12, and 16 inch. It is recommended that main sizes be a minimum of 6 inches to allow for required fire flow.[4] Should the water mains be smaller in size, the fire department should preplan accordingly to meet the required fire flows of occupancies in the area. An increased main size reduces loss of pressure due to the friction of the water against the inside of the pipe. Increased size also allows for higher flow rates at the same pressure. A simple formula used to illustrate this point is that the diameter squared of the larger pipe divided by the diameter squared of the smaller pipe equals the number of times the water flow is increased. A 12-inch pipe is expressed as 12 × 12 = 144. A 6-inch pipe is expressed as 6 × 6 = 36. The flow is determined by 144 divided by 36 = 4, illustrating that doubling the pipe size quadruples the flow at the same pressure.

If it does not exceed 600 feet in length and is connected in a grid pattern, 6-inch pipe may be used. For shopping centers and industrial areas, 8- and 12-inch mains are recommended. In heavily built-up areas of homes or other flammable construction, the main size used in industrial areas is recommended. Valves are recommended to be placed a maximum of 800 feet apart. In a properly gridded system, this distance would allow a repair to be made to the main with only 800 feet being deactivated.[5]

Several types of fire hydrants are in use today (**Figure 12-5**). The two basic types are the wet barrel (**Figure 12-5A**) and dry barrel (**Figure 12-5B**). Common hydrant construction consists of the hydrant body (barrel) above the ground with pentagon nuts (five-sided) used to remove the caps and operate the stem. The pentagon nut is designed to be turned using a specially designed hydrant wrench. The purpose of the special nut is to foil vandals who would turn on hydrants. The nut can be operated with a large pipe wrench if the need arises. The openings, usually 2½, 4, or 4½ inches in size, are situated in a horizontal position. The thread on these outlets is commonly national standard thread. The threads are protected with caps. Most of these caps, when coming from the factory, are

FIGURE 12-5A
Wet barrel hydrant common in warm climates.

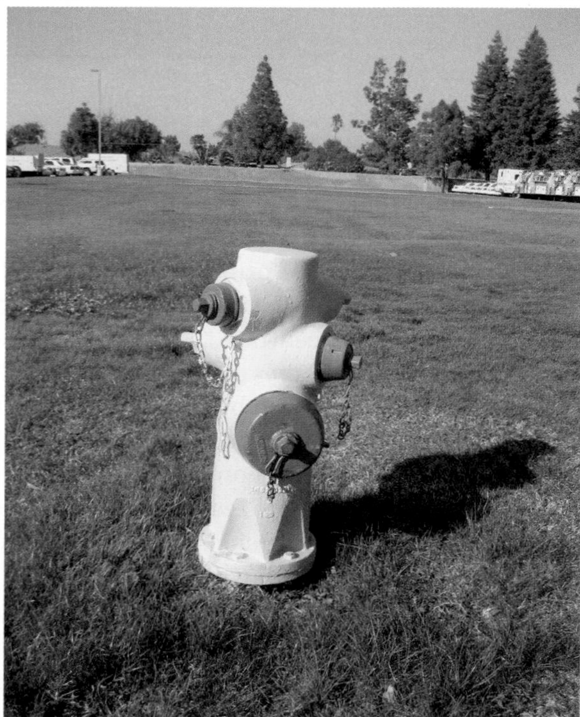

Delmar/Cengage Learning

FIGURE 12-5B
Dry barrel hydrant common in climates where sub-freezing temperatures are experienced.

Delmar/Cengage Learning

brass, making them attractive to persons wanting to sell them for their scrap value. When replacing stolen caps, plastic caps are often used. The only problem with these is that if they are installed too tightly, the nut can twist off when you try to remove them. Some departments equip their hydrants with quick connect fittings. It is important for firefighters to be acquainted with the fittings on any hydrants in their own and neighboring jurisdictions and to make sure their pumpers are equipped with the necessary adapters.

The wet barrel hydrant has water in it at all times. (**Figure 12-5A**) There is sometimes a spring-loaded valve underground that operates and shuts off the hydrant if the aboveground portion is sheared off by accident. When the caps are removed and the valve is opened, the water flows out the opening. With this type of construction, the hydrant can have several openings valved separately, which allows the pumper to be attached to the hydrant with one line. Other lines can be added later without shutting down the hydrant. Not all wet barrel hydrants are set up this way. Some types have only one valve. Whichever cap is removed is the opening the water flows from. This can be a problem if the pumper is initially attached to a smaller opening and, as the fire increases in size, it is determined that the larger opening or attaching another line to the hydrant is necessary. One way to deal with this is to always keep the water tank on the pumper full. This will give the operator a period of time, relying on tank water to supply the fire stream, to shut down the hydrant and make the necessary hookups. Depending on the tank size on the pumper and the fire flow being used, the time to perform the hookup may be very short. Pumper operators should practice this operation. This operation can be dangerous for the firefighters on the nozzles if the lines being used on the fire are supporting an interior attack. Losing the water supply to your hose line while operating on an interior attack is dangerous, to say the least.

The dry barrel hydrant is designed for use in cold climates where freezing is a problem. (**Figure 12-5B**) The water is held back by a valve underground, which is operated by the stem protruding out the top of the hydrant. The hydrant also has a drain several feet below ground level, surrounded by gravel, that allows the water to drain out of the aboveground part of the hydrant once it is turned off. This drain is set up so that water is not forced out of it when the hydrant is being operated. In a dry barrel system it may take a while for the water to rise in the pipe and water to flow out of the hydrant. This is normal but can be disconcerting when the water is needed in a hurry.

Another type of hydrant, not to be confused with the dry barrel type, is the dry hydrant. The dry hydrant is installed at a static water source for ease of setting up drafting operations. The hydrant head is positioned alongside a road at a lake or pond. The pumper hooks up to the dry hydrant and drafts water from the source. The pipe extends out into the water source, under the surface, and has a screen on the end to keep out debris (**Figure 12-6**). This greatly speeds up the operation because the operator does not need to extend heavy hard suction hoses out into the water source (see **Figure 6-26**). In areas

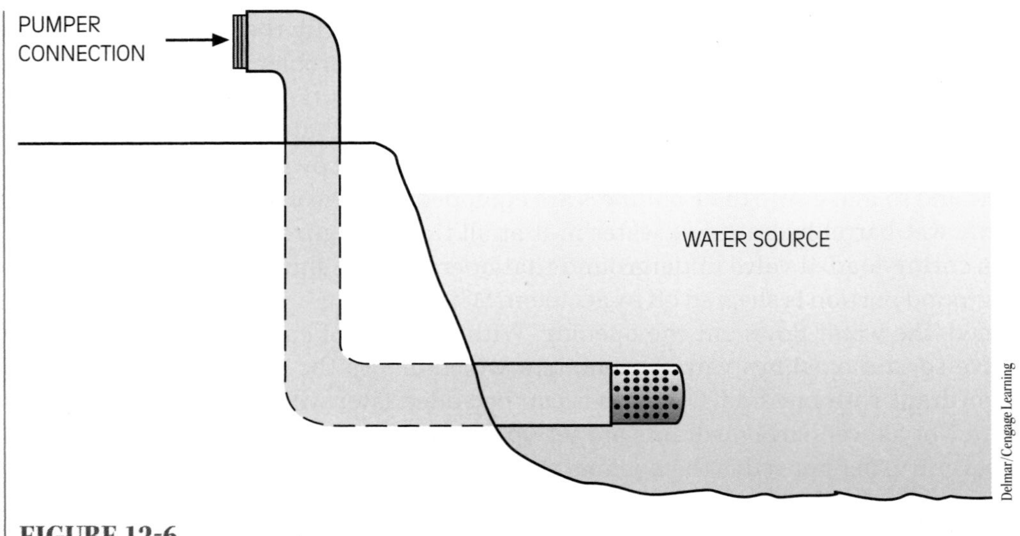

FIGURE 12-6
Dry hydrant suction source for drafting.

where ice forms on pond surfaces, it is not necessary to chop a hole in the ice to get at the underlying water. A modified version of the dry hydrant is sometimes attached to ground-level water tanks. By drafting from the tank, the operator can increase the flow into the pumper over that provided by gravity.

Hydrants are installed where required and operate off the regular water mains. The piping that extends off the water main to the hydrant is called the *bury* (**Figure 12-7**). This pipe is of a specified size determined by the type of hydrant it supplies. A concrete **thrust block** is often installed where the elbow is installed in the bury. A valve installed between the hydrant and the main allows the hydrant to be turned off for removal or maintenance. This also comes in very handy when the hydrant is knocked off in an accident and needs to be shut down. Hydrants should be installed with the openings far enough above the ground to allow easy hookup, leaving enough room to swing the handle of the hydrant wrench

thrust block A mass of concrete poured on the outside of an angle fitting and extending back to native soil. The purpose is to prevent surges in flow through a pipe from flexing the fitting and wiggling it in the ground, which would, over time, form a larger and larger underground space, possibly allowing the pipe fitting to pull apart.

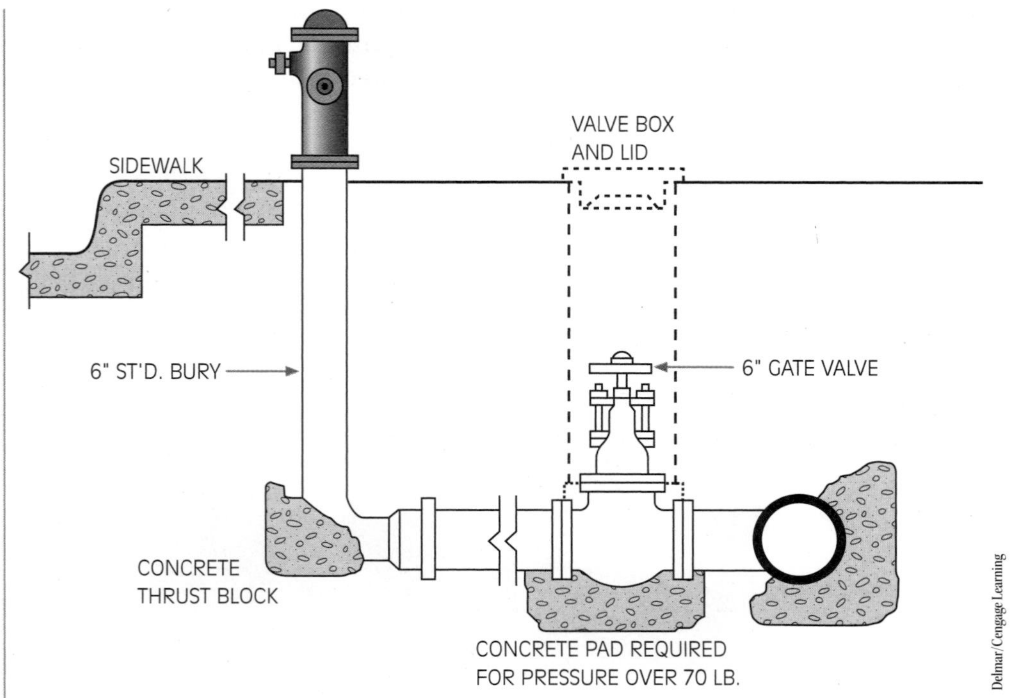

FIGURE 12-7
Schematic of hydrant with underground plumbing.

when removing the caps or opening valves. The openings should be pointed toward or parallel to the road for ease of hookup. Fences and walls should not be allowed to encroach on the area around the hydrant where they would interfere with its operation. There should be enough room between the hydrant and the street that it is not in danger of being hit by vehicles. A common way of protecting hydrants is to erect pipe barriers around them.

On airport or other special property, hydrants are installed as the situation dictates. At some airports, the hydrants are under the ground. This allows them to be available on the runway aprons without being a hazard to aircraft. Hydrants are also commonly installed on piers extending out into the ocean to provide firefighting water supply.

Hydrant spacing is specified by local ordinance. The purpose is to concentrate availability of fire flow at the blocks or groups of buildings to be protected. Some standard rules of thumb for spacing are 250 feet in compact mercantile and manufacturing districts and 500 feet in residential districts. The Insurance Services Office requires maximum spacing of 330 feet in commercial and industrial districts and 660 feet in residential areas.

NOTE Part of a complete prefire program is an annual or semiannual hydrant inspection conducted by fire department personnel in their first-in district.

Part of a complete prefire program is an annual or semiannual hydrant inspection conducted by fire department personnel in their first-in district. This program acquaints the firefighters with hydrant location and provides required maintenance (**Figure 12-8**). The maintenance should include:

- Removing all the caps
- Inspecting the condition of the outlet threads

FIGURE 12-8 Firefighters servicing hydrant.

Delmar/Cengage Learning

- Checking the hydrant barrel for foreign objects
- Replacing missing caps as necessary
- Replacing worn or missing cap gaskets
- Operating the stem of the hydrant to check for proper operation
- Lubricating and cleaning any parts as necessary
- Clearing weeds and obstructions from the area of the hydrant to provide for ease of location and access under poor visibility conditions such as fog or darkness

Many departments paint the hydrant barrels a highly visible color, such as yellow. Another way of identifying hydrant location is to affix reflective roadway markers in the street; blue is often used to distinguish them from lane markers. In areas where snow is a problem, poles are erected by the hydrant to mark their location. Any damage or repairs needed should be brought to the attention of the responsible party to place the hydrant back into serviceable condition.

The situation may arise where it is necessary to flush certain hydrants in a water system. This operation is usually performed by the water company. If the fire department wishes to flush several hydrants, the water company should be notified. When flushing the hydrants, it is important to open and close them slowly to prevent damage to the water main. If at any time a valve through which water is flowing is closed quickly, water hammer can occur. This condition is caused by a large volume of water flowing from an opening being shut down too rapidly. The large volume of water moving through the opening has great momentum. If it is suddenly stopped, the momentum of the water will cause it to send a shock wave back through the system, with resultant damage. The same rule applies when operating pumpers at hydrants. All nozzles and discharge valves should be opened and closed slowly to avoid water hammer. This can be demonstrated with a common garden hose and nozzle. Open the nozzle and let the water flow. Let go of the nozzle trigger, stopping the flow, and the hose will jump. On such a small scale no real damage is done. If the flow were 1,000 gallons a minute and you did the same thing, the forces at work would be greatly multiplied. Another problem to watch out for is causing destruction to the roadway or causing a traffic accident. A hydrant flowing 1,000 gallons a minute at 40 psi moves approximately four tons of water a minute with tremendous force. This force can easily undermine a roadway or cave in the door of a passing car, possibly even causing the driver to lose control.

If at any time a valve through which water is flowing is closed quickly, water hammer can occur.

FIGURE 12-9 Hydrant flow testing with pitot gauge.

Delmar/Cengage Learning

Hydrant testing is done on new systems to test the flow rates (**Figure 12-9**). It is also done periodically on older systems to see if the system is still performing well. The same precautions should be taken as when flushing hydrants. The test may consist of opening one hydrant to test its individual flow or opening several at once to test the ability of the water system to perform under high-demand conditions. Standard hydrant testing is performed using two hydrants, a pressure hydrant and a test hydrant. The test process consists of seven steps and can be found in the booklet, *Simplified Water Supply Testing*.[6]

For more information on servicing and testing fire hydrants, please refer to DVD clip *Servicing & Testing Fire Hydrants.*

Hydrant Painting

Hydrants are painted for visibility and because their barrels are often made from cast iron. Keeping them covered with a good coat of paint prevents corrosion of the hydrant. Hydrants are often color-coded to identify their flow capabilities. The NFPA has developed a commonly used color coding system. When the NFPA system is used, hydrants capable of flowing 1,000 gpm or more have their caps and **bonnet** painted green. Hydrants flowing 500 to 999 gpm have the caps painted orange. Hydrants with a flow capacity of 499 or less have their caps painted red.[7] An additional marking

bonnet The top of a hydrant.

FIGURE 12-10
Non-functional hydrant at roadside rest stop. The caps are painted white to indicate the hydrant is out of service.

Delmar/Cengage Learning

system sometimes used is to color-code hydrants installed on **dead-end mains** with at least one cap painted black. Hydrants that are out of service may have the bonnet painted white (**Figure 12-10**). A word of caution here is that not all fire departments follow this color coding scheme so local knowledge of the color coding system used is a must. Hydrants that are part of a private system located on a public street may be painted all one color to identify them. Private hydrants on private property are painted the color the owner wishes. The use of color coding allows the pumper operator to make a quick decision as to which hydrant to use if several are present and gives the operator some idea of the capabilities of the water supply at hand. When the color codes are included on the hydrant maps and prefire plans carried in the pumper, it is easier to pick the location of the high-flow hydrants before arriving at scene. In 1976, for the Bicentennial celebration of the U.S. Constitution, many fire hydrants were painted in patriotic motifs. Although this was attractive artistically, it was not such a good idea from the firefighter's viewpoint as the color coding was painted over in many cases.

dead-end main A water main that is not gridded into the system. Water flows into it from only one way.

WATER SYSTEMS PROGRAM

Fundamental to any fire department's ability to extinguish fires is its ability to fully utilize the available resources, one of which is water. A thorough and complete knowledge of the water systems available is required. A water system program is implemented to promote cooperation between the fire department and the water companies. Records of each hydrant should be maintained as well as complete and detailed maps of the water systems (**Figure 12-11**).

NOTE Fundamental to any fire department's ability to extinguish fires is its ability to fully utilize the available resources.

The first step in implementing a water system program is to meet with water company officials to establish a working relationship and to execute an agreement on testing and maintaining of the system. A "Letter of Working Agreement" is written and signed by the responsible officials of the water company and the fire department. The fire department's role is traditionally to service and maintain the hydrants as far as minor repairs are concerned. If required, the need for major repairs is reported to the water company.

Water company records are kept that contain a description of the water system and its capabilities, including the type of water system, its components, storage

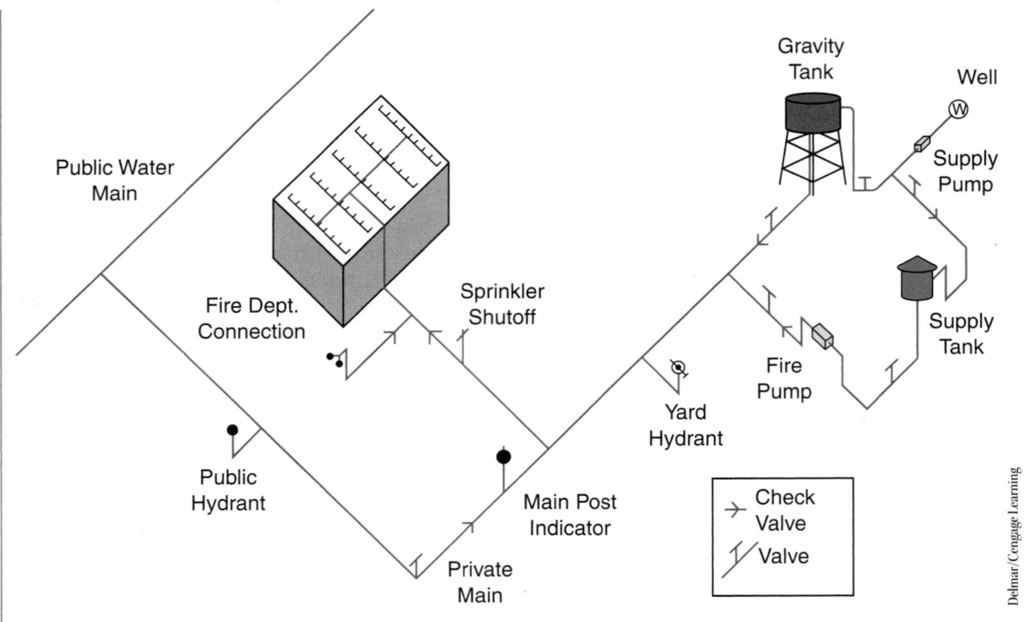

FIGURE 12-11
Water system schematic illustrating supply to fire sprinkler system in a protected occupancy.

capacity, normal pumping capacity, and emergency pumping capacity. The water company record may also contain information on the procedures needed to boost flow during a large-demand fire and emergency phone numbers for contacting water company personnel.

A grid map is maintained showing the location of hydrants and the size of mains serving them. The hydrants should be numbered on this map so individual hydrants can be referenced for record-keeping purposes. Other maps are maintained and placed in the fire apparatus for reference at the fire scene. Individual hydrants need not be numbered. Their location and color coding should be noted to aid in at-scene decision making. Valves should be noted on these maps to aid in shutting down portions of the system in case of sheared-off hydrants or broken water mains. The map should also include the emergency phone number of the water company.

Hydrant survey and service records that show the size, type, and make of the hydrant should be maintained. The flow test dates and any flushing of the hydrants should be recorded. If successive flow tests over several years show a serious decline in the hydrant's flow capability, the system may need major repair. The date the hydrant was serviced should be recorded so that no hydrants go too long without being checked.

Hydrants require very little maintenance over their life spans; however, they should be checked and serviced annually to ensure that they are in proper working order. By keeping track of their condition and when they were maintained, firefighters can identify any that are regularly in need of service due to being backed into by vehicles, or by vandalism. You may not think about them until you need them, but it sure is nice to be able to find them and have them in working order when the need arises. An additional advantage to company-level hydrant service is that it requires you to visit them and helps you remember where they are when they are needed.

In many jurisdictions where there are not built-up residential, commercial, or industrial areas, there may not be a regular water system. In other areas, the water system may not be adequate due to additional construction of structures after the water system was installed. In these situations, it is up to preplanning and the firefighters' resourcefulness to find adequate water to control fires.

Auxiliary sources of water supply can, and often do exist (**Figure 12-12**). The water may be available in holding areas in the form of reservoirs, tanks, cisterns, and swimming pools. It may be available in the flowing state in canals, rivers, or streams. There may be public or private construction equipment in the area with water-carrying capability that may be pressed into service.

By preplanning these auxiliary water supplies, they can be identified. Their locations should be shown on the maps carried in the apparatus. Often, by contacting the owner, a connection can be installed on a tank or a dry hydrant can be installed at a drafting source. Some owners may even agree to purchase a small pump to fill apparatus out of their swimming pool. In the case of helicopter bucket operations, the water source needs to be gauged for depth as well.

FIGURE 12-12
Fire water storage tank with fire department fitting, should be clearly marked. This storage tank has the marking obscured by an advertising sign for the on-site business. Fire pumper connection is circled.

Delmar/Cengage Learning

PRIVATE FIRE PROTECTION SYSTEMS

Private fire protection systems are those designed to protect individual occupancies from fire. They may be installed in private homes, businesses, manufacturing plants, or public buildings. One of their main purposes is to alert the building occupants and/or the fire department of an incipient fire, reducing loss of life and property through early detection. Other systems are designed to not only alert, but also to control or extinguish fires.

Detection Devices

Detection devices may be as basic as a smoke detector, sometimes referred to as a smoke alarm, in the home. Different types of smoke detectors are designed for use in various locations (**Figure 12-13**). The U.S. Fire Administration, in its *Facts about Fires in the U.S.*, states that a working smoke detector doubles a person's chance of surviving a fire. The report further states that approximately 90% of U.S. homes have at least one detector and 40% of the residential fires and 60% of the residential fatalities occur in homes with no detectors.[8]

A detector in common use is the ionization chamber detector. These detectors may be either wired into the building's electrical system or battery operated. The ionization chamber detector contains a small amount of radioactive material that ionizes the air entering the chamber. As ionized smoke particles enter the chamber, they disrupt the process and the alarm is triggered. The advantage of these types of detectors is they are

FIGURE 12-13
Smoke detector.

inexpensive to produce and do not need visible amounts of smoke to activate. A problem with this type of detector is that the battery is sometimes removed when it gets weak and the owner forgets to replace it. They are also subject to false alarms due to dust, steam, burning toast, and so forth.

Flame or light detectors come in two basic variations. The ultraviolet detector measures light waves and when those associated with high-intensity flames are detected, the alarm is triggered. The infrared detector measures either the flame flicker or the total infrared component of the flame.

There are detectors that rely on visible smoke to trigger the alarm. They work on the same principle as the light beams that are used in doors of stores to alert the employees that someone has entered. The light beam is aimed at the receiver and when the beam is interrupted, the alarm is triggered. Another variation has a photosensitive device that is only hit by the light when it is scattered by smoke particles. Several variations of this type of detector are available for specific applications.[9]

Rate of rise detectors are used in some occupancies. These measure the rate of temperature rise in the monitored area. If the rate exceeds specified amounts, the alarm is triggered.

Fixed temperature detectors are designed to melt at a certain temperature or measure the deflection of a bimetallic strip. Another type has a liquid in a glass bulb that breaks when the liquid expands due to heating.[10] These can be rendered ineffective due to accumulations of dirt or paint acting as insulation.

NOTE The two main purposes of detection devices are to warn building occupants of the fire start and to get the alarm to the fire department as soon as possible.

Carbon monoxide (CO) detectors are becoming more prevalent, especially in residential settings. They are designed to trigger an alarm based on an accumulation of CO over time. Sources of CO are flame-fueled devices such as ovens, furnaces, water heaters, fireplaces, and so on. According to the *Journal of the American Medical Association*, CO poisoning causes 500 nonfire-related deaths per year in the United States.[11]

The manual pull alarm station that you are probably familiar with from elementary school is another alarm-triggering device (**Figure 12-14**). By pulling down on the handle, a switch is triggered and the alarm sounds. Newer installations include a strobe light as well to alert the hearing impaired.

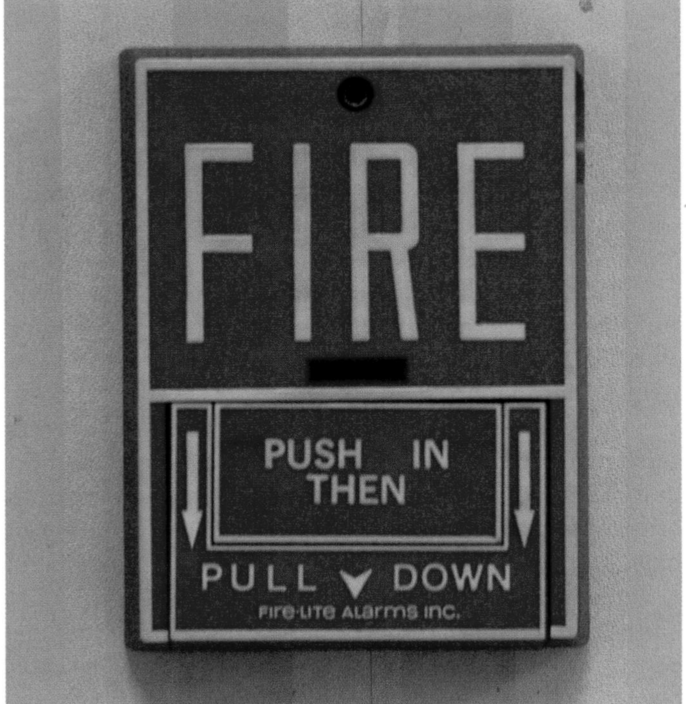

FIGURE 12-14
Manual-pull fire alarm activator.

Delmar/Cengage Learning

A water flow switch or excess flow alarm, mounted on the riser of a sprinkler system, is another means of notifying building occupants of a fire. If a sprinkler head opens, the flow of water through the system causes a switch to trigger the alarm. Most of these types of alarms also include electrical connections to an alarm company so that if no one is around to hear the alarm, it still gets transmitted to the fire department (see **Figure 12-18**).

Many facilities have a combination of these devices to sound the alarm. The system needs to be monitored at some level for the alarm to be transmitted to the fire department. There are several ways of accomplishing this, including security guards, alarm-monitoring companies, and direct tie-ins to the fire department dispatch office or fire station. Sometimes these systems are plagued with frequent false alarms, which leads to two problems. First, after a while the fire department starts to take it for granted that the alarm is false and does not respond in a timely fashion or responds without the full **first-alarm complement** of resources. Second, if the alarm is intercepted at a reception desk at the facility and not transmitted on to the fire department, the fire may have a chance to become well established before someone checks it out.

first-alarm complement The equipment normally dispatched when a fire is first reported.

EXTINGUISHING AGENTS

Water

Water is the most common fire-extinguishing agent in use today. From a mechanical standpoint, water is easily applied from many types of devices. If we were to compare water to sand as an extinguishing agent, these attributes would become immediately obvious. Water can easily be bent around corners in plumbing or flexible hoses. It can be deflected off surfaces and redirected into inaccessible spaces. By adding pressure it can be lifted to great heights and carried great distances, either inside a hose or outside as a hose stream. It will assume the shape of any container into which it is placed, making it adaptable to different shaped tanks on apparatus. It will seek the lowest level, making it run off from upper floors instead of having to be shoveled out. Being a natural product, water is readily absorbed and does not cause environmental problems when used on nonhazardous substances. In most places, water is plentiful and cheap.

Water has the ability to extinguish fires through cooling and smothering. It has one of the highest **specific heats** of any known substance, allowing it to absorb great

specific heat The ratio between the amount of heat necessary to raise the temperature of a substance compared to the amount of heat required to raise the same weight of water the same number of degrees. Water has a specific heat value of 1.

amounts of energy. One gallon will absorb approximately 1,280 BTUs in the process of raising its temperature from 62°F to 212°F. Another quality is its **latent heat of vaporization.** When this gallon of water at 212°F turns to steam, it absorbs an additional 8,080 BTUs. It requires 970 BTUs per pound to change water from a liquid to steam at 212°F.[12] The most effective absorption of heat energy is achieved when the complete volume of water applied is vaporized (converted to steam). This vaporization also produces a volume increase of 1 to over 1,700.[13] One gallon of water also produces 223 cubic feet of steam. This has the potential of displacing oxygen, especially in a confined space, such as an attic. At an efficiency rate of 90%, 50 gallons of water would produce enough steam to completely fill a one-story, 35 feet by 45 feet building. There are numerous delivery systems available to apply water to fires.

NOTE Water has one of the highest specific heats of any known substance, allowing it to absorb great amounts of energy.

latent heat of vaporization The amount of heat a material must absorb when it changes from a liquid to a vapor or gas.

Foam

Although water is the traditional extinguishing agent, the development of foam products that increase its effectiveness over that of plain water are becoming increasingly popular. Foam has three properties that make it capable of extinguishing and/or preventing fires: insulating, cooling, and forming a vapor barrier. It can cover the fuel surface, forming a layer of insulation between the heat source and the fuel supply. This is illustrated when foam is applied to grass and brush or structures prior to an advancing fire. It can cool the fuel surface below its ignition temperature. In the case of liquid fuels, it can reduce vapor production by reducing the evaporation rate through temperature reduction. Foam can also form a barrier between the vapors produced and the ignition source.

NOTE Foam has the capability of extinguishing and/or preventing ignition of fires.

Foam Components

The components of foam are water and foam concentrate. When the concentrate is proportioned into the water, at the correct rate for the application, foam solution is produced. When air is introduced into the foam solution, finished foam is produced. Some foams do not require large amounts of bubbles to work well. Others are as thick as shaving cream when applied correctly.

Types of Foam

Two basic classes of foam are available to the firefighter, all of which have particular strengths as firefighting agents. The traditional purpose of foam agents is to extinguish flammable liquid (Class B) fires by cutting off vapor production at the surface. As you will remember from Chapter 4, liquids must produce vapor to burn. If there is not enough vapor above the liquid's surface, the fire will go out. Class B Foam forms a layer above the surface of the liquid that cuts off the production of vapor. Water alone is unable to do this because it has a specific gravity greater than that of most commonly encountered flammable liquids (refer to Table 4-1). If you were to spray water on the surface of the flammable liquid, it would sink, increasing the volume of the burning spill, and cause the flammable liquid to spread, creating more of a problem. As the foam is applied to the burning liquid surface, some is sacrificed as steam as it cools the surface of the liquid.

Chemical foams were the first to be developed. They were generated using a hopper into which two chemical powders were dumped. The powders were introduced into the hose stream, the chemical reaction took place, and finished foam was produced. The problems arose when the powders would cake and not mix properly, leading to inconsistent foam quality. The gas bubbles formed by the chemical reaction would burst the hose if the nozzle were shut off prematurely. Chemical foams are not commonly used today.

Mechanical foam is another class of foams. They are formed by the introduction of foam concentrate into the hose stream. The bubbles are formed by mechanical instead of chemical action. Air is entrained in three ways to create the characteristic bubbles found in foam when it is applied: (1) The hose stream passes through special aeration devices; (2) compressed air is forced into the hose stream; or (3) as the solution leaves the nozzle, air is entrained. There are several types of mechanical foams available and in use.

Protein foam is made of natural protein solids that have been broken down chemically. Protein foam concentrate has a strong resemblance in smell and appearance to blood. Protein foam has excellent firefighting capabilities in that it is an extremely water-retentive compound with high strength and elasticity.[14] Protein foam is nontoxic despite its smell. Like most other foams, protein foam is useable in 3% and 6% concentrations in water. The stiffness of protein foams can be a detriment. The foam does not flow around and seal behind obstructions very well. This property also makes it susceptible to being blown around on the surface of the liquid by the wind.

Fluoroprotein foam is an improved version of protein foam that consists of a protein foam base with fluorinated agents added. This gives the foam produced good fuel-shedding ability, making it suitable for subsurface injection into large fuel storage tanks. Fluoroprotein foam also works well with dry chemical extinguishing agents, making it a good choice for use on three-dimensional flammable liquid fires.[15]

Alcohol-type protein foams were developed for use on the class of materials known as polar solvents.[16] **Polar solvents**, such as alcohol, are miscible in water. Hydrocarbon fuels, such as gasoline, are not miscible in water. When regular foams are used on polar solvents, the water in the foam dissolves out into the polar solvent and the foam blanket is destroyed.

The most popular type of synthetic foam is the aqueous film-forming foam (AFFF), commonly pronounced as "A triple F." This type of foam creates a sudsy blanket of trapped bubbles and then breaks down mechanically into a very thin film of water on the surface of the liquid fuel. It is self-sealing in that it has the capability of sealing itself around pipes and other structures protruding from the surface of the burning liquid. It also reseals itself when the foam blanket is disturbed if there is sufficient foam available on the surface.[17] When produced as AFFF ATC (alcohol-type concentrate), the foam can also be used on polar solvents. This type of foam typically comes in a concentrated formulation that is used at 3% on hydrocarbon fuels and 6% on polar solvents. The problem with this foam is that it is not very sudsy and it is hard to tell if effective foam is being produced. It is also difficult to tell if the foam blanket has broken down, which may endanger firefighters working in the foamed area. When used to blanket flammable liquid spills for the purpose of vapor prevention, **combustible gas indicators** (CGI) (**Figure 12-15**) should be employed. To be effective, foam must be periodically reapplied to maintain an effective vapor barrier.

polar solvents These liquids will mix readily with water as water is a polar substance. The common polar solvents are alcohols, aldehydes, esters, ketones, and organic acids.

combustible gas indicator A device that measures the percentage of lower explosive limit concentration of gas in the atmosphere. These devices must be used by trained personnel for proper interpretation of the readings.

The best plan of action is to not walk across the flammable liquid spill, even when foam is in place. If you must, be sure to drag your feet as you walk. Do not lift them for each step. This reduces the chances of breaking the foam blanket and creating a cloud of flammable vapor around you. If the foam blanket is broken and does ignite momentarily, it should reseal and the fire should go out. The natural inclination is to run, but this will only disturb the foam blanket more, increasing the intensity of the fire around you.

High expansion foams are produced by running the foam solution over specially designed netting while forcing air through the netting with a powerful fan. This creates a foam with an expansion ratio as high as 1,000 to 1.[18] The foam is introduced into a space (a basement is a good example), displacing the oxygen and insulating the unburned materials from heat. Care must be taken to use fresh air in generating the foam. Combustion products cause the foam to break down at a faster rate. It could also be possible to suck flammable vapors into the foam generator, causing a foam with flammable vapor inside the bubbles. High expansion foam will reduce visibility to nearly zero and it would be easy to become lost or trip over objects in a foam-filled area.

Delmar/Cengage Learning

FIGURE 12-15 Hazardous materials detection instruments, including combustible gas indicator.

NOTE To be effective, foam must be periodically reapplied to maintain an effective vapor barrier.

SAFETY The best plan of action is to not walk across the flammable liquid spill, even when foam is in place.

NOTE Foam designed for Class A fires is different than that designed for Class B fires. The foams are not interchangeable.

Class A Foam

Foam designed for Class A fires is different than that designed for Class B fires. The foams are not interchangeable. Class A foam is designed to be used at much lower concentrations than Class B agents. The percentages are in the 0.1% to 1% range. This is adjustable to produce foam with the desired qualities. The lower the proportion of foam concentrate, the wetter the foam will be. When used in conjunction with a compressed air foam system, Class A foam can be created that will stick to vertical surfaces and shield them from the heat of the fire. This quality allows firefighters to pretreat areas in advance of the fire. Ordinary water would just run down and end up on the ground or evaporate.[19] A type of foaming agent coming into more common use is fire blocking gel. These agents form a protective barrier which holds water suspended better than regular Class A foams, even on vertical surfaces. Gels are used in wildland firefighting to pretreat structures, trees, and other flammables before the fire approaches. "Gelling" increases firefighter safety as the firefighters are able to pretreat the structure and leave the area prior to the fire arriving. The gels may be re-wetted after they are applied to increase their effectiveness once they have dried.

Wetting Agents

Wetting agents act much the same as soap: They reduce the surface tension of the water and allow it to soak into the fuel at a faster rate. In this way, more of the water remains on the fuel and as it soaks in it tends to not evaporate as fast as surface water would. To demonstrate this, take a piece of cloth and pour some plain tap water on it. Observe how long it takes to soak in. Then take some water with dish soap in it and perform the same experiment again. The water with the soap will soak in much faster because it has less surface tension. When performing fire overhaul operations, wetting agents are especially helpful as less water is needed to accomplish the job and the deep-seated embers are extinguished more quickly.[20]

Fire Retardant

Agents applied on wildland fires are divided into two categories: short- and long-term retardants. All are water based. Water, foam, and wetting agents are classified as short-term retardants and fire suppressants. Their effectiveness is directly related to the moisture content they retain when on the fuel. They lose their effectiveness when the water evaporates and the fuel dries out after their application. Suppressants are used in direct attack and mop-up operations. They may be applied from the air or from ground-based units.

Agents that react chemically with the fuel and retain their effectiveness after they have dried out are classified as long-term retardants. Retardants are used for indirect attack. Long-term retardants are almost always applied from the air. They can be applied by air tankers or helicopters. They contain pigments so they remain visible from the air. As firefighting methods have become more politically sensitive, new types of pigments have been developed. In areas where there are rock outcroppings and structures, fugitive pigments are used. These pigments lose their color and basically disappear after a while from exposure to the sun. Regular pigments are iron oxide (rust)-based and tend to need rain to wash them off the fuel and into the soil.[21]

If you had ever thought about drinking water from a hose stream, a look at all of these possible additives should convince you not to. Many of these additives add no particular color or odor to the water. They would give you intestinal problems if you were to take them internally.

Carbon Dioxide

Carbon dioxide gas (CO_2) extinguishes fires by smothering. Carbon dioxide is an **inert** gas with the ability to dilute the oxygen in the fire area to a level where the fire will be extinguished. CO_2 systems work best in installations where air flow can be controlled, preventing premature dilution of the product. The CO_2 comes out of the extinguishing system as a mixture of vapor and dry ice particles, which is very cold, and it has a cooling effect when applied directly to the burning fuel surface. CO_2 systems are installed in areas where water is not the extinguishing agent of choice. CO_2 is selected because it can extinguish fires without leaving a residue, does not promote rust, does not create water runoff, and does not conduct electricity.[22] Some of the areas where CO_2 is preferred are telephone switching rooms, fur storage, and computer installations. In occupancies where chemicals are stored and using water would create runoff that would have to be treated as hazardous waste requiring expensive cleanup, the cost of a CO_2 system is justified (see **Figure 12-25**).

inert A substance that will not react with other substances.

Halogenated Agents

Halogenated agents extinguish the fire by breaking the chemical chain reaction, whereas CO_2 extinguishers operate by smothering. Halogenated agent systems are more effective, but there is concern about their effect on the ozone layer as they are chlorofluorocarbons (CFCs). The concentrations of halogenated agents used in fire extinguishment are not considered hazardous. However, the chemical by-products of their use can be harmful and self-contained breathing apparatus should be worn when coming into contact with these agents. An additional concern when utilizing halogenated agents is that they may not extinguish deep-seated combustion in cellulosic materials (wood, paper). Examples of these fire-extinguishing compounds are brand names such as Halon, followed by the Army Corps of Engineers numbering system based on their chemical makeup: 1211, 1301, and so forth.[23]

halogenated agents Fire-extinguishing agents containing the elements from Group 7 on the periodic table of the elements (halogens).

Halogenated agent systems are installed and operate much like the carbon dioxide systems. Their use is being phased out because of environmental concerns.

Clean Agents

So-called clean agents have been developed to replace halogenated agents. The requirements of these agents are that they have the same cleanliness (lack of residue), extinguishing capability, and low toxicity of halogenated agents but do not deplete the Earth's ozone layer.

Dry Chemical

Dry chemical extinguishing systems use a mixture of finely divided powders. These powders are treated to resist caking and are water repellant. Their effectiveness lies in their ability to break the chemical chain reaction. They also absorb some of the radiated heat from the fire and displace oxygen in a limited way. Effective on flammable liquid (Class B) and electrical fires (Class C), some formulations can also be used on ordinary combustibles (Class A). When used on ordinary combustibles, their application should always be followed up with water to extinguish deep-seated embers.[24]

NOTE When used on ordinary combustibles, the application of dry chemicals should always be followed up with water to extinguish deep-seated embers.

Dry Powder

The extinguishing agents used on combustible metals (Class D) are dry powder agents. The agent may come in a bucket, pail, or extinguisher. These agents control the fire by forming a coating on the burning surface and excluding oxygen. Some are nothing more than dry sand and others are graphite or special powders. Some contain plastic beads, which melt and help to form a coating.

Water is not commonly used on combustible metals as it may react violently, especially with magnesium and sodium, causing explosions.

Water is not commonly used on combustible metals as it may react violently, especially with magnesium and sodium, causing explosions. These explosions can spray burning material onto firefighters, leading to burn injuries. The bright light generated by the explosion can also cause eye injuries. When water is to be used, it must be applied in flooding amounts.[25]

EXTINGUISHING SYSTEMS

Sprinkler Systems

Automatic sprinkler systems have been in service for more than one hundred years. They have an excellent record of fire suppression. Sprinklers have been statistically shown to control 96% of the fires where they were activated. The remaining fires were not controlled due to improper maintenance, inadequate or shut-off water supply, incorrect installation or design, and obstructions.[26] A common misconception about fire sprinklers is that if one head is activated they all activate. This is regularly shown in television shows and movies but is incorrect. With the exception of deluge systems, only the heads that are affected by heat activate. The activation of a minimum number of heads to contain the fire reduces water damage.

NOTE Automatic sprinklers have been statistically shown to control 96% of the fires where they were activated.

One of the main causes of fires getting out of control is the lack of immediate detection. The best time to attack a fire is in its ignition stage. When there is a delay between the fire's ignition and discovery, the fire has a chance to gain headway and enter the growth stage. The primary asset of a sprinkler system is its ability to act on a fire without human intervention. Modern sprinkler systems are very reliable, are always on duty, and not busy elsewhere when a fire starts. They activate whether anyone has discovered the fire or not. There have been instances where the sprinkler system activated and extinguished a fire that was not even discovered until after it was out.

Residential Sprinklers

The efficiency of these systems has led to their adaptation for use in residences. Installed in a very simplified form, the system does not have the extensive valving required on systems installed in commercial or industrial occupancies. They also have a much-reduced pipe size in their plumbing and therefore reduced water supply. There is a shutoff valve installed on the supply side of the system for ease of turning off the flow in case of operation. There should be a water flow alarm installed so if the system activates, the fire department can be notified. The heads are designed to sit in a recess in the ceiling and drop down below ceiling level when activated.[27]

Commercial and Industrial Sprinkler Systems

A sprinkler system consists of several basic components. Sprinklers are distributed throughout the occupancy (**Figure 12-16**). There is a pipe bringing water in from the water system. The water then passes through an **open screw and yoke valve** (OS&Y) or **post indicator valve** (PI). The purpose of these valves is to make it obvious under quick

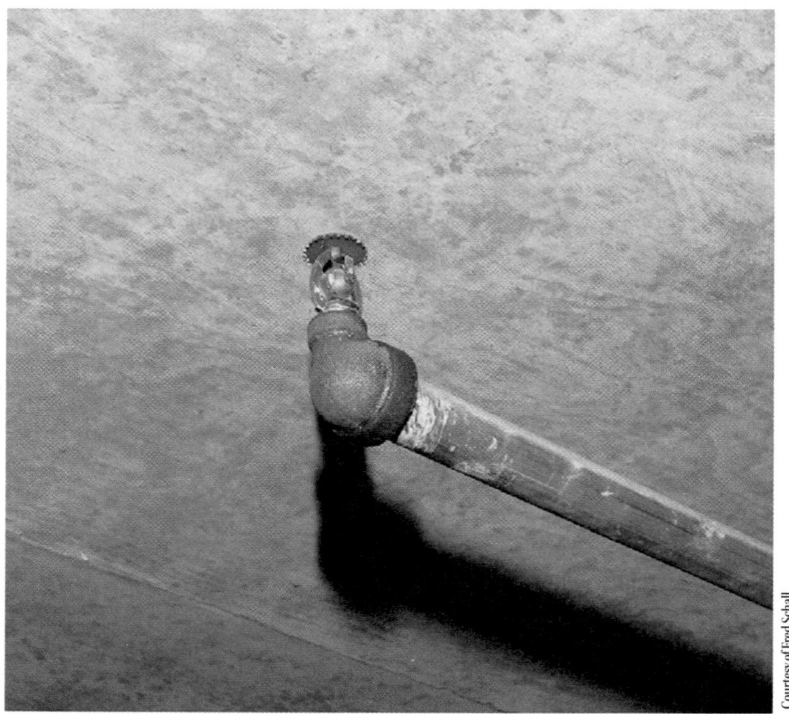

FIGURE 12-16 Heat-activated fire sprinkler.

Courtesy of Fred Schall

inspection whether they are open or closed (**Figure 12-17**). Both of these types of valves are usually kept locked in the open position to prevent tampering. The water then enters the main control valve. Main control valves differ depending on whether the system is dry pipe or wet pipe.

open screw and yoke valve A valve with a hand wheel that exposes a threaded rod when in the open position. The hand wheel looks much like a steering wheel and can be locked with a chain and padlock so it cannot turn.

post indicator valve A valve with an indicator body that sticks up out of the ground. The body has a small window that says either "shut" or "open" depending on the position of the valve.

Wet Pipe System

The wet pipe system has water under pressure behind the sprinkler heads at all times. This allows the system to operate rapidly when a head is opened. These systems are suitable for use in installations where freezing is not a problem. The control valve on a wet pipe system is a check valve to keep water from the sprinkler system from reentering the domestic supply (**Figure 12-18**). The clapper in the check valve is usually in the closed position. It only

Delmar/Cengage Learning

FIGURE 12-17 OS&Y valves chained and locked in open position.

Delmar/Cengage Learning

FIGURE 12-18 Sprinkler riser and valving located outside of building where freezing temperatures do not occur.

opens fully when a sprinkler head opens. There are pressure gauges on each side of the clapper. In a properly operating system, the pressure gauge on the sprinkler side of the valve should read the higher pressure, because as pressure surges enter the system, they raise the clapper momentarily and the higher pressure ends up trapped above the valve, keeping it in the closed position. There is also an alarm valve that activates when excess water flow is detected. Attached will be a **retard chamber** that allows pressure surges to dissipate without triggering the alarm.

There is a main system drain mounted above the clapper on the main valve body to allow draining of the system for repairs.

retard chamber A small tank attached to sprinkler systems that allows pressure surges to dissipate their energy before they enter the system and set off the water flow alarm.

NOTE Outside control valves and risers are used only in areas where freezing temperatures do not occur.

Dry Pipe System

The dry pipe system is somewhat more complicated in design. These systems are used in areas where freezing temperatures may occur, such as outside storage or unheated buildings. Because water expands as it freezes, if the system is allowed to become filled with water, the expansion will rupture the plumbing. As the temperature warms, the parts that were held in place by the ice fall to the floor. In a room with a high ceiling, falling chunks of cast iron fittings can be hazardous to your health. The dry pipe system has compressed air in the lines behind the sprinkler heads. The compressed air keeps the clapper in the main valve body from opening and admitting water. There is water on the underside of the clapper valve and as soon as it opens, the water is admitted to the system. In especially large systems, if one head were to open, there can be a one-minute delay before enough air is released to let water come out of the open sprinkler.

In areas where large amounts of water are needed immediately, deluge systems are used. They are set up with sprinkler heads that are open all the time. Fire detection devices open the valving and allow water into the system when necessary.

Once the water leaves the main valve, it enters the riser. There is a **fire department connection** attached to the riser (**Figure 12-19**). This connection is used to boost the pressure in the system by attaching a hose from a pumper. The standard operating pressure for this pumper is 150 psi, considerably more pressure than most water supply systems. The fire department connection is on the discharge side of the main valve so water from the pumper is not forced back into the domestic water supply. It is important when attaching to the sprinkler system that the pumper is not connected to a hydrant that directly supplies the water flow to the sprinkler system.

fire department connection Fittings connected to the fire protection system used by the fire department to boost the pressure and/or add water to the system.

FIGURE 12-19 Fire department connection and post indicator valve.

Delmar/Cengage Learning

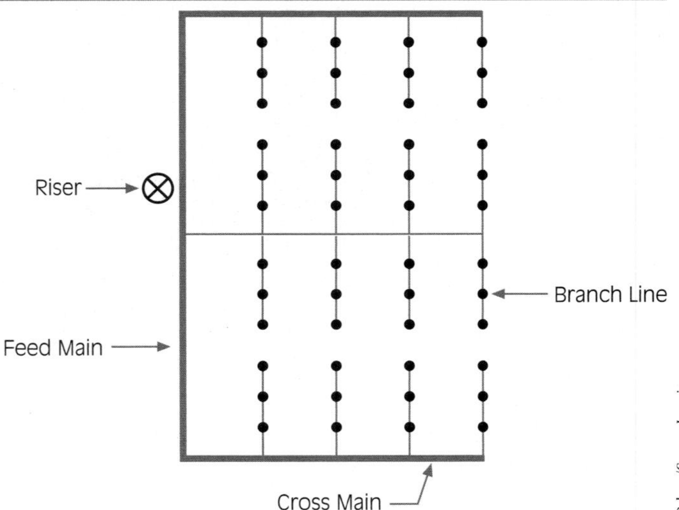

FIGURE 12-20 Sprinkler piping schematic.

Delmar/Cengage Learning

When the water leaves the riser, it enters the feed mains. These pipes are then connected to the cross mains, which are attached to the branch lines. The branch lines are where the individual sprinkler heads are attached. At the highest and furthest point in the system is the inspector test drain. This is used to verify that the required pressure is available to every sprinkler head (**Figure 12-20**).

Variations of the dry pipe system are the deluge and preaction systems. Deluge sprinkler systems are used in high fire hazard occupancies, such as plywood manufacturing. The sprinkler heads in this type of system are not heat activated and are referred to as open sprinklers. The system is activated by a supplemental fire detection system. When the deluge valve opens, water enters the plumbing and is immediately discharged through the open heads.

Preaction sprinkler systems are designed with a preaction valve holding the water back and have closed sprinkler heads. The sprinkler system is connected to a detection system. When the detection system activates, the preaction valve opens, allowing water into the piping. When heat becomes sufficient to open the sprinkler head, water is discharged.[28]

Sprinkler Heads

Sprinkler heads come in numerous styles. They are divided into three main types. The pendant type head is designed to be installed in the hanging down position. The upright head is installed with the head above the pipe (**Figure 12-21**). The sidewall sprinkler head is used in a horizontal position. Each head is equipped with a deflector that divides the water flow into a fine spray. The deflector also directs the spray. The head types are not interchangeable: if pendant heads are used in the upright position, and so forth., the heads will not function correctly.

Sprinkler heads are designed to operate at different temperatures. Either the frame of the head is painted or the glass bulb contains a colored liquid that is color coded for their operating temperature. The maximum ceiling temperatures range from 100°F to 625°F.[29] Should any painting of the ceiling or walls be done, the sprinkler heads themselves must be masked off from paint. Once the painting is completed the masking material must be removed or it may interfere with the proper activation of the head(s) during a fire. Sprinkler heads can be rendered ineffective when stock is piled too close to them, limiting their spray pattern. Another factor affecting their effectiveness is the use

FIGURE 12-21
Upright sprinkler head.

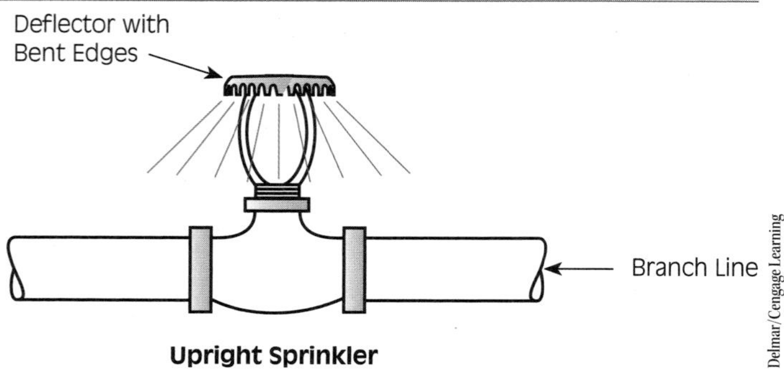

of solid shelving. The spray can be kept from penetrating to the seat of the fire by deflection. To avoid the problem experienced with solid shelving, in some instances heads are installed between the shelves. This does, however lead to a greater risk of heads being knocked off or damaged during shelf stocking operations, especially where forklifts are used to move stock.

Standpipe Systems

A fire protection system that provides a fire hose attachment station on each floor is the standpipe system. By having a plumbing system installed in the building with fire department hose connections, it is not necessary to take the time to lay hose to each floor. The Class I standpipe is designed for fire department use and has a 2½-inch hose connection (**Figure 12-22**). The Class II standpipe is designed for use by building occupants until the fire department arrives and has a 1½-inch hose connection with hose attached. The Class III system has both a 2½-inch hose connection for fire department use and a 1½-inch connection with hose for occupant use or a 2½-inch connection on a 1½-inch reducer and hose that is easily removable to allow access to the 2½-inch connection.[30]

NOTE By having a plumbing system installed in the building with fire department hose connections, it is not necessary to take the time to lay hose to each floor.

FIGURE 12-22 Standpipe outlet connection for 2½-inch fire hose.

Delmar/Cengage Learning

Foam Systems

Foam systems can be stationary or vehicle mounted. There are four ways that foam solution is created from foam concentrate and water. The first of these is through eduction using a device called an eductor (**Figure 12-23**). Eductors operate on the venturi principle. The water is forced through a small opening that expands after it passes over the orifice where the foam concentrate is admitted to the fire stream. On portable eductors, the water inlet is attached to the hose coming from the pumper. The outlet is attached to the hose going to the nozzle. There is a plastic hose attached to the eductor that is inserted into the foam concentrate container, usually a 5-gallon bucket or 55-gallon drum. Where the plastic hose attaches to the eductor body, there is a plastic ball on the inside, which is designed to act as a check valve, preventing water from being forced into the foam concentrate container. After use of the eductor the plastic ball should be removed and cleaned. If it is not, dried concentrate can cause the eductor to malfunction. These devices are also called in-line eductors. On the top there may be a dial to set the percentage of

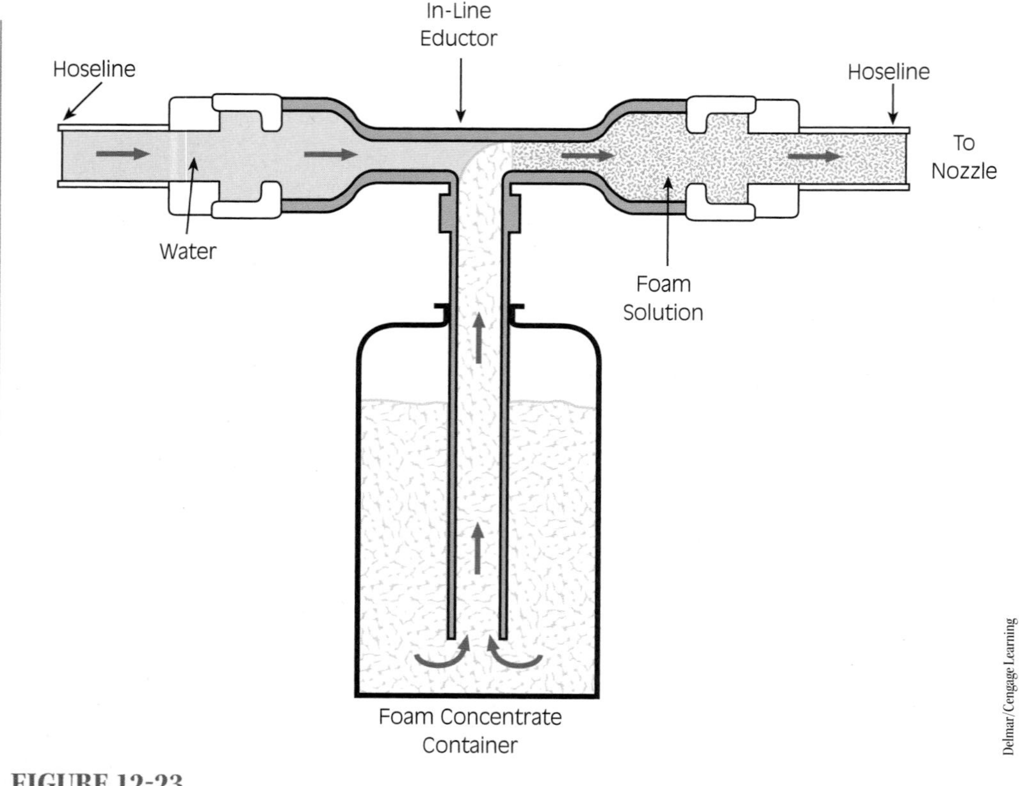

FIGURE 12-23
In-line foam eductor.

the foam solution produced, usually either 3% or 6% depending on the concentrate and application requirements.

The second way foam solution is produced is through injection. In this type of system, the foam concentrate is directly injected into the water. These systems may be set up to perform the injection before or after the pump.

The third method is called batch mixing. In batch mixing, the foam concentrate is introduced directly into the water reservoir. In wildland firefighting operations, the operators often carry premeasured amounts of Class A concentrate on the apparatus. When the alarm is received, they pour the concentrate into the tank and start the pump to circulate the solution so it is well mixed when they arrive at the scene. Wetting agents are usually added in this way. This method is also used on helicopters. The helicopter has an on-board tank of foam concentrate.

When the load of water is picked up, in either a bucket or tank, the required amount of foam concentrate is introduced into the water. The drawback to batch mixing is that to get a consistent concentration of the foam solution, the water tank must be completely emptied and refilled every time the solution is recreated. With Class A foam, careful proportioning is not that critical; with Class B foam, the concentration is very important and batch mixing is not recommended (**Figure 12-24**).

The fourth method is premixing. In premixing, the foam concentrate and water are mixed to the proper concentration and stored in this way. Foam extinguishers are designed this way. The problem with premixing is that foam solution and wetting agent solutions are corrosive on steel and cast iron. It is not a good idea to leave premixed foam

FIGURE 12-24
Class B and Class A foam concentrate containers.

Delmar/Cengage Learning

in an apparatus tank for long periods of time. As a general rule, anytime foam or wetting agents are used, the system must be drained and flushed to avoid corrosion of either the tank or plumbing. If the apparatus is equipped with a plastic tank, carrying premixed foam may seem like a good idea. However, the plumbing, such as the pump casing, is cast iron or steel and still presents a problem.

An addition to Class A foam systems that is gaining in popularity is the compressed air foam system (CAFS). The system is equipped with an air compressor that pumps air into the hose stream after the pump. The use of this type of system does not rely on the nozzle or other device to aerate the foam. It also makes the hoselines lighter as they contain air as well as foam solution. By adjusting the level of foam concentrate and air in the solution, the foam can be made for different purposes. A very dry foam will stick to vertical surfaces. A wet foam will not stick but provides good penetration into the fuel.

Class B foam is very wet and does not stick to vertical surfaces. It works well on two-dimensional fires—those that have width and length, such as a pool of liquid. On three-dimensional fires—those that have vertical surfaces as well—it does not work well. An example would be flaming liquid spraying from a flange or piping leak. In this scenario, the Class B foam would be able to extinguish the pool fire, but not the fire at the source of the leak. In this situation, a dry chemical extinguisher can be used to extinguish the fire at the site of the leak.

NOTE Any time foam concentrate is used, the manufacturer's specifications as to the concentration of the product must be closely followed.

In the case of an aircraft hangar or other special occupancy, the foam system is preplumbed and discharges out of sprinkler heads or special nozzles.[31] The system is designed to rain the foam down on the flaming material. As the foam reaches the floor, it spreads out and can seal over the surface of the burning material beneath the aircraft. Whenever Class B foam is used on flammable liquids, it is to be applied gently to the surface. Plunging the foam under the liquid surface will just disrupt the foam blanket that you are trying to create. When Class B foam is being used, plain water hose streams are not to be used on the surface that is being foamed. They will only dilute and decrease the effectiveness of the foam blanket.

NOTE When using foam eductors and other in-line devices, the manufacturer's specifications must be followed as pertains to flow rate, inlet pressure, length of line after the eductor, and elevation above the eductor.

Any time foam concentrate is used, the manufacturer's specifications as to the concentration of the product must be closely followed. Care must also be taken to determine the type of fuel involved. Regular AFFF will not work on polar solvents and the proper concentration of AFFF ATC must be used when dealing with polar solvents. Class A foam is not designed to be used on Class B fires, and vice versa.

When using foam eductors and other in-line devices, the manufacturer's specifications must be followed as pertains to flow rate, inlet pressure, length of line after the eductor, and elevation above the eductor. If the flow rate on the nozzle is set incorrectly or the nozzle is not all the way open, the foam you are expecting is not what you will get. Class B foam does not give very good visual clues as to its concentration. You may think you are forming a good sealing blanket over a product when in fact you are not.

Gas Extinguishing Systems

Other fire suppression systems have been developed to address specific problem areas. Where water will cause excessive damage to stock or electrical installations, gas type systems are installed.

Carbon Dioxide

In carbon dioxide installations, the product is stored in large cylinders or a tank (**Figure 12-25**). When a fire occurs, the gas is released into a piping system and expelled from nozzles in the area to be protected.[32] CO_2 is not poisonous but can be harmful to humans because of its ability to dilute the oxygen content of the room, leading to asphyxiation. The system is to be installed with a warning system to evacuate occupants prior to discharge. In any operation where a CO_2 system has been used to attack the fire, personnel must wear self-contained breathing apparatus to avoid the danger of low oxygen concentration.

FIGURE 12-25
Carbon dioxide fire extinguishing system at farm chemical storage facility.

Delmar/Cengage Learning

In any operation where a CO_2 system has been used to attack the fire, personnel must wear self-contained breathing apparatus to avoid the danger of low oxygen concentration.

Carbon dioxide will extinguish most types of fires but dissipates rapidly enough that immediate follow-up is necessary. When used on burning liquid fires, it will extinguish the flame, but not cool metal parts of the liquid's container. If the metal parts are at a temperature above the ignition temperature of the liquid, re-ignition can occur. In ordinary combustible fires, CO_2 will not penetrate and extinguish deep-seated smoldering. In this type of fire, follow-up with water is necessary to ensure that the fire will not reignite. The concentration of the CO_2 in the atmosphere must be sufficient to lower the oxygen content to a point where the fire is extinguished. In large areas, this would require prohibitive amounts of product.

Dry Chemical Systems

Stored in a container, the powder may or may not be under direct pressure. In a pressure extinguisher, the powder is forced out a tube that extends into the bottom of the container by the pressure of the expellant gas above it. In this type of extinguisher, the gas is usually nitrogen, which is nonreactive. In an extinguisher with the expellant gas stored in a remote reservoir, the reservoir must be punctured to release the gas into the container expelling the contents. The expellant gas in this type of extinguisher is usually carbon dioxide. This type of extinguisher is commonly carried on fire equipment because it is easy to refill the powder and attach a new pressure cartridge after use (**Figure 12-26**).

Relatively inexpensive and easy to store, dry chemical systems are very common. They are installed in range hoods in restaurant kitchens and other areas where flammable liquids need to be extinguished.[33] These systems are mounted in heavy machinery and racing vehicles due to their dependability under rough treatment and their extinguishing capability. Their ability to perform well in the presence of water is a definite plus. Foam does not perform well on three-dimensional fires. The dry chemical can be discharged into the fire stream at the nozzle and extinguish the fire when foam or water alone will not do the job. These systems are installed on aircraft rescue firefighting apparatus in a twinned system with aqueous film-forming foam. The water in the foam cools the metal parts, the foam extinguishes the pool fire, and the dry chemical can extinguish the liquid as it runs down the fuselage. This same evolution can be performed with plain water and a dry chemical extinguisher on vehicle fires.

FIGURE 12-26 Various types of fire extinguishers.

Delmar/Cengage Learning

Wet Chemical Extinguishing Systems

Wet chemical extinguishing systems—Class K—were developed to extinguish fires in commercial cooking operations, such as deep fryers. Previously dry chemical and wet chemical Class B systems performed this function. The oils used now are vegetable based, not animal based. They have a lowered ignition temperature and the cooking equipment retains heat better than the old systems. The purpose of the wet chemical system is to reduce the temperature of the burning liquid as well as to apply the extinguishing agent.[34]

Fire Extinguishers

Over the years, portable fire extinguishers have been developed using almost every type of extinguishing agent. Multipurpose dry chemical extinguishers are designed to be used on ordinary combustible (Class A), flammable liquid (Class B), and energized electrical (Class C) fires. Plain water extinguishers are suitable for use on Class A fires. There are CO_2 and halogenated agent extinguishers for Class A, B, and C fires as well. A flammable metal (Class D) extinguisher may be as simple as a bucket of dry sand with a scoop or shovel used to apply the sand. Others will have powdered graphite and plastic pellets that melt and form a coating over the burning metal. Fire extinguishers come in all sizes from 2 pounds to 250 pounds of extinguishing agent.[35] (See **Figures 12-27** and **12-28**).

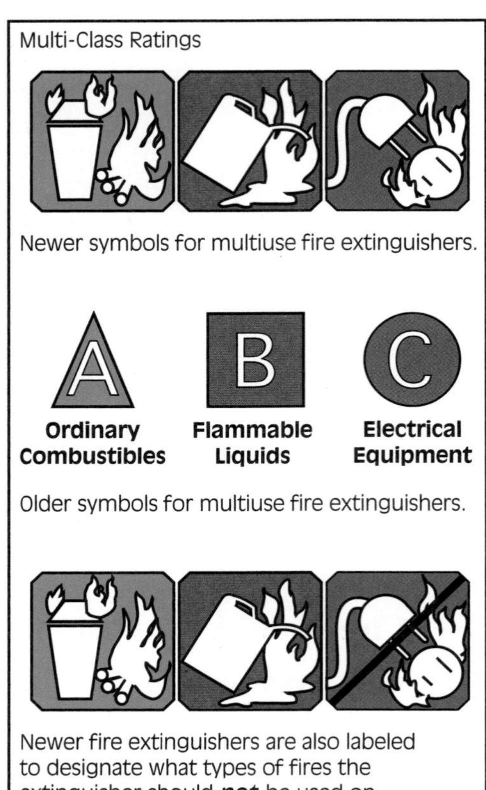

FIGURE 12-27
(A) Older versions of fire extinguishers are labeled with colored geometrical shapes with letter designations. (B) Newer fire extinguishers are labeled with a picture label system. (C) Many fire extinguishers can be used to fight more than one class of fire.

For more information on the operation of portable fire extinguishers, please refer to DVD clip *Operating Portable Fire Extinguisher.*

Obsolete Agents

It is quite possible that you may find obsolete types of fire extinguishers still in use. Some of these include soda acid and carbon tetrachloride. If you do come across these types of extinguishers, they should be removed from service. The soda acid extinguisher has a

Extinguisher Type	Class A Fire	Class B Fire	Class C Fire	Class D Fire	Class K Fire
Type A Extinguisher	×				
Type BC Extinguisher		×	×		×
Type ABC Extinguisher	×	×	×		×
Metal/Sand (Class D Extinguisher)				×	
Type K Extinguisher		×	×		×
CO_2		×	×		

Delmar/Cengage Learning

FIGURE 12-28
It is important to use the proper extinguisher for the particular class of fire.

small amount of strong acid inside and must be handled carefully. The carbon tetrachloride-type extinguisher contains a toxic substance and must be disposed of properly. If you come across an extinguisher and are not sure of its type, contact your local extinguisher service company and they can assist you.

Fire Pumps

In some occupancies fire pumps are installed either to boost pressure or to ensure a water supply to the firefighting system. Two types of fire pumps are commonly used for this purpose. One is basically the same type of pump that is installed in fire engines, a centrifugal pump. It is driven by either a gas or diesel engine or electric motors (**Figure 12-29**). For safety purposes one of each (diesel and electric driven) should be installed to provide service in case of a power failure. The other type is the vertical turbine pump (see **Figure 12-1**). Designed to operate automatically under demand, these systems also have a manual start. Firefighters should be familiar with the location and operation of these pumps in their district.[36]

Pressure-Reducing Devices

In high-rise buildings, pressure-reducing devices are installed because the amount of pressure needed to pump water to the one hundredth floor would cause entirely too much pressure on the fiftieth floor. These devices can seriously influence the effectiveness of a fire stream. As with any of the other factors that affect firefighting, it is extremely important that firefighters preplan their area, making sure that they understand the limitations and operation of any systems that will be used in firefighting. Most importantly, fire departments need to be equipped with the proper combination of hose diameter and nozzles for operating in high-rise structures. Without these, hose streams have very little reach and are rendered ineffective, creating a safety hazard for firefighting personnel.

FIGURE 12-29 Diesel- and electric-powered centrifugal fire pumps used to boost pressure in sprinkler/standpipe system.

Delmar/Cengage Learning

Fire departments need to be equipped with the proper combination of hose diameter and nozzles for operating in high-rise structures.

SUMMARY

The fire department should have a close working relationship with the local water company. Through this mutually beneficial relationship, the operation of the water system can be maintained and improved when necessary. A thorough knowledge of the strengths and weaknesses of the water system is necessary for decision-making purposes at the fire scene.

Numerous firefighting agents are in use today. To fully understand their effectiveness and uses, you must become familiar with all of them. As a safety issue, their hazards must also be known. The building owner may ask for your opinion on what type of system to install and you should at least be acquainted with the options.

The firefighting agents are all applied through some type of system. By schooling yourself in their operation and uses it is possible to intelligently preplan what is necessary to support these systems in case of fire. You will probably be called on to inspect these systems at some point in your career. If you do not have a clue as to how they operate, you will not know whether they are in working order. As more types of agents and systems are introduced, you must stay informed about their capabilities. In today's world of fire fighting, it is entirely possible that you could be placed in a situation where you are required to assist with or operate a system you do not have on your own apparatus.

It is the fire department's job to pick the best option available at the fire scene. This will assist in quick suppression of the fire and reduction of damage to what you are trying to save and the environment.

REVIEW QUESTIONS

1. What is the most common extinguishing agent used by firefighters?
2. What is the difference between a public and private water company?
3. When does the fire department get involved in planning fire flow for new construction?
4. List three components of a water supply system.
5. What are the differences between direct pumping and gravity-fed water systems?
6. Why is duplication of water supply equipment necessary?
7. List the names of the mains in a distribution system and show their relationship to each other.
8. Why is it necessary for a supply system to be gridded?
9. What is the difference between a wet barrel and dry barrel hydrant?
10. What is a dry hydrant?
11. List the steps to be performed in hydrant maintenance.
12. How are hydrants flow tested?
13. A hydrant with orange painted caps will flow a minimum of how much water (GPM)?
14. What are the component parts of a water systems program?
15. List two types of detection systems and explain their operation.
16. What are the differences between dry pipe and wet pipe sprinkler systems?
17. List two fire suppression systems that do not use water and explain how they work.
18. In terms of wildland use, list short-term and long-term retardants and explain how they are used in firefighting.
19. What is a foam eductor and how does it work?
20. What is the difference between AFFF ATC, and AFFF?
21. Why is it important to know the manufacturer's instructions when using foam concentrates and devices?

DISCUSSION QUESTIONS

1. From an engineering standpoint, what can be done to reduce the risk of damage from hostile fires in structures in your community?

2. With fire sprinklers rated as 96% effective in stopping hostile fires, why are they not included in more structures? What can be done to change this?
3. Justify buying a new pumper with a built-in foam system, even when it is more expensive.

NOTES

1. Insurance Services Office, *Grading Schedule for Municipal Fire Protection* (New York, NY: Insurance Services Office, 1974).
2. National Fire Protection Association, *Fire Protection Handbook*, Nineteenth Edition (Quincy, MA: National Fire Protection Association, 2003).
3. Insurance Services Office, *Grading Schedule for Municipal Fire Protection* (New York, NY: Insurance Services Office, 1974).
4. Ibid.
5. Ibid.
6. *Simplified Water Supply Testing* (Chicago, IL: American Mutual Insurance Alliance, 1964).
7. National Fire Protection Association, *Standard 291, Fire Flow Testing and Marking of Hydrants* (Quincy, MA: National Fire Protection Association, 2007).
8. U.S. Fire Administration, *Facts about Fires in the U.S.* (Washington, DC: Federal Emergency Management Agency, 2003).
9. National Fire Protection Association, *Fire Protection Handbook*, Nineteenth Edition (Quincy, Ma: National Fire Protection Association, 2003).
10. Ibid.
11. M. King, C. Bailey, *Carbon Monoxide—Related Deaths—United States, 1999—2004*, (CDC, retrieved from: cdc.gov/mmwr/preview/mmwrhtml/mm5650a1.htm)
12. Frank L. Fire, *The Common Sense Approach to Hazardous Materials* (New York, NY: Fire Engineering, 2004).
13. National Fire Protection Association, *Fire Protection Handbook*, (Quincy, MA: National Fire Protection Association, 1997).
14. Richard L. Tuve, *Principles of Fire Protection Chemistry* (Quincy, MA: National Fire Protection Association, 1976).
15. Ibid.
16. Ibid.
17. Ibid.
18. Ibid.
19. National Fire Protection Association, *Standard 1150, Foam Chemicals for Fires in Class A Fuels* (Quincy, MA: National Fire Protection Association, 2004).
20. National Fire Protection Association, *Standard 18, Wetting Agents* (Quincy, MA: National Fire Protection Association, 2006).
21. National Fire Equipment System Course, *S370 Intermediate Air Operations* (Boise, ID: National Interagency Fire Center, 2002).
22. National Fire Protection Association, *Fire Protection Handbook*, Nineteenth Edition, (Quincy, MA: National Fire Protection Association, 2003).
23. Ibid.
24. Ibid.
25. Ibid.
26. National Fire Protection Association, *Standard 13, Installation of Sprinkler Systems* (Quincy, MA: National Fire Protection Association, 2007).

27. National Fire Protection Association, *Standard 13D, Installation of Sprinkler Systems in One- and Two- Family Dwellings and Manufactured Homes* (Quincy, MA: National Fire Protection Association, 2007).
28. National Fire Protection Association, *Fire Protection Handbook*, Nineteenth Edition, (Quincy, MA: National Fire Protection Association, 2003).
29. Ibid.
30. National Fire Protection Association, *Standard 14, Standpipe and Hose Systems* (Quincy, MA: National Fire Protection Association, 2007).
31. National Fire Protection Association, *Standard 11, Low-, Medium-, and High-Expansion Foam* (Quincy, MA: National Fire Protection Association, 2005).
32. National Fire Protection Association, *Standard 12, Carbon Dioxide Extinguishing Systems* (Quincy, MA: National Fire Protection Association, 2008).
33. National Fire Protection Association, *Standard 10, Portable Fire Extinguishers* (Quincy, MA: National Fire Protection Association, 2007).
34. National Fire Protection Association, *Standard 17A, Wet Chemical Extinguishing Systems* (Quincy, MA: National Fire Protection Association, 2009).
35. National Fire Protection Association, *Standard 10, Portable Fire Extinguishers* (Quincy, MA: National Fire Protection Association, 2007). National Fire Protection Association, *Fire Protection Handbook*, Nineteenth Edition, (Quincy, MA: National Fire Protection Association, 2003).
36. Ibid.

CHAPTER 13

Emergency Incident Management

LEARNING OBJECTIVES

Upon completion of this chapter, you should be able to:

- Explain the need for a plan at every incident.
- Differentiate between offensive, defensive, and combination modes of attack.
- Explain the need for organized thought processes in incident assessment.
- Describe the strategic priorities at an incident.
- Explain the terms strategy, tactics, and tasks.
- Explain the need for size-up of an incident.
- Explain how a size-up is performed and what information must be communicated in a report on conditions.
- List the five major components of the National Incident Management System (NIMS).
- Describe the components of the Incident Command System (ICS).
- List the positions and their functions in the ICS.
- Explain the need for unified command on a multijurisdictional incident.

INTRODUCTION

Incidents come in all types and sizes. It is the responsibility of everyone at the incident scene to do their part to assure a swift and successful conclusion to the incident. By training yourself in the area of size-up, you can learn to assess the incident from the time the call is received until its conclusion. By applying the strategic priorities in the proper order, you can develop a feel for what has to be done and when it has to be done. As you become more skilled in these areas, you can better assist the person in command of the incident. You will be better able to give him the information he needs from your location. You can also start to become more involved in the planning process that must take place at all incidents.

To aid in the effective management of incidents, the incident command system (ICS) has been developed. The incident command system presents a structure that is adaptable to all incident types. The ICS facilitates resources from different agencies, public and private, working together in a coordinated fashion. This is accomplished through training, common terminology, and an established command structure. By learning the ICS prior to an incident, resources from different agencies and disciplines can come together at the scene and operate in an effective, coordinated manner.

MANAGEMENT RESPONSIBILITY

Emergency incident management is primarily the responsibility of the first-in fire officer, because most incidents only require the resources of one company to handle. However, not all incidents can be handled by the resources that initially arrive at the scene. Should an incident escalate, the first-in officer has the opportunity to turn the incident into a major disaster through poor management, or conversely, to have a positive effect on the outcome of the incident. Some situations require more resources and possibly even long periods of time to control. The first-in personnel must have a complete knowledge of the hazards and potential of the incident as well as their own priorities and capabilities. The only way an incident can be brought under control is through the total and coordinated efforts of all personnel at the scene.

NOTE Should an incident escalate, the first-in officer has the opportunity to turn the incident into a major disaster through poor management, or conversely, to have a positive effect on the outcome of the incident.

NOTE It is the responsibility of every firefighter at the scene to be involved in assisting in the control of the incident.

It is the responsibility of every firefighter at the scene to be involved in assisting in the control of the incident. As a new firefighter, you may feel that you are not involved in the decision-making process. This is not true; all incidents require that everyone at the scene be alert and aware of the hazards and needs at the incident as it develops. The standard rule is "victims do not arrive at the scene in fire trucks." Your actions at the incident have a direct impact on its outcome. For this reason you need to be aware of and capable of participating in the decision-making process.

"Victims do not arrive at the scene in fire trucks."

INCIDENT PLANNING

Every incident must have a plan. The plan must be formulated to achieve effective utilization of resources and to resolve the incident with as little further damage as possible. The first decision to be made in forming an incident plan is what the objectives are. Objectives are defined as something toward which effort is directed; an aim or end of action; a goal. If there is no clearly defined goal in mind for the incident, it cannot be determined how to achieve the goal, nor can progress towards the goal be measured. Objectives for different kinds of incidents will, of course, differ. The objectives in a hazardous materials incident may be only to control the spread of the material. In a wildland incident, they may be, initially, to protect structures in the fire's path and then to effect control. Objectives must be clearly stated and understood as well as being measurable and attainable.

NOTE Every incident must have a plan. The plan must be based on objectives that are clearly stated, measurable, attainable, and understood.

Strategy

The next step in the formulation of the plan is to determine the strategies necessary to achieve the objectives. Strategy is defined as the art of devising plans to achieve a goal or objective.[1] From a firefighting standpoint, **strategy** may be defined as the method used to coordinate the tactical operations of units to achieve the desired incident objectives. All of the strategies must support the objectives. Strategies are based on the size-up of the incident, objectives, strategic priorities, and mode of attack selected.

strategy The method used to coordinate the tactical operations of units to achieve the desired incident objectives.

Tactics are the actions taken to achieve the strategies. Examples of tactics would be performing an interior search, ventilation, and advancing hoselines to the interior of a structure. In a wildland situation tactics could include constructing fire control lines with bulldozers and handcrews.

tactics Actions taken to achieve strategies.

Size-Up

At this point, it is common to ask yourself "How will I ever be able to take into account all of the factors that are required to come up with a plan of operation?" In his book, *Fire Fighting Tactics*, Lloyd Layman presented the process of size-up.[2] The first-in officer is to size up the incident. Size-up is a mental process requiring the incident commander to perform the following steps:

NOTE Size-up is a mental process.

1. *Determining facts when the alarm is received.* The time of day, which will affect whether life safety is likely to be a major factor; the weather, especially important in wildland incidents; the address or location of the incident; what type of occupancy is involved, such as a school, hospital, or vehicle; what type of incident is reported—fire, traffic accident, or hazardous materials spill. These are just a representative sample of the facts the firefighter can start to consider as soon as the alarm is received. Size-up starts even before units arrive at the scene.

   Preplanning is also a part of the size-up function. Through proper preplanning and inspection, many facts about the occupancy involved will already be known. A tremendous aid in entering a business to fight a fire after dark is to have been on a preplan or inspection tour under nonemergency conditions. Knowing something of the building layout is a great help when the smoke is thick and the rooms are pitch black.
2. *Anticipating probabilities.* Probabilities are the likelihood of what may be involved in the reported incident. If it is a reported small brush fire on a hot, windy, summer day, it will probably spread quickly. A traffic accident or car fire at rush hour on the freeway will probably cause access problems for fire apparatus. Some probabilities can be anticipated through study and experience, others cannot.
3. *Assessing your own situation.* The actual situation facing the firefighter. The ladder truck is in the shop for repair and the building on fire is higher than the longest ladder available at the scene. The fire is putting up a large column of smoke, visible as you leave the station. There is no smoke showing and building occupants report that they have it under control. There are a wide variety of situations that you will be confronted with over the course of your career. You must be prepared to think clearly and act decisively at all of them.

4. *Making a decision.* The time when the firefighter, considering the facts, probabilities, and his own situation, determines what is the best course of action under the given circumstances. Are additional resources needed? Should the attack strategy be an offensive or defensive mode of attack? This is when the incident commander makes or breaks the incident. Now is the time to determine the objectives, strategies and tactics. Too conservative and the incident exceeds control measures. Too liberal and resources are wasted and unavailable if another incident occurs.
5. *Planning the operation.* This is when the tactics are put into action. The incident commander communicates the plan to the resources at scene and things start to happen. The attack is started. Resources report back with their progress. Are we winning or losing? Are we going to save the structure or create a new parking lot?

NOTE Size-up does not stop when the plan is put into operation. It is an ongoing process which revaluates incident factors and resource accomplishments in controlling the incident.

Size-up does not stop when the plan is put into operation. When resources report their progress and the incident commander observes what is happening, the process starts over again with new facts. As the facts, probabilities, and your own situation change, there will have to be new decisions and adjustments to the plan of operation. As previously mentioned, size-up is a process. It starts before resources leave the station and continues throughout the incident.

The person who wishes to become an effective incident commander will go back over this process after the incident and evaluate where improvements could be made. A group critique of the operations performed is a good idea. This is called an *after action review* (AAR). AARs should be examinations of what went right and wrong. When approached positively and not used to place blame or criticism, the people involved can be honest and the AAR can be used as a training tool.

NOTE The person who wishes to become an effective incident commander will go back over this process after the incident and evaluate where improvements could be made.

When you arrive at the scene, you may not have enough resources to have any direct effect on the incident at that time. One thing that can be done is to give a comprehensive size-up that will help your supervisor determine what the incident's resource needs are going to be. It is often better to scout the incident and give a good size-up than to get involved in taking action that will have little effect. By giving a good description, your supervisor can get the additional resources needed started to the scene. Time is critical. If he is still several minutes out, he can at least get the additional resources ordered. Most incident commanders would rather have to turn some resources around than not have enough and have the incident overwhelm the resources.

Wildland Fire Report of Conditions

One of the benefits of a complete report of conditions is that it requires the consideration of many of the elements necessary to perform a size-up. The following is an example of the requirements for a report of conditions for wildland fires.[3] This information should be given to your supervisor or the dispatch center over the radio. The other units responding should be able to hear your description as well.

1. *Correct location.* Often reports of wildland fires come in from passersby or persons who see the smoke but are some distance from the fire. It is important to make sure to give a corrected location when necessary. This may also require you to give the other responding units directions to the scene. By doing so, responding units can more easily approach the fire and the dispatch center can give the corrected information to any aircraft that are requested.
2. *Size.* The current size of the fire. This is given as the number of acres involved. No one is perfect at this, especially when the fire is in hilly terrain. Give your best estimate.
3. *Fuel type.* In determining the type of additional resources that may be necessary, the type of fuel burning is critical. In grass and light fuels, control and extinguishment can usually be accomplished with pumpers and dozers assisted by aircraft when necessary. The fuel burns clean and leaves little mop-up work. In heavy fuels, such as brush and timber, the extinguishment and mop-up require hand crews. In these fuels, helicopters are very effective. Also, the heavier fuels are more likely to produce **spot fires**.
4. *Slope and aspect.* The slope of the hill the fire is on has a lot to do with its rate of spread. A fire out in the flats is not going to spread nearly as rapidly as a fire at the toe of a slope. A fire on the very top of a hill tends to burn downhill slowly. The slope also affects the type of resources needed. A steep slope may not allow pumping equipment to be used without putting in hose lays. This tactic is much slower to place into operation than **pump and roll** tactics. Slopes that are too steep or rocky for dozers or hand crews to operate will necessitate an indirect attack.

   Aspect is the relationship of the slope to the points of the compass. A fire on a south-facing slope tends to have lighter fuels and have the sun shining on it for most of the day. The fuels will also be drier, causing them to burn more rapidly. North-facing slopes tend to have heavier fuels and do not get as much sun.

spot fires In heavier fuels, flying fire brands can land outside the fire perimeter and start new fires.

pump and roll A tactic used in grass fires utilizing pumpers that can drive while the pump is operating. Hoselines are connected to the apparatus and water is sprayed to extinguish the fire edge.

5. *Rate of spread.* Given as slow, moderate or rapid. In planning for resource needs, the rate of spread of the fire directly affects the size of the fire perimeter by the time the additional resources arrive. The higher the rate of spread, the more difficult the fire will be to contain and more control line will have to be put in due to its increased size.
6. *Exposures.* The exposures should be specified as to the number and type. Exposures consist of structures and other improvements, such as power poles. When structures are endangered, resources must be ordered to protect them as they are of the highest priority.
7. *Weather conditions.* Wind is going to have a direct effect on the spread of wildland fires. They tend to spread rapidly with the wind and slowly against it. A strong downslope wind can drive a fire at extremely high rates of spread in a downhill direction. This is contrary to the common wisdom that fires spread uphill at a faster rate.
8. *Potential of the fire.* Is the fire going to be easy to control or is this going to be a major operation? Will resources be needed for a couple of hours or a couple of days? Should the order be placed for a base/camp to be set up?
9. *Additional resource needs.* In your best estimate, what additional resources are needed to control the fire? When requesting resources from outside your own department's capabilities, they can require a four- to eight-hour lead time. If you need help, ask for it. One of the great truths of firefighting is that if you do not order what you need, it is not coming.
10. *Objectives.* All of the resources responding need to know what the plan is. Clearly stated objectives, as well as a good size-up, can aid them in getting a mental picture of what you have and what you will need them to do when they arrive.

Structure Fire Report of Conditions

In the size-up/report of conditions of a structure fire many of the considerations are the same as for wildland fires.

1. *Correct location.* Self-explanatory.
2. *Height/stories.* The height of the structure involved. Specify which floor the fire is on.
3. *Size.* The size of the structure. There is a big difference in the resource requirements for a single-family dwelling versus a large commercial building.
4. *Type of structure.* A wood frame structure is handled differently than a pre-cast concrete tilt-up structure. Factories have different resource requirements than hospitals.
5. *Location and area involved.* Is the fire in the front or the back? The size of the area involved is usually given as a percentage. For example, 10%, up to and including fully involved.
6. *Level of involvement.* Whether the fire is visible. Examples are nothing showing, smoke showing, smoke and flames showing, and through the roof.

7. *Exposures.* Number and type. This could be a vehicle parked next to the structure or a separate building in close proximity that is threatened. Remember that radiated heat is not affected by wind.
8. *Potential of fire.* Whether the fire is small or coming out of every window and threatening to spread to the adjacent structures.
9. *Additional resource needs.* Number and type. For example, one more engine, three more engines, or a ladder truck.
10. *Objectives.* State your priorities. When the structure is fully involved, all you may be able to do is protect the exposures until sufficient resources arrive and water supply is set up to attack the fire. Specify whether the attack is to be offensive, defensive, or combination.
11. *Obtain an* **"all clear."** Life safety is the primary objective at any incident and the "all clear" signifies that the **primary search** is completed and all victims who could be rescued have been. This is not going to be possible in every case but should always be one of the first considerations.

One of the best things the first arriving unit can do is to give the others responding a clear description of what is occurring. This way they get a mental picture of the incident and start their own size-up process. At this time, the responding higher-level supervisor may wish to pose questions as to what you have and what you need. This helps them to get their thought process started as well.

all clear An "all clear" is the short descriptive phrase which indicates that a primary search of the structure for victims has been completed.

primary search A rapid search of all involved and exposed areas which are affected by the fire but can be entered, to verify removal and/or safety of all occupants. Should this not be possible a secondary search is conducted as soon as it is safe to do so.

Strategic Priorities

Most of the strategic priorities in use today are based on those identified by Lloyd Layman in his book *Fire Fighting Tactics.* Layman divided the strategic priorities into seven areas:[4]

1. Rescue is the first strategic priority. Life safety is the most important consideration in any incident operation. In fire situations, this may require evacuating people from the fire's path, searching structures looking for trapped occupants, and erecting ladders to pull people from imminent danger. Another method of rescuing people from fires is to place hose lines into operation that cut off the fire spread toward the trapped persons. In hazardous materials situations, rescues may involve evacuation, sheltering in place, or actually pulling people out of tanks and contaminated areas. The first search of the incident is called the primary search. When it is completed

and all live persons are rescued, the "all clear" message is transmitted to the incident commander. Any search after that is called a secondary search. Some incident commanders are of the opinion that if the incident is so well involved that no initial search can be conducted, any search conducted at the incident is considered a secondary search.

NOTE Rescue is the first strategic priority.

2. Exposures are those items that are not yet involved in the incident, but soon will be if no action is taken to protect them. As mentioned in Chapter 4, radiated heat is a very real problem at fire scenes. When hose lines are placed into operation to protect adjoining structures or foam is applied to the exterior of structures to protect them from advancing wildland fires, these tactics are considered exposure protection. Exposures can also exist inside of a structure. When a fire spreads through a structure, it usually starts in one area. Interior hose lines are placed so spread of the fire is prevented from happening, in effect protecting interior exposures.
3. Confinement is the act of stopping the spread of the incident. Depending on the scale of the incident, confinement can have different meanings. In a hazardous materials incident, confinement may involve diking the spill to keep it out of a storm drain or waterway. In a structure fire incident, it may mean confining the fire to the building or even block of origin. A typical confinement objective on wildland fires is to confine the fire to one canyon or one side of a road.
4. Extinguishment in the case of fires is the act of putting the fire out. In terms of hazardous materials, it could be equated to stopping the source of the leak.
5. Overhaul is performed when the fire is extinguished. Firefighters must make sure that the fire is totally out, which often requires checking concealed spaces in attics and inside walls. Overhaul also includes placing the structure into a condition in which it can be turned over to the owner. This does not include structural repairs. On smaller incidents, such as a kitchen fire, it usually includes cleaning up and removing fire debris and water and evacuating smoke. This is not only good safety practice, but good public relations as well.

   At wildland fires, overhaul is commonly called mop-up. Mop-up consists of searching out embers that could threaten control lines, dropping dangerous trees, and improving the fire line.

   Although dirty and non-glamorous, overhaul/mop up are extremely important, often taking longer than extinguishment operations to accomplish. There is nothing like having to return to a fire scene to combat a new fire due to a **rekindle** caused by

rekindle A fire that reignites after it was thought to be extinguished. Commonly happens in attics, basements, and walls of structure fires and in logs on wildland fires. Usually due to incomplete overhaul/mop-up.

insufficient or improper overhaul/mop-up. During overhaul operations is no time to let down your guard. In structure fires, at this point in the fire, carbon monoxide concentrations are at their highest and SCBA must be worn until the atmosphere can be changed. Fire may have weakened the structure to the point that it is not safe to enter and operations may have to be conducted from the exterior. At wildland fires, trees that were damaged by the fire have a tendency to fall down, so a lookout must be maintained when working around them.

6. Salvage operations can be performed at any time during the incident. Salvage is the act of saving the contents of the building from damage due to incident operations. We cannot do anything about the damage that has occurred before we get at scene, but we can try our best to prevent any unnecessary further damage after we arrive. Firefighting usually involves applying water to control the fire. Salvage is the actions taken to protect building contents from water damage. Salvage covers are used to cover furniture and other objects to prevent damage; where applicable, objects are carried outside. In some areas these items must be guarded to prevent them from being stolen. After wildland fires, dozer and hand lines are **rehabilitated** to prevent erosion when the rain comes. Salvage is not just good practice to reduce dollar loss; like overhaul, it is good public relations and a part of modern firefighting professionalism.
7. Ventilation is another function that can be performed at any time during incident operations. Considered an incident support operation, ventilation aids in quick and efficient fire control as well as enhancing safety. Ventilation reduces the possibilities of flashover and back draft. It also makes the fire fighting job easier by increasing visibility and reducing heat and steam felt by the firefighters performing the attack. Ventilation, properly performed and used as part of a coordinated attack, can also act as a confinement tool on the fire ground. By venting smoke and heat buildup from an attic, the attic can be prevented from reaching its ignition temperature. Some types of ventilation operations can be used to cut off fire spread by redirecting heat to the outside of the structure.

The seven strategic priorities are listed in order but are not necessarily performed in that order. The main idea is to have a list of options to consider and to aid in the mental process of prioritizing operational objectives.

The common acronym for Layman's strategic priorities is "RECEO SV."

rehabilitate There are two common usages of this word in fire fighting. To rehabilitate personnel means that they rest, cool off, and replenish body fluids. To rehabilitate a fire line means to construct water bars to direct water runoff and prevent erosion. Under the Federal "Minimum Impact Suppression Tactics" policy, rehabilitation may mean the erasure of fire lines as much as possible. In other words, to cause as little damage as possible controlling the fire as fire is a part of the natural environment.

Modes of Attack

Offensive Mode

Strategies are based on differing modes of attack. In the offensive mode, resources are applied directly to controlling the fire or other incident. This method is often used on small fires or incidents where firefighters can make access without unreasonable risk to their safety and there is still something to save. There have been instances where rescuers have lost their lives on incidents that were essentially body recovery. The incident commander must constantly be aware of the risks versus benefits of actions taken at an incident. It is a common saying that firefighters will risk much to save much, such as the lives of viable victims, and risk little to save little. As a firefighter, it is your responsibility to be aware of what is going on around you and to constantly evaluate your situation. Direct attack on a wildland fire or interior attack on a structure fire would be considered the offensive mode.

NOTE In the offensive mode, resources are applied directly to controlling the fire or other incident.

Defensive Mode

In situations where the fire is too large or too well established, the defensive mode is used. In the constant evaluation of risk versus benefit, if it is found that the risk to personnel is too high in relation to the outcome of their actions, the incident commander should initiate actions in the defensive mode or switch from offensive to defensive mode. Structure fires where structural collapse is imminent or has already occurred would require the defensive mode. Attacking a block-sized structure fire with exterior lines and elevated streams from ladder trucks would be considered defensive mode. A wildland fire being pushed through brush with 30-foot flame lengths would also require the defensive mode. Hazardous materials incidents, which require a precautionary approach, are usually handled initially in the defensive mode until the material and its hazards can be identified.

NOTE In situations where the fire is too large or too well established, the defensive mode is used.

NOTE There are situations where both offensive and defensive modes may be used at the same time—in a carefully coordinated fashion—to control an incident.

Combination Mode

There are situations where both offensive and defensive modes may be used in a combined attack mode to control an incident. This must be done in a carefully coordinated attack. The combination mode requires good communications between incident command and the forces, and among the forces themselves, to avoid one group adversely affecting the safety or operations of the other. Combined attacks are often used on large wildland incidents in the form of direct attack on the slow-moving parts of the fire and indirect attack on the rapidly moving parts. Another example is to add structure protection. The flanks of the fire are being attacked directly or indirectly to establish **perimeter control**. Other resources are assigned within the fire perimeter protecting structures as the fire moves past. Resources assigned to perform **structure protection** do not take action to control the perimeter of the fire; their sole function is to protect the structures in its path.

Combination attacks require careful coordination and should not be used for interior-attack structural firefighting. The introduction of hose streams into a ventilation hole or other opening may reverse the outflow of smoke and heat, causing injury or death to interior attack forces.

perimeter control Controlling the edges of a wildland fire.

structure protection Protecting structures in danger of being consumed by an advancing wildland fire.

Combination attacks require careful coordination and should not be used for interior-attack structural firefighting. The introduction of hose streams into a ventilation hole or other opening may reverse the outflow of smoke and heat, causing injury or death to interior attack forces.

Tactics

Tactics are defined as the art of directing and employing resources to achieve the objectives.[5] Engine companies are usually involved at the tactical level of fire fighting. Once the objectives are determined and communicated, and the strategies are selected, the individual companies can employ the tactics necessary to achieve the objectives.

Some examples of tactics would be to perform ventilation to control fire spread or putting in a dozer line and then backfiring to slow the head of a wildland fire.

Tasks

Tasks are defined as pieces of work (jobs) to be completed in a specified period of time.[6] Tasks are those jobs necessary to achieve the tactics required by the incident plan. Most new firefighters will initially be expected to operate on the task level. An operation such as advancing a hose line into a burning structure requires the execution of several tasks. One of the first would be to don the SCBA. Another would be taking the hose line from the apparatus and stretching it out. As you can see, even a simple tactic is made up of a good number of tasks. It is very important that the firefighter master these tasks. Every chain is only as strong as its weakest link. If a firefighter does not know how to don the breathing apparatus quickly, the hose line will not be stretched in a timely manner and the whole operation will be slowed down. In firefighting, efficiency and proficiency lead to incident effectiveness.

NOTE In firefighting, efficiency and proficiency lead to incident effectiveness.

To look at this whole process and the relationship of the parts, let's look at a relatively simple, but common, fireground operation at the scene of a structure fire with one bedroom involved. The objectives would be to ensure life safety and confine the fire to the room involved, causing as little property damage as possible. The strategies would be primary search and interior attack, offensive mode. The tactics would be to ventilate the roof, advance a hose line in through the front door attacking the fire at its seat, conduct the search, and perform salvage. Some of the tasks involved would be to don SCBA, ladder the roof, cut the ventilation hole, advance the hose line, search the structure, remove any occupants, attack the seat of the fire, and so on. It is easy to see that the incident could easily require the effort of numerous personnel to bring it to a swift and successful conclusion. This incident could easily utilize two engine companies and a truck company to perform all of the operations.

Even an incident of this small size requires a plan with closely coordinated management and effective communications between all of the personnel involved. When the companies start the attack, a lack of coordination may mean that one company directs its hose stream into a window while the other company is advancing its hose through the interior. The interior attack personnel are going to get steam burns if this happens. If things do not go as planned and the incident switches to the defensive mode, it will become more complex still. Exposures will have to be protected, water supplies need to be further developed, crews need rehabilitation, and so forth.

Without a well-thought-out plan of operations, the incident we just looked at can easily turn into a minor disaster. All of the firefighters at the scene must be aware of how the operation is proceeding. If the plan is not working, the objectives, strategy, and tactics will have to be changed to take into account the new situation.

One of the key elements here is communication. The incident commander needs to communicate the plan to the company commanders, and the company commanders must communicate their portion of the plan to their companies. Operations should not proceed until all participants have a clear understanding of their responsibilities. This does not mean that they stand around for 30 minutes and discuss it. The personnel, through their knowledge of strategic priorities and tactics, should not need lengthy instructions. The incident commander tells the company commander to make an interior attack and primary search. The company commander tells her crew which size hose line to use and that they are going to make an interior attack and primary search. The firefighters know that SCBA will be required as well as what other tools to bring in case forcible entry is necessary. Beforehand, the firefighters should know whose responsibility it is to pull the hose and who will get the tools.

NOTE The incident commander needs to communicate the plan to the company commanders, and the company commanders must communicate their portion of the plan to their companies.

NOTE The incident commander not only needs to communicate decisions to the company commanders, but must also receive information from them in return.

Communication should be two-way. The incident commander not only needs to communicate decisions to the company commanders, but must also receive information from them in return. At an incident of any size much larger than a vehicle fire, the incident commander cannot see all of the incident. In a structure fire, the incident commander cannot see through the walls. At a wildland incident, the fire may be spread over many acres, even extending over the back side of a hill or into a canyon. On large wildland incidents, the incident command post may be miles from the actual incident. When an objective is met, the incident commander may have other work for the involved company and needs to know they are available. If some part of the plan is or is not working, it must be communicated back to the incident commander. When the operation is going wrong, the plan must be amended to correct the deficiencies. The incident commander cannot make these decisions if not informed of the necessity for change.

An example of this would be from the previously mentioned bedroom fire. The truck company is told to "ventilate the roof." The engine company is waiting to perform interior attack once the hole is completed. If the truck company officer does not notify the incident commander that the ventilation operation is completed, the engine company will be waiting at the door for the order to advance their line. Once the hole is completed, the incident commander may want the truck company to start salvage operations. Without two-way communication, the operations are slowed down and inefficient use is made of resources. At worst, the fire develops to the point where it is beyond the capabilities of the resources at scene or someone is seriously injured.

NATIONAL INCIDENT MANAGEMENT SYSTEM (NIMS)

In response to attacks on September 11, 2001, President George W. Bush issued Homeland Security Presidential Directive 5 (HSPD-5) in February 2003.

HSPD-5 called for a National Incident Management System (NIMS), identified steps for improved coordination of federal, state, local, and private industry response to incidents, and described the way these agencies will prepare for such a response.[7]

The Secretary of the Department of Homeland Security announced the establishment of NIMS in March 2004. One of the key features of NIMS is the Incident Command System (ICS). NIMS training is available on the Internet at www.fema.gov/nims/.

The Incident Command System (ICS)

An **incident** is an occurrence, either caused by humans or natural phenomena, that requires response actions to prevent or minimize loss of life or damage to property and/or the environment.

Examples of incidents include:

- Fire, both structural and wildland
- Natural disasters, such as tornadoes, floods, ice storms, or earthquakes
- Human and animal disease outbreaks
- Search and rescue missions
- Hazardous materials incidents
- Criminal acts and crime-scene investigations
- Terrorist incidents, including the use of weapons of mass destruction
- National Special Security Events, such as presidential visits or the Super Bowl
- Other planned events, such as parades or demonstrations

incident An incident is an occurrence, either caused by humans or natural phenomena, that requires response actions to prevent or minimize loss of life or damage to property and/or the environment.

NOTE The system's organizational structure is adaptable to any kind of incident to which a fire agency is likely to respond.

Given the magnitude of these types of events, it is not always possible for any one agency alone to handle all the management and resource needs.

Partnerships are often required among local, state, tribal, and federal agencies. These partners must work together in a smooth, coordinated effort under the same management system.

The Incident Command System, or ICS, is a standardized, on-scene, all-hazard incident management concept. ICS allows its users to adopt an integrated organizational structure to match the complexities and demands of single or multiple incidents without being hindered by jurisdictional boundaries.

ICS has considerable internal flexibility. It can grow or shrink to meet different needs. This flexibility makes it a very cost effective and efficient management approach for both small and large situations.

NOTE ICS is scalable to the size and complexity of the incident it is being used to manage.

History of the Incident Command System

The Incident Command System (ICS) was developed in the 1970s by a still-active group known as FIRESCOPE. ICS was developed following a series of catastrophic fires in California's urban interface. Property damage ran into the millions of dollars, and many people died or were injured. The personnel assigned to determine the causes of this disaster studied the case histories and discovered that response problems could rarely be attributed to lack of resources or failure of tactics. What were the lessons learned?

Surprisingly, studies found that response problems were far more likely to result from inadequate management than from any other single reason.

Weaknesses in incident management were often due to:

- Lack of accountability, including unclear chains of command and supervision
- Poor communication due to both inefficient uses of available communications systems and conflicting codes and terminology
- Lack of an orderly, systematic planning process
- No common, flexible, predesigned management structure that enables commanders to delegate responsibilities and manage workloads efficiently
- No predefined methods to integrate interagency requirements into the management structure and planning process effectively

A poorly managed incident response can be devastating to our economy and our health and safety. With so much at stake, we must effectively manage our response efforts. The Incident Command System, or ICS, allows us to do so. ICS is a proven management system based on successful business practices. This course introduces you to basic ICS concepts and terminology.

ICS Built on Best Practices

ICS is:

- A proven management system based on successful business practices
- The result of decades of lessons learned in the organization and management of emergency incidents

ICS has been tested in more than 30 years of emergency and nonemergency applications, by all levels of government and in the private sector. It represents organizational "best practices," and—as a component of NIMS—has become the standard for emergency management across the country.

NIMS requires the use of ICS for all domestic responses. NIMS also requires that all levels of government, including territories and tribal organizations, adopt ICS as a condition of receiving federal preparedness funding.

What ICS Is Designed To Do

Designers of the system recognized early that ICS must be interdisciplinary and organizationally flexible to meet the following management challenges:

- Meet the needs of incidents of any kind or size.
- Allow personnel from a variety of agencies to meld rapidly into a common management structure.
- Provide logistical and administrative support to operational staff.
- Be cost effective by avoiding duplication of efforts.

ICS consists of procedures for controlling personnel, facilities, equipment, and communications. It is a system designed to be used or applied from the time an incident occurs until the requirement for management and operations no longer exists.

Applications for the Use of ICS

Applications for the use of ICS include:

- Fire, both structural and wildland
- Natural disasters, such as tornadoes, floods, ice storms, or earthquakes
- Human and animal disease outbreaks
- Search and rescue missions
- Hazardous materials incidents
- Criminal acts and crime-scene investigations
- Terrorist incidents, including the use of weapons of mass destruction
- National Special Security Events, such as presidential visits or the Super Bowl
- Other planned events, such as parades or demonstrations

ICS may be used for small or large events. It can grow or shrink to meet the changing needs of an incident or event.

ICS Features

ICS is based on proven management principles, which contribute to the strength and efficiency of the overall system.

ICS principles are implemented through a wide range of management features including the use of common terminology and clear text, and a modular organizational structure.

ICS emphasizes effective planning, including management by objectives and reliance on an Incident Action Plan.

ICS helps ensure full utilization of all incident resources by:

- Maintaining a manageable span of control
- Establishing predesignated incident locations and facilities

- Implementing resource management practices
- Ensuring integrated communications

The ICS features related to command structure include chain of command and unity of command as well as unified command and transfer of command. Formal transfer of command occurs whenever leadership changes.

Through accountability and mobilization, ICS helps ensure that resources are on hand and ready to deploy.

And, finally ICS supports responders and decision makers by providing the data they need through effective information and intelligence management.

Common Terminology and Clear Text

The ability to communicate within the ICS is absolutely critical. An essential method for ensuring the ability to communicate is by using common terminology and clear text.

A critical part of an effective multiagency incident management system is for all communications to be in plain English. That is, use clear text. Do not use radio codes, agency-specific codes, or jargon.

NOTE Clear text must be used for communications at incidents. Do not use radio codes, agency-specific codes, or jargon.

ICS establishes common terminology allowing diverse incident management and support entities to work together. Common terminology helps to define:

- *Organizational functions.* Major functions and functional units with incident management responsibilities are named and defined. Terminology for the organizational elements involved is standard and consistent.
- *Resource descriptions.* Major resources (personnel, facilities, and equipment/supply items) are given common names and are "typed" or categorized by their capabilities. This helps to avoid confusion and to enhance interoperability.

NOTE The typing of resources is expressed through *kind* and *type*. An engine would be a kind of resource (e.g., an engine versus a water tender). The type would be determined by the resource's capability. A Type 1 resource has a greater capability than a Type 2, and so on. An example of resource typing is contained in Appendix E.

- *Incident facilities.* Common terminology is used to designate incident facilities.
- *Position titles.* ICS management or supervisory positions are referred to by titles, such as Officer, Chief, Director, Supervisor, or Leader.

Each of the above areas will be covered in more detail in the remaining sections.

Modular Organization

The ICS organizational structure develops in a top-down, modular fashion that is based on the size and complexity of the incident, as well as the specifics of the hazard environment created by the incident. As incident complexity increases, the organization expands from the top down as functional responsibilities are delegated.

The ICS organizational structure is flexible. When needed, separate functional elements can be established and subdivided to enhance internal organizational management and external coordination. As the ICS organizational structure expands, the number of management positions also expands to adequately address the requirements of the incident.

In ICS, only those functions or positions necessary for a particular incident will be filled. If one person can manage all of the major functional areas, on a small incident such as a car crash into a tree with one victim, no further organization need be implemented. Positions that are not assigned are assumed by the next higher level of management. Basically, if you are the incident commander and you do not assign an operations chief, then you are also the operations chief. In simple terms "what you do not assign you assume" as far as management responsibility. If one or more of the major functional areas reaches the point where it requires independent management, a person is assigned the responsibility of that function.

Management by Objectives

All levels of a growing ICS organization must have a clear understanding of the functional actions required to manage the incident. Management by objectives is an approach used to communicate functional actions throughout the entire ICS organization. It can be accomplished through the incident action planning process, which includes the following steps:

1. Understand agency policy and direction.
2. Assess incident situation.
3. Establish incident objectives.
4. Select appropriate strategy or strategies to achieve objectives.
5. Perform tactical direction (applying tactics appropriate to the strategy, assigning the right resources, and monitoring their performance).
6. Provide necessary follow up (changing strategy or tactics, adding or subtracting resources, etc.).

The first objective for all incidents is to "Provide for responder and public safety." The term "responder" should be used instead of "firefighter," as there are usually more than just firefighters responding to and taking action at the incident. There may be responders from law enforcement, Emergency Medical Services, and possibly others. The objective must take into account *all* responders. This objective is implied in unwritten

IAPs (Incident Action Plans) and written in formal IAPs. The chosen strategy and tactics must not disregard this objective.

The first objective for most incidents is to "Provide for responder and public safety." This objective is implied in unwritten IAPs and written in formal IAPs. The chosen strategy and tactics should not disregard this objective.

Reliance on an Incident Action Plan

In ICS, considerable emphasis is placed on developing effective Incident Action Plans (IAPs).

An IAP is an oral or written plan containing general objectives reflecting the overall strategy for managing an incident. An IAP includes the identification of operational resources and assignments and may include attachments that provide additional direction.

Every incident must have a verbal or written IAP. The purpose of this plan is to provide all incident supervisory personnel with direction for actions to be implemented during the operational period identified in the plan. Incident Action Plans include the measurable strategic operations to be achieved and are prepared around a timeframe called an **Operational Period**.

Incident Action Plans provide a coherent means of communicating the overall incident objectives in the context of both operational and support activities. The plan may be oral or written except for hazardous materials incidents, which require a written IAP.

At the simplest level, all IAPs must have four elements:

1. What do we want to do?
2. Who is responsible for doing it?
3. How do we communicate with each other?
4. What is the procedure if someone is injured?

operational period Operational periods are a predetermined time frame and vary as to length. They may be as long as 12 to 24 hours for a wildland incident or as short as an hour for a hazardous materials incident.

It is not enough to have an incident action plan. It must be provided to the responding resources in a briefing so that they are clear on the objectives and the plan to accomplish them. This is to include the identification of hazards at the incident and the action taken to mitigate them.

Manageable Span of Control

Another basic ICS feature concerns the supervisory structure of the organization. *Span of control* pertains to the number of individuals or resources that one supervisor can manage effectively during emergency response incidents or special events. Maintaining an effective span of control is particularly important on incidents where safety and accountability are a top priority.

Span of control is the key to effective and efficient incident management. The type of incident, nature of the task, hazards and safety factors, and distances between personnel and resources all influence span of control considerations. Maintaining adequate span of control throughout the ICS organization is very important.

Effective span of control on incidents may vary from three (3) to seven (7), and a ratio of one (1) supervisor to five (5) reporting elements is recommended. If the number of reporting elements falls outside of these ranges, expansion or consolidation of the organization may be necessary. There may be exceptions, usually in lower-risk assignments or where resources work in close proximity to each other.

Predesignated Incident Locations and Facilities

Incident activities may be accomplished from a variety of operational locations and support facilities. Facilities will be identified and established by the Incident Commander depending on the requirements and complexity of the incident or event.

It is important to know and understand the names and functions of the principal ICS facilities.

Incident Facilities

The *Incident Command Post*, or ICP, is the location from which the Incident Commander oversees all incident operations. There is generally only one ICP for each incident or event, but it may change locations during the event. Every incident or event must have some form of an ICP. The ICP may be located in a vehicle, trailer, tent, or within a building. The ICP will be positioned outside of the present and potential hazard zone but close enough to the incident to maintain command. The ICP will be designated by the name of the incident (e.g., Trail Creek ICP, Gulf ICP).

Staging Areas are temporary locations at an incident where personnel and equipment are kept while waiting for tactical assignments. The resources in the staging area are always in available status. Staging areas should be located close enough to the incident for a timely response, but far enough away to be out of the immediate impact zone. There may be more than one staging area at an incident. Staging areas can be colocated with the ICP, bases, camps, helibases, or helispots. Resources in the staging area are to be ready to respond within three minutes.

NOTE Available status means that the resource is staged and ready to respond in three minutes or less.

A *Base* is the location from which primary logistics and administrative functions are coordinated and administered. The base may be colocated with the ICP. There is only one base per incident, and it is designated by the incident name (e.g., Trail Creek Base). The base is established and managed by the Logistics Section.

A *Camp* is the location where resources may be kept to support incident operations if a base is not accessible to all resources or the incident is of a large enough scale as to require extended transportation times from the base to the tactical work assignments. Camps are temporary locations within the general incident area, which are equipped and staffed to provide food, water, sleeping areas, and sanitary services. Camps are designated by geographic location or number. Multiple camps may be used, but not all incidents will have camps.

A *Helibase* (**Figure 13-1**) is the location from which helicopter-centered air operations are conducted. Helibases are generally used on a more long-term basis and include such services as fueling and maintenance. The helibase is usually designated by the name of the incident (e.g., Trail Creek Helibase).

Helispots are more temporary locations at the incident, where helicopters can safely land and take off. Multiple helispots may be used.

FIGURE 13-1
Multiple helicopters operating from helibase at incident.

Courtesy of Casey Christie, The Bakersfield Californian

Resource Management

ICS resources can be factored into two categories:

1. *Tactical resources.* Personnel and major items of equipment on assignment to incidents that are available or potentially available to the operations function are called tactical resources.
2. *Support resources.* All other resources required to support the incident. Food, communications equipment, portable toilets, supplies, and fleet vehicles are examples of support resources.

Tactical resources are always classified as being in one of the following statuses:

- *Assigned* resources are working on an assignment under the direction of a Supervisor.
- *Available* resources are assembled, have been issued their equipment, and are ready for immediate assignment.
- *Out-of-Service* resources are not ready for available or assigned status.

NOTE Available status means that the resource is staged and ready to respond in three minutes or less.

Maintaining an accurate and up-to-date picture of resource utilization is a critical component of resource management.

Resource management includes processes for:

- Categorizing resources
- Ordering resources
- Dispatching resources
- Tracking resources
- Recovering resources

It also includes processes for reimbursement for resources, as appropriate.

Integrated Communications

The use of a common communications plan is essential for ensuring that responders can communicate with one another during an incident. Communication equipment, procedures, and systems must operate across jurisdictions (interoperability).

Developing an integrated voice and data communications system, including equipment, systems, and protocols, must occur prior to an incident.

Effective ICS communications include three elements:

1. *Modes.* The "hardware" systems that transfer information.
2. *Planning.* Planning for the use of all available communications resources.
3. *Networks.* The procedures and processes for transferring information internally and externally.

Chain of Command and Unity of Command

In the Incident Command System:

- Chain of command means that there is an orderly line of authority within the ranks of the organization, with lower levels subordinate to, and connected to, higher levels.
- Unity of command means that every individual is accountable to only one designated supervisor to whom they report at the scene of an incident.

These principles clarify reporting relationships and eliminate the confusion caused by multiple, conflicting directives. Incident managers at all levels must be able to control the actions of all personnel under their supervision. These principles do not apply to the exchange of information. Although orders must flow through the chain of command, members of the organization may directly communicate with each other to ask for or share information.

The command function may be carried out in two ways:

1. As a *Single Command,* in which the Incident Commander will have complete responsibility for incident management. A Single Command may be simple, involving an Incident Commander and single resources, or it may be a complex organizational structure with an Incident Management Team.
2. As a *Unified Command,* in which responding agencies and/or jurisdictions with responsibility for the incident share incident management.

Unified Command

A Unified Command may be needed for incidents involving:

- Multiple jurisdictions
- A single jurisdiction with multiple agencies sharing responsibility
- Multiple jurisdictions with multi-agency involvement

If a Unified Command is needed, Incident Commanders representing agencies or jurisdictions that share responsibility for the incident manage the response from a single Incident Command Post.

A Unified Command allows agencies with different legal, geographic, and functional authorities and responsibilities to work together effectively without affecting individual agency authority, responsibility, or accountability.

Under a Unified Command, a single, coordinated Incident Action Plan will direct all activities. The Incident Commanders will supervise a single Command and General Staff organization and speak with one voice.

NOTE The need for a unified command is based on the fact that incidents have no regard for jurisdictional boundaries. Incident planning requires unified objectives in the IAP and these are established through Unified Command.

Transfer of Command

The process of moving the responsibility for incident command from one Incident Commander to another is called transfer of command. Transfer of command may take place when:

- A more qualified person assumes command. The emphasis here is on qualification, not higher rank.
- The incident situation changes over time, resulting in a legal requirement to change command.
- Changing command makes good sense (e.g., an Incident Management Team takes command of an incident from a local jurisdictional unit due to increased incident complexity).
- There is normal turnover of personnel on long or extended incidents (i.e., to accommodate work/rest requirements).
- The incident response is concluded and incident responsibility is transferred back to the home agency.

The transfer of command process always includes a transfer of command briefing, which may be oral, written, or a combination of both. Once the predetermined time for transfer occurs, a formal transfer of command takes place, usually during an incident shift change briefing, and the transfer is announced on all incident radio frequencies. The method often used by federal incident management teams is to shadow the present command for a period of time and negotiate a time when formal transfer of command will occur. This process is referred to as a "transition of command."

Should a change of command occur at an incident it must be formally announced that a "transfer of command" has occurred.

Accountability

Effective accountability during incident operations is essential at all jurisdictional levels and within individual functional areas. Individuals must abide by their agency policies and guidelines and any applicable local, tribal, state, or federal rules and regulations. The following guidelines must be adhered to:

- *Check-In*: All responders, regardless of agency affiliation, must report in to receive an assignment in accordance with the procedures established by the Incident Commander.
- *Incident Action Plan*: Response operations must be directed and coordinated as outlined in the IAP.

- *Unity of Command*: Each individual involved in incident operations will be assigned to only one supervisor.
- *Span of Control*: Supervisors must be able to adequately supervise and control their subordinates, as well as communicate with and manage all resources under their supervision.
- *Resource Tracking*: Supervisors must record and report resource status changes as they occur.

Mobilization

At any incident or event, the situation must be assessed and response planned. Resources must be organized, assigned, and directed to accomplish the incident objectives. As they work, resources must be managed to adjust to changing conditions.

Managing resources safely and effectively is the most important consideration at an incident. Therefore, personnel and equipment should respond only when requested or when dispatched by an appropriate authority.

To provide for accountability of resources, no resource should self-dispatch to an incident. They should only respond when ordered (dispatched) to do so.

Information and Intelligence Management

The analysis and sharing of information and intelligence is an important component of ICS. The incident management organization must establish a process for gathering, sharing, and managing incident-related information and intelligence.

Intelligence includes not only national security or other types of classified information but also other operational information that may come from a variety of different sources, such as:

- Risk assessments
- Medical intelligence (i.e., surveillance of hospitals and other medical providers for occurrences of certain type of reportable diseases, such as smallpox)
- Weather information
- Geospatial data
- Structural designs
- Toxic contaminant levels
- Utilities and public works data

ICS Organization

The ICS organization is unique but easy to understand. There is no correlation between the ICS organization and the administrative structure of any single agency or jurisdiction. This is deliberate, because confusion over different position titles and organizational structures has been a significant stumbling block to effective incident management in the past.

For example, someone who serves as a Chief every day may not hold that title when deployed under an ICS structure. In departments that require incident qualification training and experience be documented and tracked, an advantage of the ICS is that all positions and the personnel trained to staff them are identified before the incident occurs.

NOTE An advantage of the ICS is that all positions and the personnel qualified to staff them can be identified before the incident occurs. Personnel are pre-qualified through training and experience.

Performance of Management Functions

Every incident or event requires that certain management functions be performed. The problem must be identified and assessed, a plan to deal with it developed and implemented, and the necessary resources procured and paid for.

Regardless of the size of the incident, these management functions will still apply.

Five Major Management Functions

There are five major management functions (**Figure 13-2**) that are the foundation upon which the ICS organization develops. These functions apply whether you are handling a routine emergency, organizing for a major non-emergency event, or managing a response to a major disaster. The five major management functions are:

1. *Incident Command* sets the incident objectives, strategies, and priorities and has overall responsibility for the incident.
2. *Operations* conducts operations to reach the incident objectives. Establishes the tactics and directs all operational resources.
3. *Planning* supports the incident action planning process by tracking resources, collecting/analyzing information, and maintaining documentation.

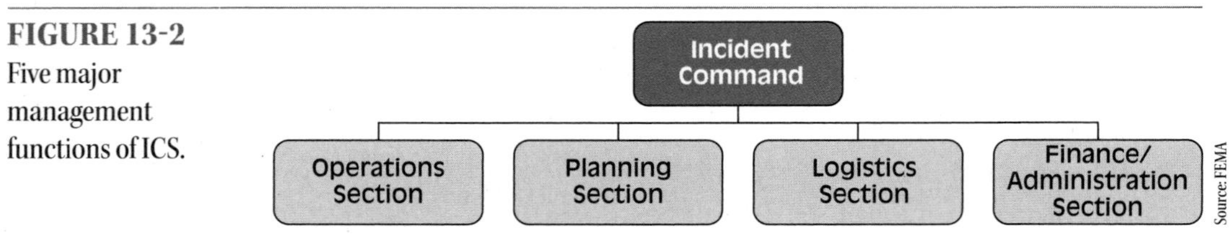

FIGURE 13-2 Five major management functions of ICS.

4. *Logistics* provides resources and needed services to support the achievement of the incident objectives.
5. *Finance/Administration* monitors costs related to the incident. Provides accounting, procurement, time recording, and cost analyses.

Organizational Structure—Incident Commander

The Incident Commander has overall responsibility for managing the incident by establishing objectives, planning strategies, and implementing tactics. The Incident Commander is the only position that is always staffed in ICS applications. On small incidents and events, one person, the Incident Commander, may accomplish all management functions. The Incident Commander is responsible for all ICS management functions until he or she delegates the function. The rule that applies here, as previously stated, is "what you do not assign you assume."

Organizational Structure—ICS Sections

Each of the primary ICS Sections may be subdivided as needed. The ICS organization has the capability to expand or contract to meet the needs of the incident.

A basic ICS operating guideline is that the person at the top of the organization is responsible until the authority is delegated to another person. Thus, on smaller incidents where these additional persons are not required, the Incident Commander will personally accomplish or manage all aspects of the incident organization.

ICS Position Titles

To maintain span of control, the ICS organization can be divided into many levels of supervision. At each level, individuals with primary responsibility positions have distinct titles (Table 13-1).

TABLE 13-1 Supervisory Position Titles

Organizational Level	Title	Support Position
Incident Command	Incident Commander	Deputy
Command Staff	Officer	Assistant
General Staff (Section)	Chief	Deputy
Branch	Director	Deputy
Division/Group	Supervisor	N/A
Unit	Leader	Manager
Strike Team/Task Force	Leader	Single Resource Boss

Using specific ICS position titles serves three important purposes:

1. Titles provide a common standard for all users. For example, if one agency uses the title "Branch Chief," another "Branch Manager," and so forth, this lack of consistency can cause confusion at the incident.
2. The use of distinct titles for ICS positions allows for filling ICS positions with the most qualified individuals rather than by seniority.
3. Standardized position titles are useful when requesting qualified personnel. For example, in deploying personnel, it is important to know if the positions needed are Unit Leaders, technical experts, etc.

Incident Commander's Overall Role

The Incident Commander has overall responsibility for managing the incident by establishing objectives, planning strategies, and implementing tactics. The Incident Commander must be fully briefed and should have a written **delegation of authority**. Initially, assigning tactical resources and overseeing operations will be under the direct supervision of the Incident Commander.

Personnel assigned by the Incident Commander have the authority of their assigned positions, regardless of the rank they hold within their respective agencies.

delegation of authority In a single jurisdiction incident the authority is already established (e.g., the local fire department is responsible for incidents within its jurisdiction). Should an incident management team be assigned to manage the incident they will require a written delegation of authority.

Incident Commander Responsibilities

In addition to having overall responsibility for managing the entire incident, the Incident Commander is specifically responsible for:

- Ensuring incident safety
- Providing information services to internal and external stakeholders
- Establishing and maintaining liaison with other agencies participating in the incident

The Incident Commander may appoint one or more Deputies, if applicable, from the same agency or from other agencies or jurisdictions. Deputy Incident Commanders must be as qualified as the Incident Commander.

Selecting and Changing Incident Commanders

As incidents expand or contract, change in jurisdiction or discipline, or become more or less complex, command may change to meet the needs of the incident.

Rank, grade, and seniority are not the factors used to select the Incident Commander. The Incident Commander is always a highly qualified individual trained to lead the incident response.

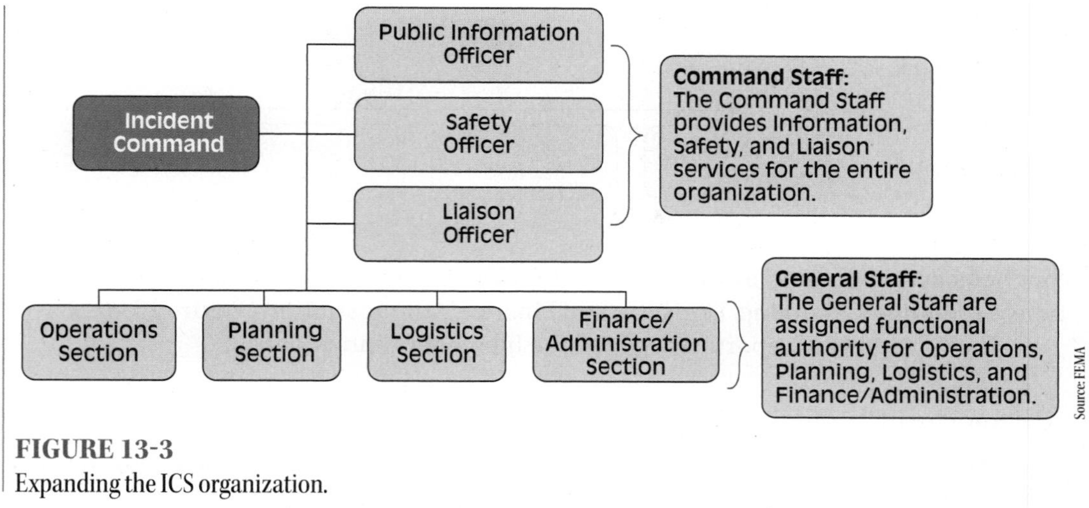

FIGURE 13-3
Expanding the ICS organization.

Expanding the Organization

As incidents grow, the Incident Commander may delegate authority for performance of certain activities to the Command Staff and to the General Staff. The Incident Commander will add positions only as needed (**Figure 13-3**).

Command Staff

Depending upon the size and type of incident or event, it may be necessary for the Incident Commander to designate personnel to provide information, safety, and liaison services for the entire organization. In ICS, these personnel make up the Command Staff and consist of the:

- Public Information Officer, who serves as the conduit for information to internal and external stakeholders, including the media or other organizations seeking information directly from the incident or event
- Safety Officer, who monitors safety conditions and develops measures for assuring the safety of all assigned personnel
- Liaison Officer, who serves as the primary contact for supporting agencies assisting at an incident

The Command Staff reports directly to the Incident Commander.

General Staff

Expansion of the incident may also require the delegation of authority for the performance of the other management functions. The people who perform the other four management

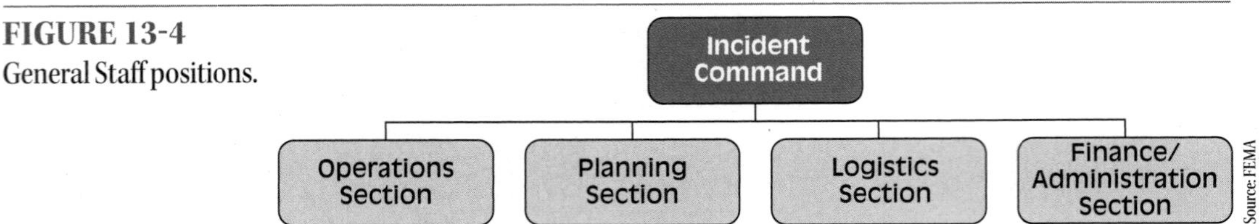

FIGURE 13-4
General Staff positions.

functions are designated as the General Staff. The General Staff is made up of four Sections: Operations, Planning, Logistics, and Finance/Administration (**Figure 13-4**).

The General Staff reports directly to the Incident Commander.

ICS Section Chiefs and Deputies

As mentioned previously, the person in charge of each Section is designated as a Chief. Section Chiefs have the ability to expand their Section to meet the needs of the situation. Each of the Section Chiefs may have a Deputy, or more than one, if necessary. The Deputy:

- May assume responsibility for a specific portion of the primary position, work as relief, or be assigned other tasks.
- Should always be as proficient as the person for whom he or she works.

At large incidents, especially where multiple disciplines or jurisdictions are involved, the use of Deputies from other organizations can greatly increase interagency coordination.

Operations Section

Until Operations is established as a separate Section, the Incident Commander has direct control of tactical resources. The Incident Commander will determine the need for a separate Operations Section at an incident or event. When the Incident Commander activates an Operations Section, she will assign an individual as the Operations Section Chief.

Operations Section Chief

The Operations Section Chief will develop and manage the Operations Section to accomplish the incident objectives set by the Incident Commander. The Operations Section Chief is normally the person with the greatest technical and tactical expertise in dealing with the problem presented by the incident.

NOTE The Operations Chief is in charge of all tactical resources assigned to the incident.

Operations Section: Maintaining Span of Control

The Operations function is where the tactical fieldwork is done and the most incident resources are assigned. Often the most hazardous activities are carried out there. The following supervisory levels can be added to help manage span of control.

- Divisions are used to divide an incident geographically.
- Groups are used to describe functional areas of operation.
- Branches are used when the number of Divisions or Groups exceeds the span of control, and can be either geographical or functional (**Figure 13-5**).

Operations Section: Divisions

Divisions are used to divide an incident geographically. The person in charge of each Division is designated as a Supervisor. How the area is divided is determined by the needs of the incident.

The most common way to identify Divisions is by using alphabet characters (A, B, C, etc.). Divisions are designated in a clockwise fashion beginning at the "front" of the incident (Division A). Other identifiers may be used as long as Division identifiers are known by assigned responders. An example of this would be to divide a multi-story building into divisions numbered by the floor they are on, such as Division 1 on the ground floor, and so forth.

The important thing to remember about ICS Divisions is that they are established to divide an incident into geographical areas of operation.

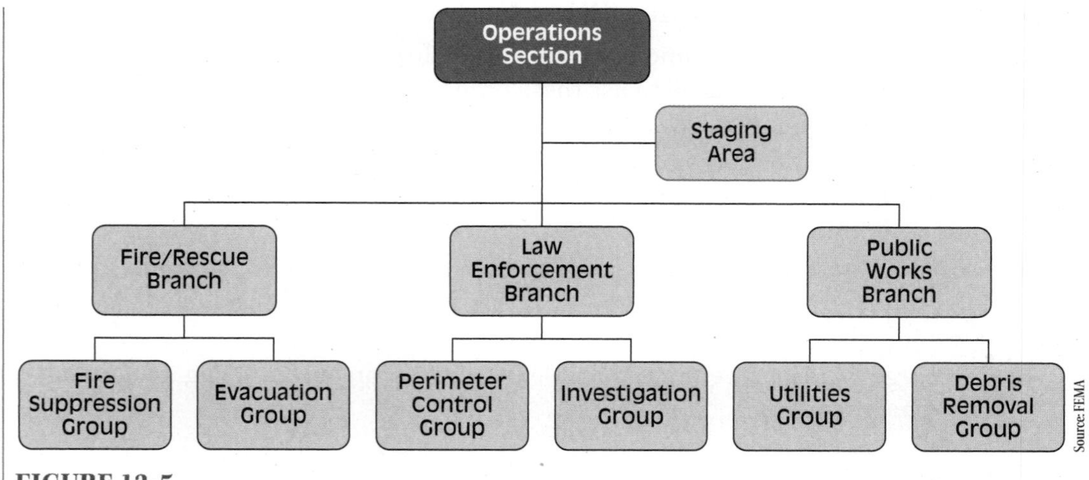

FIGURE 13-5
Operations Section: establishing branches.

Operations Section: Groups

Groups are used to describe functional areas of operation. The person in charge of each Group is designated as a Supervisor.

The kind of Group to be established will also be determined by the needs of an incident. Groups are normally labeled according to the job that they are assigned (e.g., Rescue Group, Ventilation Group, Structure Protection Group, etc.). Groups will work wherever their assigned task is needed and are not limited geographically, and they may work across division boundaries.

Operations Section: Divisions and Groups

Divisions and Groups can be used together on an incident. Divisions and Groups are at an equal level in the organization. One does not supervise the other. When a Group is working within a Division on a special assignment, Division and Group Supervisors must closely coordinate their activities. The evacuation group could be working in Division C assisting the public in evacuating a subdivision.

Operations Section: Establishing Branches

If the number of Divisions or Groups exceeds the span of control, it may be necessary to establish another level of organization within the Operations Section, called Branches. The person in charge of each Branch is designated as a Director. Deputies may also be used at the Branch level. Branches can be divided into Groups or Divisions—or can be a combination of both.

Operations Section: Branches, Other Factors

While span of control is a common reason to establish Branches, additional considerations may also indicate the need to use these Branches, including:

- *Multidiscipline Incidents.* Some incidents have multiple disciplines involved (e.g., Firefighting, Law Enforcement, Health & Medical, Hazardous Materials, Public Works & Engineering, Energy, etc.) that may create the need to set up incident operations around a functional Branch structure.
- *Multijurisdictional Incidents.* In some incidents it may be better to organize the incident around jurisdictional lines. In these situations, Branches may be set up to reflect jurisdictional boundaries.
- *Very Large Incidents.* Very large incidents may be organized using geographic or functional Branches.

Managing the Operations Section

While there are any number of ways to organize field responses, Branches and Groups may be used to organize resources and maintain span of control.

Operations Section: Expanding and Contracting

The Incident Commander or Operations Section Chief at an incident may work initially with only a few single resources or staff members.

The Operations Section usually develops from the bottom up. The organization will expand to include needed levels of supervision as more and more resources are deployed.

Task Forces are a combination of mixed resources with common communications operating under the direct supervision of a Leader (**Figure 13-6**). Task Forces can be versatile combinations of resources and their use is encouraged. The combining of resources into Task Forces allows for several resource elements to be managed under one individual's supervision, thus lessening the span of control of the Supervisor. An example of this would be two fire engines and a water tender to support them under one Task Force Leader.

Strike Teams are a set number of resources of the same kind and type with common communications operating under the direct supervision of a Strike Team Leader. Strike Teams are highly effective management units. The foreknowledge that all elements have the same capability and the knowledge of how many will be applied allows for better planning, ordering, utilization, and management.

Single Resources may be individuals, a piece of equipment and its personnel complement, or a crew or team of individuals with an identified supervisor that can be used at an incident.

It is important to maintain an effective span of control. Maintaining span of control can be done easily by grouping resources into Strike Teams, Task Forces, Divisions or Groups. Another way to add supervision levels is to create Branches within the Operations Section.

At some point, the Operations Section and the rest of the ICS organization will contract. The decision to contract will be based on the achievement of tactical objectives. Demobilization planning begins upon activation of the first personnel and continues until the ICS organization ceases operation.

Planning Section Units

The Planning Section can be further staffed with four Units (**Figure 13-7**). In addition, Technical Specialists who provide special expertise useful in incident management and response may also be assigned to work in the Planning Section. Depending on the needs, Technical Specialists may also be assigned to other Sections in the organization.

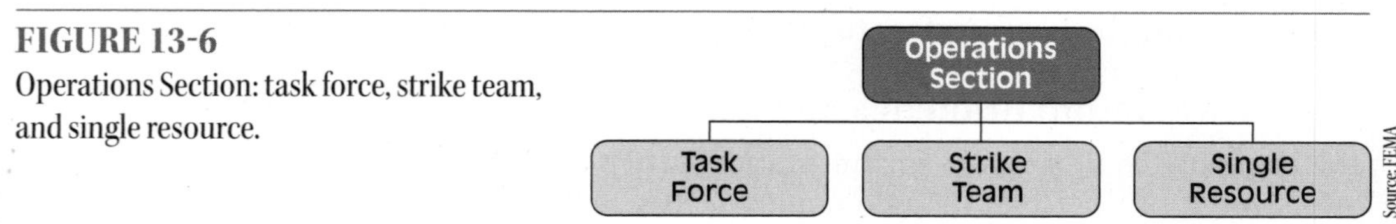

FIGURE 13-6
Operations Section: task force, strike team, and single resource.

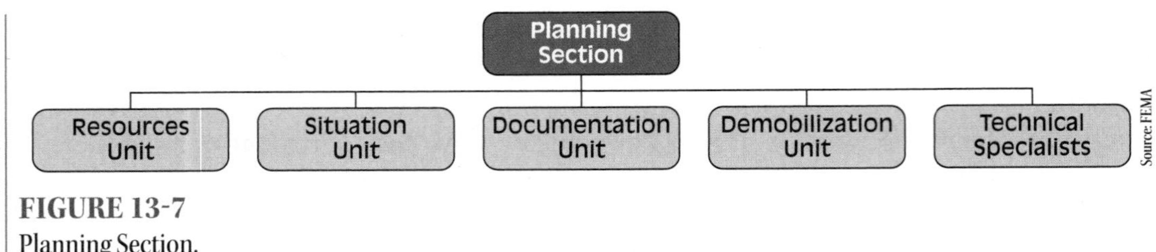

FIGURE 13-7
Planning Section.

- *Resources Unit.* Conducts all check-in activities and maintains the status of all incident resources. The Resources Unit plays a significant role in preparing the written Incident Action Plan.
- *Situation Unit.* Collects and analyzes information on the current situation, prepares situation displays and situation summaries, and develops maps and projections.
- *Documentation Unit.* Provides duplication services, including the written Incident Action Plan. Maintains and archives all incident-related documentation.
- *Demobilization Unit.* Assists in ensuring that resources are released from the incident in an orderly, safe, and cost-effective manner.

Logistics Section

The Incident Commander will determine if there is a need for a Logistics Section at the incident, and will designate an individual to fill the position of the Logistics Section Chief. If no Logistics Section is established, the Incident Commander will perform all logistical functions. The size of the incident, complexity of support needs, and the incident length will determine whether a separate Logistics Section is established. Additional staffing is the responsibility of the Logistics Section Chief.

Logistics Section: Major Activities

The Logistics Section is responsible for all of the services and support needs, including:

- Ordering, obtaining, maintaining, and accounting for essential personnel, equipment, and supplies
- Providing communication planning and resources
- Setting up food services
- Setting up and maintaining incident facilities
- Providing support transportation
- Providing medical services to incident personnel

Logistics Section: Branches and Units

The Logistics Section can be further staffed by two Branches and six Units (**Figure 13-8**).

Not all of the Units may be required; they will be established based on need. The titles of the Units are descriptive of their responsibilities.

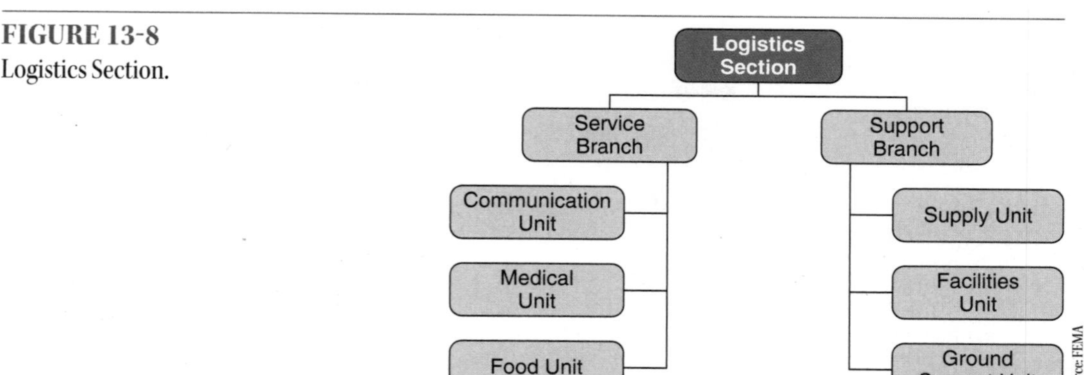

FIGURE 13-8
Logistics Section.

The Logistics Service Branch can be staffed to include:

- *Communication Unit.* Prepares and implements the Incident Communication Plan, distributes and maintains communications equipment, supervises the Incident Communications Center, and establishes adequate communications over the incident.
- *Medical Unit.* Develops the Medical Plan, provides first aid and light medical treatment for personnel assigned to the incident, and prepares procedures for a major medical emergency.
- *Food Unit.* Responsible for providing meals and drinking water for incident personnel, and obtains the necessary equipment and supplies to operate food service facilities at bases and camps.

The Logistics Support Branch can be staffed to include:

- *Supply Unit.* Determines the type and amount of supplies needed to support the incident. The Unit orders, receives, stores, and distributes supplies, and services nonexpendable equipment. All resource orders are placed through the Supply Unit. The Unit maintains inventory and accountability of supplies and equipment.
- *Facilities Unit.* Sets up and maintains incident facilities. Provides managers for the Incident Base and Camps. Also responsible for facility security and facility maintenance services: sanitation, lighting, and cleanup.
- *Ground Support Unit.* Prepares the Transportation Plan. Arranges for, activates, and documents the fueling and maintenance of assigned ground transportation. Arranges for the transportation of personnel, supplies, food, and equipment.

Finance Administration Section

The Incident Commander will determine if there is a need for a Finance/Administration Section at the incident and designate an individual to fill the position of the Finance/Administration Section Chief.

If no Finance/Administration Section is established, the Incident Commander will perform all finance functions.

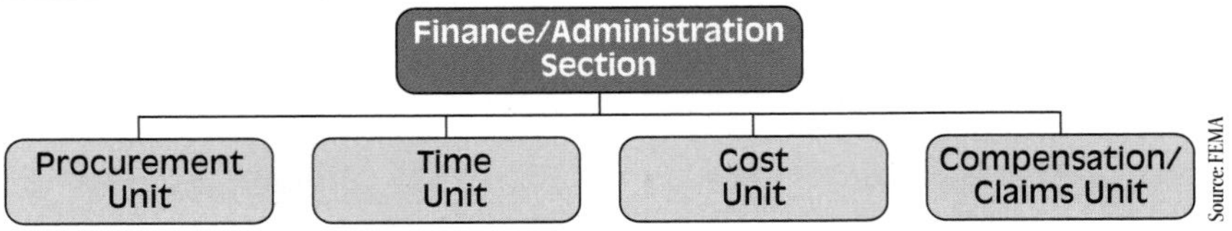

FIGURE 13-9 Finance/Administration Section.

Finance Administration Section: Major Activities

The Finance/Administration Section is set up for any incident that requires incident-specific financial management. The Finance/Administration Section is responsible for:

- Contract negotiation and monitoring
- Timekeeping
- Cost analysis
- Compensation for injury or damage to property

Finance Administration Section: Increasing Use

More and more, larger incidents are using a Finance/Administration Section to monitor costs. Smaller incidents may also require certain Finance/Administration support.

For example, the Incident Commander may establish one or more Units of the Finance/Administration Section for such things as procuring special equipment, contracting with a vendor, or making cost estimates for alternative response strategies.

Finance Administration Section: Units

The Finance/Administration Section may staff four Units (**Figure 13-9**). Not all Units may be required; they will be established based on need.

- *Procurement Unit.* Responsible for administering all financial matters pertaining to vendor contracts, leases, and fiscal agreements.
- *Time Unit.* Responsible for incident personnel time recording.
- *Cost Unit.* Collects all cost data, performs cost effectiveness analyses, provides cost estimates, and makes cost savings recommendations.
- *Compensation/Claims Unit.* Responsible for the overall management and direction of all administrative matters pertaining to compensation for injury as well as claims-related activities kept for the incident.

SUMMARY

At any time at an incident, you should be able to answer three questions: What do you have? What do you need? What is your plan? If you cannot do so, you do not have clearly defined objectives. If you ever hope to become an effective incident commander, you must think along these lines at all times when operating at emergencies. As you progress in your experience and training, the answers to these three questions will become more refined. As you are given more responsibility, the amount of information that goes into determining your answers will increase and so will the importance of your decisions.

For effective management of any type of incident, there must be an incident action plan. Plans are based on the incident objectives with strategies and tactics designed to achieve the objectives. Only through a proper size-up and logical thought processes can a plan be developed. Any time plans are developed, all members of the management team must buy into the plan and fully support it.

The incident command system is a method of placing the plan into operation. It does no good to have a complicated organization without a clearly defined set of objectives. Conversely, it is not effective to have a clearly stated set of objectives and strategies with no structure for placing the plan into operation.

An advantage of the ICS is that all positions and the trained personnel to staff them can be identified before the incident occurs. The incident command system works so well because the personnel that staff the necessary functions to support the incident are trained and prepared to assume their positions. When the ICS is implemented, the resources can be ordered and placed into action in a structure that was organized and understood before the incident occurs. Everyone in the organization should know where they fit, in terms of who they report to and who reports to them. Through a working knowledge of a position in the incident command system, any position-qualified firefighter can be assigned to an incident and carry out the duties of the assigned position.

As you can see, a major incident requires a large number of personnel just to staff the ICS positions that support the personnel and equipment operating to control the incident. This provides for an orderly development and demobilization of the organization as the demands of the incident increase or decrease. With the standardization of the ICS across the nation through NIMS, a management team or resources from another area can be brought in to assist with an incident if the need arises.

The ICS system also provides adaptability for almost any type of incident. The management staff can adapt the structure and positions of the organization within the ICS framework to meet incident needs.

REVIEW QUESTIONS

1. What ICS position is in charge at an incident?
2. Define objectives.
3. Define strategy.
4. Define tactics.
5. What are the three attack modes?
6. Interior structural attack is performed in which mode?
7. List the seven strategic priorities.

8. List the steps in a size-up. Give a listing of the information necessary for structure and wildland size-ups.
9. What are the five main components of the NIMS?
10. Whose responsibility is it to implement the ICS on incidents?
11. What are the positions of the command staff?
12. What are the positions of the general staff?
13. Which position has control of tactical resources at an incident?
14. What is the difference between a group and a division?
15. What is the difference between a strike team and a task force?
16. What does it mean to be in available status?
17. When operating an engine on an ICS-managed incident, which unit do you see for fuel?

DISCUSSION QUESTIONS

1. Why is unified command important in the case of a multiagency or multijurisdiction incident?
2. Why is it important to have a commonly understood incident command system in place before agencies are required to help each other out in times of major emergencies?
3. With the activation of resources on a nationwide basis (e.g., Oklahoma City bombing, the Pentagon and the New Yorks World Trade Center attacks, Hurricane Katrina, Gulf Oil Spill), do you think the fire service needs more or less standardization in incident command systems?
4. Using examples of various incident types (flood, fire, snow storm, hurricane, wildland fire, etc.) what would be the appropriate ICS organization to manage these incidents? Using these incident types, create an ICS organizational chart for the ones that you are most likely to encounter in your area.
5. Why should "Provide for responder and public safety" be the first objective in any Incident Action Plan?

NOTES

1. Merriam Webster, *New Collegiate Dictionary* (Springfield, MA: Merriam Webster, Inc., 2010).
2. Lloyd Layman, *Fire Fighting Tactics* (Quincy, MA: National Fire Protection Association, 1972).
3. National Wildfire Coordinating Group, *Incident Response Pocket Guide* (Boise, ID: National Wildfire Coordinating Group, 2010).
4. Lloyd Layman, *Fire Fighting Tactics* (Quincy, MA: National Fire Protection Association, 1972).
5. Merriam Webster, *New Collegiate Dictionary* (Springfield, MA: Merriam Webster, Inc., 2010).
6. Ibid.
7. FEMA, *IS-100—Incident Command System (ICS) 100 Training* (www.FEMA.gov, 2010)

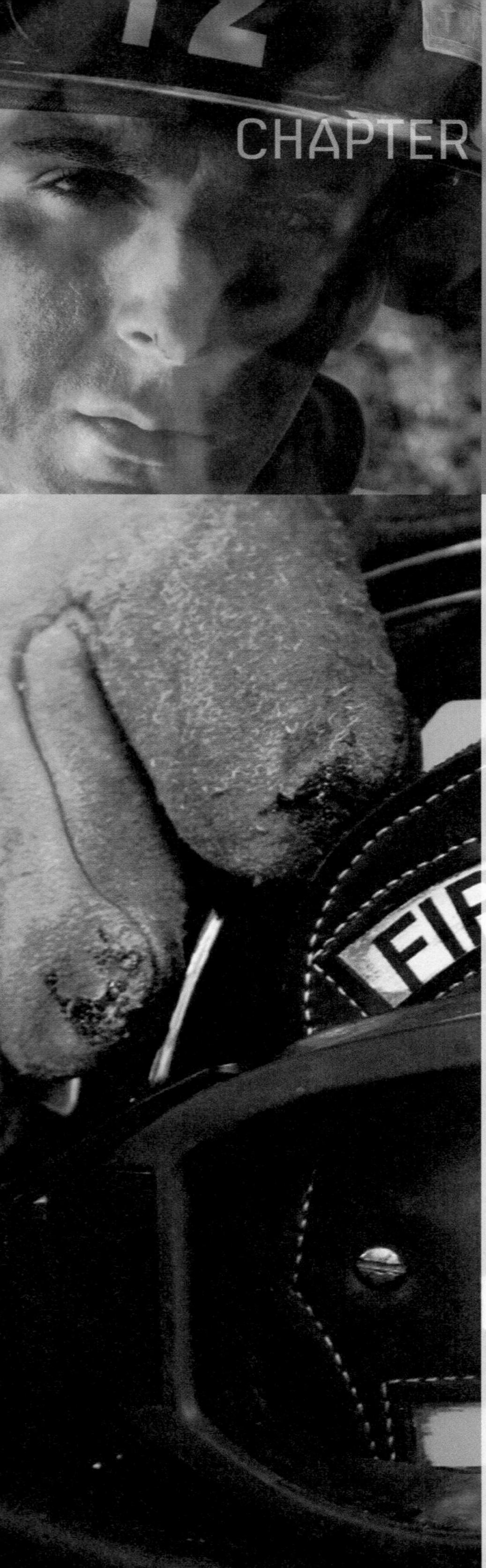

CHAPTER 14

Emergency Operations

LEARNING OBJECTIVES

Upon completion of this chapter, you should be able to:

- List the *16 Firefighter Life Safety Initiatives*.
- Identify the role of the fire department at various types of emergencies.
- List limitations of the fire department in certain types of emergencies.
- List important safety considerations when operating at different types of emergencies.
- Identify the importance of decision making in maintaining firefighter safety.

INTRODUCTION

The personnel resources of the fire department can be divided into two general areas, operations (also known as "Line") and support (also known as "Staff"). The operational personnel resources of the fire department, not to be confused with the Operations Section under ICS, consist of the people in the fire stations and their supervisors. Operations/Line personnel are the personnel charged with responding to and the **mitigation** of incidents.

Support functions are those that aid the Operations/Line personnel in performing their jobs. These can include mechanical repair services, dispatching, training, and personnel services.

One of the fundamental roles of the fire department is to respond to emergencies. Not every incident is an emergency and not every emergency is the responsibility of the fire department to manage. In many types of incidents, the fire department does what it can in light of its legal and resource limitations. A well-trained firefighter should be able to deal with most types of emergency situations. You may not be trained in that particular incident type, but many of the basic rules apply and can be used as a guide to action in unusual situations.

mitigation Reducing the hazard, making less severe.

PERSONNEL

The type of personnel who can be expected to respond to an incident depends on the incident type. Just as some doctors specialize in different types of medicine and parts of the human body, the fire department has its own specialists. These personnel have become trained in dealing with special incident types as well as being competent in regular fire department functions such as firefighting and medical aid.

NOTE The fire department is not the only agency that shows up at an incident that has a say in determining the objectives, strategy, and tactics to be used.

The fire department is not the only agency that shows up at an incident that has a say in determining the objectives, strategy, and tactics to be used (**Figure 14-1**). Law enforcement and other public agencies assist and sometimes are in command

FIGURE 14-1 Firefighters and ambulance personnel treating patient at scene of vehicle accident.

Delmar/Cengage Learning

of incidents in which the fire department is involved. In some types of incidents the number of representatives from other agencies may outnumber the firefighters who perform the necessary tasks to mitigate the incident. An incident of this type would be managed using Unified Command.

As expressed in Chapter 13, the first objective of any incident is to "Provide for responder and public safety." This chapter focuses on primarily firefighter safety, but the general rules should apply to all responders.

16 FIREFIGHTER LIFE SAFETY INITIATIVES

In 2004 the National Fallen Firefighter Foundation examined the causes of line-of-duty firefighter deaths. As a result of this meeting they put forth the 16 Firefighter Life Safety Initiatives to be followed by all firefighters.[1]

1. Define and advocate the need for a cultural change within the fire service relating to safety; incorporating leadership, management, supervision, accountability and personal responsibility.
2. Enhance the personal and organizational accountability for health and safety throughout the fire service.
3. Focus greater attention on the integration of risk management with incident management at all levels, including strategic, tactical, and planning responsibilities.
4. All firefighters must be empowered to stop unsafe practices.

5. Develop and implement national standards for training, qualifications, and certification (including regular recertification) that are equally applicable to all firefighters based on the duties they are expected to perform.
6. Develop and implement national medical and physical fitness standards that are equally applicable to all firefighters, based on the duties they are expected to perform.
7. Create a national research agenda and data collection system that relates to the initiatives.
8. Utilize available technology wherever it can produce higher levels of health and safety.
9. Thoroughly investigate all firefighter fatalities, injuries, and near misses.
10. Grant programs should support the implementation of safe practices and/or mandate safe practices as an eligibility requirement.
11. National standards for emergency response policies and procedures should be developed and championed.
12. National protocols for response to violent incidents should be developed and championed.
13. Firefighters and their families must have access to counseling and psychological support.
14. Public education must receive more resources and be championed as a critical fire and life safety program.
15. Advocacy must be strengthened for the enforcement of codes and the installation of home fire sprinklers.
16. Safety must be a primary consideration in the design of apparatus and equipment.

In terms of practical application of the 16 Firefighter Life Safety initiatives every firefighter should adhere to the following:

- *Duty and responsibility.* Make every day a training day so everyone goes home.
- *Firefighter maintenance program.* Receive regular medical checkups, get regular exercise, and eat healthy.
- *Rehab guidelines.* Stop before you drop, stay hydrated, and monitor vital signs.
- *Passengers when responding to incidents.* Wear full PPE, get in the apparatus, sit down, fasten your seatbelt, and ride with drivers who will get you there in one piece.
- *Drivers responding to incidents.* It is not a race; safe is more important than fast. Stop at all red lights and stop signs, and if others do not pull over—don't run them over.
- *Interior firefighting.* Work as a team, stay together, stay oriented, manage your air supply, and take the proper tools with you for any interior operation. Every member should have a radio, provide constant updates, and constantly assess for risk versus benefit.

To access the on-line course *Courage to be Safe® So Everyone Goes Home*, go to http://www.ctbsintructors.com. A certificate of completion is available upon successfully completing this course.

Rapid Intervention Teams

The assignment of one or more companies as rapid intervention teams (RITs, also referred to as rapid intervention companies or crews [RICs], etc.) at working incidents provides the ability to immediately initiate a rescue effort to locate, rescue, or assist firefighters who are in trouble at the scene of an incident. RIT members should be standing by wearing their full protective clothing with self-contained breathing apparatus ready for immediate use. They should have forcible entry tools, rescue rope, and any other equipment that could be needed quickly. At hazardous materials incidents or other situations where special protective equipment is required, the RIT should be ready with the same level of protective clothing and equipment that the entry team requires. Some departments have established Firefighter Assist and Search Teams (F.A.S.T.) to accomplish rapid intervention tasks.

For more information on firefighter rescue, please refer to DVD clip *Rescue Procedures—Assisting Rescue Teams.*

Two In, Two Out

OSHA has created a regulation commonly referred to as "Two In, Two Out." The regulation specifies that whenever personnel are operating in atmospheres that require SCBA, especially atmospheres that are described as **immediately dangerous to life and health (IDLH)**, the procedure of Two In, Two Out must be used. OSHA recognizes that "conditions present during an advanced interior structural fire create an IDLH atmosphere." OSHA states that this applies to fire scenes that have gone "beyond the ignition stage"—in other words, fires in the growth, fully developed or decay stages.

In some departments one, two, or three personnel will be first at scene. If they wish to initiate interior fire attack when an IDLH atmosphere is present or likely to be

IDLH Immediately dangerous to life and health. IDLH atmospheres are capable of causing death, irreversible adverse health effects, or the impairment of an individual's ability to escape from a dangerous atmosphere.

present, engine company personnel must first alter the IDLH atmosphere, with the following exception:

> If, upon arrival at the scene, firefighters find an imminent life-threatening situation where immediate action may prevent the loss of life or serious injury, such action shall be permitted with less than four firefighters on the scene, when actions are based on appropriate concepts of risk assessment and management. Such action is intended to apply only to those rare and extraordinary circumstances when, in the firefighter's professional judgment, the specific instance requires immediate action to prevent the loss of life or serious injury and four firefighters have not yet arrived on the fireground.[2]

In essence, an interior attack with less than the OSHA-required four personnel at scene can be made if there is a high probability of effecting a rescue or stopping the fire from threatening persons in imminent danger. Prior to entry by the full at-scene complement of personnel there must be an announcement over the radio and a transfer of command to the next incoming company or chief officer.

Personnel operating in hazardous areas at emergency incidents shall operate in teams of two or more, using the buddy system. Team members operating in hazardous areas are required to be in communication with each other through visual or voice contact at all times. Radios or other means of electronic contact shall not be substituted for direct visual or voice contact between members of an individual team in the hazardous area. Team members shall remain in close proximity to each other to provide assistance in case of emergency. Backup team members will monitor the status of the entry team by maintaining visual and/or voice contact.

Of the four personnel at scene when operating in IDLH or potential IDLH atmospheres, the entry team must consist of two personnel. One of those two personnel who remain outside may be engaged in other activities, as long as this does not jeopardize firefighter safety and the individual's ability to participate in any rescue operation.

STRUCTURE FIREFIGHTING

One of the basic responsibilities of any fire department is fighting structure fires (**Figure 14-2**). The equipment in the form of pumpers, hose, and nozzles is designed with this primarily in mind. Previously this was the type of emergency work that firefighters performed the most. This is no longer true with the move of the fire department into medical aid delivery, but it is still one of the most important jobs performed. Most departments spend the bulk of their time and training budget preparing for this one aspect of emergency service. A fire department that can keep its structure losses to a minimum is performing an important part of its function very well. Every once in a while there is bound to be a fire that exceeds the control capabilities of the fire department and its resources. When this happens, the public is sure to take a close look at the fire department and its budget and ask if the money is being used effectively.

FIGURE 14-2
Firefighter advancing hoseline to attack a structure fire.

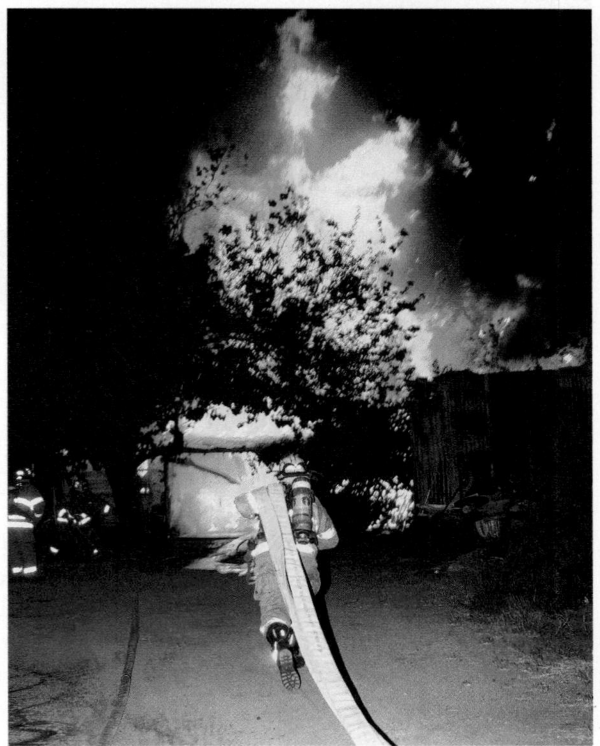

Courtesy of Edwina Davis

Firefighting operations at structure fires take one of two modes, offensive or defensive. In the offensive mode, firefighters enter the burning structure and attack the seat of the fire. In the defensive mode, the water is applied through windows or into other openings to control the fire. A key point is that hose streams should not be directed through ventilation openings when resources are inside the structure. In some cases the initial attack is made through a window to darken down the fire and then firefighters enter and complete the extinguishment. If firefighters are able to enter the structure upon arrival, a **primary search** is made for victims.

Hose streams should not be directed through ventilation openings when resources are inside the structure.

primary search A rapid search of all involved and exposed areas affected by the fire and which can be entered, to verify a removal and/or safety of all occupants. Should this not be possible a secondary search is conducted as soon as it is safe to do so.

For more information on search and rescue, please refer to DVD clip *Search & Rescue.*

For more information on ventilation, please refer to DVD clip *Ventilation.*

A firefighting operation at a structure must be a coordinated attack. The first arriving unit sizes up the fire and decides which methods will be the most effective in bringing the fire under control. Hose lines are pulled and charged, the power is cut off at the electric meter, ventilation is performed, and the firefighters attack the fire (**Figure 14-3**). If ventilation is performed before the lines are ready, the fire will intensify and may exceed the ability of the personnel at the scene to control it. Depending on the size and complexity of the fire, any number of operations are required to bring it under control. If the fire is already through the roof, ventilation may be unnecessary.

NOTE A firefighting operation at a structure must be a coordinated attack.

FIGURE 14-3
Firefighters performing roof ventilation.

Courtesy of Casey Christie, The Bakersfield Californian

For more information on interior structure fire attack, please refer to DVD clip *Fire Suppression, Coordinated Fire Attack.*

When firefighters respond to structure fires, several questions are important, no matter whether the structure is a dumpster, shed, home, or warehouse. One of these questions is, what are the contents? In the past, people were not as likely to have an assortment of household chemicals stored in the structure and firefighting was somewhat safer (**Figure 14-4**). To limit exposure to these items, firefighters have better protective equipment. Now that protective equipment has improved, firefighters sometimes think that they are invincible—which is certainly not the case. A SCBA will protect you from inhaling chemicals but will not protect you from skin contact. In a hot-structure firefighting operation, you will be sweating profusely and all of your pores will be open. If a chemical that can harm you through skin contact comes into contact with an unprotected area, you will suffer an exposure. Many of these chemicals do not show immediate effects. After repeated exposures over a 30-year career, you may find yourself suffering from cancer.

A SCBA will protect you from inhaling chemicals but will not protect you from skin contact.

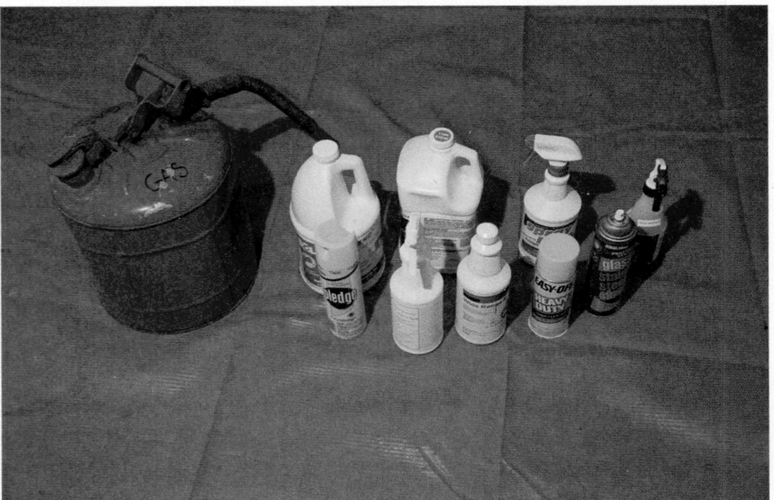

FIGURE 14-4 Hazardous household chemicals.

Delmar/Cengage Learning

Leather gloves and boots will protect you from some hazards and not from others. They are made of animal skin, which like your own, is porous. If your gloves or boots become **contaminated** with a harmful chemical, they cannot be **decontaminated**. Every time you wear them and start to sweat you are going to be exposed again. If you get a chemical on your turnout gear or uniform and take them home for washing, you can expose your whole family. A dose of a harmful chemical that will not show any ill effect on you may be lethal to small children.

Other building contents can pose a hazard. In a kitchen fire, canned goods can explode and splatter you with their boiling contents. During gasoline shortages, people were known to store trash cans full of gasoline in their homes. Some room contents, in the form of furniture, pose a hazard. Polyurethane foam can burn as hot and produce as much smoke as gasoline. Fires have occurred where the use of flammable liquids was suspected, because of heavy black smoke. After extinguishment, it was determined that the smoke was caused by burning foam, not gasoline. In areas where hunting is popular, many homes have a supply of ammunition on hand. The list of hazards in a common structure fire can go on almost indefinitely.

Interior structural components pose dangers as well. Suspended ceilings have crashed down and trapped firefighters in their framing, concealed wiring, and duct work. A ceiling fan is heavy enough to do some damage to you if you are hit. Tall book cases and high-piled stock can be knocked down by hose streams and land on firefighters. One of the worst interior structural hazards is stairs. Their use may present no problem at the beginning of the fire, but when the situation deteriorates, they can act as a chimney for smoke and flame and cut off your escape, trapping you above the fire. In heavy smoke and darkness, it is easy to fall down the stairs if you are not paying attention to where you are going. At other times the fire can weaken the stairs that you went up. As you come back down, they collapse, dropping you into a raging inferno. Curtains hanging over a window

One of the worst interior structural hazards is stairs.

For more information on interior-attack structural firefighting, please refer to DVD clip *Fire Extinguishment—Interior Structure.*

contaminated To become coated with a harmful substance.

decontaminate The physical removal of contaminants from people or equipment.

that you are using as an emergency escape route can become wrapped around you and explode into flame in a flashover situation.

Structural collapse is one of the firefighter's worst nightmares (**Figure 14-5**). A floor or roof above you can collapse and land on you, pinning you down until you run out of air or are burned to death. The floor or roof you are standing on can fail, dropping you into the fire below. Modern construction makes use of truss roof construction in many types of buildings; older construction utilized the "bow string truss" type of roof construction. Both of these types can fail dramatically when exposed to fire. **Parapet walls** and overhangs have a history of falling onto firefighters. Even whole walls have fallen, burying firefighters and equipment in the rubble.

As you walk around a burning structure, do not walk upright in front of windows. If you happen to be in front of a window at the time the fire back drafts, you will get burned. Hose lines inside the structure are quite capable of blowing the glass out of a window and cutting you badly. The old story about walking under ladders being bad luck is especially true in firefighting. If someone above you is breaking glass or drops an axe, you are likely to get hit.

As you walk around a burning structure, do not walk upright in front of windows.

parapet wall A wall that extends above the roof line.

FIGURE 14-5
Structure collapse.

Delmar/Cengage Learning

A type of structure fire that is becoming more common is the clandestine drug lab (meth lab) that has caught fire. The people cooking the drugs usually set up the operation, start the process, and then often leave for the period of time it takes to run. They may have even booby-trapped the area. In any event, they are not going to be standing out front to warn you about the dangers of what is going on inside when you arrive. Should you arrive at this type of incident, isolate and deny entry. Do not enter the structure yourself and do not touch anything. Law Enforcement has access to specially trained persons who handle this type of incident.

Not all of the hazards at a structure fire are on the inside. Heavy roof loads in the form of air conditioning units and other equipment have a tendency to come crashing through when the structure is weakened by fire (**Figure 14-6**). Chimneys pose the same hazard. A television antenna can act as a spear if it comes through the roof and pins you to the floor.

What can you do to protect yourself? Always wear your full PPE and remain alert for anything out of the ordinary. A fire that resists normal extinguishment methods may mean that there is some type of unusual fuel involved. Flame and smoke colors are an indicator. Normal structure fires burn with a reddish orange to yellow flame and black

Not all of the hazards at a structure fire are on the inside.

FIGURE 14-6 Partial roof collapse at residential structure fire.

Courtesy of Edwina Davis

smoke. If the flames are dark orange, blue, green, or some other strange color, there is something going on you should be extra careful of. If the smoke is reddish or other unusual color, it is an indicator that something is out of the ordinary.

Be suspicious. If you arrive at a structure and the windows are blown out or the doors are lying on the ground and appear to have been blown off, there is something unusual going on. If you are on your way to the fire and you see someone running down the street away from the fire with a gas can in his hand, that is a definite clue.

Leave yourself a second way out. Before entering a structure try to locate an alternate opening. Make sure there is another way out if the fire gets out of control and cuts off the path through which you entered. When working on a roof, above a fire, there should always be enough ladders placed for all of the personnel to make a quick escape. Never fewer than two ladders should be placed. If firefighters are going upstairs, ladders should be placed at windows so they can escape without using the stairs. Jumping to the ground from a second or third floor window or from the roof is not good practice and may very well end your career, if not your life.

Leave yourself a second way out.

Do not freelance. Stay with your company and your officer. Your officer has much more firefighting experience than you do and is more likely to recognize a bad situation developing before you do. By staying together, firefighters can help each other out. If one should become trapped under some rubble or drops a leg through the floor or roof, the others can lend a hand. At the least they can direct the hose stream while you disentangle yourself. Someone should always know where you have gone. If you are missing, it is better that someone recognizes that fact right away and not after they return to the station.

Do not freelance.

High-rise firefighting presents a whole list of additional hazards to the structural firefighter. Some have gone as far as to call them "elevated crematoriums." The floor areas of high-rises are often rented to different occupants. Each individual floor can have its own unique layout according to the tenant's wishes. This makes it easy to become disoriented. A practice that sometimes works is to take a quick tour of the floor below the fire floor to get an idea of the layout before entering the fire floor. The interior walls on each floor are often nothing more than partitions ending at the suspended ceiling. Above the ceiling the fire can run freely across the hidden attic space above your head. You cannot see the fire in this space and it may come out between you and your escape route. The best way to access a high-rise fire is up the stairways. When the fire is on the 110th floor, this will take a long time to accomplish and after you carry your gear to the fire floor, you may be too fatigued to be of much good. At this point the elevators sound like a good option. The trouble is, if they open onto the fire floor, you cannot be sure what you will be confronted with when the door opens. The fire department hose connections are in the stairwells. You may not be able to make it from the elevator lobby to the stairwell because of fire and heat. It is easy to get confused in heavy smoke. When you cannot make it from the elevator to the hose connection, you have no water with which to attack the fire to try to improve your situation. There are alternatives available. One is to take the elevator to the floor below the fire and use the stairs from there. This requires being absolutely sure which floor the fire is on. The elevators can be used if they are the type that opens at a **sky lobby** below the fire. This way you will not end up with the door opening onto an inferno. There are systems available on some elevators that allow firefighters to override their automatic functions. If this system is installed and the firefighters are familiar with how it works, it is another option (**Figure 14-7**).

When fighting fire in a high-rise it must first be determined exactly which floor(s) the fire is on prior to any use of elevators.

In high-rise fire situations one of the exterior hazards is falling glass (**Figure 14-8**). Most of today's high-rise buildings have glass exteriors. When there is a fire these glass panels have a tendency to fall out or get knocked out by occupants or by firefighters effecting ventilation. A large piece of glass falling from far above the street is capable of severing hose lines and doing severe damage to firefighters. Pieces of glass have the ability to sail quite far as they fall. No one should be standing out in the open within 200 feet of the fire building if possible. When underground access can be found to the structure, that is often the best way to enter.

A hazard almost always present in structure fires is electricity. One of the first things done is to detail someone to cut off the power to the structure. You should never try to pull the electrical meter. It can explode in your face, showering you with glass. Even though someone has cut the switch or main breaker to the structure, it is no guarantee that the power is off. This is especially true in multi-family, commercial, and industrial occupancies. If the occupants are stealing electricity, they may have wired around the meter. When

sky lobby A lobby on a high floor level in a high-rise building. Elevators leave from this area to service the upper floors.

FIGURE 14-7
Fire department elevator control.

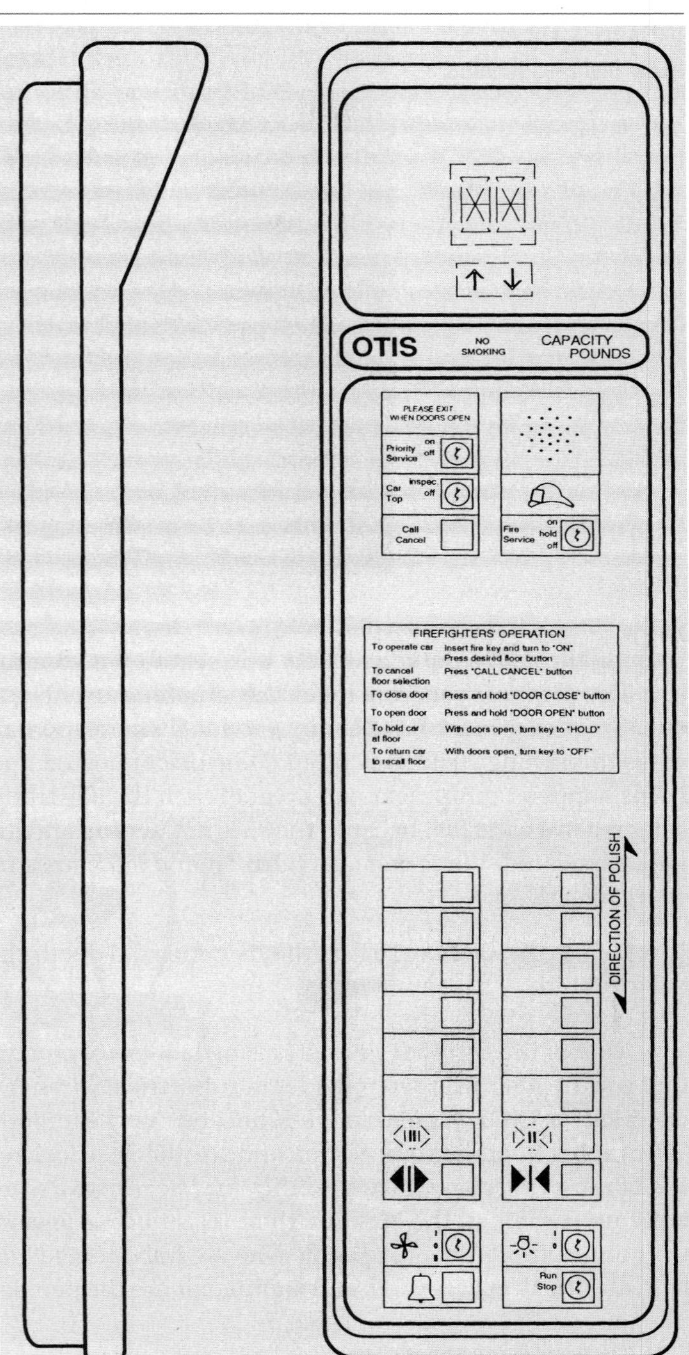

Courtesy of Otis Corp.

FIGURE 14-8
Fire in high-rise building.

Source: National Institute of Standards and Technology

walking through a darkened area, it is best to put your arms up with your palms toward you. This prevents you from reflexively grabbing an energized electrical wire if you contact one. Great care must be taken in watching where apparatus is parked because overhead wires to burning structures often come disconnected and can fall across the apparatus. If this happens, jump clear and do not touch the apparatus. Any time you see a wire, consider it a live wire. Just because they are not **arcing** and jumping does not mean that they are deactivated. Wires can also burn off and fall across metal fences, such as chain link, energizing them.

NOTE One of the first things done is to detail someone to cut off the power to the structure.

One of the cardinal rules of raising ladders of any type is to look up first. Make sure that you are not raising your ladder into electrical wires. A wooden ladder will not conduct electricity as well as an aluminum one, but they can both kill you in the right situation. Your rubber boots contain carbon and are not electrical insulators. Pike poles are not to be used for moving electrical wires. They may be resistive when new, but as they get older and build up carbon on the handles, their resistance is lowered. Do not attempt to cut wires with bolt cutters. The only people who are fully trained and carry the required equipment to deal with electrical wires and equipment are the people from the power company.

arcing Spark created when electrical contact is made.

NOTE One of the cardinal rules of raising ladders of any type is to look up first.

A hazard most people do not stop to consider at structure fires is pets. It can be very exciting to open a door or gate and see a large pit bull launching itself at you. Some people keep exotic pets, such as rattlesnakes, in their homes. Livestock can also pose a hazard to firefighters in certain situations. Entering a fenced area to extinguish a grass fire and meeting up with an enraged bull is a bad position to be in. In some situations your PPE may help to prevent injury. Your best defense is always a cautious approach.

FIRE DEPARTMENT OPERATIONS AT SPRINKLERED OCCUPANCIES

The three principle causes of unsatisfactory sprinkler performance include a closed valve in the water supply line, delivery of inadequate water supply to the system, and occupancy changes that render the installed sprinkler system unsuitable. These can be prevented through effective preincident planning and the implementation of testing and maintenance programs.

Departments should establish Standard Operating Procedures for operations at sprinklered or standpiped occupancies. One of the first actions to be taken is to ensure the supply to the system is boosted through the use of a pumper attached to the fire department connection. All of the system's valves should be checked to ensure that they are in the fully open position, unless marked closed for maintenance. This is not the case for residential systems. They are only designed to handle normal pressures provided by the water main system. Once flow from the system is ensured, hose lines can be advanced to the seat of the fire. Do not turn off the system to enhance visibility before the fire is confirmed to be under control. When a part of the system must be turned off for overhaul or salvage operations, try to turn off only the portion of the system in the affected area. The system should be placed back into service immediately before leaving the scene. If the system cannot be placed back into service, the property owner or representative should be notified.

After any fire-caused sprinkler system activation, an investigation as to cause should be conducted. Also include system effectiveness in the reporting. A full list of the components of this type of investigation are contained in NFPA Standard 13E, *Properties Protected by Automatic Sprinkler Systems.*

ELECTRICAL INSTALLATIONS

Areas that require careful preplanning and consideration are electrical substations and vaults (**Figure 14-9**). They are a hazard from a safety standpoint due to the high potential for getting electrocuted. It is also possible to cause much more damage than the fire will

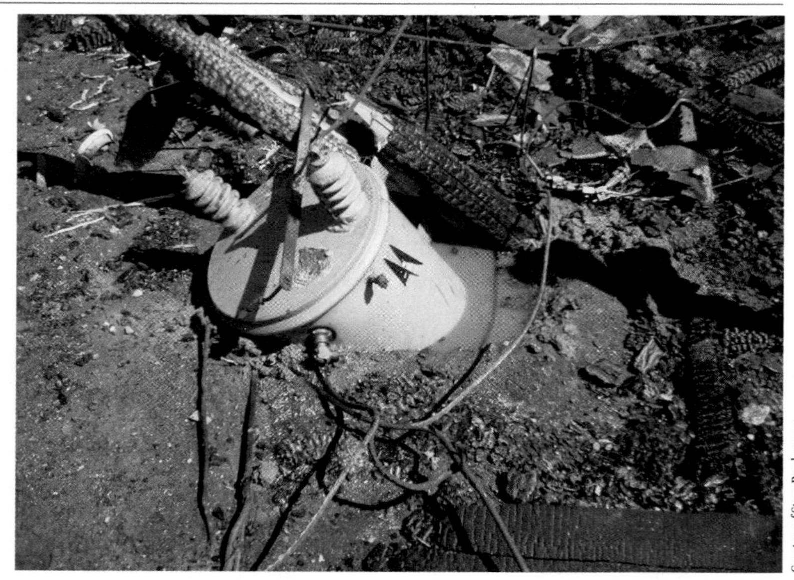

FIGURE 14-9
Results of fire in electrical equipment.

Courtesy of Steve Pendergrass

by the indiscriminate use of water or **dry chemical extinguishers**. A more common fire is the power pole fire. Water can conduct electricity. When extinguishing electrical equipment, it is better to use a fog pattern or short bursts of water to accomplish the task. This lessens the chances of getting shocked. Transformers used to contain **PCB oil**, which is considered to be carcinogenic. Power companies have, for the most part, phased this oil out. There may still be some out there and you should proceed with great caution around electrical equipment.

NOTE As a general rule firefighters should not enter or extinguish fires in electrical installations unless guided by electric company employees.

As a general rule, firefighters should not enter or extinguish fires in electrical installations unless guided by electric company employees. These personnel are trained and know where it is safe to operate and which firefighting technique should be applied. They know which switches to throw to shut down the power. Remember from Chapter 4 that a class C fire reverts to class A, B, D, or K once the power is shut down.

dry chemical extinguisher Fire extinguisher using a chemically active powder.

PCB oil Oil containing polychlorinated biphenyl, a compound that is considered to cause cancer.

WILDLAND FIREFIGHTING

In wildland firefighting operations, the basic methods of extinguishment are to apply water or fire retardant to the fire edge or to create a fire break or control line around the perimeter (**Figure 14-10**). This is done with a variety of methods. In grasslands, pumpers can be used to make a direct attack on the fire edge. Crews using hand tools can also create a **scratch line** that breaks the fuel's continuity and stops the fire spread. Dozers are used in the same manner. In heavier fuels, due to radiated heat and flame lengths, a direct attack is not possible. In these fuels, crews get well ahead of the fire and make an indirect attack, creating fire breaks with hand crews and dozers. Natural barriers, like lakes and roads, are also used as part of the fire break. When **class A foam** is available, it can be used to protect exposures by applying it as a wetting agent to raise the surface fuel moisture to where it will not burn.

A method often used to create a safe "black line" is to backfire. The backfire removes the fuel between the control line and the head of the advancing fire. The backfire burns back toward the main fire and away from the control line (**Figure 14-11**). This allows a relatively narrow control line to be used. The line is effectively widened by the removal of the fuel between the advancing fire and the control line. When burning is done to remove

scratch line A quickly created wildland fire control line, constructed using hand tools.

class A foam Foam designed for use on ordinary combustible materials.

FIGURE 14-10
Firefighters attacking wildland fire.

Courtesy of Edwina Davis

FIGURE 14-11 Engine in safety zone as fire burns past during backfiring operation.

Delmar/Cengage Learning

fuel along the **flanks** or to remove **unburned islands** left in the fire perimeter, it is called *firing out*. These firefighting methods should never be attempted by untrained personnel. When performed improperly, they can cause the fire to jump control lines.

NOTE These firefighting methods should never be attempted by untrained personnel.

The safety rules developed for wildland firefighting can be applied to all types of firefighting. The rules cover the main points that should be observed in any fire situation. The basic safety rules of operating at a wildland fire are the *Ten Standard Firefighting Orders* and the *18 Situations That Shout Watch Out*.[3]

SAFETY The safety rules developed for wildland firefighting can be applied to all types of fire fighting.

flank The sides of an advancing wildland fire.

unburned island Area of unburned fuel within a fire perimeter.

The Ten Standard Firefighting Orders

F Fight fire aggressively, but provide for safety first.
I Initiate all action based on current and expected fire behavior.
R Recognize current weather conditions and obtain forecasts.
E Ensure instructions are given and understood.
O Obtain current information on fire status.
R Remain in communication with crew members, your supervisor, and adjoining forces.
D Determine safety zones and escape routes.
E Establish lookouts in potentially hazardous situations.
R Retain control of yourself and your crew at all times.
S Stay alert, keep calm, think clearly, and act decisively.

18 Situations That Shout Watch Out

1. The fire is not scouted and sized up.
2. You are in country not seen in daylight.
3. Safety zones and escape routes are not identified.
4. You are unfamiliar with the weather and local factors influencing fire behavior.
5. You are uninformed on strategy, tactics, and hazards.
6. Instructions and assignments are not clear.
7. No communications link with crew members/supervisor.
8. Constructing line without safe anchor point.
9. Constructing fire line downhill with fire below.
10. Attempting frontal assault on fire.
11. Unburned fuel between you and the fire.
12. Cannot see main fire, not in contact with anyone who can.
13. On a hillside where rolling material can ignite fuel below.
14. Weather is getting hotter and drier.
15. Wind increases and/or changes direction.
16. Getting frequent spot fires across the line.
17. Terrain and fuels make escape to safety zones difficult.
18. You feel like taking a nap near the fire line.

All of these listed orders and situations were developed because someone was seriously injured or killed. In recent incidents where wildland firefighters were injured or killed, violations of these orders and situations were found to be contributing factors. The four most common causative factors involved in tragedy and near-miss wildland fires are (1) they usually happen on small fires or deceptively quiet sectors of large fires; (2) they happen in light fuels, such as grass or brush; (3) there is an unexpected shift in the wind direction or speed; and (4) when fires run uphill.

NOTE All of these listed orders and situations were developed because someone was seriously injured or killed.

LCES

Because it is very hard to memorize the *Ten Standard Firefighting Orders* and all of the *18 Situations That Shout Watch Out*, two more-condensed safety messages have been adopted. They cannot totally take the place of those described previously, but serve as a general reminder.

The first of these messages is **LCES**.[4]

NOTE LCES is the minimum safety standard to be established at any incident, regardless of type.

LOOKOUTS should always be posted to keep an eye on the fire and the weather. These personnel are not to become involved in actual firefighting. Their job is to keep an eye on the fire and let the crew know if the fire starts to increase in intensity or make a run at the crew's position. In areas where **hazard trees** are a problem, this may require one person watching for every two working.

COMMUNICATIONS must be maintained between the lookouts and the crews working, between the crews and their leader, between the leader and the adjoining crews, and between the crew leaders and the command personnel.

ESCAPE ROUTES should always be planned and communicated to all of the crew members in case of a change in fire behavior. They must provide access to a safety zone. As in other types of firefighting, two escape routes are better than one, in case one gets cut off.

SAFETY ZONES should be placed as often as deemed necessary. They must be large enough to accommodate all of the crews in the area in case of a burn over. The **safety zones** need to be spaced closely enough that the crew has time to travel to them if the fire dictates that they must seek refuge. They must be clearly indicated and accessible.

LCES Lookouts, Communications, Escape routes, and Safety zones.

hazard trees Trees that have burned out at the base or are liable to drop large limbs.

escape route A preplanned and understood route to a safety zone.

safety zone A place where firefighters can be safe from the incident's hazards.

F LCES Δ

When looking at the application of LCES, a modified version includes what Brad Mayhew (a former hotshot firefighter with the U.S. Forest Service) terms "F LCES Δ."[5] The F stands for "worst case scenario," also known as **WCS**. In other words, what is the worst this incident can become? In looking at the World Trade Center, the incident was handled as a structure fire with rescue in two high-rise buildings. It then became much worse. Nothing such as this had happened before—the total collapse of the structures—and it was not foreseen. Other firefighters face somewhat similar situations. They respond to an incident that is fairly commonplace and then it escalates into something much worse. This could be a dumpster fire which then turns out to contain hazardous materials. To protect ourselves from the extraordinary happening during the ordinary, we need to be sure that we are trained, equipped, and lead appropriately. This includes looking out for ourselves as well as others.

At the scene of a structure fire the Lookout is the Incident Commander or his designee remaining outside the structure to observe any changes to the structure or the fire intensity. The Communications is the link between the exterior and the interior, either by radio or other voice communication. The Escape Route is the path of egress off of the roof or out of the structure. The Safety Zone is the safe area outside of the structure, clear of smoke and falling debris. These elements of safety (LCES) should be in place prior to engaging at any incident, regardless of type.

The Greek letter **delta (Δ)** signifies change. Fire occurs in an ever-changing environment. Over the life of the fire combustibles are consumed. If the fire is in a structure the structure may be weakened. Out in the hills the fire moves over varying terrain and fuel types. Firefighters need to pay attention to what is changing now and what may change in the future. The personnel assigned to the incident are changing as well. They are becoming fatigued, possibly bored, or even overconfident. Maybe they are becoming overly stressed and entering the flight or fight mindset. All of this must be reviewed constantly in context of the LCES that is in place. Is the LCES that was put into place at the onset of the attack still valid? If not, why not and what can be done to bring it into compliance? It all goes back to situational awareness on the part of everyone assigned to the incident. They all need to be focused, alert, and vocal if something does not look right.

Remember that information is what drives decision-making. The information must be gathered and considered and appropriate action taken to respond to it. The last step in any decision-making process is to reevaluate. Once the response to the change occurs it must be reevaluated again, continuously. Firefighters must all remember to do so at every incident, every time.

WCS Worst case scenario.

delta (Δ) Change in factors affecting the incident, including personnel fatigue, time of day, structural weakening, and so forth.

Look up, Look down, Look around

The second of the condensed safety messages, "look up, look down, look around" is as follows:[6]

LOOK UP before you start to work, and every so often while working, get your eyes off the ground and look up. See what the wind, smoke, and fire are doing. Is the fire moving toward you or away from you? Are there aircraft starting to work overhead, close enough to be a danger to you? Look uphill from your location. Is there heavy equipment operating further up the hill that may roll rocks or logs down on top of your position?

LOOK DOWN, watch your footing. At night it is easy to become blinded from the light of the fire and walk into holes or fall from drop offs. Logs and pipes in tall grass can easily trip you. When carrying a razor sharp Pulaski, this could be disastrous. Look downhill and be sure that the fire has not hooked around or been ignited by rolling material below your position. The moon can illuminate the area in which you are working quite well. The problem is that moonlight does not provide you with good depth perception. It is plenty light enough to see where you are going, but easy to walk right off into a ditch. LOOK AROUND, be aware of what is going on around you. You may be working too close to someone operating a chain saw or swinging a chopping tool. Be aware of what the fire is doing. Keep track of the escape routes and safety zones. Do not become separated from your crew.

These safety messages can be applied to every type of firefighting, not just wildland. Their main focus is that you keep informed on what is happening and remain aware of how you are going to seek safe refuge. Firefighters tend to fall victim to the "candle and moth" syndrome and are drawn in to seeing only the fire and nothing else. Do not become so intent on extinguishing the fire that you do not consider where it is going and what damage it has caused to trees and structures, either of which may fall on you if you are not careful.

NOTE The main focus of these safety messages is that you keep informed on what is happening and remain aware of how you are going to seek safe refuge.

WILDLAND URBAN INTERFACE/INTERMIX FIREFIGHTING

Wildland urban interface firefighting is becoming more common. More and more people have escaped the cities by building their homes in the foothills and mountainous areas. This has given rise to the situation where the firefighters cannot afford to just back off a couple of ridges, create a fire break, and wait for the fire to get there. Now firefighters must place themselves and their equipment in the path of the advancing fire (**Figure 14-12**). In this type of firefighting, the main objective is to protect as many structures as possible. This is primarily done by placing a pumper at each structure. It would seem that a pumper with a 1,000 gallon-a-minute pump and 500 gallons of water could take on most any fire.

FIGURE 14-12 Engine strike team at wildland fire.

Delmar/Cengage Learning

This is untrue and is proved so on a regular basis, with the result being injured firefighters and a destroyed pumper. In areas of dry brush and trees, flame lengths can easily exceed 50 feet. Firefighters do not have the capability to stop this type of fire in a direct attack, frontal assault.

This firefighting method is not meant to be a suicide mission. There are several things to consider when assessing your ability to protect a structure and provide for your own safety. These are listed here as TRIAGE.

- T Take time to evaluate water needs versus availability.
- R Recon safety zones and escape routes.
- I Is the structure defendable based on construction type, topography, and anticipated fire behavior?
- A Are flammable vegetation and debris cleared within a reasonable distance?
- G Give a fair evaluation of the values at stake versus resources available and do not waste time on the losers.
- E Evaluate the safety risk to the crew and the equipment.

When the decision is made to attempt to save the structure from the advancing fire, the following safety considerations should be followed. The acronym for these is PROTECTION.

P Park engines backed in so a rapid exit can be made if necessary.
R Remember to maintain communication with your crew and adjoining resources.
O On occasions when you are overrun by fire, use apparatus or structures as a refuge.
T Tank water should not get below 50 gallons in case it is needed for crew protection.
E Engines should keep headlights on, windows closed, and outside speakers turned up.
C Coil a charged 1½-inch hose line at the engine for protection of crew and equipment.
T Try not to lay hose longer than 150 feet from your engine.
I It is important to keep apparatus mobile for maximum effectiveness.
O Only use water as needed and refrain from wetting ahead of the fire.
N Never sacrifice crew safety to save property.

As you can see, from the safety messages in the foregoing list, great emphasis is placed on maintaining communications. Without good communications among and between resource groups, a coordinated attack cannot be implemented. As a factor directly relating to safety, communications are important to keep everyone informed as to what the incident is doing and what action to take when things go wrong.

OIL FIREFIGHTING

In areas with oil production or refining, oil firefighting is a skill that needs to be developed (**Figure 14-13**). The primary objectives in oil firefighting are to extinguish the fire and control the source of any leaks. Fire props at oil firefighting schools are designed so that firefighters advance hose lines toward the burning product and push it away from them. Once at the burning flange or pipe, the valve is closed to shut off the flow. In plumbing fires, this is sometimes the technique used to control fires. Often a refinery employee can shut down the leak from a location remote from the fire and the fire department's job is to confine the fire and protect exposures until the fuel remaining in the pipe burns itself out.[7]

In refinery fires, many different products can be encountered. All of them have different burning characteristics. On the receiving side, the product is crude oil. It is very viscous; it is then refined into lighter products such as diesel fuel and gasoline. The lightest product is hydrogen gas. It burns so cleanly that the flame is often invisible. It would be possible for a firefighter to walk into the jet of burning hydrogen gas without seeing it. You should project a hose stream out in front or even use a regular broom, held out in front of you to search out the flame. Generally, firefighters should be accompanied by refinery employees any time they enter the refinery for firefighting purposes. It is dangerous to shut off valves or attempt other mitigation efforts without the proper guidance and expertise.

FIGURE 14-13
Truck wreck with petroleum fuel fire.

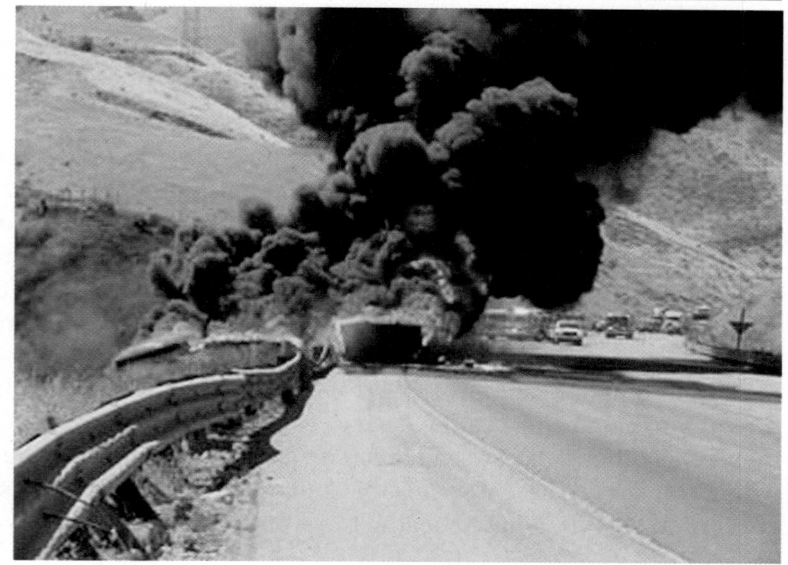

Courtesy of Edwina Davis

When the fire occurs in a storage tank, different methods are used. In storage tank fires only the surface is burning. This may appear rather easy to control because the surface area is contained by the walls of the tank. The problem is that just pouring water into the tank at a high rate will not extinguish the fire.

Much of the water will be turned to steam by the heat of the fire, and oil floats on water, leading to three main problems in this type of firefighting:[8]

1. *Boil over* occurs when water is trapped under the surface of the oil. These layers of water are called lenses, due to their shape. As the oil surface burns, a heat wave travels downward through the oil at a rate of 12 to 48 inches per hour.[9] The temperature of the heat wave is higher than the boiling point of the trapped water. When this heat wave hits a lens of water, it turns the water to steam. The water, when turned to vapor, expands at a rate of approximately 1,700 to 1. The expansion of the water to steam causes the oil to erupt upward and may cause an amount of oil to be ejected from the tank. If you were too close to the wall of the tank, you could be in the path of the falling, burning, or at least extremely hot, oil.
2. *Slop over* occurs when fire streams are applied to the burning liquid surface at such an angle that hot and/or burning oil is forced over the edge of the tank. The oil then flows down the side of the tank and you are now confronted with a fire on the ground as well as the one in the tank.
3. *Froth over* occurs when fire streams are directed at an angle that they plunge under the surface of the burning oil. This would be most likely to occur if the water were applied

from elevated stream devices such as ladder trucks. As the water enters the surface, it is turned to steam, much like the boil over, and the surface is turned to a burning oil froth. This froth can easily overflow the walls of the tank and spread as a ground fire.

In the tanks with a **cone roof** (**Figure 14-14**) and vapor space above the contents, **subsurface foam injection** plumbing is often provided. The foam generating equipment is hooked to a manifold on the tank and foam is pumped in until the fire is extinguished. Other tanks may have a system of piping that extends up the outside of the tank. The foam-generating equipment is then attached to this plumbing and the foam flows onto the surface of the product, extinguishing the fire. When neither of these is provided, the foam must be applied by aerial apparatus or from the ground through large-bore nozzles that can project a stream over the wall of the tank. In some cases a nozzle is mounted to fixed plumbing for tank protection (**Figure 14-15**). Throwing a ladder on the outside of the tank to gain access to the surface is extremely dangerous and definitely not recommended. If other tanks are endangered by the radiated heat of the fire, they may be cooled with water spray. The water must be applied directly on the tank shell. Setting up a **water curtain** to absorb radiated heat in the air is ineffective.

cone roof A style of tank construction with a vapor space over the product. The lid is connected to the tank with a weak seam that will rupture before the tank wall seams.

subsurface foam injection Plumbing installed on a tank to allow for the introduction of foam under the surface of the contained liquid.

water curtain A screen of water spray set up to protect exposures.

FIGURE 14-14
Weak seam failure in cone roof oil tank due to fire.

Courtesy of Edwina Davis

FIGURE 14-15 Fixed nozzle used for oil tank fire protection.

Delmar/Cengage Learning

Most modern storage tanks are of the floating roof type. The roof floats directly on the product, which reduces the vapor space inside the tank. The most common fire is in the seal area between the floating lid (**Figure 14-16**) and the tank shell. When large amounts of water are pumped onto the lid, it can cause it to sink, increasing the burning surface area greatly. The two methods most commonly used on a seal fire are attack lines using foam or dry chemical extinguishers. If the tank is equipped with a plumbing system for foam application, it should be designed with a dam to keep the foam applied over the area of the seal.

Gasoline Spills

The mass release of gasoline, or any other flammable liquid, requires that the vapors be controlled, if it can be done in a safe manner. Class B Foam is the agent of choice for this operation, though it is not effective on all Class B materials. When Class B foam is properly applied, it can seal the liquid surface so that the flammable vapors are contained and ignition of the vapors is prevented. When Class B foam is used, it must be periodically reapplied as it will break down after a time and become ineffective in controlling the vapors.

Any time Class B foam is used on a flammable liquid spill, firefighters should not enter the area of the spill.

FIGURE 14-16
Floating roof tanks.

Delmar/Cengage Learning

Any time Class B foam is used on a petroleum spill, firefighters should not enter the area of the spill. Walking through the foam blanket breaks the seal and releases the flammable vapors. If the vapors find an ignition source, you will be surrounded by flames. In this situation, the natural reaction is to run. This just disrupts the foam blanket further and makes the situation worse. If you must walk through the foam, shuffle your feet, do not pick them up as you walk. This helps prevent breaking the foam blanket.

Liquefied Petroleum Gas (LPG)

A commonly used by-product of oil refining is LPG. The storage containers for these products, including butane and propane, are made of a steel shell that is welded together. They are easily recognizable by their hemispherical ends. They may be of many different sizes and either mounted on the ground, carried on trains, or mounted on trucks and trailers in the larger sizes. Some other places these containers are encountered are tanks on backyard barbecues, forklifts, travel trailers, motor homes, and propane-powered vehicles. The smaller versions are sometimes encountered in garage fires.

The tanks are built to withstand high pressures, above that of the **working pressure** of the tank. They are equipped with a **relief valve** on the top that should keep them from ever reaching the tank's **burst pressure**. This works well under normal conditions. The

working pressure The pounds per square inch of pressure that a tank is designed to contain.

relief valve Device used to release unwanted pressure.

burst pressure The pounds per square of inch of pressure at which a container will fail.

problem is that not all conditions remain normal. In earthquakes, tornadoes, and floods the tanks become dislodged from their mountings. In vehicle accidents, with truck- or train-mounted tanks, the tank can end up in any position. It may have its structural integrity compromised from the tank skidding along the ground or striking another solid object. In accident situations the relief valve may end up on the bottom because the tank is upside down.

When one of these containers is involved in a fire, the recommendation is that large volumes of water be applied at each point of flame contact. The volume is determined by the size of fire and number of containers involved. When the fire is burning from the relief valve on top of the tank, there may not be any direct flame contact, but there will be a problem caused by the radiated heat. As the container is heated, the rate of gas release increases. This only compounds the problem as the fire is intensified and the container is then heated at a faster rate. Steel starts to weaken as it is heated.[10] When the pressure in the container increases to the point where it overcomes the strength of the weakened steel, the container shell will separate, releasing the contents (**Figure 14-17**).

When a release of the contents happens, it is called a **Boiling Liquid Expanding Vapor Explosion (BLEVE)**, sometimes referred to as a "Blast Leveling Everything Very Effectively." When a flammable gas is involved, the release creates a vapor cloud that ignites from the fire causing the rupture. The resultant cloud of flaming material and heat and shock waves have tremendous destructive potential. Anyone in the immediate area has very little chance of survival.

BLEVE Boiling liquid expanding vapor explosion.

FIGURE 14-17
Gas cylinders: distorted (left) and ruptured due to heat from fire (right).

When fighting fires involving containers with a flammable BLEVE potential, water should be applied with remotely supplied master stream appliances. Many facilities are set up with large nozzles just for this purpose. On the roadway these are not going to be readily available. When personnel set up appliances, they should get them set up and get out of the area as quickly as possible. Container pieces have been known to fly as far as one-half mile from the explosion site.[11]

Container pieces have been known to fly as far as one-half mile from the explosion site in a BLEVE.

For more information on gas cylinder fire attack, please refer to DVD clip *Fire Suppression, Fire Control—Gas Cylinder Fire.*

A BLEVE can also occur in a nonflammable liquid. Any container that has liquid contents and will withstand a pressure rise inside before it bursts can BLEVE. When a can of cream corn explodes in a kitchen fire, the same forces are at work. The results are just not as spectacular. An example of this type of explosion on a small scale is to make popcorn. As the water inside the kernel is heated to the burst pressure of the container, the kernel pops open.

Not all flammable gas releases are caused by or result in fires. LPG can be released due to mechanical failure of plumbing or human error. The vapor density is heavier than that of air and the LPG vapors will tend to pool in low areas. The course of action in these situations is to control ignition sources and disperse the vapors with water fog.

Natural Gas

A common flammable gas, natural gas, poses little hazard when released into the open, as it is lighter than air and will disperse itself. However, it has the potential for collecting in structures when leaks occur in interior-mounted gas meters and plumbing. There have been numerous instances where digging work is being done with either backhoe tractors or shovels and gas lines were cut. The resultant leakage of gas should be handled by the gas company. The danger of static electricity causing ignition should not be ignored and an attack line should be pulled and charged just in case. When a backhoe or other equipment ruptures the line, it should be shut down immediately and not restarted until the source of the gas is turned off and any remaining gas is dispersed.

A very real hazard to firefighters is the person bent on suicide who releases natural gas or LPG into a structure. If the neighbors smell the gas and the fire department responds before the person has expired, they may decide to take you with them by igniting the gas as you approach. Any gas smell incident should be treated as a true emergency with life-threatening potential. The worst scenario for firefighters is the cloud of flammable gas that is not yet ignited. You may find yourself in the cloud without realizing it. If at that time the gas finds its ignition source, you may become engulfed in a fireball.

HAZARDOUS MATERIALS INCIDENTS (HAZMAT)

Hazmat incidents are becoming more common, partly because of greater awareness and partly because of increased use of these materials (**Figure 14-18**). Approximately 2,000 new chemical combinations are introduced every year. Those that have no commercial value are not created in any great quantity, but those with commercial value are shipped in every mode of transportation and stored at all kinds of facilities.[12] If you were to inventory the garage in almost any home in the country, you would find some type of material that could be classified as hazardous.

NOTE If you were to inventory the garage in almost any home in the country, you would find some type of material that could be classified as hazardous.

The days of charging in and taking immediate action at all incidents are long gone. In every situation, you must be alert for the presence of materials that pose more than the ordinary hazard. This subject was already discussed to some extent in the portion of this chapter on structure fires. This section deals with incidents that are recognized as involving hazardous materials either before arrival or upon arrival.

FIGURE 14-18 Firefighters operating at a hazardous materials incident.

Delmar/Cengage Learning

The days of charging in and taking immediate action at all incidents are long gone.

The federal government has legislated, under the Superfund Amendments and Reauthorization Act of 1986 (SARA), that all employers and employees follow the requirements of OSHA 29 CFR Part 1910.20. Part 1910.20 requires training to at least the first responder level for employees responding to take action at incidents that involve hazardous materials. Annual refresher training is also required. This law also requires the establishment of an incident command system on any incident involving hazardous materials. In states without a state-level OSHA, Section 126 requires the Environmental Protection Agency to issue an identical set of regulations to cover state and local government employees.

When responding to a hazmat incident, the primary consideration is to make a precautionary approach. Slow down and think about what you are going to do before you take action. Regular PPE is not designed to protect you from hazardous materials. Many chemicals can penetrate your turnout gear and attack you. The effects may not be immediately evident and may accumulate over many years. Even SCBA is not enough protection when the chemical involved is a skin absorption hazard. Leather gloves, as previously mentioned, are no protection from chemicals. They just soak up the chemical and hold it next to your skin as you sweat and your pores open up.

NOTE When responding to a hazmat incident, the primary consideration is to make a precautionary approach.

All approaches to suspected hazmat incidents should be made from upwind, uphill, and upstream. This may require a detour from the most direct route in some situations. The approach should be made with as few personnel as possible. Firefighters should work in pairs. If it only takes two to make the approach and one to get close enough to read the label, then only two should make entry. Once at scene, any apparatus should be parked facing out. An engine can make a much faster escape from a deteriorating situation with its forward gears than it can in reverse.

All approaches to suspected hazmat incidents should be made from upwind, uphill, and upstream.

The initial responsibility of the fire department is to isolate, identify, and deny entry. The purpose of this operation is to ensure that persons who have not been exposed do not become exposed. **Perimeters** are set up to control the access of personnel to the scene.[13] The innermost perimeter is the exclusionary or "hot" zone and only properly equipped and trained personnel should be allowed into this area. Proper recognition and identification must take place before such actions are taken. Any offensive actions take place only after proper risk–benefit analysis has taken place. Should hot zone entry take place, only the minimum number of responders necessary should be allowed into the hot zone. If your department has a hazmat team, this is where their additional training, expertise, and equipment are used. Around this perimeter a secondary zone, called the contamination reduction or "warm" zone, is set up. Anyone who passes from the hot zone to the warm zone is to be properly decontaminated. This is often done by trained engine company personnel. Any nonessential personnel are to stay in the support or "cold" zone. In situations with victims, they should be brought out of the hot zone and properly decontaminated before they are turned over to EMS personnel. The EMS personnel should not be allowed to enter the hot zone to rescue them unless they have the proper training and PPE (**Figure 14-19**).

perimeter Boundary for controlled access (hazmat). The fire's edge (wildland).

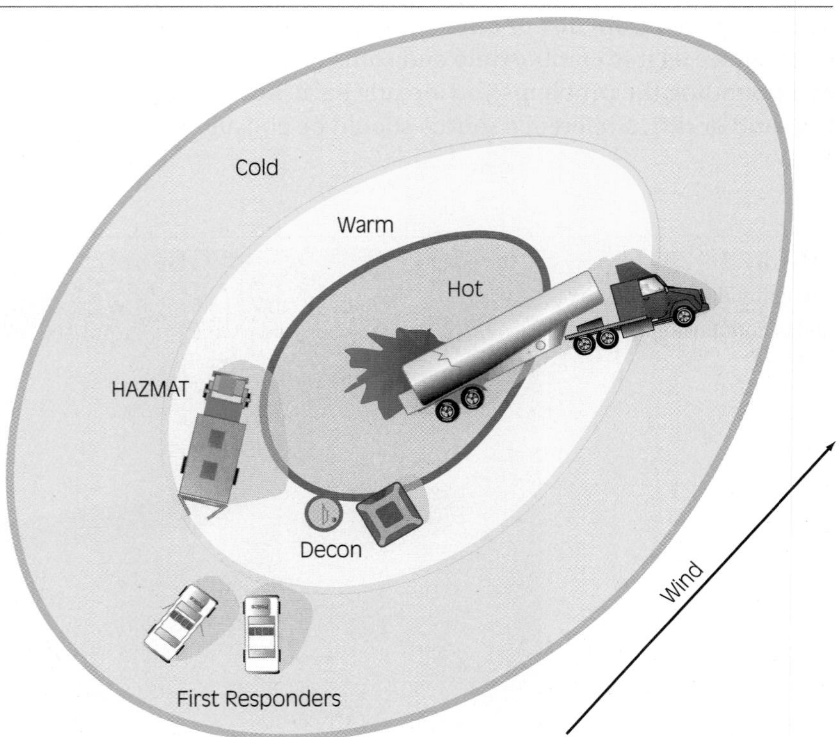

FIGURE 14-19 Zones set up at hazardous materials incident.

By identifying the material involved, a decision can be made as to the level of hazard and how it should be handled. The truck driver who says "I've handled this stuff for years" is not a reliable source of hazard information. Reference sources are available to research the hazards of many but not all substances. The problem is that in many incidents there may be two or more chemicals mixed together. This can create a combination that is not clearly defined. When in doubt, go with the worst hazard of each of the chemicals involved and act accordingly.

When the material is positively identified, information can be gathered from the U.S. Department of Transportation's *Emergency Response Guidebook*, CHEMTREC, the manufacturer, and computer databases. Many chemicals have more than one hazard. For example, many liquid pesticides are flammable as well as toxic. The *Emergency Response Guidebook* tends to describe the hazards of the materials it lists from a transportation viewpoint. It is not the definitive answer in all situations. Before any action is taken, at least three sources should be referenced as to the material's hazards. When the material has been positively identified and the proper protective clothing is determined and available, three basic actions are taken. These are diking, diverting, and controlling. Diking, or berming, is done to contain the flow of a material from the area of the leak. A dike can be constructed from dirt or other readily available materials (**Figure 14-20**). Sawdust is a poor choice as it is combustible and some chemicals can react with it, starting a fire and compounding the problems you already have. Before diking with any other material than dry sand or dirt, a reference source should be consulted.

FIGURE 14-20 Firefighter diking spill to contain liquid material.

Delmar/Cengage Learning

Before diking or berming a material with another material, it should be determined that the substances will not react with each other in a negative fashion (i.e., a strong acid and wet sand, which may result in a boiling reaction with increased release of hazardous vapor).

Diverting is done to direct the flow of a hazardous material that cannot be contained in a dike due to amount or availability of diking materials. It may also need to be done because approaching the material closely enough to dike it would pose an unnecessary safety risk. Diverting can consist of covering a storm drain grate with plastic and a ring of dirt to keep the material out of the sewer. Often in releases of large amounts of material, diverting is all that the first-in company can accomplish until more help arrives at scene.

Controlling is done to stop or reduce the flow rate of the leak. If deemed safe, a leaking 55-gallon drum can be rolled so the hole is above the liquid level. In other situations a plastic bucket can be placed where it will catch a dripping material. Spilled powder materials can be covered with a piece of plastic sheeting to keep them from being spread by the wind or draft from passing vehicles. Releases from plumbing can often be controlled by finding the valve and turning it off. All of these procedures are fairly simple, commonsense techniques that require no special equipment or training. Many specially designed leak control devices should only be used by trained personnel. Controlling usually requires a close approach to the source of the leak and is not advised until the material and its hazards are identified.

Once the incident is under control and the source of the leak is stopped, the cleanup phase begins. Many fire departments do not do cleanup. This is left to private contractors that specialize in this type of operation. Cleanup often requires specialized equipment and permits for storage and transportation of hazardous materials. The health department usually determines how clean the area needs to be and has the final say on how clean is clean.

For more information on response to hazardous materials incidents, please refer to DVD clip *Firefighter Skills: Hazardous Materials Awareness and Operations.*

WEAPONS OF MASS DESTRUCTION

As we have seen from the attacks on the Murrah Federal Building (Oklahoma City, Oklahoma, 1995), the Pentagon (Arlington County, Virginia, 2001), and the World Trade Center (New York, New York, 2001), and in other parts of the world, acts of terrorism are a very real threat. As with other incidents, firefighters are often the first responders. In the World Trade Center incident alone 343 firefighters lost their lives.

The purpose of weapons of mass destruction (WMD) attacks is to take human lives and/or to cause panic and disruption. These attacks may be chemical, biological, radioactive, nuclear, and/or explosive (CBRNE) in nature.

Some of the conditions found in a terrorist (CBRNE) incident, but not at a hazardous materials incident, are:

- *Crime scene.* Requires you to preserve as much evidence as possible.
- *Major interaction with local, state, and federal agencies*, such as the Federal Bureau of Investigation (FBI), the Bureau of Alcohol, Tobacco, Firearms and Explosives (BATF), and the Environmental Protection Agency (EPA).
- *Scene communication overload.* Radio traffic will be intense and the cellular phone systems may become overloaded with citizens, responders, and the news media.
- *Chaos.* People will be trying to escape the area and the number of responding agencies will be many more than for an ordinary hazmat incident, therefore an effective incident command structure must be implemented early on in the incident.
- *Overwhelming of resources.* The number of victims and destruction/contamination of the area may rapidly overwhelm the first responders' ability to handle, or even assess, the scope of the incident.
- *Secondary devices designed to kill responders.* A secondary device may be a bomb planted to go off after emergency personnel have arrived at the scene of a bombing or other attack.
- *Pre-incident indicators.* There may be a threat phoned or mailed, or an agency, such as the FBI, may be aware of a threat prior to the incident.
- *Deliberate attack.* Terrorist attacks are designed to cause as many casualties as possible and are done deliberately, usually in crowded places.
- *Super toxic material.* A material, such as Sarin (200 times more toxic than chlorine) may be used to increase the number of casualties.
- *Identification of material used.* In hazmat incidents there are usually some indicators as to what material is involved. In a terrorist attack the only visible indicators may be the symptoms displayed by the victims.
- *Mass casualties with many fatalities.* The purpose of a terrorist attack is to draw attention. A high death/casualty toll is certain to gain media attention.
- The *psychological effects* will go far beyond those directly involved and add casualties to the number directly affected by the incident.
- *Mass decontamination.* In an attack utilizing a toxic agent, contamination must be confined to as small an area as possible. Any persons, whether victims or responders, and equipment must be decontaminated before leaving the scene.
- *Unusual risk to emergency responders and civilians.* Regular PPE may not be sufficient to protect you from the effects of toxic materials and secondary devices may be present. Remember, the purpose of a terrorist attack is casualties.[14]

EMERGENCY MEDICAL SERVICE OPERATIONS (EMS)

The fire department of today has become much more involved in providing emergency medical services (EMS) and in doing so has moved into a new area with its own problems. Operations in the area of emergency medicine for most fire departments revolve around first aid for injuries, basic life support, and extrication of victims from vehicle accidents. Firefighters also respond to victims of assaults, including gunshot wounds, knife wounds, and rape. Medical emergencies can include diabetic problems, overdoses, and emergency childbirth. All of these scenarios pose their own types of hazards.

One of the primary problems involved in providing emergency medical assistance is avoiding exposure to bloodborne and airborne pathogens.

One of the primary problems involved in providing emergency medical assistance is avoiding exposure to bloodborne and airborne **pathogens**. The federal government has enacted legislation that addresses this problem. The legislation is part of 29 CFR 1910.1030 *Occupational Exposure to Bloodborne Pathogens*. This regulation identifies emergency response personnel as Category 1 employees. Category 1 employees are at the greatest risk of occupational exposure to communicable diseases. These bloodborne communicable diseases include, but are not limited to, **HIV** and hepatitis. The NFPA addresses this issue in NFPA 1500, *Standard on Fire Department Occupational Safety and Health Program.*[15] NFPA 1581, *Standard on Fire Department Infection Control Program*[16] defines minimum requirements and criteria. The standard lists required program components and includes recommendations on fire department facilities, personnel, PPE, and procedures for cleaning, disinfecting, and disposal.

When responding to emergency medical incidents, personnel should be provided with special PPE. The PPE required provides a liquid-proof barrier between the firefighter and the patient to prevent any of the patient's body fluids from coming into contact with your skin or **mucous membranes**. The two terms used to describe this PPE system are "universal precautions," used by the American Red Cross and Centers for Disease Control, and "body substance isolation" (BSI), from the National Fire Academy course, *Infection Control for Emergency Response Personnel* (ICERP). Both of these terms refer to the wearing of appropriate gowns or garments, properly rated gloves, and face and eye protection.

pathogen A disease-causing agent.

HIV Human Immunodeficiency Virus, cause of AIDS.

mucous membranes Inside of the nose, mouth, and covering of the eye.

All of these are available as disposable items; this way they can be bagged and disposed of properly without having to be touched again. The rule of thumb when it comes to wearing medical PPE is, "If it's wet and it isn't yours, don't get it on you."

Medical incident PPE includes appropriate gowns or garments, properly rated gloves, and face and eye protection.

Regular turnout gear is equipped with a vapor barrier inside the flame-resistant outer shell and will protect you fairly well. The problem is that once you get blood from a patient all over your turnouts, where are you going to clean them? To just take them home is extremely dangerous to your family and yourself and may be against the law. To do this properly they should be placed in a plastic bag and laundered to remove any contaminants. This is great if you have two sets of turnouts, but most people do not. There are going to be times when working in your turnouts is unavoidable. An appropriate gown or garment is no protection from the dangers of getting burned if gasoline ignites when extricating a victim from a wreck. In situations where there is no danger from fire, an appropriate gown or garment should be worn. Properly rated gloves can be worn under your leather gloves, to prevent the properly rated gloves from becoming cut on glass and sharp objects and to protect your hands from liquids soaking through.

When involved in situations that require your caring for more than one patient, it is recommended that you change gloves between patients. When the incident is concluded and you are ready to pick up and return to the station, it is a good idea to bag up any gloves and other materials that have become contaminated in a biohazard container (**Figure 14-21**). Always be sure that contaminants are not spread to the handles of the resuscitator or medical aid kit. If someone on the crew has touched a patient's body fluids and then opened the medical aid kit with the same gloves on, the handles of the kit are now contaminated. If you remove your gloves and place the kit back into the compartment on the apparatus, you have contaminated the compartment handle and yourself. In this same way contamination can be spread to the steering wheel, radios, and other equipment. Always be sure to check the bottom of your boots for contamination.

When working around paramedics, they will likely be starting IVs. After the IV is inserted, the needle must be placed in a sharps container. They should never be dropped on the ground or stuck into the mattress on the ambulance cot. If needles are dropped on the ground and you kneel down on one, you are likely to get stuck. It is your responsibility to provide for your own safety and keep an eye on where the needles are placed after use.

NOTE When involved in situations that require your caring for more than one patient, it is recommended that you change gloves between patients.

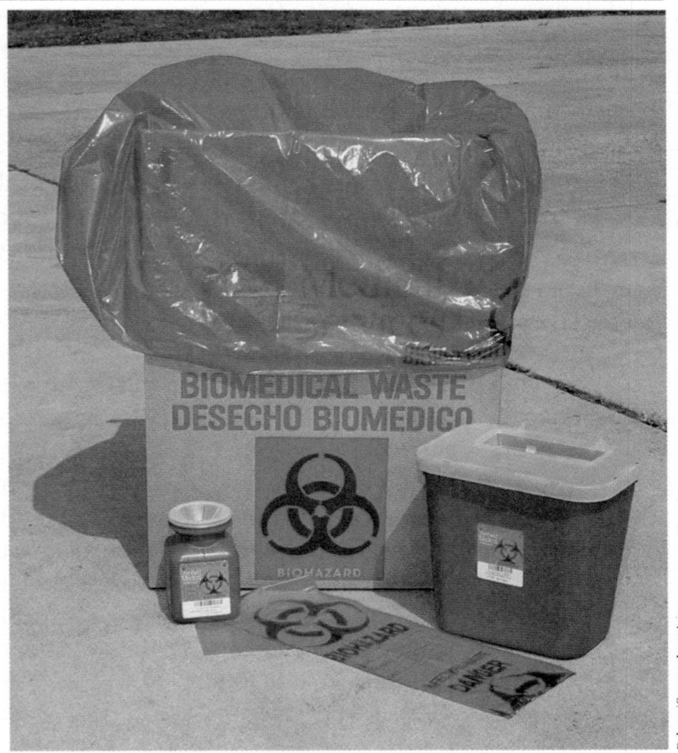

FIGURE 14-21 Biohazard containers for infectious waste.

Delmar/Cengage Learning

Some other precautions that you can take are to stay in good health: Do not report to work sick; eat a good diet; wash your hands well and often, especially after incidents; and approach every patient, regardless of age or gender, as if they have some deadly infection. This does not mean do not treat them; it means treat them with caution.

HIGHWAY INCIDENT SAFETY

A major concern of firefighters and other responders to roadside incidents is the threat of passing vehicles. Every year firefighters and other personnel are hit by passing vehicles while working at the scene of roadside incidents. These can easily be fatal to the responder. This has become a national concern and many states are now passing and enforcing traffic laws and posting signs that direct drivers to either slow down or, whenever possible, to change lanes to avoid roadside workers.

The rules for working roadside incidents of any type are summed up in the three E's of highway survival:[17]

1. *Eye on traffic.* Always be aware of your surroundings and keep an eye on the passing traffic. Position yourself so you can keep one eye on traffic and one on what you are

doing whenever possible. The best case scenario is a total shutdown of the roadway in the vicinity of the incident, but this is not always possible. A time when many responders get hit is when they exit their vehicle as they are assessing the scene and not looking at traffic. It is imperative that all personnel look first before exiting the apparatus.

2. *Exposure.* Minimize the time spent on the side of the roadway and in high-risk areas as much as possible. Keep all responders and victims out of these areas and in a place of safety. One of the ways to create a safer work area on a roadside incident is to position a large piece of apparatus, such as an engine, in the path of oncoming vehicles so they have to use another lane to avoid it. All warning lights should be on, the brakes set, wheel chocks set, and the wheels should be turned in a direction such that if the protective vehicle is rear-ended it will move in a direction away from the responders. This is usually to the right. Flares should be positioned between the protective vehicle and oncoming traffic so that passersby have ample time to respond and stop or change lanes. The average stopping distance for an auto on dry pavement at 60 miles per hour is determined at 303 feet and trucks at 361 feet. Even without stopping, an unimpaired and focused driver travelling at 60 miles per hour is going to take 132 feet to react to the situation.[18] Flares must be positioned well back from the incident scene in the direction the threat is coming from. Always remember to check for spilled fuel prior to allowing or using road flares. Reloading any hose on the apparatus should be done in a safe area, not on the roadside.
3. *Escape route.* Assess the situation, survey the area at the scene, and plan a course of action for fleeing from an approaching vehicle. As in every situation, look for two ways out should one of them be compromised. If the incident is located on an overpass or other bridge, the safest action is to close the roadway. Responders cannot safely just jump over the guard rail in an emergency.

A very real danger at roadside incidents is passing traffic. Minimize the time spent in and around the roadway as much as possible.

In addition, always don all appropriate protective equipment. This includes fire resistive clothing and vest (as shown in **Figure 11-3**) as specified in 23 CFR Rule 634, *Firefighter High-Visibility Safety Apparel* and NFPA 1901, *Standard for Automotive Fire Apparatus.*

An excellent source of information concerning traffic and roadway safety is the *Manual of Uniform Traffic Control Devices* (MUTCD) published by the Department of Transportation (DOT),[19] and the *Emergency Vehicle Safety Initiative FA-272* published by the National Highway Traffic Safety Administration (NHTSA).[20]

VEHICLE ACCIDENTS

When working vehicle accidents, it is important to always consider the dangers of spilled fuel. The apparatus should be parked uphill of the accident so spilled fuel cannot run down underneath it. If you must work in an area with spilled fuel to rescue a victim, the spill should be covered with Class B foam. A firefighter should be designated to stay with the attack line; if he gets involved in helping you or gets bored and sets down the nozzle, that may be when the fire starts.

When working vehicle accidents, it is important to always consider the dangers of spilled fuel which may ignite.

When working with power tools to extricate accident victims, caution must be exercised to protect yourself and the victims. Air bags are now common on vehicles and pose the hazard of going off as you lean into the vehicle either over them or in front of them. Power tools should not be used when there is spilled fuel in the immediate area. A car roof can be cut off with hacksaws when necessary. It is slower than a rescue tool but is not an ignition hazard. When using the **power shears** all loose trim metal should be removed first. A power rescue tool is capable of exerting in excess of 10,000 pounds of force.[21] If a door is popped off incorrectly, it could easily hit someone or fall and cut your toes off. The main point is to not use this equipment unless you and the others operating it are trained in its use.

For more information on tactics utilized at vehicle accidents, please refer to DVD clip *Rescue Procedures—Vehicle Extrication.*

power shears Cutting attachment for a rescue tool.

VEHICLE FIRES

Vehicle fires can be compared to structure fires with wheels. They can have all of, if not more than, the hazards of a structure fire. A fire in a truck tractor-trailer rig can have numerous hazards from both the large amounts of diesel fuel they carry and the cargo itself. Busses are more commonly fueled with liquified natural gas (LNG), a BLEVE hazard. Smaller vehicles are now presenting differing hazards with the use of alternative fuels such as liquefied petroleum gas (LPG), as are hybrid vehicles with their large acid-containing batteries now and hydrogen fuel cells in the near future.

Automobile fires should be approached from the side or quarter. A five-mile-an-hour bumper is mounted on two shock absorbers, and if they explode the bumper can be propelled forward with great force. It would hit you just under your knees. Exploding tires can be a shrapnel hazard as well. Full PPE, including SCBA, should always be worn. There are many materials used in today's cars that are toxic if inhaled when burning. Cars do not generally blow up in a fire as they do on television and in the movies. The seams of the gas tank can let go, spilling forth a large amount of burning fuel (**Figure 14-22**).

For more information on extinguishing vehicle fires, please refer to DVD clip *Fire Extinguishment—Passenger Vehicle.*

AIRCRAFT FIREFIGHTING

Aircraft are encountered by firefighters in numerous situations, either as an emergency incident or as a tool. We will first look at aircraft incidents.

When an aircraft crashes or catches fire, it can present many different hazards. It is carrying a certain amount of fuel and it has batteries, tires, flammable metals, and possibly oxygen cylinders. If the incident involves military aircraft, there is always the possibility of munitions on board and scattered about a crash scene. When an aircraft crashes, the fuel

FIGURE 14-22 Firefighters operating at vehicle fire. Both of these firefighters should be wearing their SCBA facemasks.

Courtesy of Edwina Davis

is often atomized and can explode into flame. This can cause a spectacular fire with the need for rescue of the occupants.

The first priority in aircraft firefighting is to create a path, which allows the apparatus and firefighting personnel to approach and the victims that can rescue themselves to escape. Next, entry is made into the body of the aircraft to rescue anyone alive inside and to complete extinguishment of the fire. The last operation is to complete overhaul of the fire scene.

Over the years there have been crashes of aircraft into structures. This scenario is further complicated by the presence of a structure fire and associated rescue problems. The aircraft are often just completing takeoff and carrying a large amount of fuel. When the crash is a distance from a large airport, specialized crash fire apparatus may not be available and the fire must be controlled by regular apparatus.

EMS AND FIREFIGHTING WITH AIRCRAFT

Aircraft are used as a tool by firefighters as well. The type of aircraft firefighters most commonly encounter at incidents are helicopters used for transportation of victims on emergency medical incidents and for tactical and logistical needs on other incident types. The rules for working around helicopters are the same no matter what the type of incident:[22]

- Approach and depart helicopters from the side or front in a crouching position, in view of the pilot.
- Approach and depart the helicopter from the downhill side to avoid the main rotor.
- Approach and depart the helicopter in the pilot's field of vision; do not go anywhere near the **tail rotor**.
- Use a chin strap or secure your head gear (hard hat) when working under the **main rotor**.
- Carry tools horizontally, beneath waist level to avoid contact with the main rotor.
- Fasten your seat belt when you enter the helicopter and refasten it when you leave. A seat belt dangling out of the door can cause major damage to the thin aluminum skin of a helicopter.
- Use the door latches as instructed. Use caution around plexiglass, antennas, and any moving parts.
- When entering or exiting the helicopter, step on the **skid**. If you place your foot next to the skid and the weight of the ship changes, it may run over your foot.

tail rotor Vertical propeller used for steering control that is installed on the tail of a helicopter.

main rotor Horizontal blades that create lift for a helicopter.

skid The long tubular shaped feet that helicopters sit on when on the ground.

- Any time you ride in a helicopter in a wildland fire situation you are required to wear full PPE.
- Do not throw articles from the helicopter as they may end up in the main or tail rotors.

When setting up a landing zone for a helicopter, there are some very important items to consider. Secure all loose articles in the area, such as boxes or other items that may become airborne. Perform dust abatement by wetting down the area, if possible. If the helicopter is to land on a roadway, make sure that vehicle and pedestrian traffic is stopped. When a main or tail rotor hits a vehicle, the helicopter is out of service and unable to fly. If a rotor hits a person, it will most likely kill him and disable the helicopter. Wear eye and hearing protection when working around helicopters. When a helicopter lands or takes off, it kicks up a tremendous amount of wind and debris. Be sure to roll up the windows on any vehicles in the area. Provide for plenty of clearance for the helicopter to land (**Figure 14-23**).

When the helicopter is coming in to land and you have established radio communications with the pilot, there are several items the pilot should be made aware of. Identify

FIGURE 14-23
Helicopter with external mounted hoist being used to medevac injured person.

Courtesy of Edwina Davis

and notify the pilot of any elevated hazards in the area, such as power lines, fences, and light poles. Either set up a wind indicator, with flagging tape on a pike pole or shovel, for example (**Figure 14-24**), or throw several shovels of dirt in the air, or stand with your back to the wind with your arms out in front of you. Notify everyone at the scene that the helicopter is about to land so they can protect their eyes from debris.

One of the main points to remember when working around any aircraft is that contact with any moving part on an aircraft is often fatal. All of their parts are extremely expensive. In most situations the aircraft carries its own crew that will assist you in operating in the vicinity of the aircraft. You should not touch anything until you are briefed on its use, nor should you approach the aircraft until given permission by a crew member or the pilot.

On wildland incidents, helicopters are used for dropping water and retardant on the fire. The larger helicopters can carry up to 2,500 gallons of water and create a very strong rotor downwash. When the helicopter comes in to make its drop, get out of the area or lie on the ground and cover up. Try to be uphill of the drop so you do not get washed down the hill or hit by rocks loosened by the water. A large helicopter can cause the tops of dead trees to break loose and become a falling hazard. The downwash from the main rotor will also cause a change in fire activity as the flames are fanned.

Contact with any moving part on an aircraft is often fatal.

FIGURE 14-24 Improvised wind indicator using shovel and flagging tape.

Delmar/Cengage Learning

Another activity is logistical support. Helicopters are used to sling load materials to crews on the fire line. When sling loads are delivered, firefighters should avoid placing themselves in the approach and departure path of the helicopter and stay clear of the landing zone. If the helicopter develops trouble in flight, one of the first things the pilot does is release the load to reduce weight.

Airplanes are used to drop fire retardant at wildland fires. The material dropped is either a chemical compound mixed with water or water alone. Some of these aircraft are capable of delivering up to 3,000 gallons at one time. If you consider that 3,000 gallons, at over nine pounds a gallon, is coming down from around 200 feet above ground level at 130 miles per hour, the dangers become evident. The turbulence caused by the plane's passage can knock the tops out of trees and fan the fire. If you are in the area where a drop is to be made, it is best to clear the area prior to the drop. If this is not possible, lie on the ground on your stomach, facing the aircraft, and place your hands on top of your head. If you are hit by a retardant drop you will be covered by a sticky, slippery coating that has a strong ammonia odor. It is not really harmful, but it is uncomfortable. Be aware that it is slippery on dry grass and rocks and will make hand tools hard to hold on to. This material is slightly corrosive and should be washed off vehicles as soon as possible.[23]

If you are in the area where a drop is to be made, it is best to clear the area prior to the drop.

DECISION MAKING

Any discussion of safety in the emergency services has to include decision making. Decisions were involved in all the places and situations in which firefighters and other responders were killed or injured, including: during training, responding to incidents, while performing at incidents, and when returning from incidents. Without a review and understanding of how we make decisions, especially under stress, we are missing one of the key elements of keeping ourselves and those around us safe from harm. As has been stated, "All firefighters deserve a safe trip home."

Decision making is being looked at and discussed across the fire service, on both the structural firefighting side and the wildland firefighting side. Many of the elements of decision making can be directly cross-applied or adapted from one side to the other. In some cases fire agencies have both responsibilities. All of those involved in emergency response have lessons they can learn from each other when it comes to responding to emergencies, which tend to be dynamic, high-stress situations.

As a new firefighter it is going to require some tact on your part should you choose to approach your supervisor on these items. Not everyone has bought in to the more recently emphasized concepts in emergency-response decision making. Some are unaware and some do not necessarily agree. So step lightly when discussing this with others.

Crew Resource Management (CRM)

In adopting the concept of *crew resource management* the fire service has followed a model developed in the aviation industry. In the early 1980s, the aviation industry recognized that human error was the prevailing cause in aviation disasters.[24]

Additional industries adopted the concept of **CRM** in the 1980s and 1990s. The resultant increase in safety and reduction in injuries is an indicator of how effective CRM can be when utilized properly. The U.S. Coast Guard reported a 74 percent reduction in its injury rate since the adoption of CRM. U.S. air disasters showed a marked reduction from approximately 20 per year to one or two per year.

CRM is based on five elements. These are:

1. *Communication.* CRM requires that people speak directly and respectfully and communicate responsibly. A model of this is the one presented in the NWCG's *Incident Response Pocket Guide* (IRPG). It states under "communications responsibilities" that firefighters must: brief others as needed, debrief your actions, communicate hazards to others, acknowledge messages, and ask if you don't know. It also states that all leaders of firefighters have the responsibility to provide complete briefings that include a clearly stated "Leader's intent." This includes:
   - *Task.* What is to be done
   - *Purpose.* Why it is to be done
   - *End state.* How it should look when done.[25]

   Implied in this is that *all* personnel at the scene should have an idea of what is going on, what they are supposed to be doing, and what is to be accomplished.
2. *Situational awareness.* This can be described as maintaining attentiveness to an event. Situations in the emergency services are particularly dynamic and require the full attention of everyone to recognize hazards and changing elements at the scene. The *IRPG* lists the elements of situational awareness as:
   - Objectives
   - Communication
   - Who's in charge
   - Previous fire (incident) behavior
   - Weather forecast
   - Local factors

CRM Is based on the assumption that human error is the primary cause of fireground fatalities and injuries and by using CRM the fire service can reduce the number of negative outcomes.

3. *Decision making.* Decision making must be based on information. This includes information from training, experience, and information gathered at the scene. The decision making is focused on risk/benefit analysis. We must look at what we are doing and decide whether it is worth it. No firefighter should die to save property, though every year some do. Too much information overloads the decision maker and too little leads to poor decision making. Somewhere in the middle is the right amount of information selectively focused on the right elements of the scene to make sound decisions. We must also recognize that there are some things we know, some things that we do not know, and some things that we do not know that we do not know. This is where experience enters the equation. A person with quality experience can focus on the important elements, pay less attention to the less important elements, and not forget that there are factors in a dynamic situation that may not be readily perceived. This reinforces the need to maintain situational awareness.
4. *Teamwork.* Teamwork is critical. Every member of the team must know their job and know where they fit in the organization. This could be a company level organization or a major operation with hundreds of personnel. This refers back to the incident command concepts of *Unity of Command* and *Chain of Command.* Failure to perform as a team at emergency incidents can lead to a reduction in operational effectiveness, excessive property loss, injury, and death.
5. *Barriers.* Barriers can affect any or all of the previous factors. Rather than go into it in this section (CRM), it will be further explained in the next section on "The 2&7 Tool."

The 2&7 Tool

Brad Mayhew, author of "F LCES Δ," is conducting ongoing research into how firefighters make decisions. His **2&7 Tool** illustrates two errors and seven barriers common to poor decision making. Though he is from a wildland fire background his tool can be applied to other types of emergency situations as well.[26]

The two errors:

1. Underestimating hazards and using inadequate safety measures.
2. Failing to notice changing conditions and adjust tactics accordingly.

If you were to apply these two errors to any number of case studies of firefighter near-misses and fatalities in all risk situations, you could see that they often apply. Firefighters commit too deeply to the situation and overextend their ability to escape. Even a rapid intervention team (RIT/RIC) cannot get you out if you are in too far or the situation deteriorates to the point that they cannot locate you or get to you. Driving too fast on the way to the incident is an example of error number 1, because you are unable to stop

The 2 & 7 Tool The 2 errors and 7 barriers common to poor decision making.

quickly enough, or avoid the hazard, because of your excessive speed. Another example is placing only one or two ladders to the roof for one engine or truck company. As the incident progresses and they are joined by other firefighters, there are too many on the roof to make a quick escape down the number of ladders available should conditions deteriorate and it becomes necessary for them to do so. In a wildland fire the example would be getting too far from the safety zone, creating too long an escape time, and then the fire flares up, cutting off your escape. Or, as in the ladder example, more personnel are assigned to the division than can safely fit in the safety zone. This sort of thing just seems to creep up on firefighters because they are often overly-focused on getting the job done and not on what else is happening (lack of situational awareness).

The seven barriers:

1. *Inexperience.* As previously noted, experience is valuable. That is if it is quality experience in performing safely, not learning cowboy tactics and thinking they are OK just because you got away with it before.
2. *Getting too comfortable.* Thinking nothing bad will happen. Or as in Barrier 1, nothing bad has happened yet, so what is the likelihood that it will this time? State legislatures would not have to pass laws against talking and texting on cell phones while driving—if people were attentive and not overly comfortable with doing it, and thinking nothing bad will happen while texting when they are supposed to be focused on their driving.
3. *Distraction from primary duty.* You can only focus fully on one thing at a time. Outside factors at the incident scene can divert your attention from what you should be focusing on. A part of the structure collapsing down the street diverts your attention and the piece you are working under comes down on top of you. The driver operator's primary duty is to get you there safely, yet we still suffer from fatal responding vehicle rollovers due to excessive speed and other accidents.
4. *Priorities out of order.* The first priority in any incident is life safety, including your own. Firefighters have been killed trying to perform "rescues" that were obviously body recoveries. Some fight fires for the thrill, danger, and adrenaline rush. These are the ones that will get you hurt or killed.
5. *Social influences.* One of the strongest of these is peer pressure. In the fire service it is common that new firefighters have to "prove themselves." This occurs at the station and on the fire ground. This does not mean that we should all get into the mode of "group think" where we stop thinking for ourselves or overextend just to prove something to others.
6. *Stress reaction.* As stress builds, the rational thinking part of the brain tends to shut down. We focus on a plan of action and stick with it. This is the "do or die" mindset and leads to tunnel vision, which reduces situational awareness. This becomes evident when the interior attack crew is ordered out of the building by the Incident Commander because the roof is about to fall in and they reply, "We almost got it,

Chief, just a few more minutes." They are not in position to see the big picture and need to do as ordered and not argue.

7. *Physical impairment.* Physical impairment can come in many forms; carbon monoxide poisoning, lack of sleep, drinking alcohol the night before coming on duty, drugs, and too many incidents in too short a time. These all lead to a general diminishment in awareness and decision-making ability.

We must all guard ourselves and those around us from making the two errors and falling victim to the seven barriers if we are to be safe and effective firefighters. This applies to every incident, every time, no matter how simple or commonplace.

SUMMARY

This chapter presented a number of the operations engaged in by firefighters. It by no means covers every situation or type of event. Numerous safety rules were presented that can be applied in a wide range of situations. There are complete training programs available for different incident types. The main point you should be left with is that every situation has its own set of hazards and it is your responsibility to make sure that you provide for your personal safety and the safety of others. The worst hazard is often the one that is not recognized. You must train yourself to evaluate situations as they arise and anticipate situations that can harm you. Hazards and the risks that firefighters take in relation to them are based on decision making. Decision-making skills are critical in maintaining firefighter safety. This requires you to remain constantly vigilant at every type of incident. When you become complacent and stop paying attention, or think that things are only routine, you just may be in the greatest danger. Never lose your situational awareness as the consequences can be deadly.

REVIEW QUESTIONS

1. List at least three interior hazards encountered at structure fires.
2. List at least three exterior hazards encountered at structure fires.
3. What is wrong with freelancing at emergency scenes?
4. Why should the IC set up a RIT at the scene?
5. Why should you not enter electrical substations without electrical company personnel?
6. Using your text, list the 10 Standard Firefighting Orders.
7. Using your text, list the 18 Situations That Shout Watch Out.
8. What are the four components of LCES?
9. When implementing LCES, why does it need to be constantly reevaluated as the incident progresses?

10. What is meant by the term WCS in F LCES Δ?
11. What are the three ways to "look"?
12. List several of the safety considerations when providing structure protection.
13. What are the three "overs" referred to in oil firefighting?
14. What is the meaning of the term "BLEVE"?
15. How should all suspected hazardous materials incidents be approached?
16. What is meant by the acronyms WMD and CBRNE?
17. What is meant by "universal precautions" in regard to EMS incidents?
18. Why should an attack line be pulled on all vehicle rescue situations?
19. What are the three priorities in an aircraft firefighting incident?
20. What action should you take if an air tanker is about to make a drop on your position?
21. What are the safety rules for approaching a helicopter?
22. What are the 2 common errors and 7 barriers to decision making as described in *The 2&7 Tool*?

DISCUSSION QUESTIONS

1. Why is the buddy system important when operating in IDLH atmospheres?
2. If a WMD/NBC event were to happen in your community, is the fire department prepared to deal with it? Which other local agencies would be involved?
3. Why is a good safety attitude as necessary as safety training in staying safe in emergency and nonemergency situations?
4. Explain, by using examples, the meaning of the following quotation: "To fight fire one must be aggressive; to survive it one must be cautious."
5. Why is it sometimes a better option to take no action in certain situations?
6. In some states emergency lights are not to be left on when working roadside incidents if all vehicles and victims are safely out of the roadway. What are your thoughts on this practice?
7. Why is decision making a skill that needs to be highly developed by firefighters?

NOTES

1. National Fallen Firefighters Foundation, *Everyone Goes Home - 16 Firefighter Life Safety Initiatives,* www.everyonegoeshome.com, 2010
2. OSHA, *Standard 29 CFR 1910.120: Hazardous Waste Operations and Emergency Response* (Washington, DC: OSHA, 1986).
3. National Fire Equipment System Course, *S 336: Tactical Decision Making in Wildland Fire* (Boise, ID: National Interagency Fire Center, 2004).
4. National Fire Equipment System Course, *134: Lookouts, Communications, Escape Routes and Safety Zones* (Boise, ID: National Interagency Fire Center, 2003).

5. Mayhew, Brad, "The Intent of LCES," www.firelinefactors.com, 2009.
6. National Fire Equipment System Course, *S 336: Tactical Decision Making in Wildland Fire* (Boise, ID: National Interagency Fire Center, 2004).
7. Shell Oil Co., *Reno Fire School Training Manual* (Reno, NV: Shell Oil Co., 1991).
8. Ibid.
9. National Fire Protection Association, *Fire Protection Handbook* (Quincy, MA: National Fire Protection Association, 2008).
10. Ibid.
11. Ibid.
12. Frank L. Fire, *The Common Sense Approach to Hazardous Materials* (New York, NY: Fire Engineering, 2009).
13. International Fire Service Training Association, *Hazardous Materials for First Responders* (Stillwater, OK: Fire Protection Publications, 2004).
14. SBCCOM, *Domestic Preparedness Training Program* (U.S. Army Edgewood Research, Development, and Engineering Center: 1999).
15. National Fire Protection Association, *Standard 1500, Fire Department Occupational Safety and Health Program* (Quincy, MA: National Fire Protection Association, 2007).
16. National Fire Protection Association, *Standard 1581, Fire Department Infection Control Program* (Quincy, MA: National Fire Protection Association, 2005).
17. Automobile Club of Southern California, *ERS Tow and Service Instruction* (Los Angeles, CA: Automobile Club of Southern California, 2003).
18. James Madison University, "Ratio of Stopping Speed to Stopping Distance" http://www.jmu.edu/safetyplan/vehicle/generaldriver/stoppingdistance.shtml, 2010
19. Department of Transportation, *Manual of Uniform Traffic Control Devices* (Washington, DC: Department of Transportation, 2009).
20. National Highway Traffic Safety Administration, *Emergency Vehicle Safety Initiative FA-272* (Washington, DC: National Highway Traffic Safety Administration, 2004).
21. Hurst Performance, *Model S511 Instruction Manual* (Huntingdon Valley, PA: 2009).
22. National Fire Equipment System, *Basic Aviation Safety Guide* (Boise, ID: National Interagency Fire Center, 2004).
23. National Fire Equipment System Course, *S-270 Basic Air Operations* (Boise, ID: National Interagency Fire Center, 2003).
24. International Association of Fire Chiefs, *Crew Resource Management* (Fairfax VA, 2003).
25. National Wildfire Coordinating Group, *Incident Response Pocket Guide*, (Boise, ID: National Interagency Fire Center, January, 2006).
26. Mayhew, Brad, "The 2&7 Tool," www.firelinefactors.com, 2009.

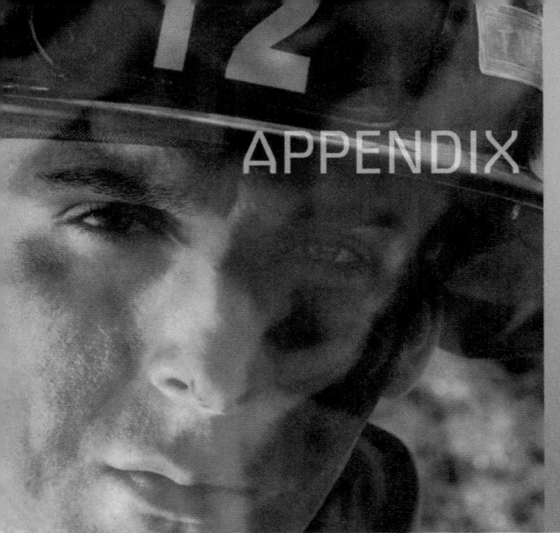

APPENDIX A

The U.S. Fire Problem

TABLE A-1

Year	Fires	Civilian Deaths	Civilian Injuries	Firefighter Deaths	Firefighter Injuries	Direct Property Damage (in billions) As Reported	Direct Property Damage (in billions) In 2009 Dollars
1977	3,264,000	7,395	31,190	157	112,540	$4.7	$16.6
1978	2,817,500	7,710	29,825	173	101,100	$4.5	$14.8
1979	2,845,500	7,575	31,325	125	95,780	$5.8	$17.2
1980	2,988,000	6,505	30,200	138	98,070	$6.3	$16.4
1981	2,893,500	6,700	30,450	136	103,340	$6.7	$15.8
1982	2,538,000	6,020	30,525	128	98,150	$6.4	$14.2
1983	2,326,500	5,920	31,275	113	103,150	$6.6	$14.2
1984	2,343,000	5,240	28,125	119	102,300	$6.7	$13.8
1985	2,371,000	6,185	28,425	128	100,900	$7.3	$14.5
1986	2,271,500	5,850	26,825	119	96,450	$6.7	$13.1
1987	2,330,000	5,810	28,215	132	102,600	$7.2	$13.6
1988	2,436,500	6,215	30,800	136	102,900	$8.4	$15.2
1989	2,115,000	5,410	28,250	118	100,700	$8.7	$15.1
1990	2,019,000	5,195	28,600	108	100,300	$7.8	$12.8
1991	2,041,500	4,465	29,375	108	103,300	$9.5[1]	$15.0[1]
1992	1,964,500	4,730	28,700	75	97,700	$8.3	$12.7
1993	1,952,500	4,635	30,475	79	101,500	$8.5[2]	$12.6[2]

Delmar/Cengage Learning

(continues)

TABLE A-1 Continued

Year	Fires	Civilian Deaths	Civilian Injuries	Firefighter Deaths	Firefighter Injuries	Direct Property Damage (in billions) As Reported	Direct Property Damage (in billions) In 2009 Dollars
1994	2,054,500	4,275	27,250	105	95,400	$8.2	$11.9
1995	1,965,500	4,585	25,775	98	94,500	$8.9	$12.5
1996	1,975,000	4,990	25,550	96	87,150	$9.4	$12.9
1997	1,795,000	4,050	23,750	99	85,400	$8.5	$11.4
1998	1,755,500	4,035	23,100	91	87,500	$8.6	$11.3
1999	1,823,000	3,570	21,875	112	88,500	$10.0	$12.9
2000	1,708,000	4,045	22,350	103	84,550	$11.2	$14.0
2001	1,734,500	6,196[3]	21,100[4]	443[5]	82,250	$44.0[6]	$53.3[6]
2002	1,687,500	3,380	18,425	97	80,800	$10.3	$12.3
2003	1,584,500	3,925	18,125	106	78,750	$12.3[7]	$14.4[7]
2004	1,550,500	3,900	17,875	105	75,840	$9.8	$11.1
2005	1,602,000	3,675	17,925	87	80,100	$10.7	$11.8
2006	1,642,500	3,245	16,400	89	83,400	$11.3	$12.0
2007	1,557,500	3,430	17,675	103	80,100	$14.6[8]	$15.1[8]
2008	1,451,500	3,320	16,705	104	79,700	$15.5[9]	$15.5[9]
2009	1,348,500	3,010	17,050	82	78,150	$12.5	$12.5

[1]Includes $1.5 billion in damage caused by the Oakland fire storm.

[2]Includes $809 million in damage caused by Southern California wildfires.

[3]Includes 2,451 civilian deaths that occurred from the events of September 11, 2001.

[4]Includes 800 civilian injuries that occurred from the events of September 11, 2001.

[5]Includes 340 firefighters at the World Trade Center, September 11, 2001.

[6]Includes $33.44 billion in property loss that occurred from the events of September 11, 2001.

[7]Includes the Southern California Wildfires (Cedar and Old Wildfires) with an estimated total property loss of $2,040,000,000. Loss by specific property type for this fire was not available.

[8]This includes the 2007 California Fire Storm with an estimated property damage of $1.8 billion.

[9]This figure includes the 2008 California wildfires with an estimated property loss of $1.4 billion.

Direct property damage figures do not include indirect losses, such as business interruption. Inflation adjustment to 2007 dollars is done using the consumer price index.

Source: "Fire Loss in the United States 2009," Michael J. Karter, Jr., August 2010 and previous reports in the series; "Firefighter Fatalities in the Unites States," Rita F. Fahy, Paul R. LeBlanc, Joseph L. Molis, June 2010 and previous reports in the series; "U.S. Firefighter Injuries," Michael J. Karter, Jr., Joseph L. Molis, October 2009 and previous reports in the series.

TABLE A-2

Year	Fires	Civilian Deaths	Civilian Injuries	Direct Property Damage (in billions) As Reported	Direct Property Damage (in billions) In 2009 Dollars
1977	750,000	6,135	22,600	$2.2	$7.7
1978	730,500	6,185	21,260	$2.2	$7.2
1979	721,500	5,765	20,450	$3.0	$7.5
1980	757,500	5,446	221,100	$5.5	$7.9
1981	733,000	5,540	20,375	$3.3	$7.7
1982	676,500	4,940	21,100	$3.3	$7.2
1983	641,500	4,820	21,450	$3.3	$7.1
1984	623,000	4,240	19,275	$3.4	$7.1
1985	622,000	5,025	19,825	$3.8	$7.5
1986	581,500	4,770	19,025	$3.6	$7.0
1987	551,500	4,660	20,440	$3.7	$7.0
1988	552,500	5,065	22,600	$4.0	$7.3
1989	513,500	4,435	20,750	$4.0	$6.9
1990	467,000	4,115	20,650	$4.3	$7.0
1991	478,000	3,575	21,850	$5.6[1]	$8.7[1]
1992	472,000	3,765	21,600	$3.9	$5.9
1993	470,000	3,825	22,600	$4.8[2]	$7.2[2]
1994	451,000	3,465	20,025	$4.3	$6.3
1995	425,500	3,695	19,125	$4.4	$6.1
1996	428,000	4,080	19,300	$5.0	$6.8
1997	406,500	3,390	17,775	$4.6	$6.1
1998	381,500	3,250	17,175	$4.4	$5.8
1999	383,000	2,920	16,425	$5.1	$6.6
2000	379,500	3,445	17,400	$5.7	$7.1
2001	396,500	3,140	15,575	$5.6	$6.8
2002	401,000	2,695	14,050	$6.1	$7.2
2003	402,000	3,165	14,075	$6.1[3]	$7.1[3]
2004	410,500	3,225	14,175	$5.9	$6.8
2005	396,000	3,055	13,825	$6.9	$7.6
2006	412,500	2,620	12,925	$7.0	$7.4

Delmar/Cengage Learning

(continues)

TABLE A-2 Continued

Year	Fires	Civilian Deaths	Civilian Injuries	Direct Property Damage (in Billions) As Reported	Direct Property Damage (in Billions) In 2009 Dollars
2007	414,000	2,895	14,000	\$7.5[4]	\$7.8[4]
2008	403,000	2,780	13,560	\$8.6[5]	\$8.5[5]
2009	377,000	2,590	13,050	\$7.8	\$7.8

[1]Includes \$1.5 billion in damage caused by the Oakland Fire Storm, most of which was lost to homes but for which no detailed breakdown by property type was available.

[2]Includes \$809 million in damage caused by Southern California wildfires.

[3]This does not include the Southern California wildfires with an estimated property damage of \$2 billion.

[4]This does not include the 2007 California Fire Storm with an estimated property damage of \$1.8 billion.

[5]Does not include the 2008 California wildfires.

Residential structures include homes, hotels and motels, dormitories, barracks, and the like but do not include hospitals, nursing homes, residential schools, jails or prisons, among other properties that provide sleeping accommodations.

Direct property damage figures do not include indirect losses, such as business interruption. Inflation adjustment to 2007 dollars is done using the consumer price index.

Source: "Fire Loss in the United States 2009," Michael J. Karter, August 2010, and previous reports in the series.

TABLE A-3

Year	Fires	Civilian Deaths	Civilian Injuries	Direct Property Damage (in Billions)[1] As Reported	Direct Property Damage (in Billions) In 2009 Dollars
1977	348,000	370[2]	3,710	\$1.9	\$6.9
1978	331,500	165	3,725	\$1.8	\$6.0
1979	315,000	205	4,400	\$2.4	\$7.2
1980	307,500	229	3,625	\$2.4	\$6.3
1981	294,500	220	5,325	\$2.7	\$6.4
1982	270,000	260	4,475	\$2.5	\$5.5
1983	227,000	270	4,700	\$2.5	\$5.4
1984	225,000	285	3,750	\$2.5	\$5.1
1985	237,500	240	3,525	\$2.7	\$5.3

(continues)

TABLE A-3 Continued

Year	Fires	Civilian Deaths	Civilian Injuries	Direct Property Damage (in billions)[1] As Reported	Direct Property Damage (in billions) In 2009 Dollars
1986	218,500	215	3,725	$2.3	$4.5
1987	206,500	220	3,375	$2.5	$4.8
1988	192,500	215	3,675	$3.2	$5.8
1989	174,500	220	3,275	$3.5	$6.1
1990	157,000	285[3]	3,425	$2.5	$4.1
1991	162,500	190	3,125	$2.8[4]	$4.4[4]
1992	165,500	175	2,725	$3.1	$4.7
1993	151,500	155	3,950[5]	$2.6[6]	$3.8[6]
1994	163,000	125	3,100	$2.6	$3.7
1995	148,000	290[7]	2,600	$3.3[8]	$4.6[8]
1996	150,500	140	2,575	$3.0	$4.1
1997	145,500	120	2,600	$2.5	$3.3
1998	136,000	170	2,250	$2.3	$3.1
1999	140,000	120	2,100	$3.4	$4.4
2000	126,000	90	2,200	$2.8	$3.5
2001[9]	125,000	80	1,650	$3.2	$3.9
2002	118,000	80	1,550	$2.7	$3.2
2003	117,500	220[10]	1,525	$2.6[11]	$3.0[11]
2004	115,500	80	1,350	$2.4	$2.7
2005	115,000	50	1,500	$2.3	$2.5
2006	111,500	85	1,425	$2.6	$2.8
2007	116,500	105	1,350	$3.1[12]	$3.2[12]
2008	112,000	120	1,400	$3.8[13]	$3.8[13]
2009	103,500	105	1,690	$3.0	$3.0

[1]Individual incidents with large loss can affect the total for a given year. Note the following: The 1988 figure includes a Norco, Louisiana petroleum refinery fire with a loss of $330 million. The 1989 figure includes a Pasadena, Texas, polyolefin plant fire with a loss of $750 million.

[2]Includes 165 deaths at the Beverly Hills Supper Club fire in Southgate, Kentucky.

[3]Includes 87 deaths at the Happy Land social club firein New York City (N.Y.C.).

[4]Does not include $1.5 billion in damage caused by Oakland Fire Storm.

(continues)

[5]Includes 1,024 injuries that occurred at the World Trade Center explosion and fire in N.Y.C.

[6]Does not include the Southern California Wildfire with a loss of $809 million.

[7]Includes 168 deaths that occurred at the federal office building fire in Oklahoma City, OK.

[8]Includes an Oklahoma City, OK office building with a loss of $135 million, a Georgia manufacturing plant fire with a loss of $200 million, and a Massachusetts industrial complex fire with a loss of $500 million.

[9]Does not include the events of 9/11/01, where there were 2,451 civilian deaths, 800 civilian injuries, and $33.44 billion in property loss.

[10]This includes 100 fire deaths in the Station Nightclub Fire in Rhode Island and 31 deaths in two nursing home fires in Connecticut and Tennessee.

[11]This does not include the Southern California Wildfires with an estimated property damage of $2 billion.

[12]This does not include the 2007 California Fire Storm with an estimated property damage of $1.8 billion.

[13]Does not include the 2008 California wildfires.

The term non-residential includes public assembly, educational, institutional, store and office, industry, utility, storage and special structure properties.

Direct property damage figures do not include indirect losses, such as business interruption. Inflation adjustment to 2008 dollars is done using the consumer price index.

Source: "Fire Loss in the United States 2009," Michael J. Karter, Jr., August 2010, and previous reports in the series.

TABLE A-4

Year	Fires	Civilian Deaths	Civilian Injuries	Direct Property Damage[1] (in billions) As Reported	Direct Property Damage[1] (in billions) In 2009 Dollars
1977	1,098,000	6,505	26,310	$4.1	$14.6
1978	1,062,000	6,350	24,985	$4.0	$13.2
1979	1,036,500	5,970	24,850	$5.0	$14.7
1980	1,065,000	5,675	24,725	$5.5	$14.2
1981	1,027,500	5,760	25,700	$6.0	$14.1
1982	946,500	5,200	25,575	$5.7	$12.7
1983	868,500	5,090	26,150	$5.8	$12.5
1984	848,000	4,525	23,025	$5.9	$12.1
1985	859,500	5,265	23,350	$6.4	$12.8
1986	800,000	4,985	22,750	$5.8	$11.4
1987	758,000	4,880	23,815	$6.2	$11.8
1988	745,000	5,280	26,275	$7.2[2]	$13.0[2]
1989	688,000	4,655	24,025	$7.5[3]	$13.0[3]

(continues)

TABLE A-4 Continued

Year	Fires	Civilian Deaths	Civilian Injuries	Direct Property Damage (in billions)[1] As Reported	Direct Property Damage (in billions) In 2009 Dollars
1990	624,000	4,400	24,075	$6.7	$11.0
1991	640,500	3,765	24,975	$8.3[4]	$13.1[4]
1992	637,500	3,940	24,325	$7.0[5]	$10.6[5]
1993	621,500	3,980	26,550	$7.4[6]	$11.0[6]
1994	614,000	3,590	23,125	$6.9	$9.9
1995	573,500	3,985[7]	21,725	$7.6	$10.7
1996	578,500	4,220	21,875	$7.9	$10.9
1997	552,000	3,510	20,375	$7.1	$9.5
1998	517,500	3,420	19,425	$6.7	$8.9
1999	523,000	3,040	18,525	$8.5	$10.9
2000	505,500	3,535	19,600	$8.5	$10.6
2001[8]	521,500	3,220	17,225	$8.9	$10.8
2002	519,000	2,775	15,600	$8.7	$10.4
2003	519,500	3,385[9]	15,600	$8.7[10]	$10.1[10]
2004	526,000	3,305	15,525	$8.3	$9.5
2005	511,000	3,105	15,325	$9.2	$10.1
2006	524,000	2,705	14,350	$9.6	$10.3
2007	530,500	3,000	15,350	$10.6[11]	$11.0[11]
2008	515,000	2,900	14,960	$12.4[12]	$12.3[12]
2009	480,500	2,695	14,740	$10.8	$10.8

[1]Individual incidents with large losses can affect the total for a given year.

[2]The1988 figure includes a Norco, Louisiana petroleum refinery fire with a loss of $330 million.

[3]The1989 figure includes a Pasadena, Texas, polyolefin plant fire with a loss of $750 million.

[4]The 1991 figure includes theOakland Fire Stormwith a loss of $1.5 billion and the Meridian Plaza high-rise fire in Philadelphia with a loss of $325 million.

[5]The 1992 figure includes the Los Angeles Civil Disturbance with a loss of $567 million

[6]The 1993 figure includes Southern California wildfires with a loss of $809 million.

[7]Includes 168 deaths that occurred at the federal office building fire in Oklahoma City, OK.

[8]Does not include the events of 9/11/01, where there were 2,451 civilian deaths, 800 civilian injurie [illegible] losses.

[9]Includes 100 fire deaths in the Station Nightclub Fire in Rhode Islandand 31 deaths in two nursing home fires in Connecticut and Tennessee.

[10]Does not include the Southern California Wildfires with an estimated property damage of $2 billion.

[11]This does not include the 2007 California Fire Storm with an estimated property damage of $1.8 million.

[12]Does not include 2008 California wildfires.

Direct property damage figures do not include indirect losses, such as business interruption. Inflation adjustment to 2008 dollars is done using the consumer price index.

Source: "Fire Loss in the United States 2009," Michael J. Karter, Jr., August 2010, and previous reports in the series.

TABLE A-5

Year	Fires	Civilian Deaths	Civilian Injuries	Direct Property Damage (in billions) As Reported	Direct Property Damage (in billions) In 2009 Dollars
1977	723,500	5,865	21,640	$2.0	$9.6
1978	706,500	6,015	20,400	$2.1	$6.8
1979	696,500	5,500	18,825	$2.4	$7.0
1980	734,000	5,200	19,700	$2.8	$7.4
1981	711,000	5,400	19,125	$3.1	$7.4
1982	654,500	4,820	20,450	$3.1	$7.0
1983	625,500	4,670	20,750	$3.2	$6.9
1984	605,500	4,075	18,750	$3.4	$6.9
1985	606,000	4,885	19,175	$3.7	$7.4
1986	565,500	4,655	18,575	$3.5	$6.8
1987	536,500	4,570	19,965	$3.6	$6.8
1988	538,500	4,955	22,075	$3.9	$7.1
1989	498,500	4,335	20,275	$3.9	$6.7
1990	454,500	4,050	20,225	$4.2	$6.8
1991	464,500	3,500	21,275	$5.5[1]	$8.6[1]
1992	459,000	3,705	21,100	$3.8	$5.8
1993	458,000	3,720	22,000	$4.8[2]	$7.1[2]
1994	438,000	3,425	19,475	$4.2	$6.1
1995	414,000	3,640	18,650	$4.3	$6.0
1996	417,000	4,035	18,875	$4.9	$6.7
[illegible]7	395,500	3,360	17,300	$4.5	$6.0

(continues)

TABLE A-5 Continued

Year	Fires	Civilian Deaths	Civilian Injuries	Direct Property Damage (in Billions)[1] As Reported	Direct Property Damage (in Billions) In 2009 Dollars
1998	369,500	3,220	16,800	$4.3	$5.6
1999	371,000	2,895	16,050	$5.0	$6.4
2000	368,000	3,420	16,975	$5.5	$6.9
2001	383,500	3,110	15,200	$5.5	$6.7
2002	389,000	2,670	13,650	$5.9	$7.1
2003	388,500	3,145	13,650	$5.9[3]	$6.9[3]
2004	395,500	3,190	13,700	$5.8	$6.6
2005	381,000	3,030	13,300	$6.7	$7.4
2006	396,000	2,580	12,500	$6.8	$7.6
2007	399,000	2,865	13,600	$7.4[4]	$7.7[4]
2008	386,500	2,755	13,160	$8.2	$8.2[5]
2009	362,500	2,565	12,650	$7.6	$7.6

[1]Includes $1.5 billion in damage caused by the Oakland Fire Storm, most of which was lost to homes but for which no detailed breakdown by property type was available.

[2]Includes $809 million in damage caused by the Southern California Wildfires.

[3]Does not include the Southern California Wildfires.

[4]Does not include the 2007 California Fire Storm with an estimated property damage of $1.8 billion.

[5]Does not include the 2008 California wildfires with an estimated property damage of $1.4 billion.

"Homes" are dwellings, duplexes, manufactured homes (also called mobile homes), apartments, row houses, town houses and condominiums. Other residential properties, such as hotels and motels, dormitories, barracks, rooming and boarding homes, and the like, are not included.

Direct property damage figures do not include indirect losses, such as business interruption. Inflation adjustment to 2009 dollars is done using the consumer price index.

All are estimates of losses in fires reported to fire departments, based on data reported to NFPA's annual National Fire Exp Survey.Direct property damage figures have not been adjusted for inflation.

Sources: "Home Structure Fires," Marty Ahrens, March 2010; "Fire Loss in the United States During 2008," Michael J. August 2009; "Catastrophic Multiple-Death Fires for 2008," Stephen G. Badger, September 2009.

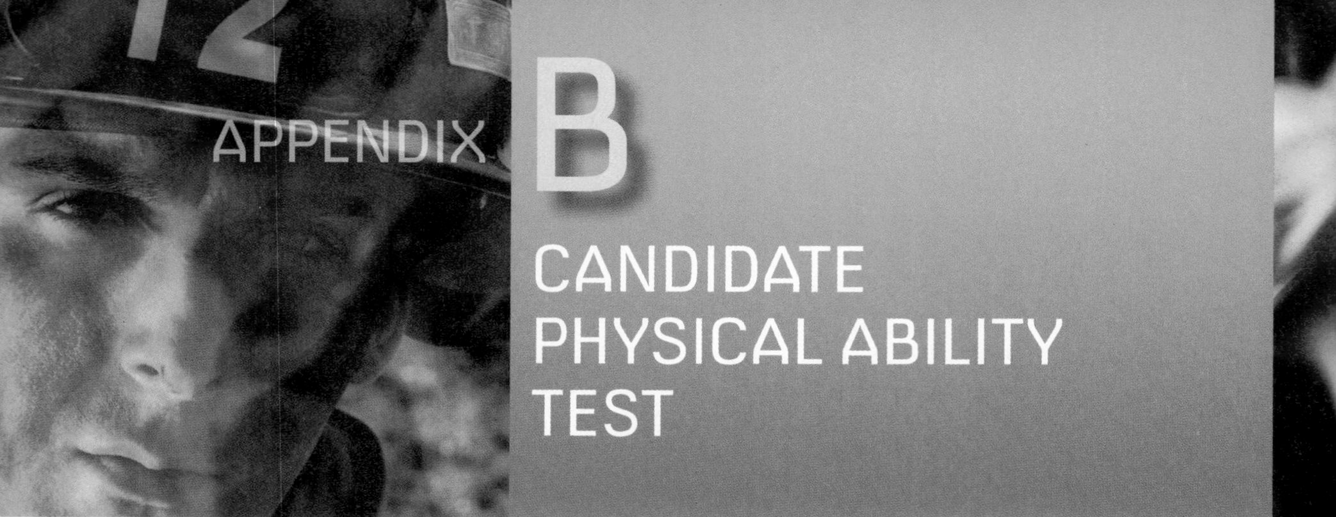

APPENDIX B

CANDIDATE PHYSICAL ABILITY TEST

The Candidate Physical Ability Test (CPAT) was developed as a joint effort of the International Association of Firefighters (IAFF), the International Association of Fire Chiefs (IAFC), 10 professional fire departments, and their union locals from across the United States and Canada. The fire departments participating in the development of the CPAT were Austin, Tex.; Calgary, Alberta; Charlotte, N.C.; Fairfax County, Va.; Indianapolis, In.; Los Angeles County, Ca.; Miami Dade, Fl.; New York City, N.Y.; Phoenix, Ariz.; and Seattle, Wash.

The test was designed to simulate activities performed by firefighters at incidents. The CPAT was developed to allow a fire department to obtain a pool of trainable potential candidates for employment who are physically able to perform essential job tasks at fire scenes.

The CPAT consists of eight separate sequential events that require the candidate to progress along a predetermined path from event to event in a continuous manner. This is a pass/fail test based on a validated maximum total time of 10 minutes and 20 seconds.

[illegible] all events, the candidate must wear long pants, a hard hat with chin- [illegible] nd footwear with no open heel or toe. Watches and loose or restrictive [illegible] itted.

[illegible] the candidate wears a 50-pound (22.68-kg) vest to simulate the [illegible] breathing apparatus and firefighter protective clothing. An addi- [illegible] g), using two 12.5-pound (5.67-kg) weights that simulate a high- [illegible] added for the stair climb event.

[illegible] to obtain the necessary information regarding the candidate's [illegible]

[illegible] were chosen to provide the highest level of consistency, [illegible] the candidate's physical abilities.

The props are placed in a sequence that best simulates their use in a fire scene while requiring an 85-foot (25.91-m) walk between events. To ensure the highest level of safety and to prevent candidates from exhaustion, no running is allowed between events. This walk allows the candidate approximately 20 seconds to recover and regroup before each event.

The CPAT includes eight sequential events as follows:

1. Stair Climb
2. Hose Drag
3. Equipment Carry
4. Ladder Raise and Extension
5. Forcible Entry
6. Search
7. Rescue
8. Ceiling Breach and Pull

The CPAT is not only a physical ability testing process. It also includes information on recruiting and mentoring personnel to prepare for the testing; on diversity and its importance in the fire service; and on recruiting methods through community outreach, community organizations, the Internet, the media, brochures, public announcements, and colleges and high schools.

The program has a Preparation Guide for candidates and information as to why a fire department must provide a Preparation Guide. Included in the Preparation Guide is information on who should receive it and what other activities fire departments can provide to prepare applicants.

For further information on the Candidate Physical Ability Test, contact an IAFF union local or check the IAFF Web site at www.IAFF.org.

APPENDIX C

Numerical Index—2010 National Fire Codes Contents

70E	Electrical Safety in the Workplace—2009	Vol 5
72	National Fire Alarm and Signaling Code—2010	Vol 6
73	Electrical Inspection for Existing Dwellings—2006	Vol 6
75	Protection of Information Technology Equipment—2009	Vol 6
76	Telecommunications Facilities—2009	Vol 6
77	Static Electricity—2007	Vol 16
79	Electrical Standard for Industrial Machinery—2007	Vol 6
80	Fire Doors and Other Opening Protectives—2010	Vol 6
80A	Exterior Fire Exposures—2010	Vol16
82	Incinerators, Waste and Linen Handling Systems and Equipment—2009	Vol 6
85	Boiler and Combustion Systems Hazards—2007	Vol 6
86	Ovens and Furnaces—2007	Vol 6
88A	Parking Structures—2007	Vol 7
90A	Installation of Air-Conditioning and Ventilating Systems—2009	Vol 7
90B	Installation of Warm Air Heating and Air-Conditioning Systems—2009	Vol 7
91	Exhaust Systems for Air Conveying of Gases, etc.—2010	Vol 7
92A	Smoke-Control Systems—2009	Vol 7
92B	Smoke Management Systems in Malls, Atria, Large Spaces—2009	Vol 7
96	Ventilation Control and Fire Protection of Commercial Cooking Operations—2008	Vol 7
99	Health Care Facilities—2005	Vol 7
99B	Hypobaric Facilities—2010	Vol 7
101®	Life Safety Code®—2009	Vol 7
101A	Alternative Approaches to Life Safety—2010	Vol 16
102	Grandstands, Folding and Telescopic Seating, Tents, and Membrane Structures—2006	Vol 8
105	Smoke Door Assemblies and Other Opening Protectives—2010	Vol 8
110	Emergency and Standby Power Systems—2010	Vol 8
111	Stored Electrical Energy Emergency and Standby Power Systems—2010	Vol 8
115	Laser Fire Protection—2008	Vol 8
120	Fire Prevention and Control in Coal Mines—2010	Vol 8
122	Fire Prevention and Control in Metal/Nonmetal Mining and Metal Mineral Processing Facilities—2010	Vol 8
130	Transit and Passenger Rail Systems—2010	Vol 8
140	Motion Picture and TV Production Facilities and Locations—2008	Vol 8
150	Fire and Life Safety in Animal Housing Facilities—2009	Vol 8
160	Use of Flame Effects Before an Audience—2006	Vol 8
170	Fire Safety and Emergency Symbols—2009	Vol 8
204	Smoke and Heat Venting—2007	Vol 8
211	Chimneys, Fireplaces, Vents, and Solid Fuel-Burning Appliances—2010	Vol 8
214	Water-Cooling Towers—2005	Vol 8
220	Types of Building Construction—2009	Vol 8
221	High Challenge Fire Walls, Fire Walls, and Fire Barrier Walls—2009	Vol 8
225	Model Manufactured Home Installation Standard—2009	Vol 8
232	Records, Protection of—2007	Vol 8
241	Construction, Alteration, and Demolition Operations—2009	Vol 8
251	Fire Tests of Building Construction and Materials—2006	Vol 8
252	Fire Tests of Door Assemblies—2008	Vol 8
253	Test for Critical Radiant Flux of Floor Covering Systems Using a Radiant Heat Energy Source—2006	Vol 8
257	Fire Tests of Window and Glass Block Assemblies—2007	Vol 8
259	Test Method for Potential Heat of Building Materials—2008	Vol 8
260	Cigarette Ignition Resistance of Components of Upholstered Furniture—2009	Vol 8
261	Method of Test for Determining Resistance of Mock-Up Upholstered Furniture Material Assemblies to Ignition by Smoldering Cigarettes—2009	Vol 8
262	Method of Test for Flame Travel and Smoke of Wires and Cables—2007	Vol 8
265	Evaluating Room Fire Growth Contribution of Textile Coverings on Full Height Panels and Walls—2007	Vol 8
268	Exterior Wall Assemblies, Test for Ignitability—2007	Vol 8
269	Toxic Potency Data, Test for Developing—2007	Vol 8
270	Measurement of Smoke Obscuration Using a Conical Radiant Source—2008	Vol 8
271	Heat and Visible Smoke Release Rates for Materials and Products Using an Oxygen Consumption Calorimeter—2009	Vol 9
274	Pipe Insulation Test to Evaluate Fire Performance Characteristics—2009	Vol 9
275	Evaluation of Thermal Barriers Used Over Foam Plastic Insulation—2009	Vol 9
285	Evaluation of Flame Propagation Characteristics of Exterior Non-Load-Bearing Wall Assemblies—2006	Vol 9

286 Wall and Ceiling Interior Finish—2006 Vol 9
287 Flammability of Materials in Cleanrooms—2007 Vol 9
288 Floor Fire Door Assemblies, Fire Tests of—2007 Vol 9
289 Individual Fuel Packages Fire Test—2009 Vol 9
290 Passive Protection Materials for Use on LP-Gas Containers, Fire Testing—2009 Vol 9
291 Fire Flow Testing and Marking of Hydrants—2010 Vol 16
301 Merchant Vessels—2008 Vol 9
302 Pleasure and Commercial Motor Craft—2010 Vol 9
303 Marinas and Boatyards—2006 Vol 9
306 Control of Gas Hazards on Vessels—2009 Vol 9
307 Marine Terminals, Piers, and Wharves—2006 Vol 9
312 Fire Protection of Vessels During Construction, Conversion, Repair, and Lay-Up—2006 Vol 9
318 Protection of Semiconductor Fabrication Facilities—2009 Vol 9
326 Safeguarding Tanks and Containers—2010 Vol 9
329 Handling Releases of Flammable and Combustible Liquids and Gases—2010 Vol 16
385 Tank Vehicles for Flammable and Combustible Liquids—2007 Vol 9
400 Hazardous Materials Code—2010 Vol 9
402 Aircraft Rescue and Fire Fighting Operations—2008 Vol 16
403 Aircraft Rescue and Fire Fighting Services at Airports—2009 Vol 9
405 Recurring Proficiency of Airport Fire Fighters—2010 Vol 9
407 Aircraft Fuel Servicing—2007 Vol 9
408 Aircraft Hand Portable Fire Extinguishers—2010 Vol 9
409 Aircraft Hangars—2004 Vol 9
410 Aircraft Maintenance—2010 Vol 9
412 Evaluating Aircraft Rescue and Fire-Fighting Foam Equipment—2009 Vol 9
414 Aircraft Rescue and Fire-Fighting Vehicles—2007 Vol 9
415 Airport Terminal Buildings, Fueling Ramp Drainage, Loading Walkways—2008 Vol 9
418 Heliports—2006 Vol 9
422 Aircraft Accident/Incident Response Assessment—2010 Vol 16
423 Construction and Protection of Aircraft Engine Test Facilities—2010 Vol 9
424 Airport/Community Emergency Planning—2008 Vol 16
450 Emergency Medical Services and Systems—2009 Vol 16
472 Competence of Responders to Hazardous Materials/Weapons of Mass Destruction Incidents—2008 Vol 10
473 Competencies for EMS Personnel Responding to Hazardous Materials/Weapons of Mass Destruction Incidents—2008 Vol 10
484 Combustible Metals—2009 Vol 10
495 Explosive Materials Code—2010 Vol 10
496 Purged and Pressurized Enclosures for Electrical Equipment—2008 Vol 10
497 Classification of Flammable Liquids, Gases, or Vapors and of Hazardous (Classified) Locations for Electrical Installations in Chemical Process Areas—2008 Vol 16
498 Safe Havens for Vehicles Transporting Explosives—2010 Vol 10
499 Classification of Combustible Dusts and of Hazardous (Classified) Locations for Electrical Installations in Chemical Process Areas—2008 Vol 16
501 Manufactured Housing—2010 Vol 10
501A Fire Safety Criteria for Manufactured Home Installations, Sites, and Communities—2009 Vol 10
502 Road Tunnels, Bridges, and Other Limited Access Highways—2008 Vol 10
505 Powered Industrial Trucks Including Type Designations, Areas of Use, Maintenance, and Operations—2006 Vol 10
520 Subterranean Spaces—2010 Vol 10
550 Fire Safety Concepts Tree—2007 Vol 16
551 Evaluation of Fire Risk Assessments—2010 Vol 16
555 Evaluating Potential for Room Flashover—2009 Vol 16
560 Ethylene Oxide for Sterilization and Fumigation, Storage, Handling, and Use of—2007 Vol 10
600 Industrial Fire Brigades—2010 Vol 10
601 Security Services in Fire Loss Prevention—2010 Vol 10
610 Motorsports Venues—2009 Vol 17
654 Prevention of Fire and Dust Explosions from Manufacturing Combustible Particulate Solids—2006 Vol 10
655 Sulfur Fires and Explosions—2007 Vol 10
664 Fires and Explosions in Wood Processing and Woodworking Facilities—2007 Vol 10
701 Methods of Fire Tests for Flame Propagation of Textiles and Films—2010 Vol 10
703 Fire Retardant-Treated Wood and Fire-Retardant Coatings for Building Materials—2009 Vol 10
704 Identification of the Hazards of Materials—2007 Vol 10

705	Field Flame Test for Textiles and Films—2009	Vol 17
720	Carbon Monoxide Warning Equipment in Dwellings—2009	Vol 10
730	Premises Security—2008	Vol 17
731	Electronic Premises Security Systems—2008	Vol 10
750	Water Mist Fire Protection Systems—2010	Vol 10
780	Installation of Lightning Protection Systems—2008	Vol 10
801	Facilities Handling Radioactive Materials—2008	Vol 11
804	Fire Protection for Advanced Light Water Reactor Electric Generating Plants—2010	Vol 11
805	Light Water Reactor Electric Generating Plants—2010	Vol 11
806	Performance-Based Standard for Fire Protection for Advanced Nuclear Reactor Electric Generating Plants Change Process—2010	Vol 11
820	Fire Protection in Wastewater Treatment and Collection Facilities—2008	Vol 11
850	Electric Generating Plants—2010	Vol 17
851	Hydroelectric Generating Plants—2010	Vol 17
853	Stationary Fuel Cell Power Systems—2010	Vol 11
900	Building Energy Code—2010	Vol 11
901	Standard Classifications for Incident Reporting and Fire Protection Data—2006	Vol 17
909	Protection of Cultural Resource Properties—2010	Vol 11
914	Fire Protection in Historic Structures—2010	Vol 11
921	Fire and Explosion Investigations, Guide for—2008	Vol 17
1000	Fire Service Professional Qualifications Accreditation and Certification System—2006	Vol 11
1001	Fire Fighter Professional Qualifications—2008	Vol 11
1002	Fire Apparatus Driver/Operator Professional Qualifications—2009	Vol 11
1003	Airport Fire Fighter Professional Qualifications—2010	Vol 11
1005	Marine Fire Fighting for Land-Based Fire Fighters	Vol 11
1006	Technical Rescuer Professional Qualifications—2008	Vol 11
1021	Fire Officer Professional Qualifications—2009	Vol 12
1026	Incident Management Personnel Professional Qualifications—2009	Vol 12
1031	Professional Qualifications for Fire Inspector and Plan Examiner—2009	Vol 12
1033	Fire Investigator Professional Qualifications—2009	Vol 12
1035	Fire and Life Safety Educator, Public Information Officer, and Juvenile Firesetter Intervention Specialist Professional Qualifications—2010	Vol 12
1037	Professional Qualifications for Fire Marshal	Vol 12
1041	Fire Service Instructor Professional Qualifications—2007	Vol 12
1051	Wildland Fire Fighter Professional Qualifications—2007	Vol 12
1061	Public Safety Telecommunicator Qualifications—2007	Vol 12
1071	Emergency Vehicle Technicians—2006	Vol 12
1081	Industrial Fire Brigade Member Professional Qualifications—2007	Vol 12
1122	Code for Model Rocketry—2008	Vol 12
1123	Fireworks Display—2010	Vol 12
1124	Fireworks and Pyrotechnic Articles—2006	Vol 12
1125	Model Rocket and High Power Rocket Motors—2007	Vol 12
1126	Use of Pyrotechnics Before a Proximate Audience—2006	Vol 12
1127	High Power Rocketry—2008	Vol 12
1141	Infrastructure for Land Development in Suburban and Rural Areas—2008	Vol 12
1142	Water Supplies for Suburban and Rural Fire Fighting—2007	Vol 12
1143	Wildland Fire Management—2009	Vol 12
1144	Reducing Structure Ignition Hazards from Wildland Fire—2008	Vol 12
1145	Class A Foams—2006	Vol 17
1150	Foam Chemicals for Fires in Class A Fuels—2004	Vol 12
1192	Recreational Vehicles—2008	Vol 12
1194	Recreational Vehicle Parks and Campgrounds—2008	Vol 12
1201	Providing Fire and Emergency Services to the Public—2010	Vol 12
1221	Communications, Emergency Services—2010	Vol 12
1250	Fire and Emergency Services Organization Risk Management—2010	Vol 17
1401	Fire Protection Training Reports and Records—2006	Vol 17
1402	Building Fire Service Training Centers—2007	Vol 17
1403	Live Fire Training Evolutions—2007	Vol 13
1404	Fire Service Respiratory Protection Training—2006	Vol 13
1405	Land-Based Fire Fighters Who Respond to Marine Vessel Fires—2006	Vol 17
1407	Training Fire Service Rapid Intervention Crews—2010	Vol 13

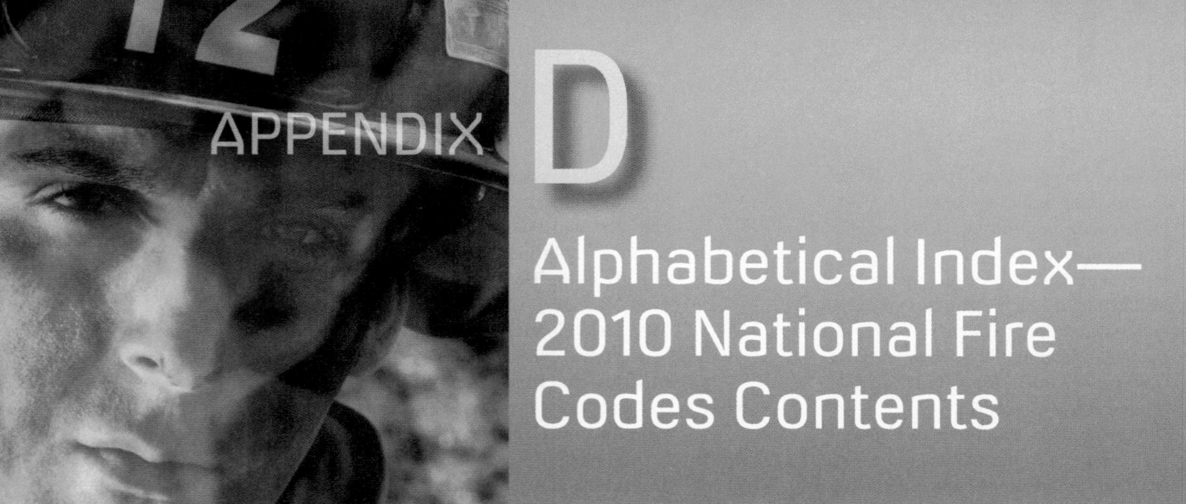

APPENDIX D

Alphabetical Index—2010 National Fire Codes Contents

APPENDIX E

FIRESCOPE ICS Resource Typing and Strike Team Requirements

Kind	Strike Team Types	Number/Type	Minimum Equipment Standards							Minimum Personnel		
			Pump Capac	Water Capac	2-1/2" Hose	1-1/2" Hose	1" Hose	Ladder	Master Stream	Strike Team Leader	Per Single Resource	Total Personnel
Engines	A	5 - Type 1	1000 GPM	400 Gal	1200 Ft	400 Ft	200 Ft	20 Ft	500 GPM	1	4	21
	B	5 - Type 2	500 GPM	400 Gal	1000 Ft	500 Ft	300 Ft	20 Ft	N/A	1	3	16
	C	5 - Type 3	120 GPM	300 Gal	N/A	1000 Ft	800 Ft	N/A	N/A	1	3	16
	D	5 - Type 4	50 GPM	200 Gal	N/A	300 Ft	800 Ft	N/A	N/A	1	3	16
Crews	G	Hand crew combinations consisting of a minimum of 35 persons (Do not mix Type 1 and Type 2 crews)	Type 1 Handcrews have no restrictions on use.							1	N/A	36
	H		Type 2 Handcrews may have use restrictions.							1	N/A	36
Dozers	K	2 - Type 1 1 - Dozer Tender	Heavy Dozer Min. 200 HP (D-7, D-8, or equivalent)							1	1 1	4
	L	2 - Type 2 1 - Dozer Tender	Medium Dozer Min. 100 HP (D-5, D-6, or equivalent)							1	1 1	4
	M	2 - Type 3 1 - Dozer Tender	Light Dozer Min. 50 HP (D-4, or equivalent)							1	1 1	4

Source: Field Operations Guide ICS 420-1 FIRESOPE Incident Command System Publication, June, 2004.

APPENDIX F
Common Acronyms

ACLS	advanced cardiac life support
AED	Automatic external defibrillator
AFFF	aqueous film-forming foam
AFFF ATC	aqueous film-forming foam alcohol type concentrate
ARFF	aircraft rescue firefighting apparatus
AVL	automatic vehicle location
BC	battalion chief
BLEVE	boiling liquid expanding vapor explosion
BLM	Bureau of Land Management
BLS	basic life support
CAD	computer-aided dispatch
CAFS	compressed air foam system
CBRNE	chemical, biological, radioactive, nuclear, explosive
CP	command post (see ICP)
CRT	cathode ray tube
DHS	Department of Homeland Security
DOT	Department of Transportation
DOT ERG	DOT Emergency Response Guidebook
EMS	emergency medical services
EMT	Emergency Medical Technician
EMTP	Paramedic
EOC	Emergency Operations Center
FAST	Firefighter Assist and Search Team
FD	fire department
FF	firefighter
FOG	ICS Field Operations Guide
FPB	Fire Prevention Bureau
FRO	first responder operations
GED	general equivalency diploma
GIS	geographic information system
GPS	global positioning system
HT	handy talkie portable radio
IAFC	International Association of Fire Chiefs
IAFF	International Association of Fire Fighters
IC	Incident Commander
ICS	Incident Command System
ICP	incident command post (see CP)
IDLH	immediately dangerous to life and health
ISO	Insurance Services Office
LDH	large diameter hose
LNG	liquefied natural gas
LPG	liquefied petroleum gas
MDC	mobile data computer
NBC	nuclear/biological chemical
NFES	National Fire Equipment System
NFPA	National Fire Protection Association
NIFC	National Interagency Fire Center
NIMS	National Incident Management System

NWCG	National Wildfire Coordinating Group
ODP	Office of Domestic Preparedness
OES	Office of Emergency Services
OOS	out of service
OSC	Operations Section Chief
OSHA	Occupational Safety and Health Administration
PAL	personal alarm locator
PASS	personal alerting safety system
PESO	Public Education/Safety Officer
PIO	Public Information Officer
PPE	personal protective equipment
psi	pounds per square inch
PTO	power takeoff
RIC	rapid intervention company
RIT	rapid intervention team
SCBA	self-contained breathing apparatus
SOG	standard operating guideline
SOP	standard operating procedure
ST	strike team
TF	task force
TO	training officer
USAR	urban search and rescue
USFS	United States Forest Service
WMD	weapons of mass destruction

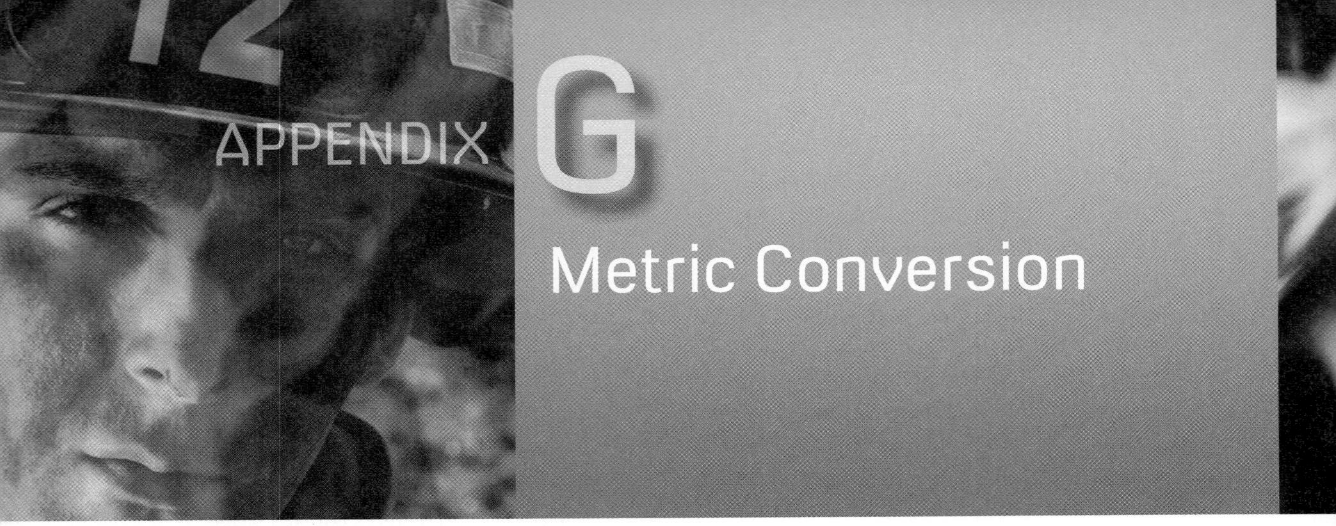

APPENDIX G

Metric Conversion

Conversion of length:

feet to meters, multiply by .305

inches to millimeters, multiply by 25.4

meters to feet, multiply by 3.28

millimeters to inches, multiply by .039

Conversion of speed:

miles (per hour) to kilometers (per hour), multiply by 1.609

kilometers (per hour) to miles (per hour), multiply by .62

Conversion of volume:

gallons (per minute) to liters (per minute), multiply by 3.79

liters (per minute) to gallons (per minute), multiply by .264

Conversion of weight:

pounds to kilograms, multiply by .4536

kilograms to pounds, multiply by 2.205

Conversion of temperature:

degrees Fahrenheit to degrees Celsius, subtract 32 and multiply by 5/9 (.555)

degrees Celsius to degrees Fahrenheit, multiply by 9/5 (1.8) and add 32

degrees Kelvin to degrees Celsius, add 273.15 to the Celsius reading

degrees Fahrenheit to degrees Rankine, add 459.69 to the Fahrenheit reading

APPENDIX H

Fire-Related Web Sites

These sites can be used by themselves or as starting points to link to other sites:

usfa.dhs.gov/fireservice/fatalities—National Institute for Occupational Safety and Health Web site. Source for statistics and investigative reports on firefighter line-of-duty deaths.

disastersafety.org—The Institute for Business & Home Safety's Web site with information on protecting property from natural disasters

emergencyvehicleresponse.com—Site focused on fire apparatus safety.

everyonegoeshome.com—Firefighter Life Safety Initiatives site

firebooks.com—Firefighters' Bookstore Web site.

firedawg.com—Information on fire-related education and training.

firehouse.com—Firehouse magazine and links to more than 8,000 Web sites.

firefighterclosecalls.com—Site dedicated to firefighter safety and survival.

firefighternearmiss.com—National Firefighter Near-Miss Reporting System

firescope.org—ICS information

firenuggets.com—Fire photos and firefighting information.

delmarfire.cengage.com—Cengage Learning's fire service products Web site.

firstgov.gov—Directory to U.S. government Web sites.

IAFC.org—International Association of Fire Chiefs Web site.

IAFF.org—International Association of Fire Fighters Web site.

ICMA.org—International City/County Management Association

NIFC.gov—National Interagency Fire Center Web site. Listing of wildland fire information.

NOAA.gov—National Oceanographic and Atmospheric Administration Web site. Site for information on fire weather, tornadoes, and other weather-related disasters.

NFPA.org—National Fire Protection Association Web site. Source of statistical, code, training, and other fire-related data.

respondersafety.com—Source for firefighter safety resources.

training.fema.gov—National Fire Academy site.

USFA.fema.gov—United States Fire Administration source for fire data, National Fire Academy course and registration information, and Learning Resources Center library access.

Wildland Fire-Related Web-Sites

fireleadership.gov—Wildland fire leadership program

iawfonline.org—International Association of Wildland Firefighters site

myfirecommunity.net—Urban/wildland interface fire prevention site.

wfap.net—USFS Apprentice program

wildfirelessons.net—Site for wildland fire safety information.

wffoundation.org—Wildland Firefighters Foundation site

nifc.gov/wfstar/index.htm—Wildland fire safety annual refresher training site

firecareers.com

firecareerassist.com

fireemployment.com

firefighter-jobs.com

firerecruit.com

firerescue1.com

recruitfirefighter.com

USAjobs.opm.gov—U.S. Government jobs site

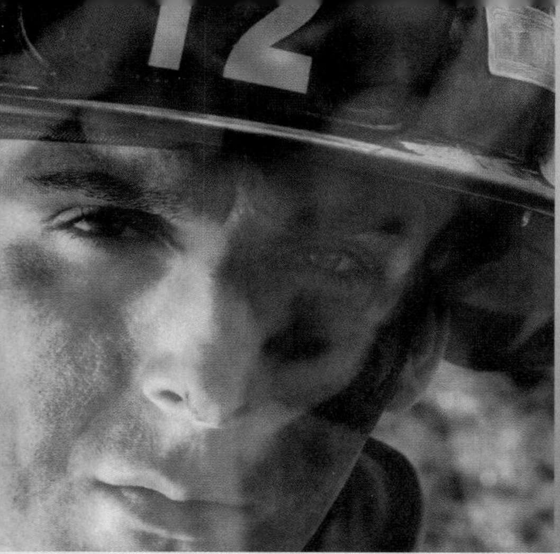

GLOSSARY

A

administrative procedures Written procedures for performing staff-related functions such as reports and other paperwork.

all clear An "all clear" is the short descriptive phrase which indicates that a primary search of the structure for victims has been completed.

alley lights Lights mounted in a light bar that shine to the side of the vehicle, commonly used for spotting addresses on structures at night.

ambient temperature The temperature surrounding an object. Air temperature.

ambulance gurney The wheeled cot that patients are placed on prior to transport in an ambulance.

arcing Spark created when electrical contact is made.

articulated boom Elevating device consisting of a boom that is hinged in the middle.

atmospheric pressure The pressure of the atmosphere exerted on any point. 14.7 psi at sea level.

automatic aid Under this system departments assist each other without regard to jurisdictional boundaries. Often used in areas where there are county islands within city limits or in interagency areas where a fire starting in one agency's area is a direct threat to another's jurisdiction.

B

back draft A type of explosion caused by the sudden influx of air into a mixture of gases, which have been heated above the ignition temperature of at least one of them.

backfires A fire lit in front of an advancing fire to remove fuel and widen control lines.

baffles Partitions placed in tanks that prevent the water from sloshing and making the vehicle unstable when turning corners.

BLEVE Boiling liquid expanding vapor explosion.

bonnet The top of a hydrant.

buddy breathing A technique in which two people share the same SCBA air supply to avoid breathing smoke or toxic fumes.

burning out Lighting a fire to remove fuel along the flanks of a fire. Also used to remove unburned islands that remain as the fire advances.

burst pressure The pounds per square of inch of pressure at which a container will fail.

C

call-back A recall of personnel to on-duty status, usually due to an emergency situation.

cascade system A system of large compressed gas cylinders connected to a manifold.

class A foam Foam designed for use on ordinary combustible materials.

combustible construction The use of unprotected wood and wood byproducts in building construction.

combustible gas indicator A device that measures the percentage of lower explosive limit concentration of gas in the atmosphere. These devices must be used by trained personnel for proper interpretation of the readings.

command presence The ability to maintain composure in situations that are stressful.

company officer The first line supervisor in the fire department. Depending on the jurisdiction involved, this position may be identified by various titles, such as captain, lieutenant, sergeant, station manager, module leader, or unit manager.

composition roofing Tar paper and shingles or tar paper covered with roofing asphalt.

cone roof A style of tank construction with a vapor space over the product. The lid is connected to the tank with a weak seam that will rupture before the tank wall seams.

confined space A space that is not designed to be occupied on a regular basis that is lacking in natural ventilation.

conflagration Very large fires that defy control efforts and cause extensive damage over a large area.

contaminated To become coated with a harmful substance.

control lines Removing fuel, applying water, or using natural barriers to stop a wildland fire from spreading.

cover To move resources into a fire station when the regular crew is assigned to an incident. In some departments this is called a *move up.*

critical incident stress debriefing A discussion in which personnel are encouraged to express their feelings after responding to and operating on particularly stressful events that result in high loss of life or other significant conditions. Conducted to help personnel better deal with their emotions.

CRM Is based on the assumption that human error is the primary cause of fireground fatalities and injuries and by using CRM the fire service can reduce the number of negative outcomes.

crown fire Fire in the tops of trees. These fires move very rapidly and defy control efforts.

curriculum A particular course of study.

D

dead-end main A water main that is not gridded into the system. Water flows into it from only one way.

decontaminate The physical removal of contaminants from people or equipment.

delegation of authority In a single jurisdiction incident the authority is already established (e.g., the local fire department is responsible for incidents within its jurisdiction). Should an incident management team be assigned to manage the incident they will require a written delegation of authority.

Delta (Δ) Change in factors affecting the incident, including personnel fatigue, time of day, structural weakening, and so forth.

demographics The statistical characteristics (e.g., age, race, gender, income) of the population of an area.

drafting pit An open-topped underground tank that is used for drafting operations and pump testing.

drill The practicing of tasks and jobs to improve performance.

driver/operator The primary responsibility of this position is to operate the pumping or aerial apparatus assigned to the fire department. Depending on the jurisdiction involved, this position may be identified by various titles, such as engineer, chauffeur, or truck operator.

dry chemical extinguisher Fire extinguisher using a chemically active powder.

duff Leaves, pine needles, and other dead forest material.

E

eligible list A certified list of persons who have successfully completed the testing process.

emergency medical dispatch A system where dispatchers are trained to give medical advice to the persons at the scene until emergency help arrives.

emergency medical technician A specified level of medical training that usually consists of approximately 100 hours of classroom and practical training.

energy The capacity for doing work.

escape route A preplanned and understood route to a safety zone.

evaporation The changing of liquid to a vapor.

Explorers A program of the Boy Scouts of America for persons 15 to 21 years of age. The Explorers work in conjunction with a professional organization such as the fire or police department to learn the operation and job requirements.

extrication The act of removing trapped victims. Term usually used in reference to vehicle accidents.

F

felony A serious crime such as murder, arson, or rape for which the punishment is either imprisonment in a state prison for more than one year, or death.

fire department connection Fittings connected to the fire protection system used by the fire department to boost the pressure and/or add water to the system.

fire flow Amount of water supply required for fire extinguishment expressed in gallons per minute (gpm).

fire-resistive construction Construction that has been designed to resist the effects of heat from fire.

first-alarm complement The equipment normally dispatched when a fire is first reported.

fiscal Financial.

flank The sides of an advancing wildland fire.

foam The finished product of water combined with certain agents that aid in the water's ability to extinguish fires. These agents are further discussed in Chapter 12.

forest litter The components of duff including tree limbs.

free radicals An atom or group of atoms that is unstable and must combine with other atoms to achieve stability.

freelance The act of performing operations without a coordinated effort or the knowledge of one's superior officer.

fuels management A program where naturally growing fuels, such as brush, are reduced to lessen fire intensity or open up areas for wildlife and cattle.

fully encapsulated suit A suit that includes total body protection. When worn with gloves, it gives head-to-toe protection from certain chemicals. The interior is sealed from the outside air.

fusees Another term for road flares.

G

generalist A person with general knowledge of no great depth in many subject areas.

ground strikes When lightning bolts reach objects on the ground, often starting fires in trees when there is no rain received with the storm.

ground sweep nozzles Nozzles mounted underneath apparatus to sweep fire from under the vehicle.

H

halogenated agents Fire-extinguishing agents containing the elements from Group 7 on the periodic table of the elements (halogens).

hazard trees Trees that have burned out at the base or are liable to drop large limbs.

helispots Unimproved areas large enough to land a helicopter.

helitack Personnel whose primary means of transportation to fires is by helicopter. They also assist in helicopter operations when the helicopter is being used for water drops or for crew and equipment transportation.

high-angle rescue Rescue utilizing ropes and other equipment.

highly protected risk A designation used by the insurance industry to describe business properties that meet certain special requirements as described in the footnote below.

high-rise building Multistory buildings in which the upper floors are beyond the reach of aerial equipment.

HIV Human Immunodeficiency Virus, cause of AIDS.

hose lay The method of laying out hose at a fire scene.

hydrant hookups Attaching the suction hose from the pumper to the hydrant.

hydraulics The computation of the required pressure to be applied to water to overcome the effects of pressure loss due to friction in piping and fire hose.

I

IDLH Immediately dangerous to life and health. IDLH atmospheres are capable of causing death, irreversible adverse health effects, or the impairment of an individual's ability to escape from a dangerous atmosphere.

ignition temperature The minimum temperature to which a substance must be raised before it will ignite. The piloted ignition temperature is usually much lower than the autoignition temperature. Piloted ignition may be provided by a spark or flame or by raising the general temperature.

Incendiary device An incendiary device is a device used to light a fire. This can be as simple as a lit cigarette folded into a matchbook or more complicated, involving chemical mixtures and a timer or cell phone to activate.

incendiary A fire that is deliberately set.

incident An incident is an occurrence, either caused by humans or natural phenomena, that requires response actions to prevent or minimize loss of life or damage to property and/or the environment.

inert A substance that will not react with other substances.

infrared sensing devices and GPS Devices that can detect heat energy through smoke and clouds. Used for aerial mapping of fire edges and locating hot spots.

interoperability The ability of different departments to communicate on common radio frequencies at incidents is called interoperability. This includes assisting fire departments and other departments, such as public works and law enforcement.

inventory The act of accounting for all of the equipment and tools assigned to a piece of apparatus.

inverter An electrical device that converts 12-volt current to 110 volt. Used to operate lights and tools from vehicle's charging system.

J

Jake brake Common name for the Jacobs Engine Brake and other brakes of that type. Used on diesel motors.

L

latent heat of vaporization The amount of heat a material must absorb when it changes from a liquid to a vapor or gas.

lateral transfer The act of changing jobs from one fire department to another without coming on at the bottom of the rank structure.

LCES Lookouts, Communications, Escape routes, and Safety zones.

liaison officer A contact person for outside agencies.

light bar Roof-mounted unit containing emergency warning lights.

local area network A system linking the computer terminals of a department on a local scale—for example, at the different desks at headquarters. A wide area network links the computers of several different geographic locations, such as between headquarters and the fire stations.

logging slash The remnants of logging operations, including limbs trimmed from downed trees and broken tree trunks.

M

main rotor Horizontal blades that create lift for a helicopter.

manipulative training Training in the operation of tools and equipment.

master stream appliance Large-bore nozzle equipped with a base. Not designed for hand-held use.

McLeod A tool for fighting wildland fires with a scraping blade on one side of the head and a rake on the other.

mechanical aptitude The ability to figure out the operation and construction of equipment from drawings.

mentor A person who guides and directs toward a goal.

miscible Capable of mixing without separation.

misdemeanor A crime punishable by up to one year in a county jail or by a fine usually not to exceed one thousand dollars or both.

mitigation Reducing the hazard, making less severe.

mixture A substance made up of two or more substances physically mixed together.

Mobile Data Computer A computer mounted in the apparatus and connected to an antenna to provide and receive CAD information.

molecule Combined groups of atoms. Molecules composed of two or more different kinds of atoms are called compounds.

motor block heater An electrical device that keeps oil in the motor warm and makes for easier starting and helps prevent damage when the motor is started in cold weather.

mucous membranes Inside of the nose, mouth, and covering of the eye.

mutual aid When departments draw up an agreement that they will assist each other upon request.

N

nonmiscible Not capable of mixing, will separate.

O

on duty The time firefighters spend performing their jobs.

one-hour fire-rated separation A fire-rated assembly that should resist breakthrough for a period of one hour. An example of this type of construction is the use of 5/8-inch-thick fire-rated gypsum wallboard or a combination of wallboard and plaster. All of the electrical boxes must be metal and not plastic. Any penetrations through the assembly must be properly protected to prevent the spread of fire.

open screw and yoke valve A valve with a hand wheel that exposes a threaded rod when in the open position. The hand wheel looks much like a steering wheel and can be locked with a chain and padlock so it cannot turn.

operational period Operational periods are a predetermined time frame and vary as to length. They may be as long as 12 to 24 hours for a wildland incident or as short as an hour for a hazardous materials incident.

operational procedures Written procedures for performing operational functions.

oral interview panel An interview technique in which the interviewers ask questions and evaluate the answers given by job candidates. They assign a score to the candidate's responses for ranking purposes during the selection process.

overhaul The operation performed to ensure that all embers are extinguished after a fire is controlled. The final extinguishment actions.

oxidation The chemical combination of any substance with an oxidizer.

oxidizer A substance that gains electrons in a chemical reaction.

P

paramedic An advanced level of medical training. Paramedics can perform invasive procedures on the patient, such as starting intravenous lines.

parapet wall A wall that extends above the roof line.

pathogen A disease-causing agent.

PCB oil Oil containing polychlorinated biphenyl, a compound that is considered to cause cancer.

perimeter control Controlling the edges of a wildland fire.

perimeter Boundary for controlled access (hazmat). The fire's edge (wildland).

perjury Making false statements in a sworn document or testimony.

petrochemical Dealing with oil refining and production facilities.

polar solvents These liquids will mix readily with water as water is a polar substance. The common polar solvents are alcohols, aldehydes, esters, ketones, and organic acids.

positive pressure mode SCBA regulator function that keeps positive pressure in the mask face piece at all times.

post indicator valve A valve with an indicator body that sticks up out of the ground. The body has a small window that says either "shut" or "open" depending on the position of the valve.

power shears Cutting attachment for a rescue tool.

prescribed burning Planned application of fire under specified conditions in a predetermined area to achieve management objectives, which include: removal or modification of fuels; clearing paths through brush; and killing unwanted plant growth.

primary search A rapid search of all involved and exposed areas which are affected by the fire but can be entered, to verify removal and/or safety of all occupants. Should this not be possible a secondary search is conducted as soon as it is safe to do so.

probationary firefighter A person hired by the fire department who has not been granted permanent status.

Pulaski A tool for fighting wildland fires with an axe on one side of the head and a grub hoe on the other.

pump and roll A tactic used in grass fires utilizing pumpers that can drive while the pump is operating. Hoselines are connected to the apparatus and water is sprayed to extinguish the fire edge.

pyrolysis The chemical decomposition of matter through the action of heat.

R

rappel To descend by means of a rope.

reforestation Planting seedling trees in areas destroyed by fire or logging.

rehab Short for rehabilitation.

rehabilitated There are two common usages of this word in fire fighting. To rehabilitate personnel means that they rest, cool off, and replenish body fluids. To rehabilitate a fire line means to construct water bars to direct water runoff and prevent erosion. Under the Federal "Minimum Impact Suppression Tactics" policy, rehabilitation may mean the erasure of fire lines as much as possible. In other words, to cause as little damage as possible controlling the fire as fire is a part of the natural environment.

rekindle A fire that reignites after it was thought to be extinguished. Commonly happens in attics, basements, and walls of structure fires and in logs at wildland fires. This is usually due to incomplete overhaul/mop-up.

relief valve Device used to release unwanted pressure.

repeaters Serve the purpose of receiving radio transmissions and boosting the signal. Used in areas where topography or tall buildings interrupt clear communications.

reserve/cadet programs Organized programs sponsored by paid fire departments that provide training in return for personnel volunteering their time.

Resource designator An identification system using numbers and letters to identify resources by agency and type.

resume A listing of a person's areas of experience and education.

resuscitator A mechanical device that can perform forced ventilation or provide oxygen to a victim or patient.

retard chamber A small tank attached to sprinkler systems that allows pressure surges to dissipate their energy before they enter the system and set off the water flow alarm.

retardant A material spread on fuels that inhibits their burning.

retro-reflective A surface, material, or device (retroreflector) that reflects light or other radiation back to its source; reflective.

S

safety section A body of law that sets the retirement benefit rate for certain professions, primarily those that are high hazard and deal with public safety, namely, fire and police.

safety zone A place where firefighters can be safe from the incident's hazards.

salvage Operations performed to prevent or reduce fire, smoke, or water damage to items of value.

scratch line A quickly created wildland fire control line, constructed using hand tools.

skid The long tubular shaped feet that helicopters sit on when on the ground.

sky lobby A lobby on a high floor level in a high-rise building. Elevators leave from this area to service the upper floors.

sling load Material transported by being placed in a net suspended underneath a helicopter.

smoke jumper Highly trained personnel who parachute in to suppress fires in remote areas.

specialist A person with extensive training in one area of operations or information.

specific heat The ratio between the amount of heat necessary to raise the temperature of a substance compared to the amount of heat required to raise the same weight of water the same number of degrees. Water has a specific heat value of 1.

spot fires In heavier fuels, flying fire brands can land outside the fire perimeter and start new fires.

standpipe system Plumbing system installed in multistory buildings for fire department use with outlets on each floor for attaching fire hose.

static water source Pond, lake, or tank used to supply fire engines.

strategy The method used to coordinate the tactical operations of units to achieve the desired incident objectives.

structure protection Protecting structures in danger of being consumed by an advancing wildland fire.

subsurface foam injection Plumbing installed on a tank to allow for the introduction of foam under the surface of the contained liquid.

T

tactical support A vehicle equipped to provide the needs of firefighters at the emergency scene. See rehab.

tactics Actions taken to achieve strategies.

tail rotor Vertical propeller used for steering control that is installed on the tail of a helicopter.

technical training Training in the specifications and limitations of equipment or calculation of information necessary to operate the equipment.

The 2 & 7 Tool The 2 errors and 7 barriers common to poor decision making.

thrust block A mass of concrete poured on the outside of an angle fitting and extending back to native soil. The purpose is to prevent surges in flow through a pipe from flexing the fitting and wiggling it in the ground, which would, over time, form a larger and larger underground space, possibly allowing the pipe fitting to pull apart.

toxicology The science of materials that are poisonous to living things, especially humans and animals.

turret nozzle Roof or bumper mounted nozzle remotely controlled from inside the cab.

U

unburned island Area of unburned fuel within a fire perimeter.

unprotected vertical shafts Laundry chutes, elevator shafts. When not enclosed in fire-resistive construction, these openings act as chimneys, allowing rapid fire spread in multilevel buildings.

upper division courses College-level courses that are applicable to a degree program for a bachelor's degree or higher. Lower division courses are those taken on the college level that are either prerequisites for higher levels of study or are used to receive an associate of arts or sciences degree.

urban interface The area where built-up areas of homes and businesses have little separation from the natural-growing wildland area.

urban intermix Buildings interspersed in wildland areas. Homes built in the woods.

V

vacuum truck Tank truck equipped with a pump that evacuates the air from inside the tank causing it to draw a vacuum. Used for picking up liquids from spills or tanks, commonly used at crude oil production facilities.

validate To make sure that the items included are actual requirements of the job.

veteran's points Points added to a person's final score on a competitive examination process. Given to persons who have satisfactorily performed military service.

voice-over A presentation style where video is shown and a person's voice is heard without seeing the person. Commonly used in documentaries.

volunteer firefighting Performing firefighting services without pay. In some areas a variation of this is the Paid Call Firefighter program. Under this program firefighters are paid a specified sum when they respond to incidents or attend training.

W

water curtain A screen of water spray set up to protect exposures.

watersheds Complex geographic, geologic, and vegetative components that control runoff of rain water and support varied ecosystems.

WCS Worst case scenario.

wildland Open land in its natural state.

worker's compensation Money paid to persons who have been injured in the course of their employment and are unable to work either temporarily or permanently.

working pressure The pounds per square inch of pressure that a tank is designed to contain.

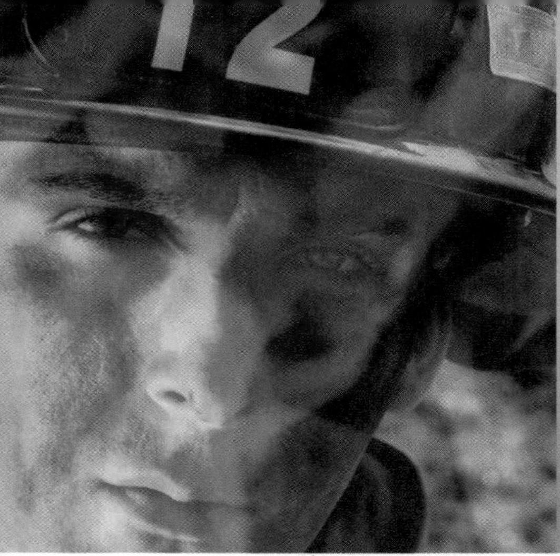

INDEX